DOMINO AND INTRAMOLECULAR REARRANGEMENT REACTIONS AS ADVANCED SYNTHETIC METHODS IN GLYCOSCIENCE

DOMINO AND INTRAMOLECULAR REARRANGEMENT REACTIONS AS ADVANCED SYNTHETIC METHODS IN GLYCOSCIENCE

Edited by

ZBIGNIEW J. WITCZAK
ROMAN BIELSKI

WILEY

Published by John Wiley & Sons, Inc., Hoboken, New Jersey.
Published simultaneously in Canada.

Library of Congress Cataloging-in-Publication Data:

Names: Witczak, Zbigniew J., 1947– , editor. | Bielski, Roman, 1946– , editor.
Title: Domino and intramolecular rearrangement reactions as advanced synthetic methods in glycoscience / edited by Zbigniew J. Witczak, Roman Bielski.
Description: Hoboken, New Jersey : John Wiley & Sons Inc., [2016] | Includes bibliographical references.
Identifiers: LCCN 2015039879 | ISBN 9781119044208 (cloth)
Subjects: | MESH: Glycosides–chemistry–Laboratory Manuals. | Chemistry Techniques, Synthetic–methods–Laboratory Manuals.
Classification: LCC QP702.G577 | NLM QU 25 | DDC 572/.567–dc23 LC record available at http://lccn.loc.gov/2015039879

Printed in the United States of America

10 9 8 7 6 5 4 3 2 1

We dedicate this book to our wives, Wanda and Barbara.

CONTENTS

FOREWORD

The development of novel methods for coupling molecules in the area of carbohydrate chemistry provides access to complex oligosaccharides, glycopeptides, and even glycoproteins of modest size, using synthetic methods. Attaching small molecules such as mono- or oligosaccharides to other carbohydrates, amino acids (or proteins), and other natural products is a challenging and important target in today's glycosciences. Another challenging task is that of a carefully designed modification of monosaccharides before introducing them into a larger structure.

Both of these capabilities require chemistry that must be performed with a high level of regio- and stereocontrol. In recent years, several novel strategies enabling these delicate manipulations have been introduced. One of these takes advantage of various rearrangement reactions and another utilizes domino reactions. Both processes offer substantial atom economy and hopefully stereo- and regiocontrol. The combination of domino/cascade/rearrangement reactions has become a very powerful tool for constructing complex structures in enantiomerically pure forms. The methodology is also very productive in the synthesis of many targets in which a well-defined steric outcome at specific positions is desired. The Editors did well by combining domino and cascade reactions with rearrangements. Indeed, one or even several rearrangements are often involved in the domino/cascade. Furthermore, a rearrangement reaction is one that takes place within a limited number of atoms belonging to one molecule. In a domino reaction, the number of available atoms is also strictly limited and mechanisms of the rearrangement and domino components are often similar.

Another very interesting category of the domino/cascade reactions worth mentioning is that of three component reactions. Its application to the synthesis of many molecules of biological interest and templates for natural products synthesis is particularly appealing. This strategy is well exemplified and detailed in Chapter 7.

While there are excellent books devoted to domino reactions and rearrangements in organic chemistry in general, there have been no corresponding comprehensive works describing "the State of the Art" in domino/cascade reactions and rearrangements in the carbohydrate field. Witczak and Bielski have recognized the strategic importance of designing/preserving chirality of molecules and applying tools of modern synthetic methodology to carbohydrate chemistry. Happily, the Editors were able to recruit a group of outstanding scientists to contribute excellent chapters addressing important aspects of the molecular transformations with a combination of rearrangements and cascade/domino options to cover most topics of value in glycosciences. Given its very careful selection of subjects and their sequential presentation in a collection of 15 chapters, this book constitutes a valuable source of practical information to all carbohydrate chemists.

I am sure that there will be a great interest in this book.

SAMUEL DANISHEFSKY

Professor of Chemistry
Memorial Sloan-Kettering Cancer Center

PREFACE

Synthesis of organic molecules designated to fulfill special requirements or exhibit specific properties has belonged and will belong to the most important targets of organic chemistry. This area of synthesis, particularly when applied to synthesis of constructs containing carbohydrates, has experienced a dramatic acceleration in recent years. One factor explaining the observed new developments has been a better understanding of the function and structure of glycoproteins and other naturally occurring glycorandomized derivatives. The present book attempts to offer an insight into these new developments created by coupling domino reactions with rearrangement reactions to achieve specifically designated/functionalized carbohydrate chemistries.

Carbohydrates represent a unique family of polyfunctional molecules, which can be chemically or enzymatically modified in a multitude of ways. They have been extensively used as starting materials in enantioselective syntheses of many, complex natural products with a plurality of chirality centers. Synthetic organic chemistry that utilizes these carbohydrate building blocks continues to spawn revolutionary discoveries in glycobiology, medicinal chemistry, pharmacology, molecular biology, and medicine, simply by providing not only the raw materials but also the accurate mechanistic insight of modem molecular sciences.

The coupling of synthetic approaches of domino reactions combined with rearrangement steps is always complex with multiple possible outcomes. In synthetic carbohydrate chemistry, it is even more difficult as it is often unpredictable and always requires very careful selection of starting templates, reagents, and reaction conditions to achieve the best approach to the target compound.

The book originates from the symposium "Domino Reactions and Rearrangement in Glycosciences," which we organized during the 248th Meeting of the American Chemical Society in San Francisco, CA, in Fall 2014. It attracted several prominent speakers, had a relatively large attendance, and was met with a lot of interest. Some of the chapters in this book are based on the presentations delivered at the symposium. Other contributions are from leading experts in the field of carbohydrate chemistry. Some of the chapters are reviews of the recent literature; some describe recent experimental data from the author's laboratories. We believe that all chapters are of a very high standard and offer a novel perspective on the discussed subjects.

Thus, it is not surprising that almost all of the chapters in the book are, to some extent, concerned with applications. It confronts the editors with a dilemma that is impossible to address satisfactorily—how to divide the book into consistent segments. By no means are we satisfied with our choice, but some kind of division had to be introduced.

Each chapter in the book covers issues related to the title reactions and discusses multiple synthetic methodologies and potential applications of the synthesized constructs.

The introductory chapter, written by internationally recognized expert in domino reactions Pellissier, discusses various aspects of selected examples of carbohydrate domino chemistry methodologies and proposes a novel strategy applicable to synthesizing certain molecular targets. It also describes the definition of domino reaction that was originally introduced by Thietze.

Chapter 2, authored by Mahrwal, deals with organocatalyzed cascade reaction strategies employed to the synthesis of important carbohydrate templates. The most efficient methodologies compiled in this review were developed before the birth of official combined domino/cascade reaction chemistry.

Chapter 3, authored by Soengas, describes intriguing results of the various elimination reaction strategies in domino reactions.

The development and application of original framework strategy of domino reaction for synthesis of biological targets were thoroughly explored by Rauter in Chapter 4. The chapter discusses most types of natural molecules, including complex carbohydrates.

In Chapter 5, Guaragna reviews an important topic of original strategy of syntheses of thioenol ether containing systems and their multistep cyclization mediated by DDQ.

In Chapter 6, Witczak and Bielski offer a different perspective on closely related subject of thio-click and domino approach to the synthesis of carbohydrate heterocycles.

Chapter 7, by Santra and coworkers, addresses microwave-assisted multicomponent cascade coupling of various compounds, including carbohydrates.

Chapter 8, by Werz, is a very broad review of spiroannulation domino reactions for the attachment of additional carbohydrate functionalities. The review discusses processes belonging to domino reactions as well as other important annulation coupling reactions traditionally used for the construction of various natural product targets.

In the second section of the book, titled "Rearrangements Reactions of Functional Sugar Templates," Witczak and Bielski explore, in Chapter 9, selected topics to introduce the reader to the subject of rearrangement reactions in carbohydrate chemistry.

Jarosz and coworkers contribute in Chapter 10 with a discussion of novel synthetic applications of rearrangement of carbohydrate backbone discovered en route to higher carbon sugars.

Chapter 11, by Bielski and Witczak, describes a new chemistry of levoglucosenone derivatives prepared via Overman and related rearrangements.

Marzabadi and coworkers describe in Chapter 12 an important study of the rearrangement products of 3,6-anhydro-D-glycals. The review highlights a synthesis of prototypes of simple functionalized molecules with high chemotherapeutic potential.

Chapter 13, by Narayanaswamy and coworkers, provides a general overview of the potential applications of ring expansion and rearrangement reactions in the synthesis of highly valuable, septanoses as bioactive, carbohydrate-based ligands with multiple biomedical applications.

Chapter 14, by Herradón and coworkers, describes rearrangements of carbohydrate templates as the pathway to functionalized heterocycles and peptide–scaffold hybrids.

The last chapter, by Nguyen and coworkers, explores an extremely important topic of a stereoselective transformation of glycals and glycosyl trichloroacetimidates and their application to the synthesis if important urea-linked glycosides.

With the increasing complexity of modern sciences in the 21st century, a need to educate industrial leaders, public, and governmental funding agencies about the intellectual and technical potential and economic importance of specific areas of life sciences has become more and more crucial. One such area is the part of glycoscience emerging as a result of a marriage between the concept of domino and rearrangement approaches in the synthetic carbohydrate chemistries. We hope that this book will fill this need, at least to some extent.

In conclusion, we are convinced that the present collection of diverse chapters offers an insight into the present stage of domino and rearrangement approaches in glycoscience and will help steer future discoveries to fulfill the enormous potential in the area of synthetic carbohydrate chemistry.

ACKNOWLEDGMENTS

We thank all the authors for excellent contributions to this volume. We also thank the peer reviewers of the chapters for their high level of expertise and helpful efforts to improve the overall quality of all the manuscripts.

We dedicate this book to our wives, Wanda and Barbara.

Zbigniew J. Witczak, PhD

Department of Pharmaceutical Sciences
Nesbitt School of Pharmacy
Wilkes University, Wilkes-Barre, PA 18766, USA

Roman Bielski

Value Recovery, Inc.
510 Heron Drive
Suite 301, Bridgeport, NJ 08014, USA
Department of Pharmaceutical Sciences
Wilkes University
Wilkes-Barre, PA 18766, USA

LIST OF CONTRIBUTORS

Peter R. Andreana, Department of Chemistry and Biochemistry *and* School of Green Chemistry and Engineering, University of Toledo, 2801 W. Bancroft St., Wolfe Hall 2232B, Toledo, OH 43606-3390, USA; phone: 409-530-1930; e-mail: peter.andreana@utoledo.edu

Tonja Andreana, Department of Chemistry and Biochemistry *and* School of Green Chemistry and Engineering, University of Toledo, 2801 W. Bancroft St., Wolfe Hall 2232B, Toledo, OH 43606-3390; USA

Vikram Basava, Department of Chemistry and Biochemistry, Seton Hall University, 400 South Orange Ave., South Orange, NJ 07079, USA

Roman Bielski, Department of Pharmaceutical Sciences, Wilkes University, 84 W. South Street, Wilkes-Barre, PA, USA; phone: 610-573-4804; e-mail bielski1@verizon.net

Jean-Paul Bourgault, Department of Chemistry and Biochemistry *and* School of Green Chemistry and Engineering, University of Toledo, 2801 W. Bancroft St., Wolfe Hall 2232B, Toledo, OH 43606-3390, USA

Vasco Cachatra, Universidade de Lisboa, Faculdade de Ciências, Departamento de Química e Bioquímica, Centro de Química e Bioquímica, Carbohydrate Chemistry Group, Ed. C8, Piso 5, Campo Grande, 1749-016 Lisboa, Portugal

Maciej Cieplak, Institute of Organic Chemistry, Polish Academy of Sciences, Kasprzaka 44/52, 01-224 Warsaw, Poland

Supriya Dey, Department of Organic Chemistry, Indian Institute of Science, Bangalore, Karnataka, India

Daniele D'Alonzo, Department of Chemical Sciences, University of Napoli Federico II, via Cintia 21, I-80126 Napoli, Italy

Agnieszka Gajewska, Institute of Organic Chemistry, Polish Academy of Sciences, Kasprzaka 44/52, 01-224 Warsaw, Poland

Annalisa Guaragna, Department of Chemical Sciences, University of Napoli Federico II, via Cintia 21, I-80126 Napoli, Italy; phone: +39-081-674119; e-mail: annalisa.guaragna@unina.it

Emi Hanawa, Department of Chemistry and Biochemistry, Seton Hall University, 400 South Orange Ave., South Orange, NJ 07079, USA

Bernardo Herradón, Instituto de Química Orgánica General, CSIC, c/ Juan de la Cierva 2, 28006 Madrid, Spain; phone: +34915618806; e-mail: b.herradon@csic.es

Sławomir Jarosz, Institute of Organic Chemistry, Polish Academy of Sciences, Kasprzaka 44/52, 01-224 Warsaw, Poland; phone: +4822-3432320; fax: +4822-6326681; e-mail: slawomir.jarosz@icho.edu.pl

N. Jayaraman, Department of Organic Chemistry, Indian Institute of Science, Bangalore, Karnataka, India

Rainer Mahrwald, Institut für Chemie der Humboldt-Universität zu Berlin, Brook-Taylor-Str. 2, 12 489 Berlin, Germany; phone: +49-30-2093-8397; e-mail: rainer.mahrwald@rz.hu-berlin.de

Enrique Mann, Instituto de Química Orgánica General, CSIC, c/ Juan de la Cierva 2, 28006 Madrid, Spain

Cecilia H. Marzabadi, Department of Chemistry and Biochemistry, Seton Hall University, 400 South Orange Ave., South Orange, NJ 07079, USA; phone: (973)761-9032; fax: (973)761-9772; e-mail: cecilia.marzabadi@shu.edu

Matthew J. McKay, Department of Chemistry, University of Iowa, Iowa City, Iowa 52242, USA

Irene de Miguel, Instituto de Química Orgánica General, CSIC, c/ Juan de la Cierva 2, 28006 Madrid, Spain

Hien M. Nguyen, Department of Chemistry, University of Iowa, Iowa City, IA 52242, USA; phone: (+1) 319-384-1887; e-mail: hien-nguyen@uiowa.edu

Anna Osuch-Kwiatkowska, Institute of Organic Chemistry, Polish Academy of Sciences, Kasprzaka 44/52, 01-224 Warsaw, Poland

Giovanni Palumbo, Department of Chemical Sciences, University of Napoli Federico II, via Cintia 21, I-80126 Napoli, Italy

Nathaniel H. Park, Department of Chemistry, University of Iowa, Iowa City, IA 52242, USA

Hélène Pellissier, Aix Marseille Université, Centrale Marseille, CNRS iSm2 UMR 7313, 13397 Marseille, France

Amélia P. Rauter, Universidade de Lisboa, Faculdade de Ciências, Departamento de Química e Bioquímica, Centro de Química e Bioquímica, Carbohydrate Chemistry Group, Ed. C8, Piso 5, Campo Grande, 1749-016 Lisboa, Portugal; phone: +351 217500952; e-mail: aprauter@fc.ul.pt or aprauter@gmail.com

Soumava Santra, Department of Chemistry and Biochemistry *and* School of Green Chemistry and Engineering, University of Toledo, 2801 W. Bancroft St., Wolfe Hall 2232B, Toledo, OH 43606-3390, USA

Artur M. S. Silva, Department of Chemistry & QOPNA, University of Aveiro, 3810-193 Aveiro, Portugal

Eric T. Sletten, Department of Chemistry, University of Iowa, Iowa City, IA 52242, USA

Raquel G. Soengas, Department of Chemistry & QOPNA, University of Aveiro, 3810-193 Aveiro, Portugal; phone: +351-234-370084; e-mail: rsoengas@ua.pt

Sara M. Tomé, Department of Chemistry & QOPNA, University of Aveiro, 3810-193 Aveiro, Portugal

N. Vijaya Ganesh, Department of Organic Chemistry, Indian Institute of Science, Bangalore, Karnataka, India

Benjamin Voigt, Institut für Chemie der Humboldt-Universität zu Berlin, Brook-Taylor-Str. 2, 12 489 Berlin, Germany

Daniel B. Werz, Institut für Organische Chemie, Technische Universität Braunschweig, Hagenring 30, 38106 Braunschweig, Germany; phone: (+) 49-531-391 5266; e-mail: d.werz@tu-braunschweig.de

Zbigniew J. Witczak, Department of Pharmaceutical Sciences, Wilkes University, 84 W. South Street, Wilkes-Barre, PA, USA; phone: 570-408-4276; e-mail: zbigniew.witczak@wilkes.edu

ABBREVIATIONS

1D	one-dimensional
2D	two-dimensional
A	adenine
Ac	acetyl
AIBN	azobisisobutyronitrile
Ar	aryl
BINAP	2,2′-bis(diphenylphosphino)-1,1′-binaphthyl
Bn	benzyl
Boc	*tert*-butoxycarbonyl
BSA	bis(trimethylsilyl) acetamido
Bz	benzoyl
C	cytosine
CAN	ceric ammonium nitrate
Cbz	carboxybenzyl
Cp	cyclopentanyl
D-A	donor–acceptor
DABCO	1,4-diazabicyclo[2.2.2]octane
DBU	1,8-diazabicyclo[5.4.0]undec-7-ene
DCC	*N*,*N*′-dicyclohexylcarbodiimide
DCE	dichloroethane
DCM	dichloromethane
DDQ	2,3-dichloro-5,6-dicyano-1,4-benzoquinone
DHP	dihydropyran
DIAD	di(isopropyl) azodicarboxylate
DMAP	4-dimethylaminopyridine

DME	dimethoxyethane
DMF	*N*,*N*-dimethylformamide
DMF	dimethylformamide
DMP	dimethoxypropane
DMS	dimethylsulfide
DMSO	dimethysulfoxide
DOS	diversity-oriented synthesis
dppb	1,4-bis(diphenylphosphino)butane
dppe	1,2-bis(diphenylphosphino)ethane
dppp	1,3-bis(diphenylphosphino)propane
DTBP	2,6-di-*tert*-butylphosphine
EDC	1-ethyl-3-(3-dimethylaminopropyl)carbodiimide
EDTA	ethylenediaminetetraacetic acid
Et	ethyl
Gal	galactose
Glc	glucose
HCS	higher carbon sugars
HMPA	hexamethylphosphoramide
HNA	1′,5′-anhydro-*arabino*-hexitol nucleic acids
homoDNA	2′,3′-dideoxy-β-*erythro*-hexopyranosyl nucleic acids
IBX	2-iodoxybenzoic acid
IMDA	intramolecular Diels-Alder reaction
i-Pr	isopropyl
LAH	lithium aluminum hydride
LDA	lithium diisopropylamide
m-CPBA	*m*-chloroperoxybenzoic acid
MCR	multicomponent reaction
Me	methyl
MS	molecular sieves
MTBE	methyl *tert*-butyl ether
mw	microwave
MWI	microwave irradiation
NIS	*N*-iodosuccinimide
NMMO	*N*-methylmorpholine-*N*-oxide
NMO	*N*-methylmorpholine-*N*-oxide
NMP	*N*-methylpyrrolidone
NMR	nuclear magnetic resonance
NOE	Nuclear Overhauser effect
P-3CR	Passerini three-component reaction
PCC	pyridinium chlorochromate
PDC	pyridinium dichromate
PG	protecting group
Ph	phenyl
PMB	4-methoxybenzyl
PPTS	pyridinium *p*-toluenesulfonate

Py	pyridine
RCM	ring-closing metathesis
rt	room temperature
SET	single-electron transfer
T	thymine
TBAF	tetrabutylammonium fluoride
TBDPS	*tert*-butyldiphenylsilyl
TBSOTf	*tert*-butyldimethylsilyl trifluoromethanesulfonate
*t*Bu	*tert*-butyl
TEA	triethylamine
TEBAC	benzyltriethylammonium chloride
TFDO	methyl(trifluoromethyl)dioxirane
THF	tetrahydrofuran
THP	tetrahydropyran
TIBAL	triisobutylaluminum
TLC	thin layer chromatography
TMEDA	tetramethylethylenediamine
TMS	trimethylsilyl
TMSCN	trimethylsilyl cyanide
TMSOTf	trimethylsilyl trifluoromethanesulfonate
Tosyl	*para*-toluenesulfonate ester
TPAP	tetrapropylammonium perruthenate
TS	*para*-toluenesulfonate ester
Ts	4-methylphenylsulfonyl (tosyl)
TTMPP	tris(trimethoxyphenyl)phosphine
U-4CR	Ugi four-component reaction

1

INTRODUCTION TO ASYMMETRIC DOMINO REACTIONS

HÉLÈNE PELLISSIER
Aix Marseille Université, Centrale Marseille, CNRS iSm2 UMR 7313, 13397 Marseille, France

1.1 INTRODUCTION

A domino reaction has been defined by Tietze as a reaction that involves two or more bond-forming transformations, taking place under the same reaction conditions, without adding additional reagents and catalysts, and in which the subsequent reactions result as a consequence of the functionality formed by bond formation or fragmentation in the previous step [1]. It must be recognized that a relatively narrow distinction exists between domino and consecutive cascade or tandem reactions. From the point of view of an operator, the only difference between the two lies in the point along the sequence at which one or more catalysts or reagents had to be added to effect either the initiation of a sequence (i.e., domino reaction) or propagation to the next step (i.e., consecutive reaction). It should be noted that the descriptors domino, cascade, and tandem are often used indistinguishably from one another in the literature [2], and various opinions exist on how such reactions should be classified. According to Tietze, a domino reaction is strictly defined as a process in which two or more bond-forming transformations occur on the basis of functionalities formed in the previous step and, moreover, no additional reagents, catalysts, or additives can be added to the reaction vessel, nor can reaction conditions be changed [1]. Denmark further posited, however, that most domino reactions, as defined by Tietze, fell under the broader category of tandem processes [3]. Other tandem reactions that are not cascades involve the isolation of intermediates, a change in reaction conditions, or

Domino and Intramolecular Rearrangement Reactions as Advanced Synthetic Methods in Glycoscience, First Edition. Edited by Zbigniew J. Witczak and Roman Bielski.

the addition of reagents or coupling partners. Furthermore, other authors classified domino reactions with even stricter conditions [4, 5]. The quality and importance of a domino reaction can be correlated to the number of bonds generated in such a process and the increase in molecular complexity. Its goal resembles nature in its highly selective sequential transformations. The domino reactions can be performed as one-, two-, and multicomponent transformations. Multicomponent reactions are defined as domino reactions involving at least three substrates and, consequently, constitute a subgroup of domino reactions [4, 6]. It is worth noting that there are some confusing ideas among chemists about the definition of a multicomponent reaction. According to Yus [6c], this type of reactions should be clearly differentiated from other one-pot processes, such as tandem or cascade reactions, and in general from all processes that involve the reaction between two reagents to yield an intermediate that is captured by the successive addition of a new reagent (sequential component reactions). Multicomponent reactions are convergent chemical processes that involve the well-defined condensation or coupling of more than two reactants to form a product that contains significant portions of all reactants, ideally all atoms [6b]. Moreover, no additional reagents, catalysts, or additives can be added to the reaction vessel, nor can reaction conditions be changed as in domino processes, according to Tietze. The use of one-, two-, and multicomponent domino reactions in organic synthesis has been increasing constantly, since they allow to economically synthesize a wide range of complex molecules, including natural products and biologically active compounds by employing a methodology that requires neither costly and time-consuming protection-deprotection processes nor the purification of intermediates [7]. Indeed, decreasing the number of laboratory operations and the quantities of chemicals and solvents has made domino and multicomponent reactions unavoidable processes [8]. The proliferation of these reactions is evidenced by the number of recent reviews covering the literature through 1992 [1, 4, 6c, 9]. Although asymmetric synthesis is sometimes viewed as a subdiscipline of organic chemistry, actually this topical field transcends any narrow classification and pervades essentially all chemistry [10]. Indeed, the preparation of chiral compounds is an important and challenging area of contemporary synthetic organic chemistry, mainly due to the fact that most natural products are chiral and their physiological or pharmacological properties depend upon their recognition by chiral receptors, which will interact only with molecules of the proper absolute configuration. The use of chiral drugs in enantiopure form is now a standard requirement for virtually every new chemical entity, and the development of new synthetic methods to obtain enantiopure compounds has become a key goal for pharmaceutical companies. The growing economic importance of chiral compounds has spurred major research efforts toward the selective preparation of chiral compounds. The synthesis of optically active chiral compounds, which play an important role in medicine and materials, is one of the most fascinating aspects of modern organic synthesis. Over the last three decades, an explosive growth of research in the field of asymmetric synthesis has occurred. Asymmetric synthesis constitutes one of the main strategies to gain access to enantioenriched compounds, involving the use of either chiral substrates and auxiliaries or catalysts derived preferentially from cheap chiral pool sources. In particular, asymmetric catalysis of organic reactions to provide enantiomerically enriched products is of central importance to modern synthetic and pharmaceutical chemistry. In this context, the possibility of preparing chiral compounds by joining two or more reactions into one asymmetric domino process has rapidly become a challenging goal for chemists, due to economical advantages, such as the avoidance of costly protecting groups and time-consuming purification procedures after each step. In particular, the combination of these fascinating one-pot reactions with asymmetric catalysis has quickly become one of the most rapidly growing areas of the current organic chemistry.

1.2 ASYMMETRIC DOMINO REACTIONS USING CHIRAL CARBOHYDRATE DERIVATIVES

In spite of the explosive development of asymmetric domino processes including multicomponent reactions, domino reactions involving chiral carbohydrate derivatives still remain rare in the literature. It is particularly true with regard to carbohydrate-derived organocatalysts. This section highlights several important works dealing with stereocontrolled domino reactions of chiral carbohydrate derivatives in a first part, and a rare enantioselective domino reaction catalyzed by a chiral carbohydrate derivative in a second part. It must be noted that all the discussed domino reactions are multicomponent ones.

1.2.1 Stereocontrolled Domino Reactions of Chiral Carbohydrate Derivatives

One of the challenges in organic synthesis is to implement various reaction strategies in a multicomponent reaction, which is a domino reaction involving at least three substrates [1, 4–6] to achieve multibond formation. Moreover, the process is to take place in a single-reaction vessel and the newly formed product must contain portions of all the employed reactants. Multicomponent reactions convert more than two educts directly into their product by one-pot reactions. The starting materials for this kind of chemical transformation may be rich in functional groups. Typically, multicomponent reactions lead to very complex products by reacting relatively simple starting materials. The multicomponent reaction proceeds according to the domino principle, since subsequent transformations are a consequence of the functionalities produced in the previous transformation. These processes are highly flexible, (chemo)-selective, convergent, atom efficient, and of high exploratory power. Inspired by the mode of action of nature, they have brought the concept of ideal synthesis closer to reality [11]. Indeed, the discussed strategy is atom economical and avoids the necessity of protecting groups and isolation of intermediates. Its goal resembles nature in its highly selective sequential transformations. Even though the history of multicomponent reactions dates back to the second half of nineteenth century with the reactions of Strecker, Hantzsch, and Biginelli, it was only in the last decades with the work of Ugi that the concept of the multicomponent reaction has emerged as a powerful tool in synthetic chemistry [12]. In particular, one should mention a venerable and old multicomponent reaction, the so-called Hantzsch reaction, which was first reported in 1882 [13]. It allows the synthesis of 1,4-dihydropyridines through the reaction of enamines, aldehydes, and 1,3-dicarbonyl compounds. Dihydropyridines are well known for their activity against calcium channels, multidrug inflammatory targets in addition to their usefulness as tools for reducing imines to amines. The harsh reaction conditions usually applied to carry out the Hantzsch reaction have significantly decelerated the development of asymmetric versions of this reaction. In comparison with other asymmetric multicomponent reactions, the asymmetric Hantzsch process has probably observed the greatest progress during the last 6 years. In 2006, Dondoni et al. reported the synthesis of chiral 1,4-dihydropyridines through the three-component Hantzsch reaction between chiral aldehydes bearing a *N*-Boc benzyl glycinate group, β-keto esters, and enamino esters, which led to the corresponding chiral 1,4-dihydropyridines [14]. The scope of this methodology was later extended by the same authors to the synthesis of various enantiopure *C*-glycosylmethyl pyridylalanines starting from the corresponding chiral aldehydes [15]. A collection of eight novel *C*-glycosylmethyl pyridine amino acids was achieved in 55–68% yields with total preservation of the stereocenter and protective group's integrity. Furthermore, in 2009, the same authors developed the first organocatalyzed three-component Hantzsch reaction occurring between a chiral *C*-glycosyl aldehyde, a β-diketone, and an enamine, to provide the corresponding enantiopure substituted 1,4-dihydropyridine *C*-glycoconjugate in a 50% yield and excellent diastereoselectivity of >98% de, as shown in Scheme 1.1 [16]. This method, based on a double asymmetric induction, arising from the use of a chiral auxiliary associated to a chiral organocatalyst, allowed the synthesis of biologically relevant *C*-nucleosides, which were not accessible through uncatalyzed procedures.

The modern concept of multicomponent reaction is intimately related to the reactions developed with isocyanide reagents [17]. The Ugi four-component reaction is the reaction of a carbonyl compound (usually an aldehyde), an amine, an isocyanide, and a carboxylic acid (or an alcohol) to yield α-amino acid derivatives. In general, the mechanism involves *in situ* formation of an imine from the aldehyde or ketone and the primary amine, followed by α-addition of the isocyanide component to the imine and carboxylic acid and subsequent rearrangement to furnish diversely substituted α-amino acid derivatives. This reaction, first described in 1959, has been more widely studied and used than any other multicomponent reaction [12]. In their early work, Ugi et al. determined that the use of a chiral acid or isonitrile in the reaction did not provide any degree of stereoselectivity [18]. In contrast, chiral ferrocenylamine inputs resulted in the synthesis of nonracemic amino acid derivatives with low to modest levels of diastereoselectivity [19]. Later, Kunz et al. developed more versatile chiral auxiliaries for the Ugi reaction, using carbohydrate derivatives [20]. High diastereoselectivities of up to 94% de of (*R*)-amino acids were obtained in reactions employing a galactosylamine derivative (Scheme 1.2, first equation) [20a]. A drawback of this asymmetric Ugi reaction was that high levels of stereoselectivity were observed only for reactions using *tert*-butyl isonitrile. In addition, the asymmetric synthesis of (*S*)-amino acids with excellent diastereoselectivities of up to 96% de

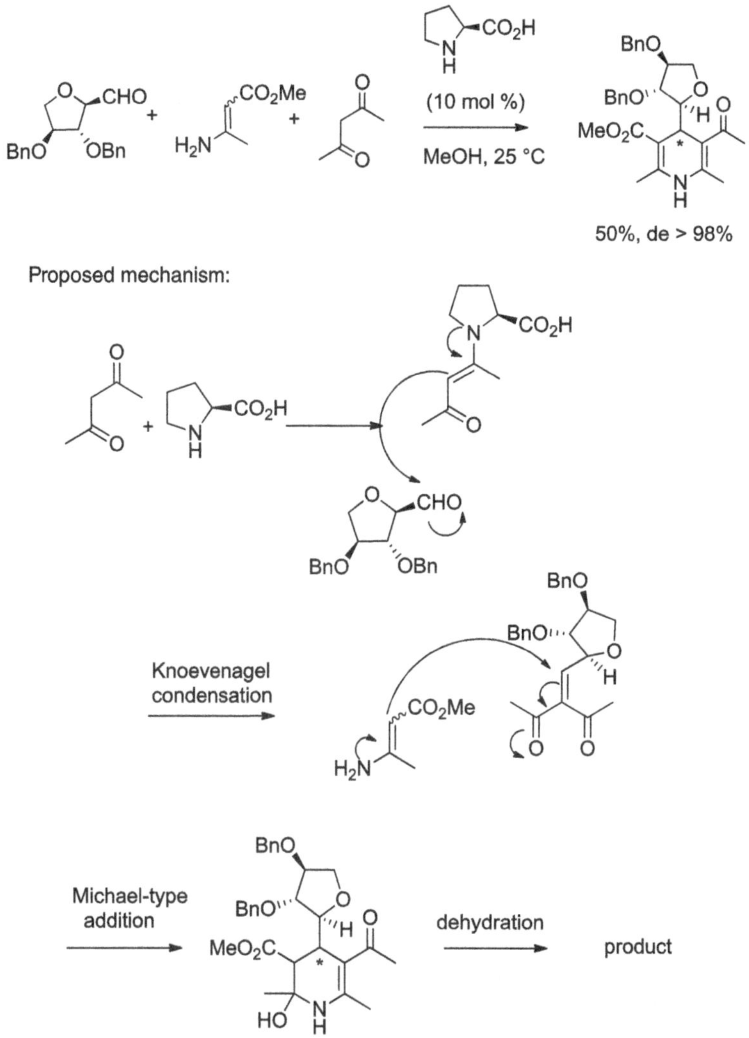

SCHEME 1.1 Three-component Hantzsch reaction of a chiral *C*-glycosyl aldehyde.

via the Ugi reaction was achieved using an arabinosylamine derivative (Scheme 1.2, second equation) [20b]. A single variant of the chemistry developed by Kunz was reported by Goebel et al. in 1991 [21]. These authors showed that 2,3,4,6-tetra-*O*-alkyl-β-D-glucopyranosylamines used as chiral amine components were even more efficient than the *O*-acyl-aldopyranosylamides used by Kunz et al. since diastereoselectivities of up to >99% de were reached with these chiral substrates.

The Passerini three-component reaction, discovered in 1921, involves the condensation of carbonyl compounds, carboxylic acids, and isocyanides to afford the corresponding α-acyloxy carboxamides [22]. Among several advantages are the

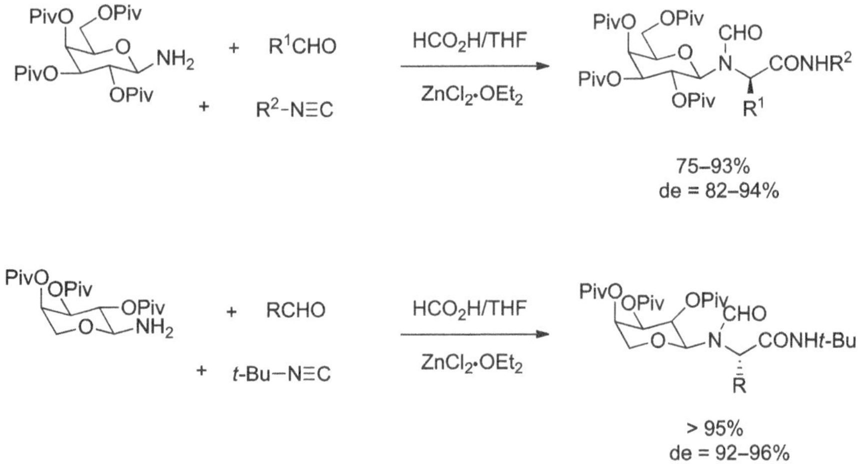

SCHEME 1.2 Three-component Ugi reactions of chiral carbohydrate-derived amines.

mildness of the reaction conditions, the broad scope, and high variability of the inputs [23]. Several chiral auxiliaries and substrates have been applied to control the stereochemical outcome of the diastereoselective Passerini reaction [24]. A particularly interesting Passerini reaction was reported by Krishna et al. in 2006 [25], who used (*p*-toluenesulfonyl)methylisocyanide (TosMIC) for the first time in this reaction as the isonitrile component. As shown in Scheme 1.3, the process employed carbohydrate-derived aldehydes as the chiral auxiliaries to react with TosMIC and carboxylic acids, providing the corresponding products as mandelamides in moderate yields (35–45%) and diastereoselectivities (50–52% de). The use of a chiral carboxylic acid in this reaction did not enhance the stereoselectivity.

The classic direct Mannich reaction, discovered in 1912 [26], is an aminoalkylation of carbonyl compounds involving ammonia (or a primary or secondary amine derivative), a nonenolizable aldehyde (usually formaldehyde) or a ketone, and an enolizable carbonyl compound, leading to β-aminocarbonyl derivatives [27]. In addition to enantioselective catalytic Mannich reactions [28], all of the possibilities of using chiral starting materials for this asymmetric multicomponent reaction have been reported. In their aim of discovering novel glycopeptide-based drugs for the treatment of bacterial and viral infections, cancer, and inflammatory processes [29], Dondoni et al. have employed a range of chiral *C*-glycosyl aldehydes as chiral auxiliaries in the Mannich reaction with *p*-methoxybenzylamine, and commercially available ketene silyl acetal, 1-methoxy-2-methyl-1-trimethylsilyloxypropene [14]. This process was promoted by $InCl_3$ and provided the corresponding chiral *C*-glycosyl α,α-dimethyl β-amino esters in high yields (60–90%) and complete diastereoselectivity in all cases of substrates studied, as shown in Scheme 1.4.

Diastereoselective nickel-catalyzed asymmetric four-component domino Reformatsky-type reactions have been developed by Dondoni et al. with the aim of achieving chiral *C*-glycosyl β-amino esters [14]. As shown in Scheme 1.5, an

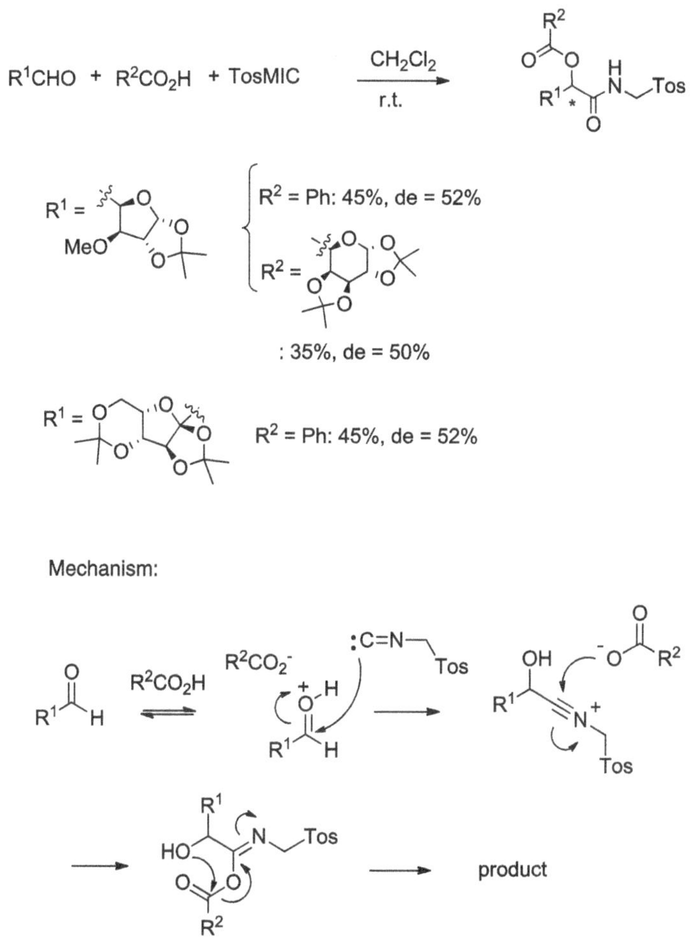

SCHEME 1.3 Three-component Passerini reaction of chiral carbohydrate-derived aldehydes.

in situ initial coupling of chiral *C*-glycosyl aldehydes with *p*-methoxybenzylamine led to the corresponding imines, which reacted with bromozinc enolate (*in situ* generated from dimethylzinc and ethyl bromoacetate). The resulting chiral Reformatsky products were obtained in moderate to good yields and as single diastereomers in all cases of substrates studied.

More recently, Beau and Norsikian reported remarkable diastereoselective domino Petasis/Diels–Alder reactions of unprotected carbohydrates with allylamines and boronic acids [30]. As shown in Scheme 1.6, free sugars, such as D-ribose and D-fucose, reacted through a three-component reaction to give the corresponding complex heterocyclic chiral domino products in good yields and often complete diastereoselectivity. The domino process started with a Petasis reaction occurring

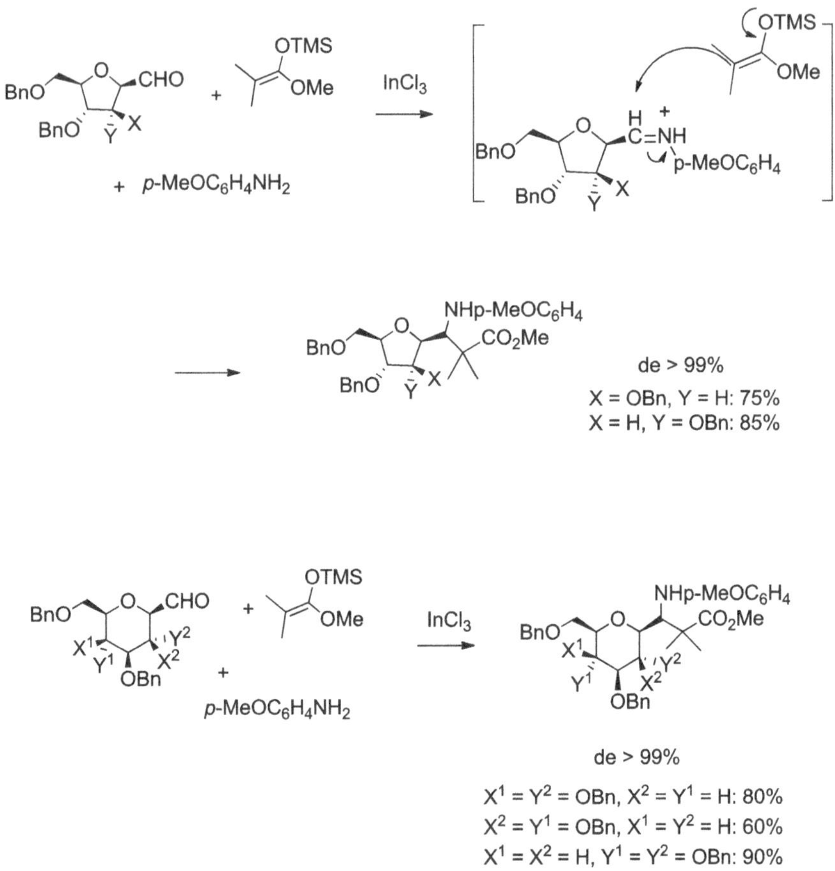

SCHEME 1.4 Three-component Mannich reactions of chiral *C*-glycosyl aldehydes.

between the free sugar and allylamine to give the corresponding acyclic Petasis product, which subsequently underwent an intramolecular Diels–Alder reaction to provide the final bicyclic chiral domino product. The process was performed in 9:1 dichloromethane/hexafluoroisopropanol as a solvent by heating the mixture at 120°C under microwave irradiation. In almost all cases studied, the reaction afforded a single diastereomer, except for the reaction of D-ribose with diallylamine and 2-furanylboronic acid, which led to a 7:3 mixture of two diastereomers, as shown in Scheme 1.6.

1.2.2 Enantioselective Domino Reactions Catalyzed by Chiral Carbohydrate Derivatives

The economical interest in combining asymmetric catalytic processes with domino reactions is obvious. The use of metal [10, 31] and biocatalysis [32] dominated the

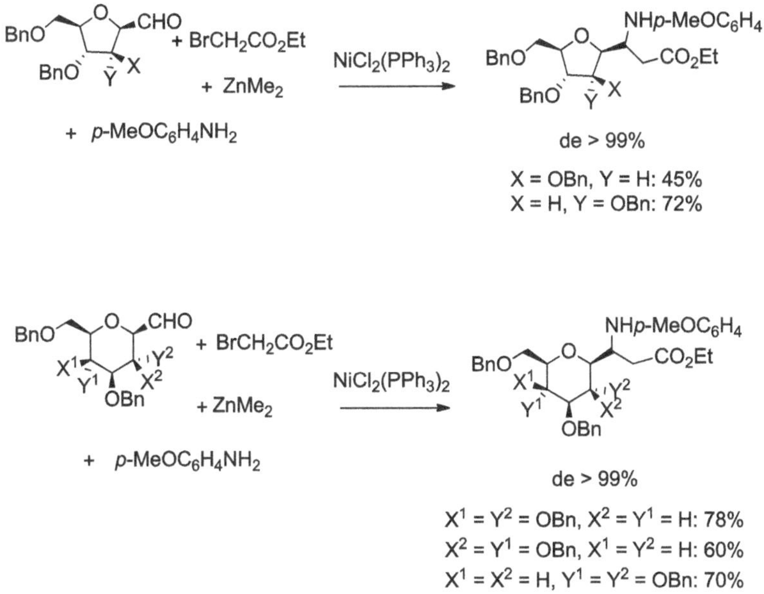

SCHEME 1.5 Four-component domino Reformatsky-type reactions of chiral *C*-glycosyl aldehydes.

catalytic processes at the end of the last century. However, a change in perception occurred during the last decade when several reports confirmed that relatively simple organic molecules, such as proline, could be highly effective and remarkably enantioselective catalysts of various fundamentally important transformations [33]. This rediscovery has initiated an explosive growth of research activities in organocatalysis, both in industry and in academia. Organocatalysts have several important advantages, since they are usually robust, inexpensive, readily available, and nontoxic. Enantioselective organocatalytic processes have reached maturity in recent years with an impressive and steadily increasing number of publications describing applications of this type of reactions, which paint a comprehensive picture for their real possibilities in organic synthesis [9l-m,9o,9q] [34]. The application of chiral organocatalysts has permitted the preparation of a number of highly valuable chiral products containing no traces of hazardous metals. The methodology offers several advantages from an economical and environmental point of view [35]. The ability of organocatalysts to promote a wide range of reactions by different activation modes makes organocatalysis ideal for applications in domino reactions, processes taking place in one pot and building complex frameworks from simple starting compounds. These organocatalyzed domino reactions are often highly efficient and follow, in some way, different biomimetic pathways, and principles that are found in biosynthetic processes in nature. In this context, the chemists have devoted more and more effort into the development of new and powerful strategies in domino reactions that avoid the use of costly and time-consuming protection–deprotection processes, as well as purification procedures of intermediates. Among enantioselective

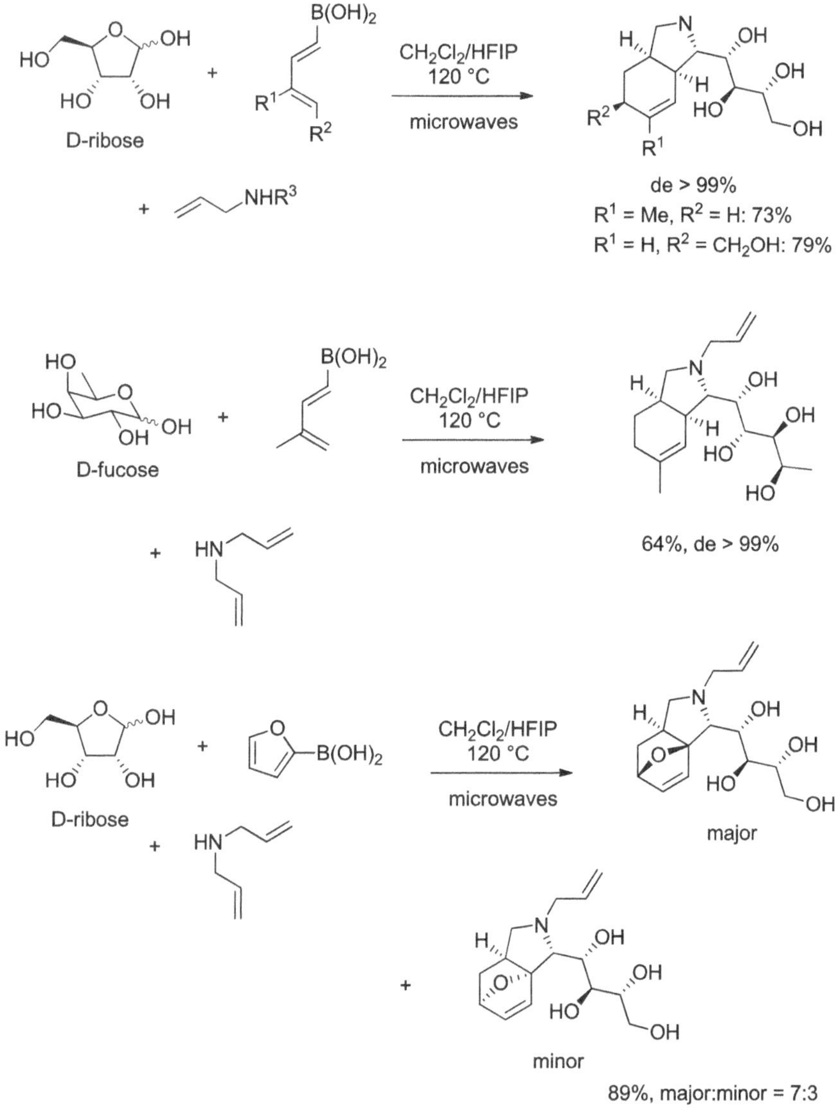

SCHEME 1.6 Three-component domino Petasis/Diels–Alder reactions of unprotected carbohydrates.

organocatalyzed reactions, the Biginelli reaction [36], which is a three-component reaction between urea or thiourea, a 1,3-dicarbonyl derivative, and an aldehyde providing medicinally relevant 3,4-dihydropyrimidin-2-(1*H*)-ones or -thiones, is known to give high enantioselectivities when induced by organocatalysts [37]. In 2009, chiral bifunctional primary amine-thiourea catalysts derived from D-glucose were demonstrated by Chen et al. to be very efficient to promote highly enantioselective

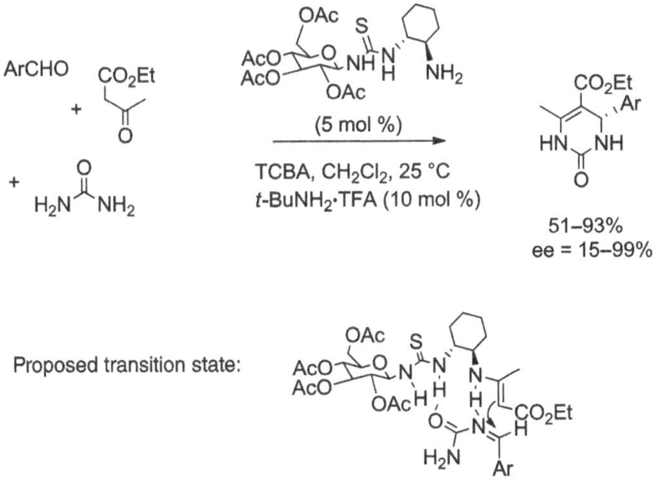

SCHEME 1.7 Carbohydrate-derived primary amine-thiourea-catalyzed Biginelli reaction.

Biginelli reaction of a range of benzaldehydes, (thio)urea, and ethyl acetoacetate [38]. The best results were obtained when using gluco-2-aminocyclohexylthiourea depicted in Scheme 1.7 in combination with 2,4,6-trichlorobenzoic acid (TCBA) and *tert*-butylammonium trifluoroacetate as additives in dichloromethane as a solvent. Under these conditions, the corresponding dihydropyrimidines were formed in moderate to high yields (51–93%) with enantioselectivities of up to 99% ee, as shown in Scheme 1.7. In order to explain the stereoselectivity of the reaction, the authors proposed that both hydrogen-bonding interactions and enamine activation of the β-ketoester occurred in the transition states (Scheme 1.7). It must be noted that a low catalyst loading of only 5 mol % was sufficient to reach the discussed results. However, poor enantioselectivities were obtained for aliphatic aldehydes, such as *n*-butanal, which provided the corresponding product in only 15% ee. The asymmetric synthesis of dihydropyrimidines can also be catalyzed by chiral bifunctional primary amine-thiourea catalysts, as reported by the same authors [39]. The activation of the intermediate imine and carbonyl component toward nucleophilic attack was effectively catalyzed by the primary amine and thiourea moieties of the catalyst. The incorporation of a carbohydrate fragment in the catalyst was particularly attractive for the well-defined stereogenic centers and ubiquity. The combination of the primary amine-thiourea catalyst depicted in Scheme 1.7, TCBA, and t-BuNH$_2$-TFA as an additive proved superior in terms of reactivity and selectivity for the catalytic enantioselective Biginelli reaction depicted in this scheme. The scope of the reaction was investigated with various aromatic aldehydes producing the corresponding dihydropyrimidines with good to high enantioselectivities, while the reaction of aliphatic aldehyde was less efficient (51% yield) and offered lower stereoselectivity (15% ee). The authors proposed a transition state, in which the *Si*-face of the intermediate imine was predominately attacked by the enamine intermediate to afford the (*S*)-configured dihydropyrimidine, because the *Re*-face was shielded by the cyclohexane ring of the catalyst. More recently, the same authors investigated the same reaction under phase transfer conditions [40]. Indeed, by using the same catalyst at 15 mol % of catalyst loading combined with 15 mol % of TfOH and 10 mol % of t-BuNH$_2$·TFA as an additive, the reaction could be performed in aqueous media, providing the Biginelli products in good to high yields (62–93%) and high to excellent enantioselectivities (87–99% ee).

1.3 CONCLUSIONS

This chapter illustrates the power of asymmetric domino reactions, which have quickly become a powerful, fascinating, and highly efficient tool in organic chemistry. Indeed, these beautiful one-pot reactions can be considered as one of the most influential reaction classes of the last century. It demonstrates that this type of reactions can be successfully applied to the field of carbohydrate chemistry. Indeed, various stereocontrolled multicomponent domino reactions involving chiral carbohydrate derivatives as substrates have been developed in the last decade. For example, remarkable diastereoselectivities of up to >99% de were reported by Dondoni in various types of asymmetric multicomponent reactions, such as three-component Hantzsch reactions, three-component Mannich reactions, and four-component Reformatsky reactions. The same level of diastereoselectivity (>99% de) was also recently achieved by Beau and Norsikian in novel asymmetric three-component domino Petasis/Diels–Alder reactions. Moreover, other types of asymmetric multicomponent reactions of carbohydrate derivatives have provided excellent diastereoselectivities. For example, three-component Ugi reactions reported by Kunz offered up to 96% de. Other diastereoselectivities such as those described by Krishna for three-component Passerini reactions were moderate (52% de). When discussing the use of chiral substrates and auxiliaries in asymmetric synthesis, it should be noted that catalytic, particularly organocatalytic, asymmetric synthesis has attracted most attention. Its combination with the concept of domino sequences has allowed reaching high-molecular complexity with very often excellent levels of stereocontrol under environmentally friendly conditions, with simple operational procedures, and advantages of savings in solvents, time, energy, and costs. Major progress has been achieved in the last years through the creation of highly enantioselective versions of existing and novel organocatalytic domino and multicomponent reactions. For example, in the area of carbohydrate-derived organocatalysts, Chen et al. [40] reported remarkable enantioselective Biginelli reactions that allowed accomplishing enantioselectivities of up to 99% ee when catalyzed by chiral bifunctional primary amine-thiourea catalysts derived from D-glucose. These significant improvements are expected to influence the development of novel asymmetric domino reactions to a great extent and will make them even more useful tools for the total synthesis of natural and biologically active products. Undoubtedly, the future direction in this emerging field is to continue expanding the scope of asymmetric domino reactions especially to inexpensive carbohydrate derivatives through the identification of novel modes of reactivity and apply these powerful, simple, and easy-to-perform strategies to the synthesis of important, biologically interesting molecules including natural products.

REFERENCES

[1] (a) Tietze, L. F.; Beifuss, U. *Angew. Chem., Int. Ed.* **1993**, *32*, 131–163. (b) Tietze, L. F. *Chem. Ind. (London)* **1995**, *12*, 453–457. (c) Tietze, L. F. *Chem. Rev.* **1996**, *96*, 115–136. (d) Tietze, L. F.; Lieb, M. E. *Curr. Opin. Chem. Biol.* **1998**, *2*, 363–371. (e) Tietze, L. F.; Brasche, G.; Gericke, K. *Domino Reactions in Organic Synthesis*; Wiley-VCH: Weinheim, 2006. (f) Tietze, L. F. *Domino Reactions–Concepts for Efficient Organic Synthesis*; Wiley-VCH: Weinheim, 2014.

[2] Nicolaou, K. C.; Edmonds, D. J.; Bulger, P. G. *Angew. Chem., Int. Ed.* **2006**, *45*, 7134–7186.

[3] Denmark, S. E.; Thorarensen, A. *Chem. Rev.* **1996**, *96*, 137–165.

[4] Chapman, C. J.; Frost, C. G. *Synthesis* **2007**, 1–21.

[5] Fogg, D. E.; dos Santos, E. N. *Coord. Chem. Rev.* **2004**, *248*, 2365–2379.

[6] (a) Tietze, L. F.; Modi, A. *Med. Res. Rev.* **2000**, *20*, 304–322. (b) *Multicomponent Reactions*; Zhu, J., Bienaymé, H., Eds.; Wiley-VCH: Weinheim, 2005. (c) Ramon, D. J.; Yus, M. *Angew. Chem., Int. Ed.* **2005**, *44*, 1602–1634. (d) Synthesis of Heterocycles via Multicomponent Reactions. In *Topics in Heterocyclic Chemistry*, Vols. I and II; Orru, R. V. A., Ruijter, E., Eds.; Springer: Berlin, 2010.

[7] (a) Grondal, C.; Jeanty, M.; Enders, D. *Nat Chem.* **2010**, *2*, 167–178. (b) Nicolaou, K. C.; Chen, J. S. *Chem. Soc. Rev.* **2009**, *38*, 2993–3009. (c) Touré, B. B.; Hall, D. G. *Chem. Rev.* **2009**, *109*, 4439–4486. (d) Colombo, M.; Peretto, I. *Drug Discovery Today* **2008**, *13*, 677–684. (e) Hulme, C.; Gore, V. *Curr. Med. Chem.* **2003**, *10*, 51–80. (f) Padwa, A.; Bur, S. K. *Tetrahedron* **2007**, *63*, 5341–5378.

[8] *Chem. Soc. Rev.* **2009**, *38*(11). Special Issue on Rapid formation of molecular complexity in organic synthesis.

[9] (a) Posner, G. H. *Chem. Rev.* **1986**, *86*, 831–844. (b) Waldmann, H. *Nachr. Chem. Tech. Lab.* **1992**, *40*, 1133–1140. (c) Fukumoto, K. *Synth. Org. Chem. Jpn.* **1994**, *52*, 2–18. (d) Bunce, R. A. *Tetrahedron* **1995**, *51*, 13103–13159. (e) Parsons, P. J.; Penkett, C. S.; Shell, A. J. *Chem. Rev.* **1996**, *96*, 195–206. (f) Balaure, P. C. F.; Filip, P. I. A. *Rev. Roum. Chim.* **2002**, *46*(8), 809–833. (g) Capdevila, E.; Rayo, J.; Carrion, F.; Jové, I.; Borrell, J. I.; Teixido, J. *Afinidad* **2003**, *506*, 317–337. (h) Tietze, L. F.; Rackelmann, N. *Pure Appl. Chem.* **2004**, *76*, 1967–1983. (i) Dalko, P. I.; Moisan, L. *Angew. Chem., Int. Ed.* **2004**, *43*, 5138–5175. (j) Pellissier, H. *Tetrahedron* **2006**, *62*, 2143–2173. (k) Pellissier, H. *Tetrahedron* **2006**, *62*, 1619–1665. (l) Enders, D.; Grondal, C.; Hüttl, M. R. M. *Angew. Chem., Int. Ed.* **2007**, *46*, 1570–1581; (m) Guillena, G.; Ramon, D. J.; Yus, M. *Tetrahedron: Asymmetry* **2007**, *18*, 693–700; (n) D'Souza, D. M.; Müller, T. J. J. *Chem. Soc. Rev.* **2007**, *36*, 1095–1108. (o) Alba, A.-N.; Companyo, X.; Viciano, M.; Rios, R. *Curr. Org. Chem.* **2009**, *13*, 1432–1474. (p). Biggs-Houck, J. E; Younai, A.; Shaw, J. T. *Curr. Opin. Chem. Biol.*, **2010**, *14*, 371–382. (q) Ruiz, M.; Lopez-Alvarado, P.; Giorgi, G., Menéndez, J. C. *Chem. Soc. Rev.* **2011**, *40*, 3445–3454. (r) Albrecht, L.; Jiang, H.; Jorgensen, K. A. *Angew. Chem., Int. Ed.* **2011**, *50*, 8492–8509. (s) Pellissier, H. *Adv. Synth. Catal.* **2012**, *354*, 237–294. (t) De Graaff, C.; Ruijter, E.; Orru, R. V. A.

Chem. Soc. Rev. **2012**, *41*, 3969–4009. (u) Volla, C. M. R.; Atodiresei, I.; Rueping, M. *Chem. Rev.* **2014**, *114*, 2390–2431.

[10] *Comprehensive Asymmetric Catalysis*; Jacobsen, E. N., Pfaltz, A., Yamamoto, H., Eds.; Springer: Berlin, 1999.

[11] (a) Hall, N. *Science* **1994**, *266*, 32–34. (b) Wender, P. A.; Handy, S. T.; Wright, D. L. *Chem. Ind. (London)* **1997**, 765–769.

[12] (a) Ugi, I.; Meyr, R.; Fetzer, U.; Steinbrückner, C. *Angew. Chem.* **1959**, *71*, 386–388. (b) Orru, R. V. A.; de Greef, M. *Synthesis* **2003**, 1471–1499.

[13] (a) Hantzsch, A. *Justus Liebigs Ann. Chem.* **1882**, *215*, 1–82; (b) Jones, G. Pyridines and Their Benzo Derivatives: Synthesis. In *Comprehensive Heterocyclic Chemistry II*, Vol. V; Katritzky, A. R., Rees, C. W., Scriven, E. F. V., Eds.; Pergamon: Oxford, 1996; p 167.

[14] Dondoni, A.; Massi, A. *Acc. Chem. Res.* **2006**, *39*, 451–463.

[15] Dondoni, A.; Massi, A.; Aldhoun, M. *J. Org. Chem.* **2007**, *72*, 7677–7687.

[16] Ducatti, D. R. B.; Massi, A.; Noseda, M. D.; Duarte, M. E. R.; Dondoni, A. *Org. Biomol. Chem.* **2009**, *7*, 1980–1986.

[17] Dömling, A.; Ugi, I. *Angew. Chem., Int. Ed.* **2000**, *39*, 3168–3210.

[18] (a) Ugi, I. *Angew. Chem., Int. Ed. Engl.* **1982**, *21*, 810–819; (b) Ugi, I. *J. Prakt. Chem.* **1997**, *339*, 499–516.

[19] Sigmuller, F.; Herrmann, R.; Ugi, I. *Tetrahedron* **1986**, *42*, 5931–5941.

[20] (a) Kunz, H., Pfrengle, W. *Tetrahedron* **1988**, *44*, 5487–5494. (b) Kunz, H.; Pfrengle, W.; Sager, W. *Tetrahedron Lett.* **1989**, *30*, 4109–4110. (c) Kunz, H.; Pfrengle, W. *J. Am. Chem. Soc.* **1988**, *110*, 651–652. (d) Kunz, H.; Pfrengle, W.; Rück, K.; Sager, W. *Synthesis* **1991**, 1039–1042.

[21] (a) Goebel, M.; Ugi, I. *Synthesis* **1991**, 1095–1098. (b) Lehnhoff, S.; Goebel, M.; Karl, R. M.; Klösel, R.; Ugi, I. *Angew. Chem., Int. Ed. Engl.* **1995**, *34*, 1104–1107.

[22] Passerini, M. *Gazz. Chim. Ital.* **1921**, *51*, 126–129.

[23] Yue, T.; Wang, D.; Masson, G.; Zhu, J. *Angew. Chem., Int. Ed.* **2009**, *48*, 6717–6721.

[24] (a) Bock, H.; Ugi, I. *J. Prakt. Chem.* **1997**, *339*, 385–389. (b) Gryko, D.; Chalko, J.; Jurczak, J. *Chirality* **2003**, *15*, 514–541. (c) Frey, R.; Galbraith, S. G.; Guelfi, S.; Lamberth, C.; Zeller, M. *Synlett* **2003**, 1536–1538.

[25] Krishna, P. R.; Dayaker, G.; Reddy, P. V. N. *Tetrahedron Lett.* **2006**, *47*, 5977–5980.

[26] Mannich, C.; Krosche, W. *Arch. Pharm.* **1912**, *250*, 647–667.

[27] (a) Arend, M.; Westermann, B.; Risch, N. *Angew. Chem., Int. Ed.* **1998**, *37*, 1044–1070. (b) Cordova, A. *Acc. Chem. Res.* **2004**, *37*, 102–112. (c) Verkade, J. M. M.; Van Hemert, L. J. C.; Quaedflieg, P. J. L. M.; Rutjes, F. P. J. T. *Chem. Soc. Rev.* **2008**, *37*, 29–41.

[28] Arrayas, R. G.; Carretero, J. C. *Chem. Soc. Rev.* **2009**, *38*, 1940–1948.

[29] *Carbohydrate-based Drug Discovery*; Wong, C.-H., Ed.; Wiley-VCH: Weinheim, 2003.

[30] Cannillo, A.; Norsikian, S.; Huu Dau, M.-E. T.; Retailleau, P.; Iorga, B. I.; Beau, J.-M. *Chem. Eur. J.* **2014**, *20*, 1–12.

[31] (a) Noyori, R. In *Asymmetric Catalysts in Organic Synthesis*; Wiley: New York, 1994. (b) *Catalytic Asymmetric Synthesis,* 2nd edition; Ojima, I., Ed.; Wiley-VCH: New York, 2000. (c) *Transition Metals for Organic Synthesis*, 2nd edition; Beller, M., Bolm, C. Eds.; Wiley-VCH: Weinheim, 2004. (d) Ramon, D. J.; Yus, M. *Chem. Rev.* **2006**, *106*, 2126–2208.

[32] (a) *Biocatalysts for Fine Chemicals Synthesis*; Roberts, S. M., Ed.; Wiley-VCH: New York, 1999. (b) *Enzyme Catalysis in Organic Synthesis*, 2nd edition; Drauz, K., Waldmann, H., Eds. Wiley-VCH: Weinheim, 2002. (c) Bommarius, A. S.; Riebel, B. R. *Biocatalysis*; Wiley-VCH: Weinheim, 2004.

[33] (a) Knoevenagel, E. *Chem. Ber.* **1896**, *29*, 172–174. (b) Eder, U.; Sauer, G.; Wiechert, R. *Angew. Chem., Int. Ed. Engl.* **1971**, *10*, 496–497. (c) Hajos, Z. G.; Parrish, D. R. *J. Org. Chem.*, **1974**, *39*, 1615–1621. (d) Ahrendt, K. A., Borths, C. J., MacMillan, D. W. C. *J. Am. Chem. Soc.* **2000**, *122*, 4243–4244. (e) List, B.; Lerner, R. A.; Barbas, C. F. *J. Am. Chem. Soc.* **2000**, *122*, 2395–2396.

[34] (a) Dalko, P. I.; Moisan, L. *Angew. Chem., Int. Ed.* **2001**, *40*, 3726–3748. (b) Berkessel, A.; Gröger, H. In *Asymmetric Organocatalysis–From Biomimetic Concepts to Powerful Methods for Asymmetric Synthesis*; Wiley-VCH: Weinheim, 2005. (c) Seayad, J.; List, B. *Org. Biomol. Chem.* **2005**, *3*, 719–724. (d) Taylor, M. S.; Jacobsen, E. N. *Angew. Chem., Int. Ed.* **2006**, *45,* 1520–1543. (e) Dalko, P. I. In *Enantioselective Organocatalysis*; Wiley-VCH: Weinheim, 2007. (f) Dalko, P. I. *Chimia* **2007**, *61,* 213–218. (g) Pellissier, H. *Tetrahedron* **2007,** *63,* 9267–9331. (h) Doyle, A. G.; Jacobsen, E. N. *Chem. Rev.* **2007**, *107*, 5713–5743. (i) Gaunt, M. G.; Johansson, C. C. C.; McNally, A.; Vo, N. C. *Drug Discovery Today*, **2007**, *2,* 8–27. (j) *Chem. Rev.* **2007**, *107*(12), 5413–5883, Special Issue on Organocatalysis (Ed.: List, B.). (k) MacMillan, D. W. C. *Nature* **2008**, *455,* 304–308. (l) Yu, X.; Wang, W. *Chem. Asian. J.* **2008**, *3*, 516–532. (m) Dondoni, A.; Massi, A. *Angew. Chem., Int. Ed.* **2008**, *47*, 4638–4660. (n) Melchiorre, P.; Marigo, M.; Carlone, A.; Bartoli, G. *Angew. Chem., Int. Ed.* **2008**, *47*, 6138–6171. (o) Peng, F.; Shao, Z. *J. Mol. Catal. A* **2008**, *285*, 1–13. (p) Barbas, C. F. *Angew. Chem., Int. Ed.* **2008**, *47*, 42–47. (q) Palomo, C.; Oiarbide, M.; Lopez, R. *Chem. Soc. Rev.* **2009**, *38*, 632–653. (r) Bertelsen, S.; Jorgensen, K. A. *Chem. Soc. Rev.* **2009**, *38*, 2178–2189. (s) Bella, M.; Gasperi, T. *Synthesis* **2009**, 1583–1614. (t) Gong, L. In *Special Topic: Asymmetric Organocatalysis*, In *Chin. Sci. Bull.* **2010**, *55*(17); (u) List, B. In *Asymmetric Organocatalysis*, In *Top. Curr. Chem.* **2010**, *291*; (v) Pellissier, H. In *Recent Developments in Asymmetric Organocatalysis*; Royal Society of Chemistry: Cambridge, 2010; (w) *Enantioselective Organocatalysed Reactions*, Vols. I and II, Mahrwald, R., Ed.; Springer: Berlin, 2011; (x) Giacalone, F.; Gruttadauria, M.; Agrigento, P.; Noto, R. *Chem. Soc. Rev.* **2012**, *41*, 2406–2447. (y) Pellissier, H. *Adv. Synth. Catal.* **2012**, *354*, 237–294.

[35] (a) de Figueiredo, R. M.; Christmann, M. *Eur. J. Org. Chem.* **2007**, 2575–2600. (b) Marquès-Lopez, E.; Herrera, R. P.; Christmann, M. *Nat. Prod. Rep.* **2010**, *27*, 1138–1167.

[36] (a) Biginelli, P. *Ber. Dtsch. Chem. Ges.* **1891**, *24*, 1317–1319. (b) Biginelli, P. *Gazz. Chim. Ital.* **1893**, *23*, 360–413.

[37] Chen, X.-H.; Yu, X.-Y.; Liu, H.; Cun, L.-F.; Gong, L.-Z *J. Am. Chem. Soc.* **2006**, *128*, 14802–14803.

[38] Wang, Y.; Yang, H.; Yu, J.; Miao, Z.; Chen, R. *Adv. Synth. Catal.* **2009**, *351*, 3057–3062.

[39] Li, X.; Deng, H.; Luo, S. Z.; Cheng, J. P. *Eur. J. Org. Chem.* **2008**, 4350–4356.

[40] Wang, Y.; Yu, J.; Miao, Z.; Chen, R. *Org. Biomol. Chem.* **2011**, *9*, 3050–3054.

2

ORGANOCATALYZED CASCADE REACTION IN CARBOHYDRATE CHEMISTRY

Benjamin Voigt and Rainer Mahrwald

Institut für Chemie der Humboldt-Universität zu Berlin, Brook-Taylor-Str. 2, 12 489 Berlin, Germany

2.1 INTRODUCTION

Due to the great biological importance of carbon chain-elongated carbohydrates, the synthesis of this structural diverse compound class has attracted high synthetic interest from the scientific community. Among the elongated carbohydrates, C-glycosides have gained considerable importance over the last few decades, since they are configuratively stable under enzymatic conditions and less prone to cleavage at the anomeric carbon. As such, they are appealing substrates for chemical biology and medicinal chemistry.

The result of this development is a constantly increasing demand for methods for the stereoselective synthesis of this class of glycosides. Although many methods exist for the synthesis of these compounds from carbohydrate derivatives or analogues [1], their applications are hampered by low yields and selectivities. This is caused by necessary complex and extensive manipulations of protective groups in combination with the activation of the anomeric carbon atom.

The diversity of the developed glycosylation protocols is mainly driven by the different requirements of aglycons deployed, the availability of substrates and their complexity, as well as the achieved selectivities. A general solution for this synthetic

Domino and Intramolecular Rearrangement Reactions as Advanced Synthetic Methods in Glycoscience, First Edition. Edited by Zbigniew J. Witczak and Roman Bielski.

problem does not exist as yet. This holds true especially for the common metal-organic reactions as they are Reformatsky reaction [2], Knoevenagel reaction [3], allylation [4], or aldol reaction [5].

A big simplification of accessing C-glycosides has been achieved by reactions of unprotected and unactivated carbohydrates with phosphorylides. Examples for Wittig reactions with unprotected carbohydrates are summarized in Reference 6, whereas examples for Horner olefination (even in aqueous reaction medium) can be found in Reference 7. The reactions were carried out mostly at higher temperature. Also, allylations of even unprotected glycals have been reported in the presence of equimolar amounts of trimethylsilyl trifluoromethanesulfonate [8].

The aggressive development of organocatalysis over the last 15 years offers new strategies for C-glycosylation. In contrast to the reactions described above, these new methods not only give access to the target structural motif from unprotected and unactivated carbohydrates, but also enable catalytic procedures.

2.2 C-GLYCOSIDES

This development is briefly illustrated by the synthesis of C-glycosides of acetone and several pentoses. For reasons of comprehensibility and comparability, these reactions are discussed with unprotected and unactivated ribose in Scheme 2.1. The methods depicted in Scheme 2.1 represent direct C-glycosylation processes of unprotected and unactivated carbohydrates (with the exception of Eq. 1; Scheme 2.1).

In addition, there are many methods for C-glycosylation reactions using protected and activated carbohydrates in multistep sequences [9]. However, these transformations are not the matter of the discussion here.

Twenty-five years ago, glycosides of acetone were first synthesized by the Mukaiyama reaction. Protected and activated carbohydrates were reacted with silyl enol ether of acetone in the presence of catalytic amounts of Lewis acids ($BF_3{\cdot}Et_2O$). Under these conditions, the expected products were isolated in partially quantitative yields (Eq. 1; Scheme 2.1) [10].

Ten years later, Lubineau et al. reported the direct reaction of unprotected carbohydrates with acetylacetone in aqueous alkali media [11]. By application of this method, an access to different mixtures of furanoid and pyranoid structures of α- and β-configured C-glycosides [12] was obtained. The ratio of the products depends on the conditions of execution for this reaction and the carbohydrates deployed (Eq. 2, Scheme 2.1). This Knoevenagel/Michael/retro-Claisen-aldol cascade is carried out at high temperature (60–90°C) and is associated with the loss of a C2 fragment of the starting 1,3-dicarbonyl component (when used with acetylacetone).

Applications of this direct C-glycosylation process can be found as a key step in the synthesis of glucosidase inhibitors [13], exogene gene inductors [14], inhibitor on neoglycoprotein/cell surface binding [15], amphophilic carbohydrates [16], carbohydrates of pyrazoles [17] and antiproliferative agents [18], derivatives of barbituric acid [19], activators of glycosaminoglycan synthesis [20], surface-active carbohydrates [21], fluorescent labels for glycomics [22], or glycoconjugates [23]. Furthermore,

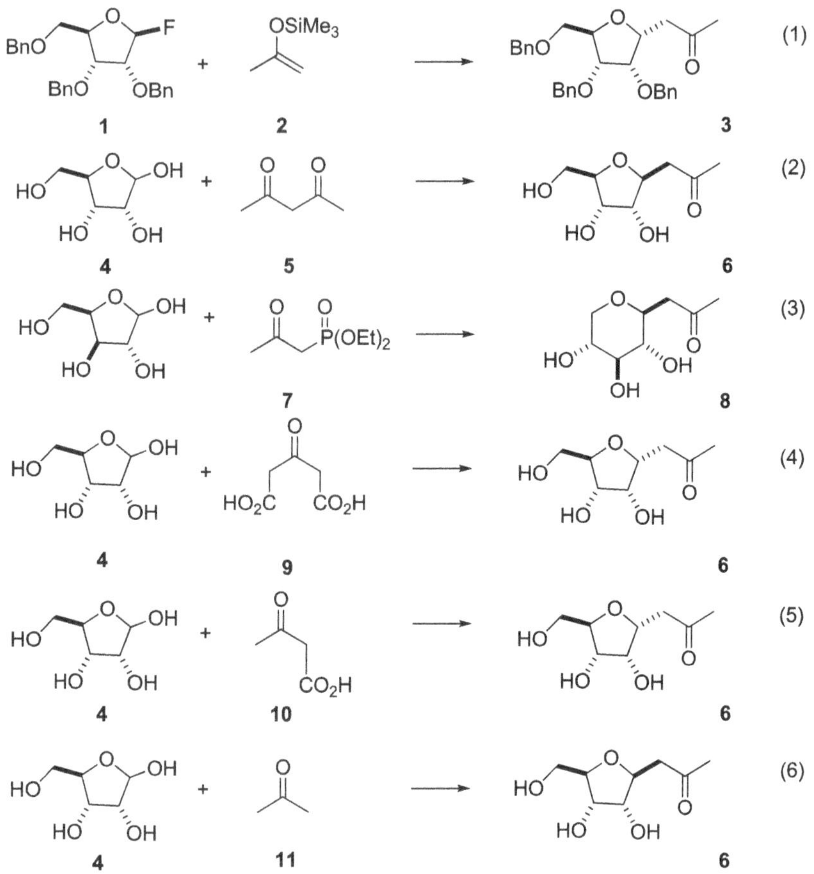

SCHEME 2.1 Syntheses of C-glycosides of acetone.

this transformation has been used for the synthesis of L-fucose and L-galactose [24]. For an overview of Knoevenagel reactions of unprotected and unactivated carbohydrates, see Reference 25.

An alternative access to C-glycosides is given by Wittig or Horner olefination of unprotected carbohydrates with the corresponding phosphonate. Subsequent intramolecular oxa-Michael reaction gives an access to glycoside **8** of acetone. A big excess of base is necessary for the reaction (see the example of xylose in Eq. 3; Scheme 2.1) [7].

We successfully developed several amine-catalyzed decarboxylative aldol/oxa-Michael cascade reactions deploying unprotected and unactivated carbohydrates to access glycosides of acetone. Useful substrates for this transformation are acetonedicarboxylic acid **9** or acetoacetic acid **10** (Eqs. 4 and 5; Scheme 2.1). These reactions are catalyzed by tertiary amines at room temperature. The corresponding C-glycoside **6** was isolated with good yields (60%) and selectivity (β/α: 3/1) [26].

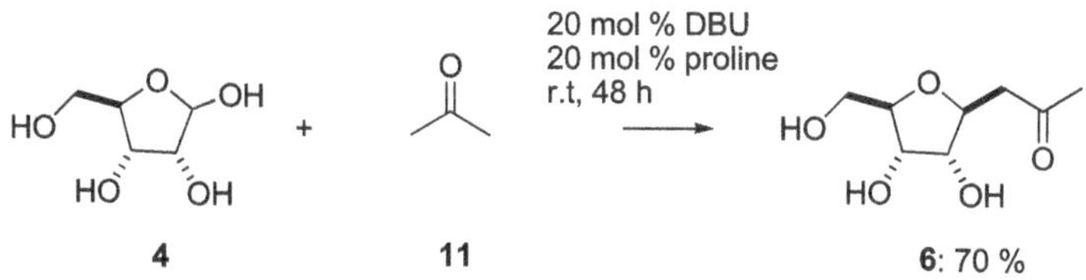

SCHEME 2.2 Amine-catalyzed cascade reactions on ribose and acetone.

During these investigations, we observed an amine-catalyzed formation of C-glycosides of unprotected carbohydrates with acetone (Eq. 6; Scheme 2.1). This reaction is catalyzed by a combination of tertiary amines and proline. By means of this methodology, an easy, catalytic access to C-glycosides can be realized under mild conditions. This operationally simple and efficient base-catalyzed transformation initiated further investigations.

In a first series, an optimization of the catalytic system was conducted. A reaction was not observed when used with amine bases. The same is true for the deployment of proline; proline does not induce a reaction. In contrast to these results, a clear reaction was detected when used with an equimolar mixture of proline and 1,8-diazabicyclo[5.4.0]undec-7-ene (DBU). Thus, an access to the unprotected acetone glycoside **6** is given with good yields of approximately 70% (Scheme 2.2).

This observation was generalized by extending the substrate scope. To this end, several carbohydrates and methyl ketones were tested in this transformation. For achieving quantitative yields, proline and DBU should be used in equimolar amounts at room temperature. Under these conditions, not only acetone, but several unsymmetrical methyl ketones can be transformed into their corresponding C-glycosides (**12**–**17**) (Scheme 2.3) [27].

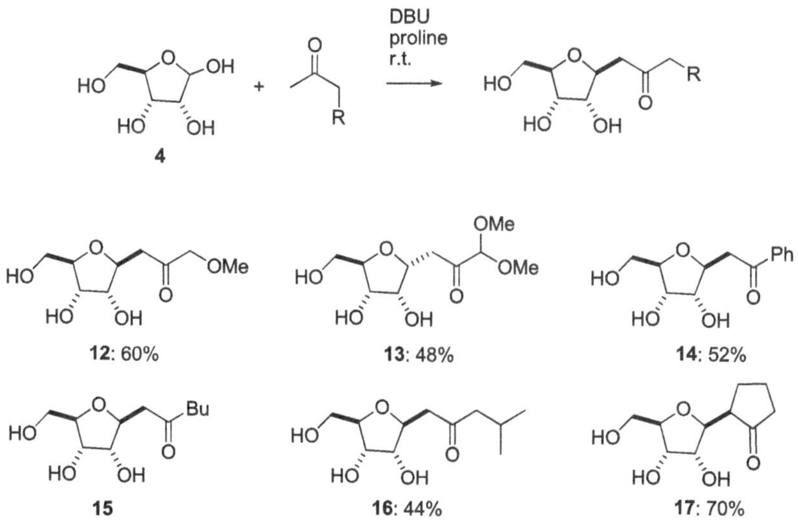

SCHEME 2.3 Synthesis of C-glycosides by cascade reactions of ribose with methyl ketones.

As the working model for the reaction mechanism, an aldol-condensation followed by an intramolecular oxa-Michael process is assumed. Considerable degrees of β-diastereoselectivity at the former anomeric carbon atom (C-1) were detected in the isolated products (dr ~7/3). Under the described reaction conditions, the furanoid structure of C-glycosides was detected as the major product. When used with oxygen-containing ketones in this cascade reaction, the favored formation of hemiketals was observed. The corresponding C-glycosidic ketones can be easily obtained by subsequent treatment with acidic ion-exchanger.

2.3 AMINE-CATALYZED KNOEVENAGEL-ADDITIONS

By further expanding the variety of substrates—from methyl ketones to 1,3-dicarbonyl compounds—a chain elongation of the employed carbohydrates instead of a C-glycosylation process was observed.

The deployed 1,3-dicarbonyl compounds were incorporated into the carbon skeleton of the carbohydrates, although with low yields. Initial experiments with deoxyribose **18** and ethyl acetoacetate **19** were conducted in the presence of catalytic amounts of diisopropylethylamine. Chain elongated 1-deoxy ketose **20** resulted from a Knoevenagel-addition/ketalization cascade reaction and was isolated in only moderate yields (29%) but with an extremely high degree of stereoselectivity (Scheme 2.4) [28]. Only one single stereoisomer was detected by NMR experiments. The C–C bond formation proceeds with an extremely high degree of *syn*-diastereoselectivity. The corresponding *anti*-configured product was not detected. These observations are consistent with results we obtained in amine-catalyzed direct aldol additions with dihydroxyacetone [29]. Further extension of this observation to other pentoses yields the corresponding Knoevenagel-addition adducts, albeit with lower yields.

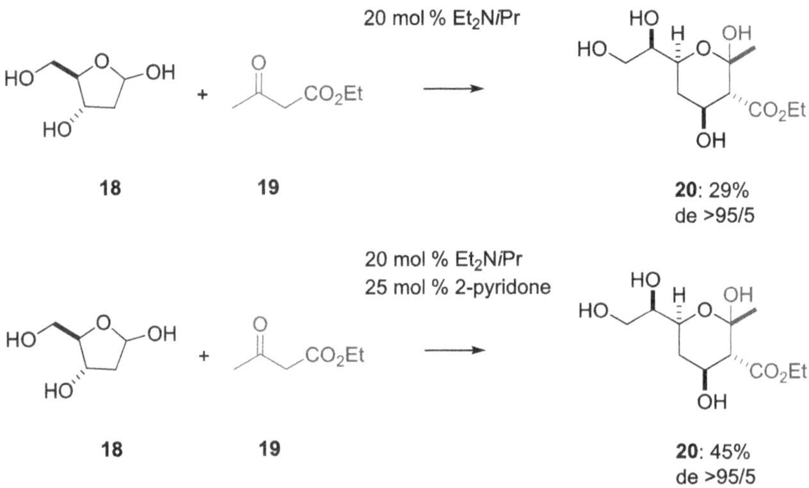

SCHEME 2.4 Base-catalyzed Knoevenagel reaction of ribose with ethyl acetoacetate. For a color version of this figure, see the color plate section.

Moreover, long reaction times are required for these conversions. When used with deoxyribose and ribose, highest yields were obtained, whereas lowest yields were found by the employment of xylose. The observed yields, in association with the configuration of the corresponding pentoses, clearly indicate a correlation with the existence of acyclic structures of the pentoses during the course of the reaction [30]. To overcome the problem of low yields of Knoevenagel addition products, we tested 2-hydroxypyridine as an additional catalyst, which is known to have a large effect on the anomerization equilibration of carbohydrates by formation of hydrogen bonds [31]. By increasing the equilibrium constant, a higher amount of active acyclic form is available per defined time unit.

And indeed, an addition of catalytic amounts of 2-hydroxypyridine increased the yields in the reaction of deoxyribose with ethyl acetoacetate **19** (**20**: 29% → 45%). The results of these optimization works are depicted in Scheme 2.4.

To use consistent stereodescriptors in further discussions, the configurative results of the reaction of ribose **4** with acetoacetic ester **19** are representatively depicted in Scheme 2.5. The proposed intermediately formed acyclic structures are shown to underline the nomenclature used (in bracket, not detected, or isolated).

The degree of internal diastereoselectivity is dictated by the existence and by the relative configuration of the 2,3-hydroxy groups of pentoses deployed. Results of

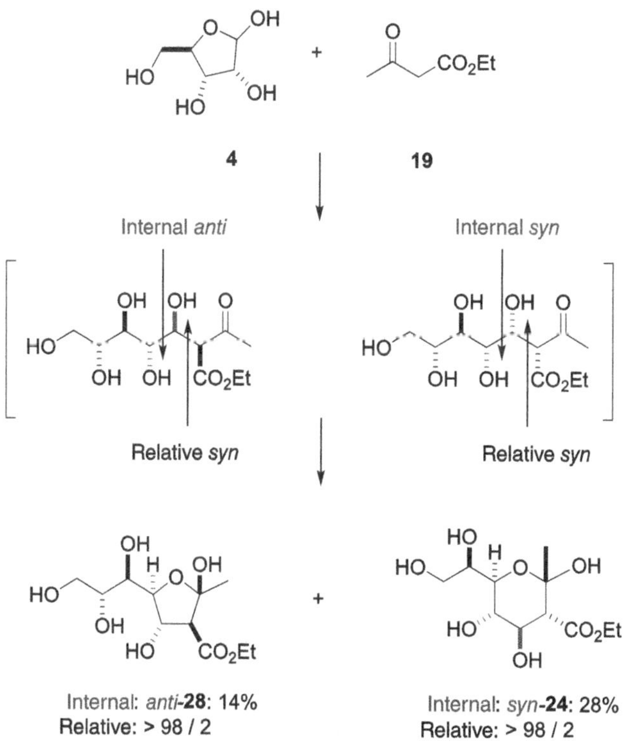

SCHEME 2.5 Configurative course of aldol reaction of D-ribose with ethyl acetoacetate. For a color version of this figure, see the color plate section.

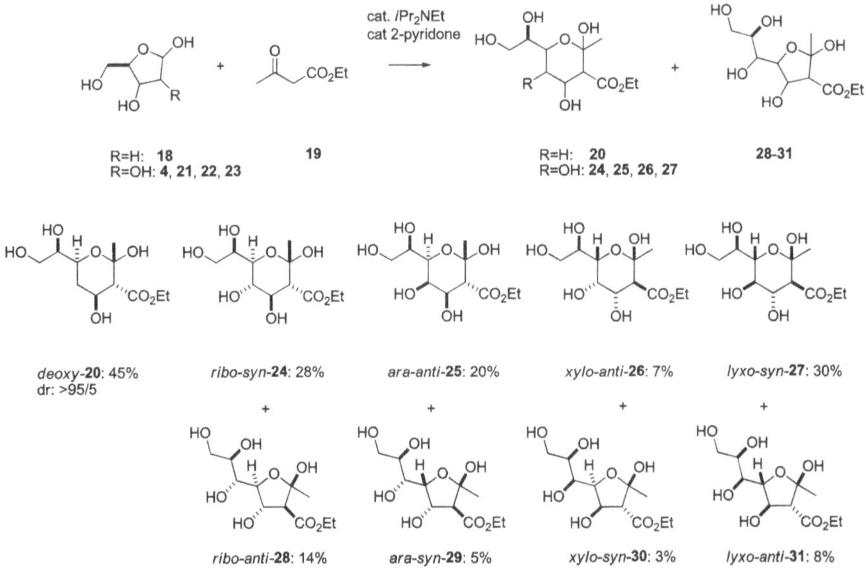

SCHEME 2.6 Chain elongation of D-pentoses with ethyl acetoacetate.

reactions of deoxyribose **18** impressively demonstrate this statement. In the deoxyribose-series only, one single stereoisomer **20** is detected (missing hydroxy group in 2 positions). This fact contrasts results that were obtained in reactions of ribose **4** with acetoacetate **19**. Two products, **24** and **28**, were isolated with extremely high degrees of relative *syn*-configuration. But they differ by their relative, internal configuration (dr ~7/3). Moreover, the relative 2,3-configuration of the deployed pentoses does not simply determine this internal diastereoselectivity. The installation of internal configuration also dictates the formation of pyranoid or furanoid structures of the elongated carbohydrates.

Several different unprotected carbohydrates have been successfully deployed in the cascade reactions with ethyl acetoacetate **19** by subsequent systematic studies. The results of these investigations are depicted in Schemes 2.6 (pentoses) and 2.7 (hexoses).

When used with ribose **4**, arabinose **21**, xylose **22,** and lyxose **23**, two diastereoisomers were detected in reactions with acetoacetate **19**. As discussed above, these findings contrast reactions with deoxyribose **18**. Again, extremely high degrees of *syn*-diastereoselectivity were observed during the C—C bond-formation process (relative diastereoselectivity). *Anti*-configured products were not detected under these reaction conditions.

An additional asymmetric induction was observed concerning the relative, internal configuration. Remarkably, relative 2,3-*anti*-configured pentoses (ribose **4** or lyxose **23**) yield internal *syn*-configured **24** and **27** as the main products. In contrast, when used with 2,3-*syn*-configured arabinose **21** or xylose **22**, internal *anti*-configured products **25** and **26** were obtained as main products [32].

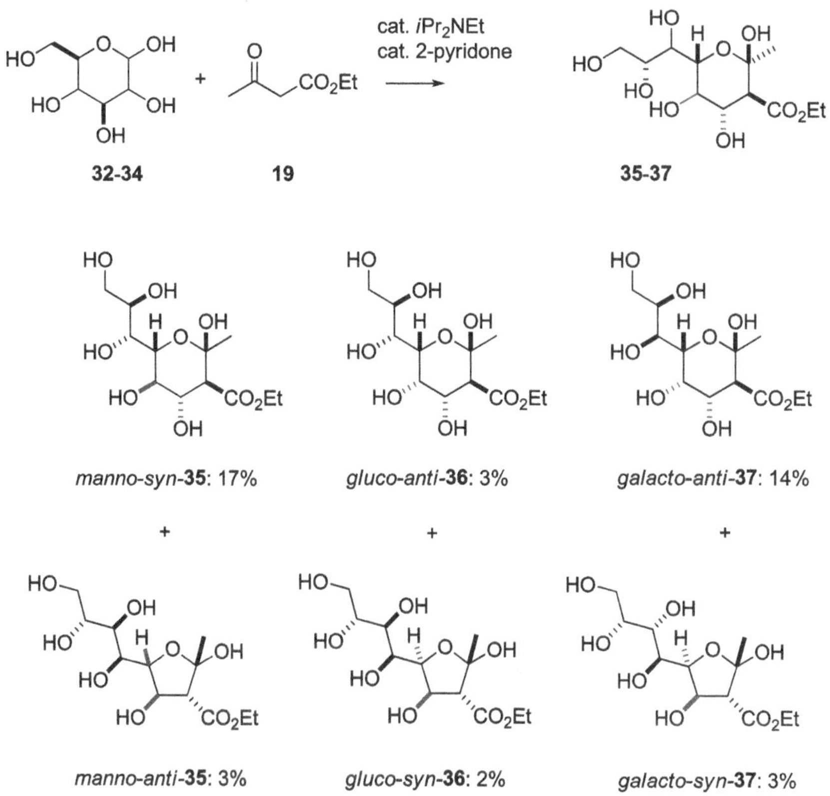

SCHEME 2.7 Chain elongation of D-hexoses with ethyl acetoacetate. For a color version of this figure, see the color plate section.

Also, this transformation was expanded to reactions with unprotected hexoses, albeit with lower yields (mannose **32**, glucose **33**, and galactose **34**; Scheme 2.7). The observed diastereoselectivity, as well as the formation of furanoid and pyranoid products, is consistent with the results obtained in the pentose series.

At this point of the investigations, we reasoned that yields could be further increased by enhancing the effective amount of reactive acyclic species within the mutarotation equilibrium. This can easily be accomplished by a selective protection of the 5-hydroxy group of starting pentoses. In this way, the thermodynamically favored pyranoid structures of the starting pentoses are no longer available to participate in the mutarotation equilibrium [33]. Hence, by removal of this structure from the mutarotation equilibrium, the effective amount of acyclic carbonyl species should be increased.

To proof this principle, readily available 5-tritylethers of pentoses **38–42** were deployed as substrates in the Knoevenagel addition [34]. As expected, Knoevenagel-addition products **43–51** were obtained with higher yields under these conditions (compare results of Scheme 2.6 with those of Scheme 2.8). A strong enhancement

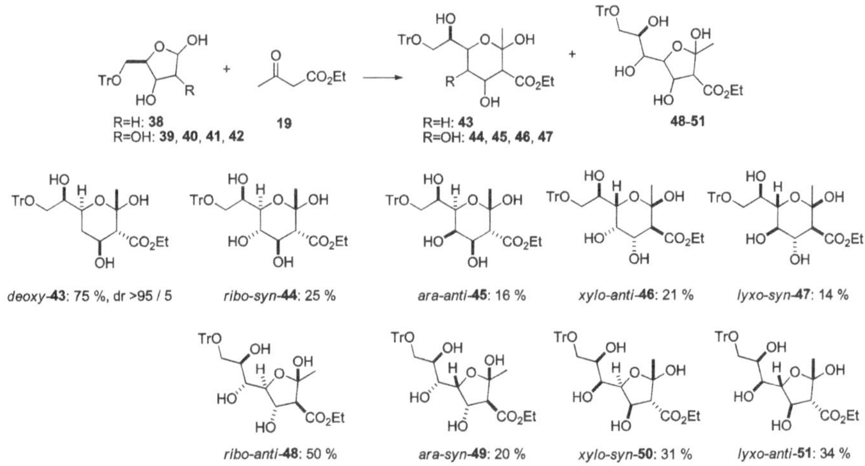

SCHEME 2.8 Amine-catalyzed reactions of ethyl acetoacetate with 5-protected D-pentoses.

of yields was observed in all described reactions. Especially yields of the expected Knoevenagel reaction product of notoriously difficult carbohydrate xylose **22** were quintupled (**26/30**: 10% → **46/50**: 52%).

Remarkably, a general inversion of internal diastereoselectivity is observed by deployment of 5-tritylated pentoses **39–42** in these transformations (compare internal diastereoselectivities of Scheme 2.6 with those of Scheme 2.8).

For an overview and a better understanding for the further discussion, the results of Schemes 2.6 and 2.8 are summarized in Table 2.1.

Explanations for these different configurative outcomes are given representatively for reactions of lyxose **23** and tritylated lyxose **42** with acetoacetic ester **19** in Scheme 2.9 and for arabinose **21** and tritylated arabinose **40** with acetoacetic ester **19** in Scheme 2.10. The extremely high relative *syn*-diastereoselectivity that is observed in

TABLE 2.1 Amine-catalyzed chain elongation of unprotected carbohydrates

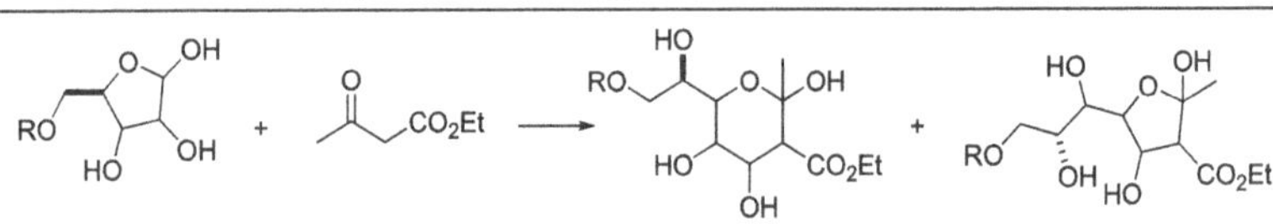

Entry	Carbohydrate	Yield [%][a] R=H	*syn/anti*	R=Tr	*syn/anti*[b]
1	Ribose	**24/28**	28/14	**44/48**	25/50
2	Arabinose	**25/29**	20/5	**45/49**	16/20
3	Xylose	**26/30**	7/3	**46/50**	21/31
4	Lyxose	**27/31**	30/8	**47/51**	14/34

[a] Isolated yields.
[b] *syn/anti* ratios were evaluated from the ^{1}H NMR spectra by integration of significant signals.

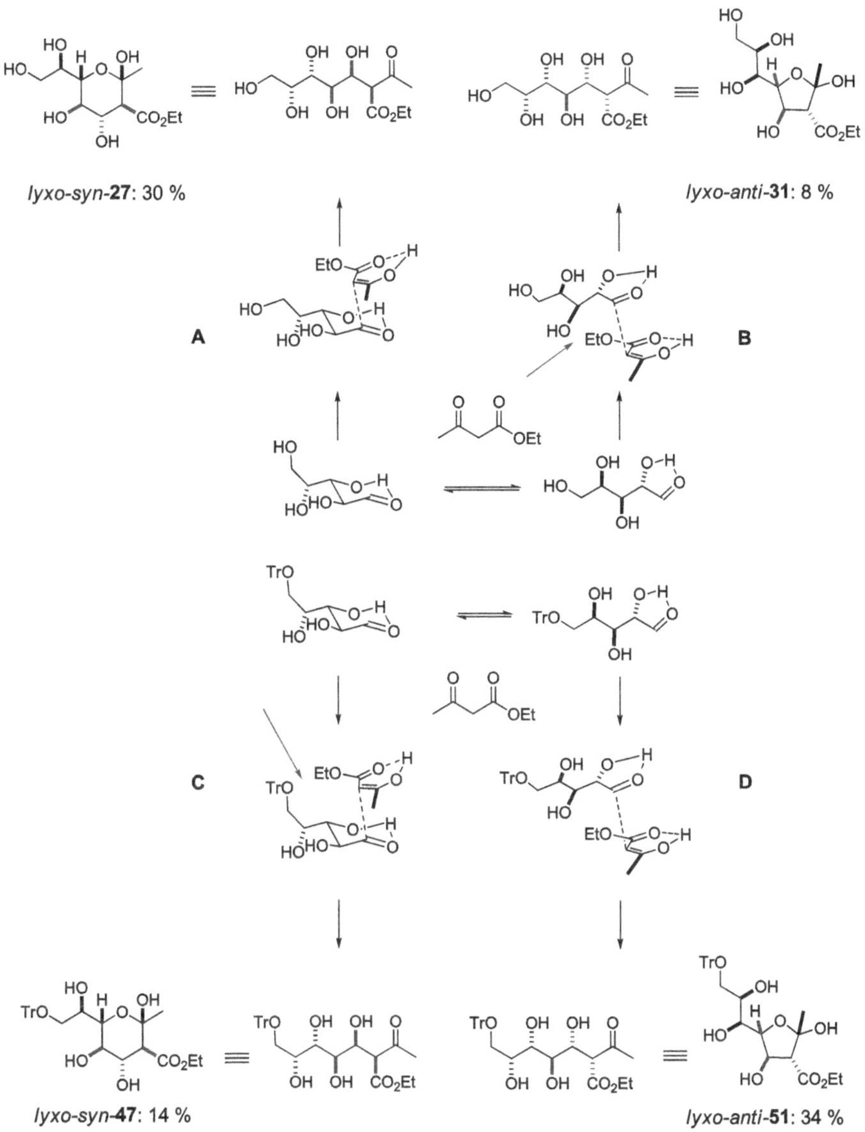

SCHEME 2.9 Configurative course of chain-elongation of lyxose **23** and 5-tritylated lyxose **42** with acetoacetic ester **19**. For a color version of this figure, see the color plate section.

all reactions during the C—C bond-formation process is based upon two configurative requirements:

- A *Z*-enolate of acetoacetic ester **19** is the only reactive species. This configuration is stabilized by a hydrogen bond.

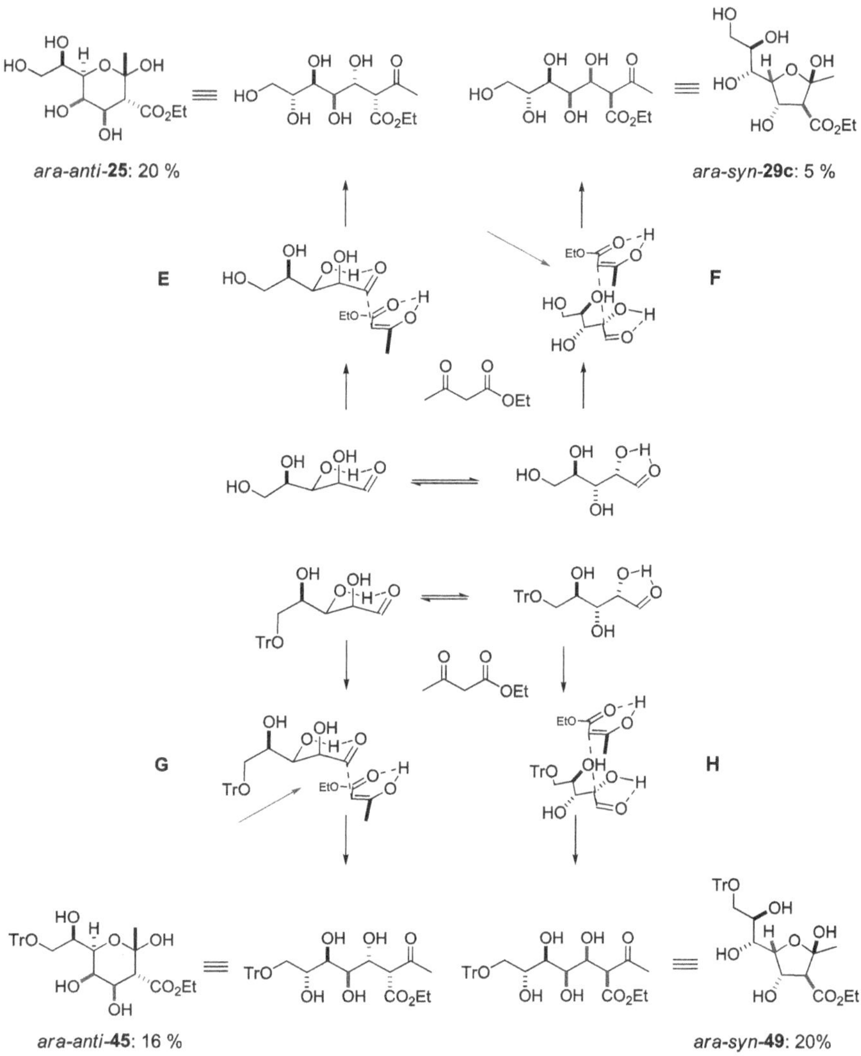

SCHEME 2.10 Configurative course of chain elongation of arabinose **21** and 5-tritylated arabinose **40** with ethyl acetoacetate **19**. For a color version of this figure, see the color plate section.

- Furthermore, the installation of *syn*-configuration during the C–C bond formation is a consequence of a *like*-approach (*Re*-side attack of acetoacetic ester to the *Re*-side of carbohydrate aldehyde or *Si*-side attack of acetoacetic ester **19** to the *Si*-side of carbohydrate aldehyde).

The correlated proposed stereochemical models are shown in Scheme 2.9 (**A** and **B** for lyxose) and in Scheme 2.10 (**E** and **F** for arabinose). The transition state models

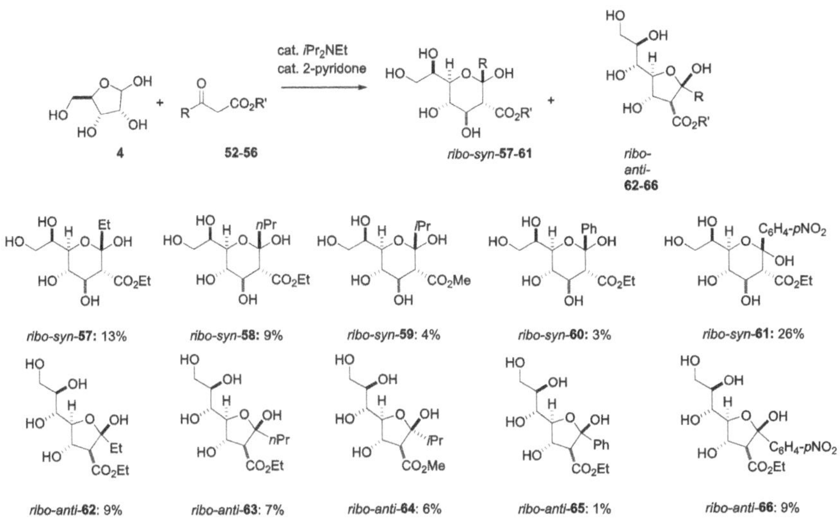

SCHEME 2.11 Chain elongation of D-ribose with different substituted acetoacetates. For a color version of this figure, see the color plate section.

(**A**/**B** and **E**/**F**), which result in the different internal stereochemical outcome, differ in the ring size of hydrogen bond stabilized cycles of pentoses (**A** and **E** → 6-membered rings; **B** and **F** → 5-membered rings). As a result of that, two transition state models are proposed. For reasons of stereochemical repulsion, models **A** and **E** are assumed to be the favored ones (follow red arrows in models **B** and **F**).

As mentioned above, an overall reversal of the internal diastereoselectivity was observed by the deployment of 5-tritylated pentoses (deoxyribose **38**, ribose **39**, arabinose **40**, xylose **41**, and lyxose **42**). These findings indicate a general steric shielding caused by the trityl group, which works in addition to and independently of possible hydrogen bonds. In contrast to results of unprotected carbohydrates (Scheme 2.6), the formation of the 5-membered ring is favored due to the bulky trityl group, which assembles in a maximum distance to the C–C bond-formation site. This is demonstrated in Schemes 2.9 and 2.10. When used with 5-tritylated arabinose **40**, the bulky trityl substituent directs the acetoacetic ester **19** to a *Si*-side access. As a consequence, *syn*-configured product **49** was detected as the main product (model **H** in Scheme 2.10). Similar considerations hold true for the preference of model **D** in the lyxose series (Scheme 2.9).

With further experiments, we have expanded this amine-catalyzed reaction of ribose **4** to several different β-keto esters. The results of this investigation are depicted in Scheme 2.11. The findings suggest that the internal diastereoselectivity and yield are influenced by the steric demand of the substituents of the acetoacetic esters deployed (**52–56**). Moreover, the results demonstrate that this new transformation can be extended to further 1,3-dicarbonyl compounds.

By tweaking the reaction conditions slightly—20 mol % proline instead of 20 mol % 2-pyridone—a completely different reaction sequence is observed. The

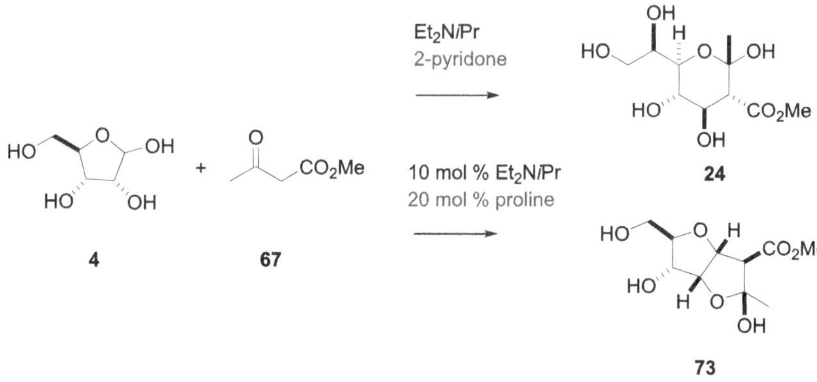

SCHEME 2.12 Different cascade reactions of D-ribose with acetoacetate. For a color version of this figure, see the color plate section.

corresponding C-glycoside **73** was isolated as a single stereoisomer instead of elongated carbohydrate **24**, isolated in reactions of ribose with acetoacetate. This compound was formed by a Knoevenagel condensation/ketalization/oxa-Michael cascade reaction. These different results are demonstrated in Scheme 2.12 [35].

A subsequent optimization yielded the following protocol. When used with 10 mol % DBU and 20 mol % proline the highly functionalized hemiketal, fused tetrahydrofuran **73** was isolated in good to high yields (Scheme 2.13). The reactions were carried out in DMF (*N*,*N*-dimethylformamide) at room temperature.

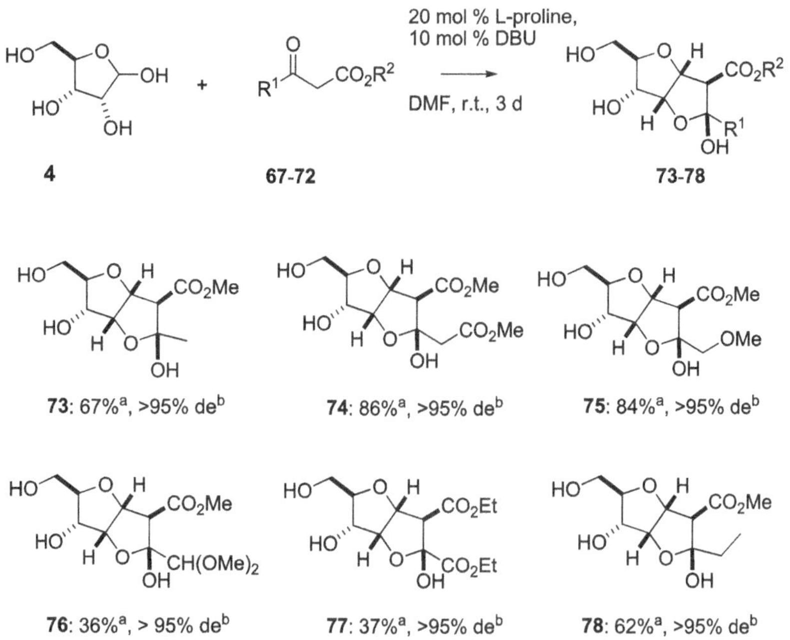

SCHEME 2.13 Amine-catalyzed reactions of different β-keto esters with D-ribose.

Inspired by these results, achieved by a simple changing of the catalytic system, we have tested the scope and limitations of this new organocatalyzed cascade reaction. In a first series, we reacted unprotected ribose with different substituted β-keto esters **67–72**. The results of these investigations are depicted in Scheme 2.13. All products, **73–78**, were isolated as single stereoisomers in their furanoid form and were found to be α-configured at the anomeric carbon atom. By-products or other stereoisomers were not detected.

The yields of the reaction depend on both steric demand of the substituent R^1 of the 1,3-dicarbonyl compounds and the electronic nature of this substituent. Very good yields were obtained by employing dimethyl 3-oxoglutarate **68** or methyl 4-methoxy-3-oxobutanoate **69**. The lowest yields were observed while using with methyl 4,4-dimethoxy-3-oxobutanoate **70** or diethyl 2-oxaloacetate **71**.

Only a single diastereoisomer was detected in all reactions we performed. Remarkably, the configurative course of this reaction is exclusively dictated by the configuration at C-2 position of the carbohydrate deployed. This is achieved by a stereoselective formation of the corresponding hemiketal **K** of the intermediately formed Knoevenagel-condensation product **I** during the reaction (see Scheme 2.16). A support for this conclusion is not only obtained from NMR characterization of intermediate **K**, but also given by the reaction of ribose with substituted acetoacetates. When used with methyl 4-methyl-3-oxovalerate **79,** three products, **80** and two epimers of *keto*-**81**, were obtained as an inseparable mixture with an overall yield of 46%. The formation of this equilibrium, due to a retro-ketalization process, cannot be prevented. The laboriously isolated product **80** yields the same equilibrium within 2 h at room temperature. Independent of this observation, all three products (**80** and two epimeric *keto*-**81**; Scheme 2.14) are formed with the same furanoid α-anomeric configuration.

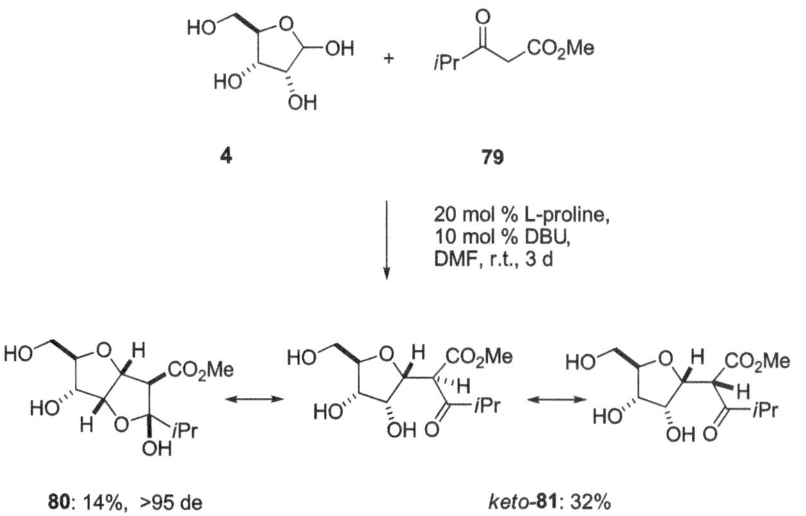

SCHEME 2.14 Reaction of D-ribose with acetoacetate **79**.

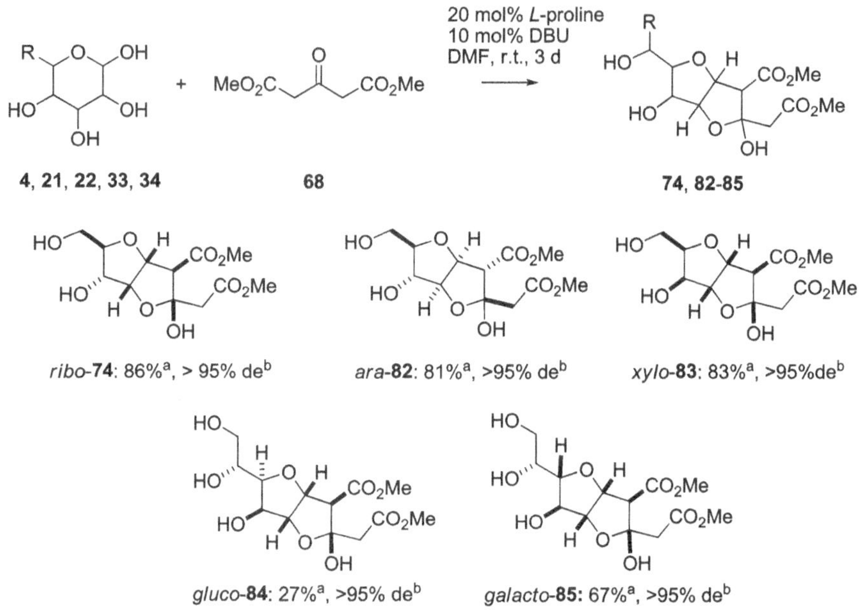

SCHEME 2.15 Reactions of dimethyl 3-oxoglutarate with ribose **4**, arabinose **2**, xylose **22**, glucose **33**, or galactose **34**.

In further experiments, we tested various pentoses and hexoses in these reactions. Dimethyl 3-oxoglutarate **68** was used as a standard 1,3-dicarbonyl compound in these transformations. The results of these studies are depicted in Scheme 2.15. The expected tetrahydrofurans **74** and **82–85** were isolated with good to high yields (with the exception of **84**: 27%). Again, only a single stereoisomer was detected in all our reactions (de >95%). Independent of the structure of the carbohydrates deployed—pentoses (ribose **4**, arabinose **21,** or xylose **22**) or hexoses (glucose **33** or galactose **34**), products were isolated exclusively in their furanoid form (compare structures of **74**, **82**, and **83** with those of **84** and **85**). On the contrary, the installation of configuration at the anomeric carbon atom and the two new stereogenic centers is dictated by the C-2 configuration of carbohydrates deployed. When used with ribose, xylose, glucose, and galactose α-configured anomers were obtained. In contrast to that, in reaction of arabinose **21** (β-configured hydroxyl group at C-2), the β-configured product **82** with inversed configurations at the two new created stereogenic centers was isolated as a single stereoisomer.

Based on these findings obtained, a Knoevenagel/ketalization/oxa-Michael cascade reaction can be assumed (Scheme 2.16). In the first step of the reaction cascade, an *E*-selective Knoevenagel reaction of the carbohydrate takes place (**I** in Scheme 2.16). This condensation process is followed by a stereoselective ketalization dictated by the configuration of the 2-hydroxy group of the carbohydrate (**K**). Subsequent *Re*-side attack of the 4-hydroxy group gives an access to fused tetrahydrofuran structure **L**. Structure **L** is identical with that of compound **74**, which is obtained when used

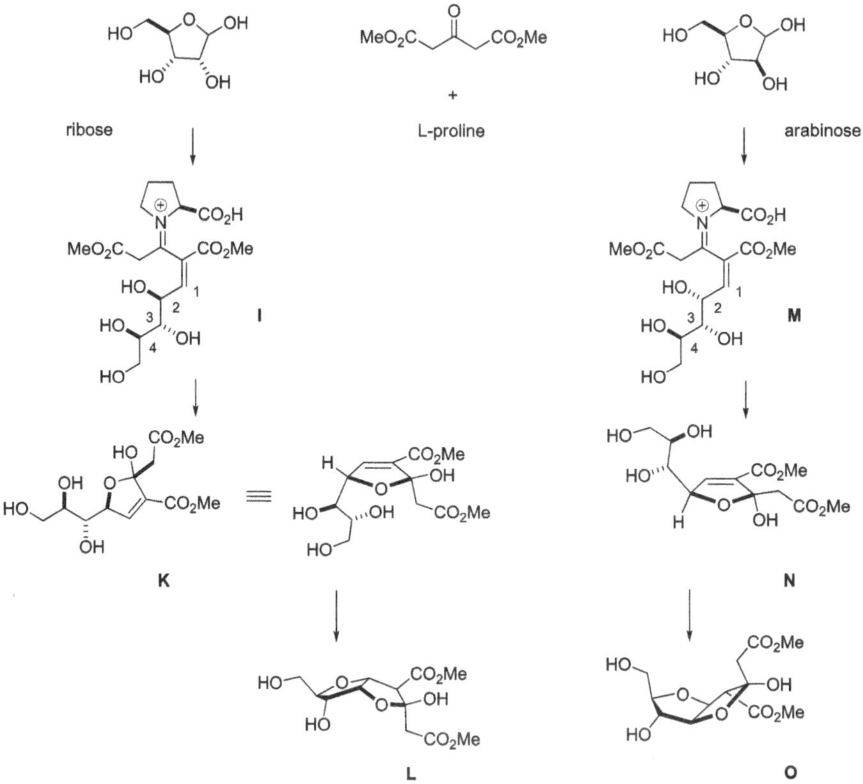

SCHEME 2.16 Configurative course of Knoevenagel condensation/ketalization/oxa-Michael cascade reactions.

with ribose. In reactions with arabinose, a *Si*-side attack of 4-hydroxy group determines the configuration of structure **O**, which is identical to the configuration of compound **82**. During this process, two new stereogenic centers are formed with defined configuration in a highly selective manner.

To date, several protocols for Knoevenagel additions to protected carbohydrates or carbohydrate-derived molecules have been reported in the literature [36]. For a manganese(III)- or cerium(IV)-mediated radical addition of malonates to protected carbohydrates, see Reference 37. For early reactions of protected carbohydrates with malonic acid in basic medium (pyridine), see Reference 6j.

Furthermore, investigations into reacting even unprotected carbohydrates with 1,3-dicarbonyl compounds have been published in the literature. Controlled by the reaction conditions, the formation of substituted furans or C-glycosides was observed. By application of acidic reaction conditions, multicondensation processes were obtained. As a consequence of that, the formation of polyhydroxylated furans was noticed [38]. These transformations are connected with the loss of stereogenic centers of the carbohydrates deployed. Under basic conditions, a Knoevenagel

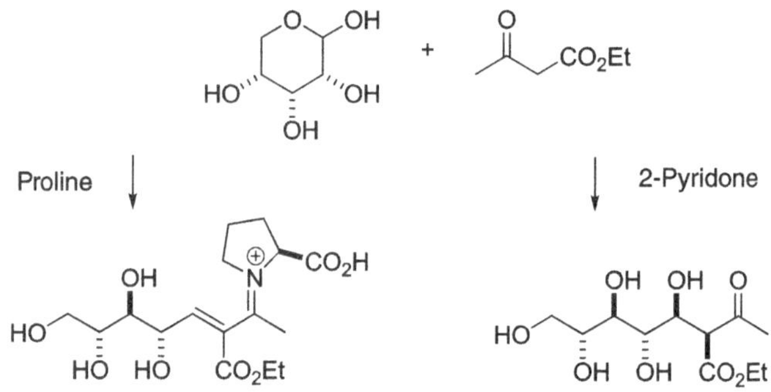

SCHEME 2.17 Different key intermediates in base-catalyzed Knoevenagel cascade reactions of ribose.

condensation occurs. Subsequent retro-Claisen aldol reaction yields the corresponding C-glycosides. This transformation involves the elimination of an acetate unit of the starting 1,3-dicarbonyl compounds, when used with acetylacetone (Lubineau reaction) [11,39].

A completely different situation is observed under the reaction conditions we describe herein. The two new methods described above enable an organocatalytic, stereoselective synthesis of the desired structural motifs without the loss of functionalities or stereogenic centers of the deployed substrates.

Both new cascade reactions were carried out in basic medium (tertiary amines). However, the key difference of this two cascade processes is as follows. The initial reaction step in the C-glycoside formation process is a Knoevenagel condensation, whereas the chain elongation cascade starts with a Knoevenagel-addition reaction of the acyclic form of the carbohydrates. This difference is dictated by the application of either 2-hydroxypyridine (Knoevenagel-addition reaction) or proline (Knoevenagel condensation). The intermediately formed Knoevenagel addition or condensation products determinate the subsequent reaction steps. Based on that, a different outcome of these cascade reactions is observed (Scheme 2.17).

2.4 MULTICOMPONENT REACTIONS

A further small tweaking of the reaction conditions involves great changes; a further new cascade channel is breaking. This is accomplished by modulation of the nucleophilicity of the activated methylene compounds applied. When used with isocyanoacetate **87** instead of cyanoacetates **86** as a methylene-activated component, an Ugi-like multicomponent cascade reaction is observed. This reaction was first detected by deployment of ethyl isocyanoacetate **87** in reactions of ribose and proline. In these experiments, proline—once the catalyst in the Knoevenagel condensation/ketalization/oxa-Michael cascade reactions—is directly incorporated into the product. As a result of that, seven-membered lactone **89** is formed. This sharp difference is demonstrated in Scheme 2.18 [40].

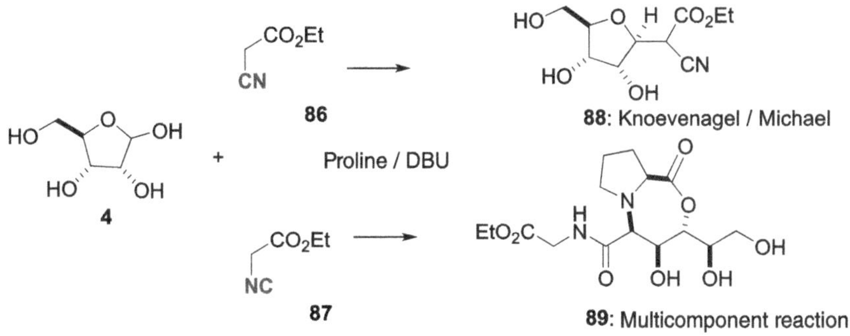

SCHEME 2.18 Comparison between organocatalyzed reactions of ethyl isocyanoacetate or cyanoacetate with ribose in the presence of proline.

It is assumed that both reactions start with the formation of the imine of the acyclic structure of carbohydrates with proline. After this initial step, the cyanoacetate **86** dictates a Knoevenagel/Michael cascade, whereas by employment of isocyanoacetate **87** a nucleophilic addition of the carbon atom of the isocyano group and subsequent rearrangement reaction are observed. An explanation for this different behavior of the isostructural compounds (ethyl esters **86** and **87**) depicted in Scheme 2.18 is given by the different sites of highest nucleophilicity as described by Mayr et al. [41]. These mechanistic considerations are depicted in Scheme 2.19.

Further investigation of this Ugi-type cascade reaction and optimization revealed a broad application of this multicomponent cascade. This transformation works without any catalyst or reagent at room temperature. For shortening reaction times and increasing yields, catalytic amounts of tertiary amines and working in boiling methanol proved to be the most successful conditions. These reaction conditions do not only influence the selectivity of the reaction but also increase the yield slightly in much shorter reaction time.

To test the general application of this transformation, we expanded these preliminary findings in a first series to reactions of D-ribose with ethyl isocyanoacetate **87**

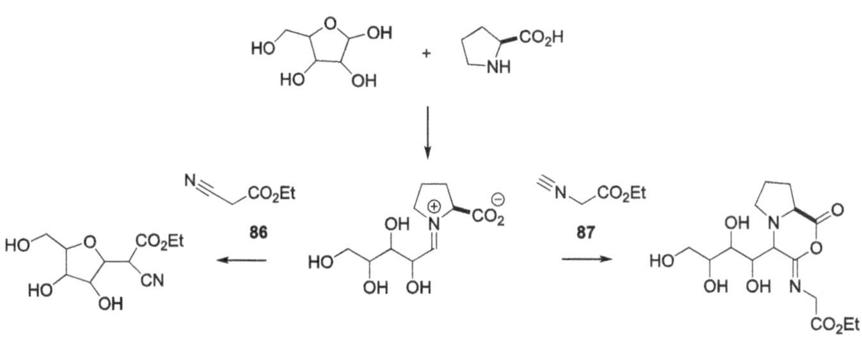

SCHEME 2.19 Mechanistic similarities and differences of Knoevenagel cascades and multicomponent cascades.

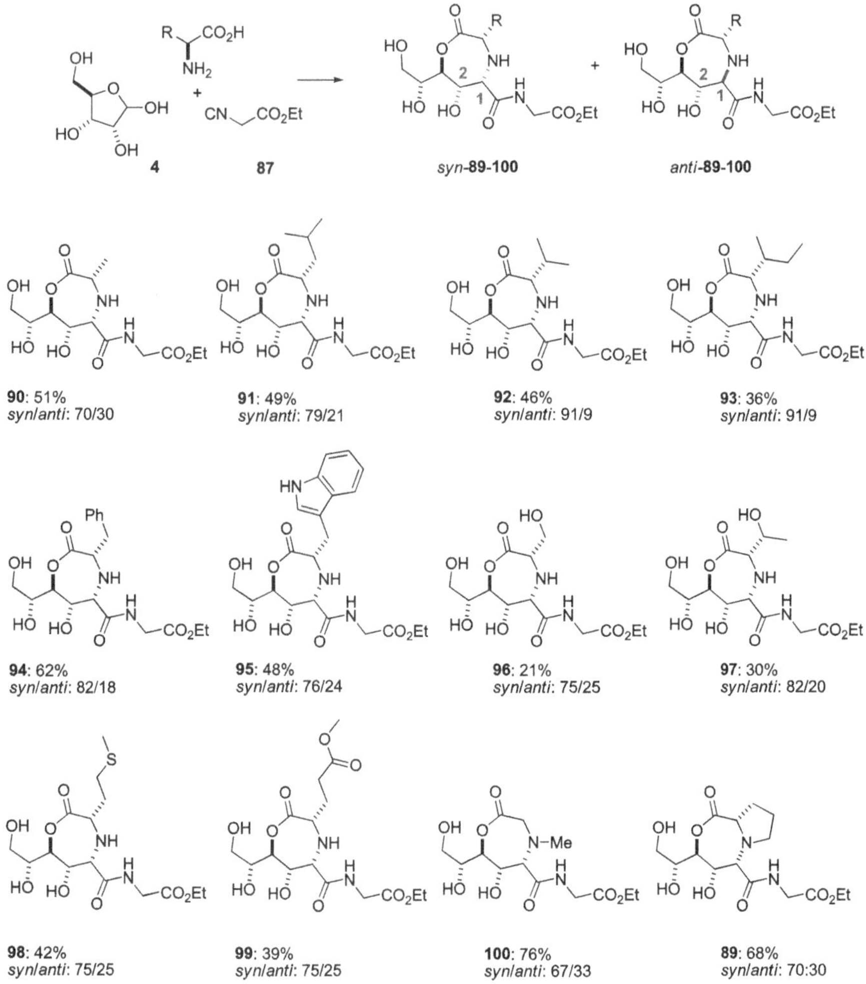

SCHEME 2.20 Multicomponent reactions of D-ribose and ethyl isocyanoacetate with L-amino acids. For a color version of this figure, see the color plate section.

and a wide range of proteinogenic amino acids (Scheme 2.20). During these investigations, the isoelectric point of the amino acid proved to be essential. This reaction works best with neutral amino acids. When used with acidic amino acids (aspartic acid or glutamic acid) or basic amino acids (histidine, arginine, or lysine), a reaction was not observed under these conditions. Furthermore, a general trend to installed configuration and diastereoselectivity was observed. By application of L-configured amino acids, 1,2-*syn*-configured structures were isolated as major products with moderate to high degrees of diastereoselectivity. The diastereoselectivity observed depends on the steric demand of the amino acids deployed. Highest degrees of diastereoselectivity resulted from the use of valine or isoleucine (**92** and **93**, *syn*/*anti*: 91/9). It is assumed

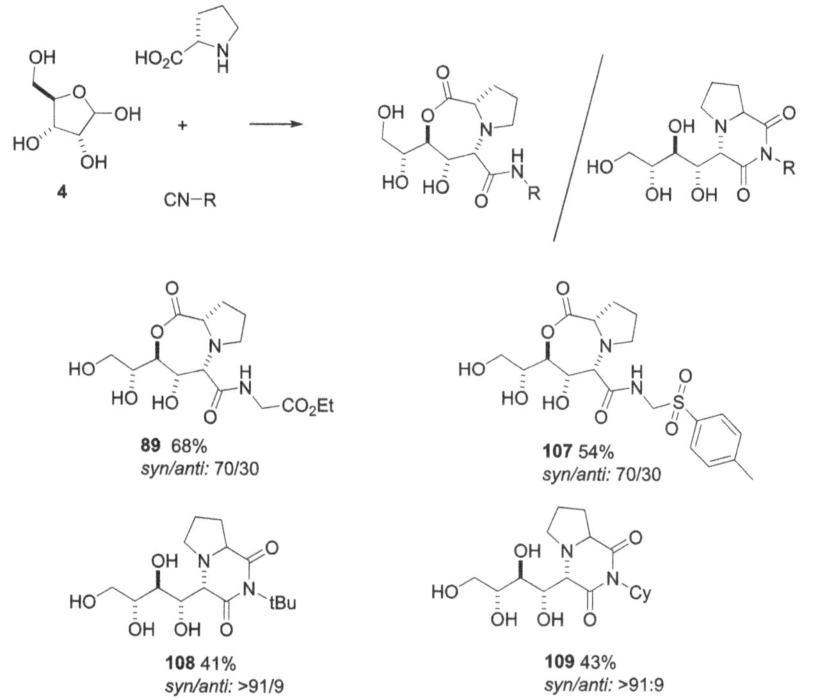

SCHEME 2.21 Multicomponent reaction of ribose and proline with different isocyanides.

that this high selectivity is caused by the β-branching of valine or isoleucine. Hence, incorporating unbranched amino acids (R=H; Scheme 2.20) should lower the selectivity. These considerations were supported by results of reactions with sarcosine. Product **100** was found with a ratio of 67/33 (*syn/anti*), though with good yields (76%). Results of these studies are depicted in Scheme 2.20.

This multicomponent sequence can also be conducted with several other isocyanoacetates, for example, *tert*-butylisocyanoacetate, *p*-toluenesulfonylmethyl isocyanoacetate, or cyclohexylisocyanoacetate. Similar results with regard to yields and selectivities were obtained when using *p*-toluenesulfonylmethyl isocyanoacetate (compare **107** with **89**; Scheme 2.21). When employing sterically demanding, electron rich isocyanides (*tert*-butylisocyanoacetate or cyclohexylisocyanoacetate) the corresponding 2,6-diketopiperazines **108** and **109** were obtained. It is assumed that these compounds are intermediates in the formation of the corresponding lactones. The electron rich, bulky substituents of compounds **108** and **109** stabilize these intermediate products and prevent the nucleophilic attack of the hydroxyl function, which results in the lactone formation as this is observed for products **89** and **107**. This consideration is supported by a separate, amine-catalyzed rearrangement of ketopiperazines **108** and **109** into the corresponding lactones.

The observation of different diastereoselectivities with L-amino acids with varying steric demand of the side chain suggests that the configuration of the amino acid

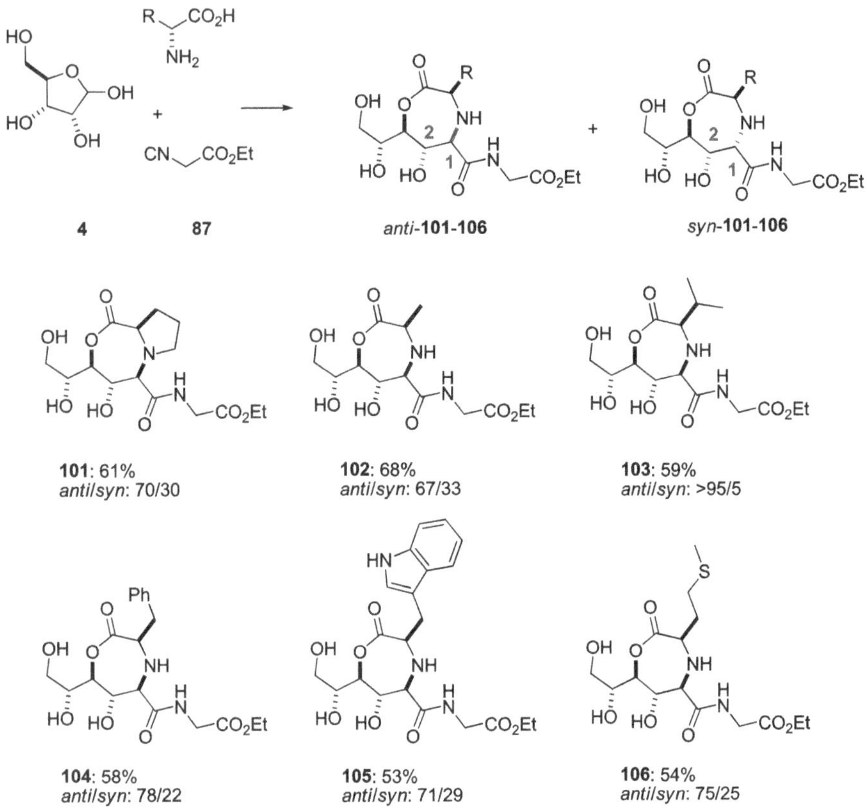

SCHEME 2.22 Multicomponent reactions of D-ribose and ethyl isocyanoacetate with D-configured amino acids. For a color version of this figure, see the color plate section.

employed dictates the installation of configuration at the anomeric carbon atom. Thus, when used with naturally configured L-amino acids, 1,2-*syn*-configured products are detected. In contrast, by applying D-configured amino acids, 1,2-*anti*-configured products are observed (Scheme 2.22).

Comparing the results of reactions of naturally configured amino acids (Scheme 2.20) with those of D-configured amino acids (Scheme 2.22), offers the obvious conclusion: nearly identical results with regard to yields and selectivities were obtained in both series, but with an inverted configuration at the former anomeric carbon atom.

In a further series, we have tested several different pentoses or hexoses in these multicomponent reactions with L-configured proline and ethyl isocyanoacetate. The results of these investigations demonstrate the outstanding role of the configuration of both, the relative and absolute configuration of the hydroxyl groups at C-2 and C-3 position of the carbohydrates incorporated. Seemingly independent of the configuration of the amino acids utilized, the configuration of the deployed carbohydrates

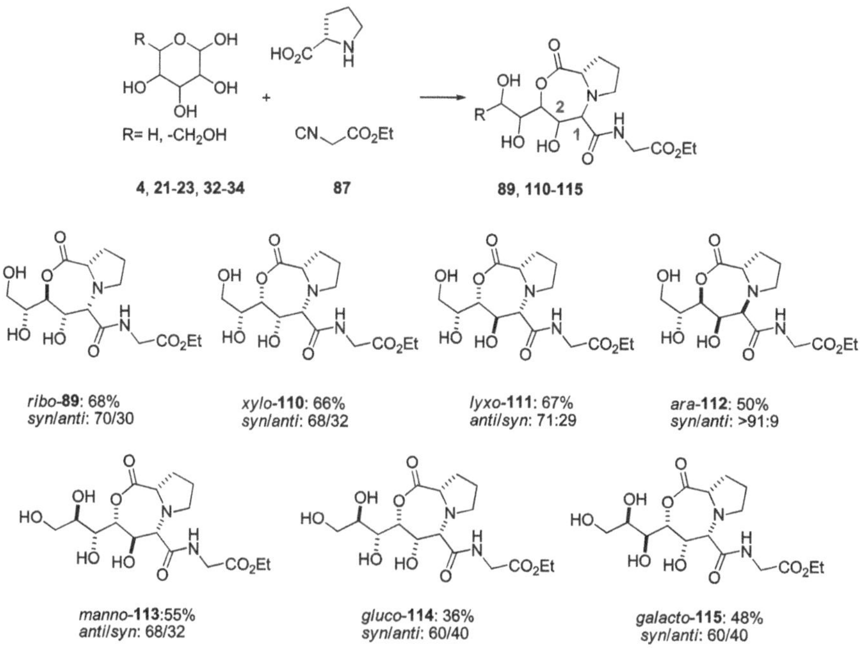

SCHEME 2.23 Multicomponent reactions of different pentoses or hexoses with ethyl isocyanoacetate and L-proline.

dictates the installation of the configuration at the carbon atom C-1 of the former carbohydrates (former anomeric carbon atom). Thus, the installation of configuration at C-1 dictates the formation of the relative 1,2-configuration of the products **89, 110–115** (Scheme 2.23). Support for these considerations is given by results of reactions with deoxyribose. In these transformations, complicated and unselective mixtures of all possible stereoisomers were detected. These results impressively demonstrate the importance of the existence and of a defined configuration of hydroxy groups at C-2 and C-3 for the configurative outcome of this multicomponent reaction.

To evaluate the influences of the configuration of carbohydrates and amino acids on the stereoselective course of this multicomponent reaction, different pentoses were reacted with both D- and L-proline in the presence of *p*-toluenesulfonylmethyl isocyanoacetate **116**. Results of these investigations are depicted in Scheme 2.24.

From inspection of these results, a correlation of the formation of 1,2-*syn*- or 1,2-*anti*-configured product with the absolute and relative configuration of the carbohydrate and the configuration of the amino acids deployed can be identified.

In reaction with 2,3-*anti*-configured pentoses (ribose or lyxose), the configurative outcome at the C-1 atom is controlled by the configuration of L- or D-proline. The reaction of D-ribose with L-proline yields 1,2-*syn*-configured product **107** as the major diastereoisomer, whereas by reaction with D-proline the 1,2-*anti*-configured product **120** is identified as the major product. As the 2,3-*anti*-configuration of lyxose is

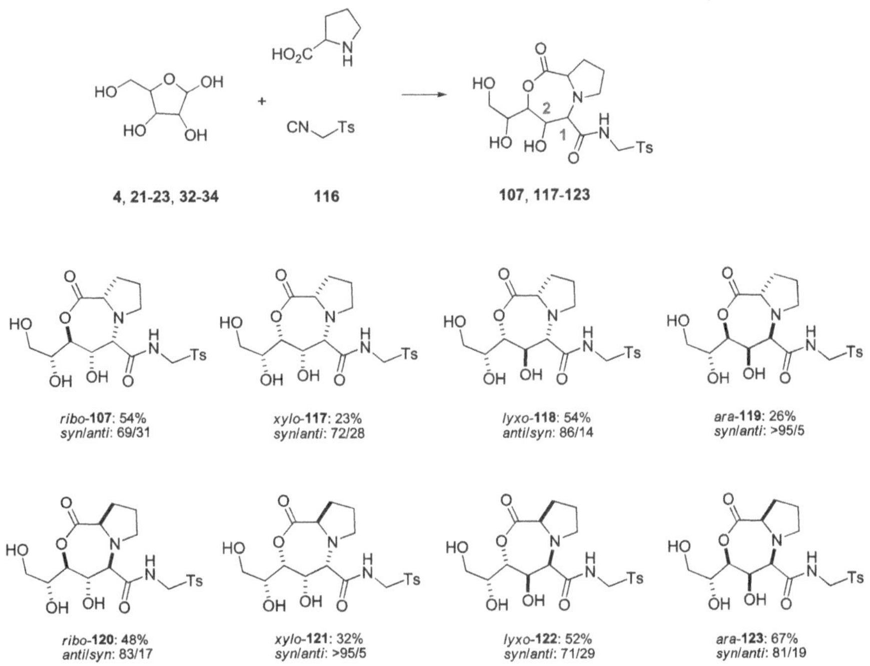

SCHEME 2.24 Multicomponent reaction with D- and L-proline.

inverted at C-2 and C-3 (in comparison with ribose), the reactions with L- or D-proline yield the same absolute configuration at C-1, but an inverted relative configuration at C-1 and C-2. Based on these results and considerations in conjunction with the observed diastereoselectivities, a mismatched situation for *ribo*-**107**/*lyxo*-**122** and a matched case for *lyxo*-**118**/*ribo*-**120** can be discussed.

In contrast to that, the installation of the configuration at the former anomeric center (C-1) is not influenced by the configuration of the amino acid deployed, when used with 2,3-*syn*-configured pentoses (arabinose or xylose). Thus, the configuration at C-1 cannot be controlled by the configuration of the amino acid. Extremely high degrees of relative *syn*-selectivity at C-1 and C-2 were observed. The formation of this extremely high diastereoselectivity in this series is solely influenced by the 2,3-*syn*-configuration of carbohydrate deployed.

In both series, 2,3-*anti*- (ribose or lyxose) and 2,3-*syn*-configured carbohydrates (arabinose or xylose), a stereochemical trend for the existence of a matched or mismatched case is detected and can be discussed. However, the origin for the installation of absolute and relative configuration is based on different modes and influences as discussed above.

In general, a mismatched case is observed if both the configuration of the carbon atom C-2 of the carbohydrate and the configuration of the amino acid are the same (**107**/**122** and **117**/**123**). In reactions where the absolute configuration of the amino acid contrasts the configuration at C-2 of the deployed carbohydrates, a matched case

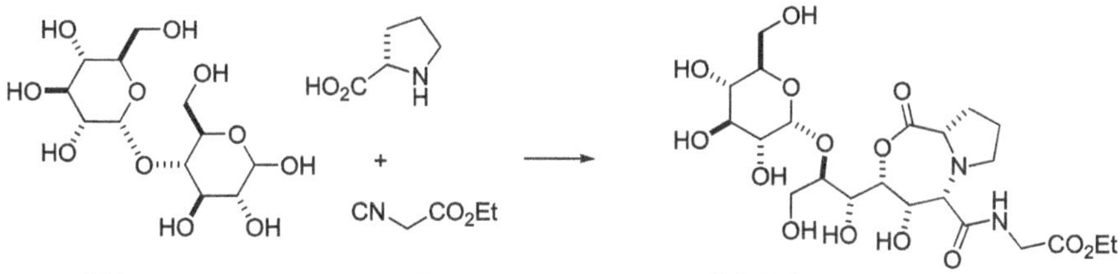

124 87 **125**: 33%
syn/*anti*: 83/17

SCHEME 2.25 Cascade reaction of maltose with L-proline and ethyl isocyanoacetate.

can be discussed (**119/121** and **118/120**). This rule of thumb holds true for reactions of proline with several different carbohydrates.

This elaborated methodology should provide an easy access to glycopeptide mimetics. To demonstrate that we reacted maltose **124** as an example for the deployment of disaccharides in these cascade reactions (Scheme 2.25), the expected lactone **125** was isolated with 33% yields after 3 h in refluxing methanol. These yields are comparable with those obtained in reactions of glucose with L-proline and ethyl isocyanoacetate (**114**: 36%; Scheme 2.23). But an increase in *syn*-selectivity was detected, indicating further supporting hydrogen bonds (compare **114**: 60/40 with **125**: 83/17).

In a further experiment, we demonstrated the utility of dipeptides in these reactions. To this end, we reacted ß-aspartame **126** with ribose and ethyl isocyanoacetate **87**. The expected lactone *syn*-**127** was isolated with a 42% yield and high levels of diastereoselectivity (Scheme 2.26).

These multicomponent reactions are novel. Multicomponent reactions of aldehydes and isonitriles with amino acids have been described in the literature [42], but the utilization of the carbonyl function of carbohydrates instead of aldehydes has not been reported previously.

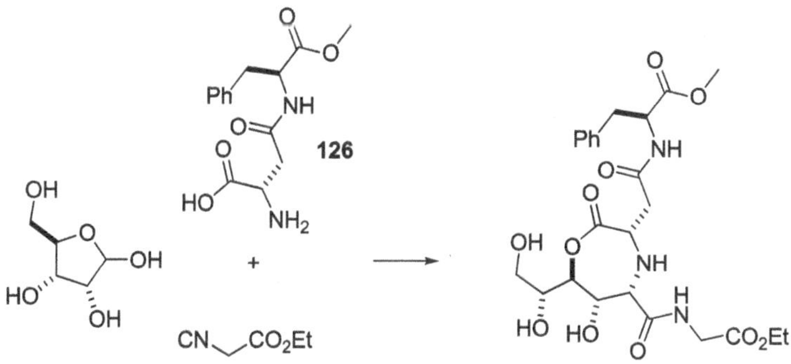

4 87 **127**: 42%
syn/*anti*: 88/12

SCHEME 2.26 Multicomponent reaction of ribose with aspartame and ethyl isocyanoacetate.

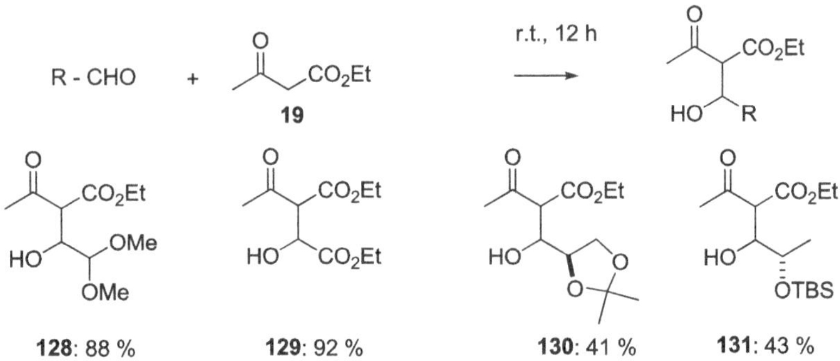

SCHEME 2.27 Catalyst-free aldol additions of aldehydes with 1,3-dicarbonyl compounds.

So far, we have demonstrated that the carbonyl functionality of aldoses can successfully be used in several cascade reactions (aldehyde in the acyclic structure). Also, ketoses proved to be useful substrates in cascade reactions with 1,3-dicarbonyl compounds (ketone in the acyclic structure). In these reactions, several unusual and highly substituted carbohydrates were formed with high degrees of chemoselectivity as well as stereoselectivity.

2.5 AMINE-CATALYZED CASCADE REACTIONS OF KETOSES WITH 1,3-DICARBONYL COMPOUNDS

The starting points of this study were observations that have been made when studying the aldol addition of aldehydes to 1,3-dicarbonyl compounds [43]. These reactions proceed without any catalyst. A condensation was not observed under these conditions. The products were isolated with quantitative yields in part. Some examples are depicted in Scheme 2.27.

At this point we wondered whether ketones, instead of aldehydes, can be deployed in these reactions. As expected, a reaction is not observed in the absence of a catalyst or reagent. After a systematical optimization, the following cascade reaction was detected. For reasons of simplification and comparability, these transformations were conducted with acetylacetone **5** as the enol component and hydroxyacetone **132** (Scheme 2.28). After 24 h at room temperature, acetate **133** was isolated in a good yield.

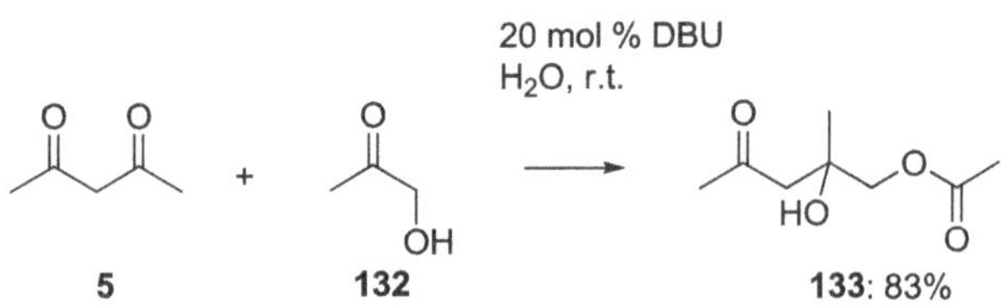

SCHEME 2.28 Amine-catalyzed cascade reactions of hydroxyacetone with acetylacetone.

The best results with regard to solvents were obtained in water. The reaction is catalyzed by tertiary amines. The most efficient catalyst concerning yields, clear reaction, and selectivity is 1,8-diazabicyclo[5.4.0]undec-7-ene (20 mol %). This transformation was expanded to several 1,3-dicarbonyl compounds. Due to the symmetry of acetylacetone **5** or dibenzoylacetone **134**, identical products were formed by the ketalization step (**133**: R_1=R_2=Me; **136**: R_1=R_2=Ph; Scheme 2.29). When used with unsymmetrical 1.3-dicarbonyl compounds, different products were formed via a regioisomeric ketalization step (reactions with benzoylacetone **138** in Scheme 2.29, **139** and **140**: R_1=Me, R_2=Ph) [44].

Similar to the mechanism described in Scheme 2.16, a Knoevenagel reaction/ketalization cascade of hydroxyacetone with 1,3-dicarbonyl compounds is assumed. In Scheme 2.16, a Knoevenagel condensation/ketalization reaction is depicted. This sequence allows a subsequent oxa-Michael addition, which yields the corresponding C-glycosides. In contrast, a Knoevenagel addition/ketalization occurs under the reaction condition described in Scheme 2.28, which is followed by an intramolecular retro-Claisen step. As a result of that, the corresponding esters were obtained (Scheme 2.29).

This transformation was successfully extended to reactions of unprotected dihydroxyacetone **141** with acetylacetone **5** or benzoylacetone **138**. The expected products derived from a Knoevenagel/ketalization/retro-Claisen cascade were isolated in high yields (Scheme 2.30).

In further experiments, we expanded this cascade to the deployment of optically active hydroxyketones. This was successfully realized by reactions of acetylacetone

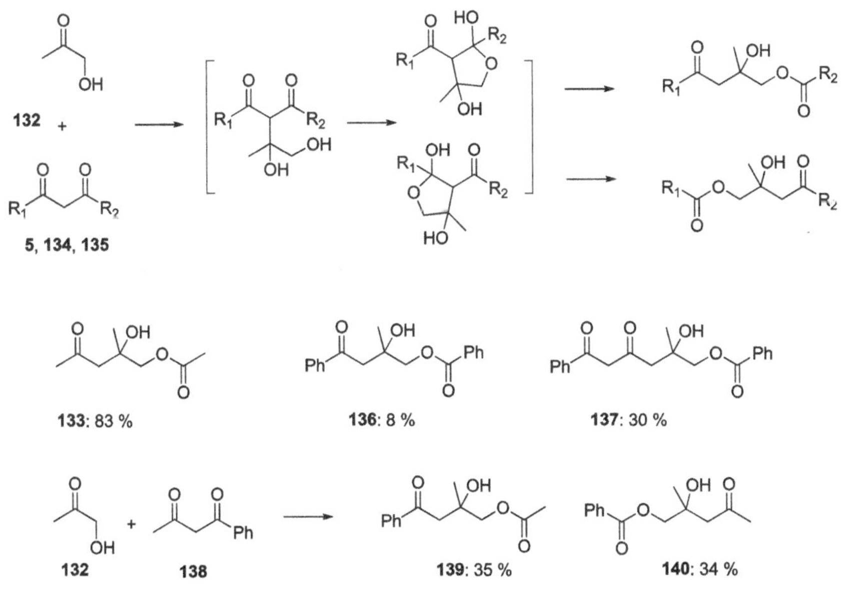

SCHEME 2.29 Amine-catalyzed cascade reactions of hydroxyacetone with 1,3-dicarbonyl compounds.

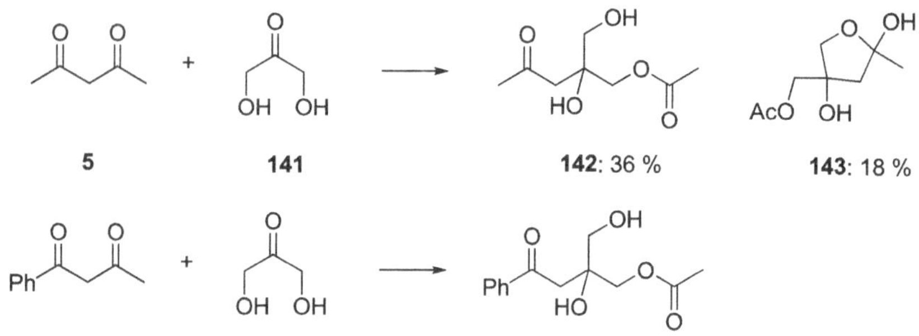

5 141 142: 36 % 143: 18 %

138 141 145: 75 %

SCHEME 2.30 Cascade reactions of dihydroxyacetone with 1,3-dicarbonyl compounds.

with L-erythrulose **146**. Again a Knoevenagel/ketalization/retro-Claisen cascade was observed to give 1-deoxyketose 147 with an extremely high degree of stereoselectivity. Thus, an access to unusually substituted, optically pure ketoses is given (Scheme 2.31).

To explain the high selectivity, which is detected in these cascade reactions, the possible products and reaction paths for reaction with L-erythrulose **146** and acetylacetone **5** are depicted in Scheme 2.32. The stereoselective Knoevenagel addition to intermediate **Q** is favored and determines simultaneously the configuration of the tertiary alcohol. A subsequent ketalization step to ketone **S** followed by the final intramolecular retro-Claisen step yields the isolated product **147** as a single stereoisomer.

These results differ considerably from those obtained by Lubineau reactions. In the Lubineau series, a Knoevenagel condensation followed by an oxa-Michael/intermolecular retro-Claisen process is proposed. As a result of the Lubineau sequence, the loss of an acetate-fragment occurs in reactions with acetylacetone.

In the amine-catalyzed reactions, a Knoevenagel addition/ketalization/ intramolecular retro-Claisen cascade is detected (Scheme 2.32). The retro-Claisen step is enabled by the ketalization of the Knoevenagel addition product **Q**. The ketalization of the Knoevenagel product (**Q**→**S**) is initiated by the hydroxyl groups of the carbohydrate moiety, as in-house NMR-experiments suggest (formation of intermediate ketal structure **K** in Knoevenagel condensation/ketalization/oxa-Michael cascade reaction Scheme 2.16). Products derived from this reaction sequence (Scheme 2.16)

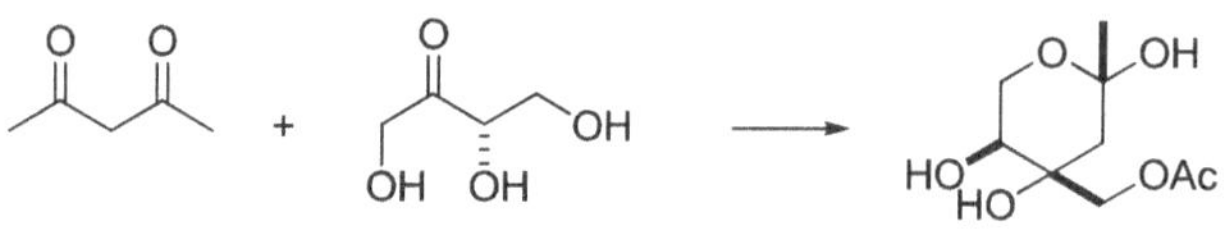

5 146 147: 51%, dr > 98/2

SCHEME 2.31 Cascade reactions of L-erythrulose with 1,3-dicarbonyl compounds.

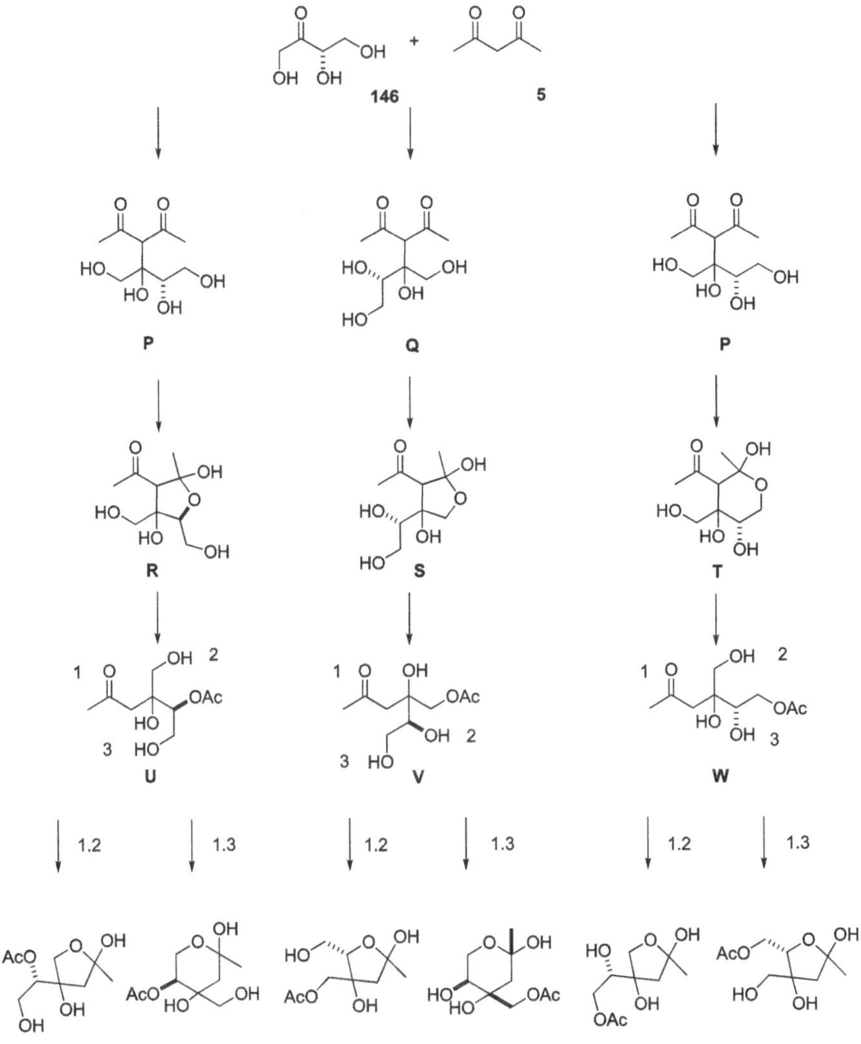

147

SCHEME 2.32 Amine-catalyzed reaction pathway of acetylacetone and L-erythrulose.

were not obtained in these amine-catalyzed transformations. The fundamental difference between these two cascade transformations is the Knoevenagel condensation, which is observed as the initial reaction in the Lubineau series and in Knoevenagel condensation/ketalization/oxa-Michael cascade reactions. In contrast, a Knoevenagel addition is observed as the starting reaction in the amine-catalyzed cascade reactions (Scheme 2.32). This main difference dictates the course of these two cascade channels (Scheme 2.33).

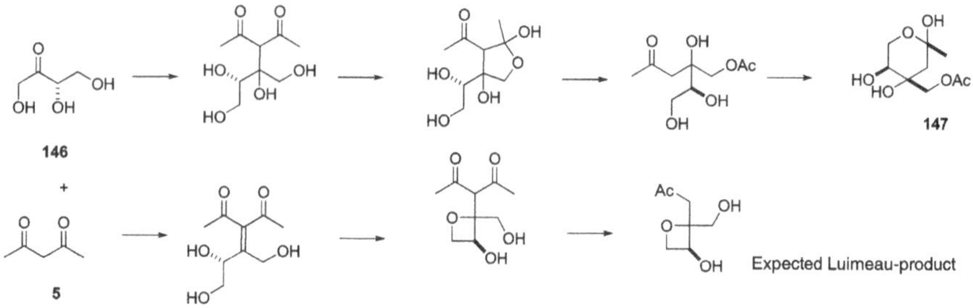

SCHEME 2.33 Comparison of amine-catalyzed cascade reaction with Lubineau reaction.

2.6 CONCLUSIONS

The extensive development of organocatalysis over the last years provided versatile methodologies for convenient utilization of the carbonyl function of unprotected carbohydrates in C—C bond-formation processes.

These biomimetic amine-activated mechanisms enable multiple cascade protocols for the synthesis of a large scope of carbohydrate-derived compound classes.

Several, only slightly different protocols have been developed for the application of 1,3-dicarbonyls in stereoselective chain elongation of unprotected carbohydrates and the synthesis of highly functionalized C-glycosides of defined configuration.

Furthermore, the latter class of compounds can also be accessed by the use of methyl ketones. So a high substrate scope is available for the installation of desired functionalities in C-glycosides.

By a one-pot, operationally simple cascade reaction of isocyanides with unprotected aldoses and amino acids, an access to a broad range of defined glycosylated pseudopeptides is given.

The elaborated organocatalyzed cascade reactions provide defined access to highly functionalized carbohydrate derivatives.

Depending on the reaction conditions, different sources of the configuration control during the bond-formation process were observed.

The demonstrated organocatalyzed cascade sequences indicate the great potential of unprotected carbohydrates in the synthesis of highly functionalized biomimetic structure motifs by operationally simple, one-pot protocols.

REFERENCES

[1] Soengas, R. G.; Otero, J. M.; Estévez, A. M.; Rauter, A. P.; Cachatra, V.; Estévez, J. C.; Estévez, R. J. *Carb. Chem.* **2012**, *38*, 263–302.

[2] (a) Hanessian, S.; Girard, C. *Synlett* **1994**, 865–867. (b) Soengas, R. S. *Tetrahedron Lett.* **2010**, *51*, 105–108.

[3] (a) Hanessian, S.; Pernet, A. G. *Can. J. Chem.* **1974**, *52*, 1280–1293. (b) Hanessian, S.; Pernet, A. G. *Can. J. Chem.* **1974**, *52*, 1266–1279. (c) Ogawa, T.; Pernet, A. G.;

Hanessian, S. *Tetrahedron Lett.* **1973**, *14*, 3543–3546. (d) Pernet, A. G.; Ogawa, T.; Hanessian, S. *Tetrahedron Lett.* **1973**, *14*, 3547–3550.

[4] (a) Gao, J.; Martichonok, V.; Whitesides, G. M. *J. Org. Chem.* **1996**, *61*, 9538–9540. (b) Spevak, W.; Dasgupta, F.; Hobbs, C. J.; Nagy, J. O. *J. Org. Chem.* **1996**, *61*, 3417–3422. (c) Kim, E.; Gordon, D. M.; Schmid, W.; Whitesides, G. M. *J. Org. Chem.* **1993**, *58*, 5500–5507. (d) Gordon, D. M.; Whitesides, G. M. *J. Org. Chem.* **1993**, *58*, 7937–7938.

[5] Haines, A. H.; Lamb, A. J. *Carb. Res.* **2000**, *325*, 323–339.

[6] (a) van Kalkeren, H. A.; van Rootselaar, S.; Haasjes, F. S.; Rutjes, F. P. J. T.; van Delft, F. L. *Carb. Res.* **2012**, *362*, 30–37. (b) Lievre, C.; Humez, S.; Frechou, C.; Demailly, G. *Tetrahedron Lett.* **1997**, *38*, 6003–6006. (c) Railton, C. J.; Clive, D. L. J. *Carb. Res.* **1996**, *281*, 69–77. (d) Horton, D.; Koh, D. *Carb. Res.* **1993**, *250*, 231–247. (e) Lee, H. W.; Lee, I. Y. C.; Park, I. H. *Bull. Korean. Chem. Soc.* **1992**, *13*, 222–223. (f) Henk, T.; Giannis, A.; Sandhoff, K. *Liebigs. Ann. Chem.* **1992**, 167–168. (g) Sun, K. M.; Dawe, R. D.; Fraser-Reid, B. *Carb. Res.* **1987**, *171*, 35–47. For reviews in this field, see (h) Cachatra, V.; Rauter, A. P. *Curr. Org. Chem.* **2014**, *18*, 1731–1748. (i) Zhdanov, Y. A.; Alekseev, Y. E.; Alekseeva, V. G. *Adv. Carb. Chem.* **1972**, *27*, 227–299. (j) Kotchekow, N. K.; Dimitriev, B. A. *Tetrahedron* **1965**, *21*, 803–815.

[7] (a) Ranoux, A.; Lemiegre, L.; Benoit, M.; Guegan, J. P.; Benvegnu, T. *Eur. J. Org. Chem.* **2010**, 1314–1323. (b) Ranoux, A.; Lemiegre, L.; Benvegnu, T. *Sci. China. Chem.* **2010**, *53*, 1957–1962.

[8] Toshima, K.; Ishizuka, T.; Matsuo, G.; Nakata, M. *Tetrahedron Lett.* **1994**, *35*, 5673–5676.

[9] (a) Lalitha, K.; Muthusamy, K.; Prasad, Y. S.; Vemula, P. K.; Nagarajan, S. *Carb. Res.* **2015**, *402*, 158–171. (b) Merino, P.; Tejero, T.; Marca, E.; Gomollon-Bel, F.; Delso, I.; Matute, R. *Heterocycles*. **2012**, *86*, 791–820. (c) Ansari, A. A.; Lahiri, R.; Vankar, Y. D. *ARKIVOC* **2013**, *2*, 316–362. (d) Stambasky, J.; Hocek. M.; Kocovsky, P. *Chem. Rev.* **2009**, *109*, 6729–6764. (e) Taylor, R. J. K.; McAllister, G. D.; Franck, R. W. *Carb. Res.* **2006**, *341*, 1298–1311. (f) Bililign, T.; Griffith, B. R.; Thorson, J. S. *Nat. Prod. Rep.* **2005**, 22, 742–760. (g) Levy, D. E. In *Organic Chemistry of Sugars;* Levy, D. E., Fuegedi, Eds.; CRC Press LLC: Boca Raton, 2006; pp 269–348. (h) Robina, I.; Vogel, P. *Synthesis* **2005**, 675–748. (i) Taillefumier, C.; Chapleur, Y. *Chem. Rev.* **2004**, *104*, 263–292. (j) Meo, P.; Osborn, H. M. I. in *Carbohydrates;* Osborn, H. M. I., Ed.; Elsevier, 2003; pp 337–384. (l) Gyorgydeak, Z.; Pelyvas, I. F. in *Glycoscience*, Vol. 1; Fraser-Reid, B. O., Tatsuta, K., Thiem, J. Eds.; Springer, 2001; pp 691–748.

[10] (a) Araki, Y.; Kobayashi, N.; Ishido, Y. *Carb. Res.* **1987**, *171*, 125–139. (b) Csuk, R.; Schaade, M. *Tetrahedron* **1994**, *50*, 3333–3348.

[11] (a) Rodridues, F.; Canac, Y.; Lubineau, A. *Chem. Comm.* **2000**, 2049–2050. Early cascade reactions (Knoevenagel/intramolecular oxa-Michael addition) of unprotected and unactivated carbohydrates with barbituric acid derivatives were reported by (b) Gonzalez, M. A.; Requejo, J. L. J.; Albarran, J. C. P.; Perez, J. A. G. *Carb. Res.* **1986**, *158*, 53–66. (c) Wulff, G.; Clarkson, G. *Carb. Res.* **1994**, *257*, 81–95.

[12] (a) Feng, W.; Fang, Z.; Yang, J.; Zheng, B.; Jiang, Y. *Carb. Res.* **2011**, *346*, 352–356. (b) Wang, J.; Li, Q.; Ge, Z.; Li, R. *Tetrahedron* **2012**, *68*, 1315–1320.

[13] (a) Mugunathan, B. *Bioorg. Med. Chem. Lett.* **2011**, *21*, 3947–3950. (b) Yan, L.; Xu, H.; Liu, F.; Zhao, J.; Liu, H. *Chin. J. Chem.* **2012**, *30*, 914–918.

[14] Liu, L.; Abdel, M. B.; Schmidt-Supprian, M.; Pohl, N. L. B. *J. Org. Chem.* **2012**, *77*, 1539–1546.

[15] Giguere, D.; Andre, S.; Bonin, M.-A.; Bellefleur, M.-A.; Provencal, A.; Cloutier, P.; Pucci, B.; Roy, R.; Gabius, H.-J. *Bioorg. Med. Chem.* **2011**, *19*, 3280–3287.

[16] Hersant, Y.; Abou-Jneid, R.; Canac, Y.; Lubineau, A.; Philippe, M.; Semeria, D.; Radisson, X.; Scherrmann, M.-C. *Carb. Res.* **2004**, *339*, 741–745.

[17] Hemamalini, A. *Carb. Res.* **2011**, *346*, 1814–1819.

[18] Riafrecha, L. E.; Rodriguez, O. M.; Vullo, D.; Supuran, C. T.; Colinas, P. A. *Bioorg. Med. Chem.* **2014**, *22*, 5308–5314.

[19] Palasz, A.; Kalinowska-Tluscik, J.; Jablonski, M. *Tetrahedron* **2013**, *69*, 8216–8227.

[20] Cavezza, A. *Bioorg. Med. Chem. Lett.* **2009**, *19*, 845–849.

[21] (a) Foley, P. M.; Phimphachanh, A.; Beach, E. S.; Zimmerman, J. B.; Anastas, P. T. *Green. Chem.* **2011**, *13*, 321–325. (b) Hemamalini, A.; Das, T. M. *New J. Chem.*, **2014**, *38*, 3015–3021.

[22] (a) Price, N. P. J. *Anal. Chem.* **2010**, *82*, 2863. (b) Carpenter, C. A.; Kenar, J. A.; Price, N. P. J. *Green. Chem.* **2010**, *12*, 2012–2018.

[23] Ajay, A.; Sharma, S.; Gupt, M. P.; Bajpai, V.; Hamidullah; Kumar, B.; Kaushik, M. P.; Konwar, R.; Ampapathi, R. S.; Tripathi, R. P. *Org. Lett.* **2012**, *14*, 4306–4309.

[24] Xia, T.-Yu.; Li, Y.-B.; Yin, Z.-J.; Meng, X.-B.; Li, S.-C.; Li, Z.-J. *Chin. Chem. Lett.* **2014**, *25*, 1220–1224.

[25] Scherrmann, M. C. *Top. Curr. Chem.* **2010**, *295*, 1–18.

[26] Voigt, B.; Mahrwald, R. unpublished results.

[27] Witte, N.; Voigt, B.; Mahrwald, R. *Synthesis*, **2015**, *47*, 2249–2255.

[28] Voigt, B.; Mahrwald, R. *Chem. Comm.* **2012**, *48*, 5304–5306.

[29] (a) Markert, M.; Mulzer, M.; Schetter, B.; Mahrwald, R. *J. Am. Chem. Soc.* **2007**, *129*, 7258–7259. (b) Rohr, K.; Mahrwald, R. *Org. Lett.* **2011**, *13*, 1878–1880.

[30] Capon, B. *Chem. Rev.* **1969**, *69*, 407–498.

[31] (a) Morpurgo, S.; Bossa, M. *Phys. Chem. Chem. Phys.* **2003**, *5*, 1181–1189. (b) Arai, K. *J. Polym. Sci., Pt A: Polym. Chem.* **1993**, *31*, 193–197. (c) Kjaer, A. M.; Nielsen, H.; Soerensen, P. E.; Ulstrup, J. *Acta Chem Scand Ser. A: Phys Inorg. Chem.* **1980**, *A34*, 281–288. (d) Morgan, D. M. L.; Neuberger, A. *Proc. R. Soc. Lond., Ser. A: Math. Phys. Eng. Sci.* **1974**, *337*, 317–332. (e) Smith, W. T.; Hearn, T. L. *Bioorg. Chem.* **1972**, *2*, 39–43. (f) Rony, P. R. *J. Am. Chem. Soc.* **1968**, *90*, 2824–2831. (g) Bentley, R.; Bhate, D. S. *J. Biol. Chem.* **1960**, *235*, 1225–1233. (h) Swain, C. G.; Brown, J. F. *J. Am. Chem. Soc.* **1952**, *74*, 2534–2537. (i) Swain, C. G.; Brown, J. F. *J. Am. Chem. Soc.* **1952**, *74*, 2538–2543.

[32] Ribose and lyxose are 2,3-*anti*-configured in the acyclic structure, whereas arabinose and xylose are 2,3-*syn*-configured pentoses in their acyclic structures.

[33] Angyal, S. J. *Angew. Chem. Int. Ed.* **1969**, *8*, 157–226.

[34] Kam, B. K.; Oppenheimer, N. J. *Carb. Res.* **1979**, *69*, 308–310.

[35] Voigt, B.; Mahrwald, R. *Chem. Comm.* **2014**, *50*, 817–819.

[36] (a) Zinner, H.; Wittenburg, E.; Rembarz, G. *Chem. Ber.* **1959**, *92*, 1614–1617. (b) Hanessian, S.; Pernet, A. G. *Can. J. Chem.* **1974**, *52*, 1280. (c) Hanessian, S.; Pernet, A. G. *Can. J. Chem.* **1974**, *52*, 1266. (d) Ogawa, T.; Pernet, A. G.; Hanessian, S. *Tetrahedron Lett.* **1973**, 3543. (e) Pernet, A. G.; Ogawa, T.; Hanessian, S. *Tetrahedron Lett.* **1973**, 3547. (f) Butcher, M. E.; Lee, J. B. *Tetrahedron Lett.* **1974**, 2663–2664. (g) Lopez Herrera, F. J. *Tetrahedron Lett.* **1980**, *21*, 4963–4966. (h) Lopez, A. F. J.; Izquierdo C. I.; Portal,

O. M. D. *Carb. Res.* **1982**, *103*, 158–164. (i) Lopez, A. F. J.; Zorrilla, B. F.; Garcia, M. P.; Santoyo, G. F. *Carb. Res.* **1985**, *135*, 303–311. (j) Lopez, H. F. J.; Pino, G. M. S. *Carb. Res.* **1986**, *152*, 283–291. (k) Tadano, K.; Maeda, H.; Hoshino, M.; Iimura, Y.; Suami, T. *J. Org. Chem.* **1987**, *52*, 1946–1956. (l) Dondoni, A.; Massi, A.; Minghini, E.; Bertolasi, V. *Helv. Chim. Acta* **2002**, *85*, 3331–3348. (m) Gallos, J. K.; Koumbis, A. E. *ARKIVOC* **2003**, 135–144. (n) Yadav, J. S.; Reddy, B. V. S.; Narsimhaswamy, D.; Lakshmi, P. N.; Narsimulu, K.; Srinivasulu, G.; Kunwar, A. C. *Tetrahedron Lett.* **2004**, *45*, 3493–3497. (o) Sabitha, G.; Reddy; E. V.; Fatima, N.; Yadav, J. S.; Krishna, K. V. S. R.; Kunwar, A. C. *Synthesis* **2004**, 1150–1154. (p) Yadav, J. S.; Reddy, B. V. S.; Gopal, A. V. H.; Rao, R. N.; Somaiah, R.; Reddy, P. P., Kunwar, A. C. *Tetrahedron Lett.* **2010**, *51*, 2305–2308. (q) Hemamalini, A.; Nagarajan, S.; Das, T. N. *Carb. Res.* **2011**, *346*, 1814.

[37] (a) Yin, J.; Linker, T. *Org. Bioorg. Chem.* **2009**, *7*, 4829–4831. (b) Linker, T.; Hartmann, K.; Sommermann, T.; Scheutzow, D.; Ruckdeschel, E. *Angew. Chem. Int. Ed. Engl.* **1996**, *35*, 1730–1732. (c) Linker, T.; Sommermann, T.; Kahlenberg, F. *J. Am. Chem. Soc.* **1997**, *119*, 9377–9384.

[38] (a) Nagarapu, L.; Chary, M. V.; Satyender, A.; Supriya, B.; Bantu, R. *Synthesis* **2009**, 2278–2282. (b) Moreno-Vargas, A. J.; Molina, L.; Carmona, A. T.; Ferrali, A.; Lambelet, M.; Spertini, O.; Robina, I. *Eur. J. Org. Chem.* **2008**, 2973–2973. (c) Price, N. P. J.; Momany, F. A.; Adeuya, A. *J. Mass. Spectr.* **2008**, *43*, 53–62. (d) Yadav, J. S.; Reddy, B. V. S.; Srinivas, M.; Divyavani, C.; Basha, S. J.; Sarma, A. V. S. *J. Org. Chem.* **2008**, *73*, 3252–3254. (e) Yan, L.; Dai, G. F.; Yang, J. L.; Liu. F. W.; Liu, H. M. *Bioorg. Med. Chem. Lett.* **2007**, *17*, 3454–3457. (f) Molina, L.; Moreno-Vargas, A. J.; Carmona, A. T.; Robina, I. *Synlett* **2006**, 1327–1330. (g) Yadav, J. S.; Reddy, B. V. S.; Sreenivas, M.; Satheesh, S. *Synthesis* **2007**, 1712–1716. (h) Bartoli, G.; Fernández-Bolaños, J. G.; Di Antonio, G.; Foglia, G.; Giuli, S.; Gonnella, R.; Mancinelli, M.; Marcantoni, E.; Paletti, M. *J. Org. Chem.* **2007**, *72*, 6029–6036. (i) Sato, S.; Naito, Y.; Aoki, K. *Carb. Res.* **2007**, *342*, 913–918. (j) Misra, A. K.; Agnihotri, G. *Carb. Res.* **2004**, *339*, 1381–1387. (k) Moreno-Vargas, A. J.; Jimenez-Barbero, U. J.; Robina, I. *J. Org. Chem.* **2003**, *68*, 4138–4150. (l) Robina, I.; Moreno-Vargas, A. J.; Fernández-Bolaños, J. G.; Fuentes, J.; Demange, R.; Vogel, P. *Bioorg. Med. Chem. Lett.* **2001**, *11*, 2555–2559. (m) Gonzalez, F. G. *Adv. Carb. Chem.* **1956**, *11*, 97–143.

[39] (a) Price, N. P. J.; Bowman, M. J.; Le Gall, S.; Berhow, M. A.; Kendra, D. F.; Lerouge, P. *Anal. Chem.* **2010**, *82*, 2893–2899. (b) Cavezza, A.; Boulle, C.; Gueguiniat, A.; Pichaud, P.; Trouille, S.; Ricard, L.; Dalko-Csiba, M. *Bioorg. Med. Chem. Lett.* **2009**, *19*, 845. (c) Nagarajan, S.; Das, T. M. *Carb. Res.* **2009**, *344*, 1028–1031. (d) Giguere, D.; Bonin, M. A.; Cloutier, P.; Patnam, R.; St-Pierre, C.; Sato, S.; Roy, R. *Bioorg. Med. Chem.* **2008**, *16*, 7811–7823. (e) Norsikian, S.; Zeitouni, J.; Rat, S.; Gerard, S.; Lubineau, A. *Carb. Res.* **2007**, *342*, 2716–2728. (f) Bragnier, N.; Scherrmann, M. C. *Synthesis* **2005**, 814–818. (g) Graziani, A.; Amer, H.; Zamyatina, A.; Hofinger, A.; Kosma, P. *Tetrahedron: Asymmetry.* **2005**, *16*, 167–175. (h) Zeitouni, J.; Norsikian, S.; Lubineau, A. *Tetrahedron Lett.* **2004**, *45*, 7761–7763. (i) Riemann, I.; Papadopoulos, M. A.; Knorst, M.; Fessner, W. D. *Aust. J. Chem.* **2002**, *55*, 147–154. (j) Moreno-Vargas, A. J.; Fernández-Bolaños, J. G.; Fuentes, J.; Robina, I. *Tetrahedron: Asymmetry.* **2001**, *12*, 3257–3266. (k) González, F. G.; Sánchez, A. G.; Goñi De Rey, M. I. *Carb. Res.* **1965**, *1*, 261–273. (l) For an overview, see Scherrmann, M. C.; Lubineau, A.; Queneau, Y. in *Handbook of Green Chemistry*, Vol. 5; Anastas, P. T., Ed.; 2010; pp 291–330.

[40] Voigt, B.; Linke, M.; Mahrwald, R. *Org. Lett.* **2015**, *17*, 2606–2609.

[41] (a) Phan, T. B.; Mayr, H. *Eur. J. Org. Chem.* **2006**, 2530–2537. (b) Tumanov, V. V.; Tishkov, A. A.; Mayr, H. *Angew. Chem. Int. Ed.* **2007**, *46*, 3563–3566.

[42] For reviews, see the following references: (a) Koopmanschap, G.; Ruijter, E.; Orru, R. V. A. *Beilst. J. Org. Chem.* **2014**, *10*, 544–598. (b) van Berkel, S. S.; Boegels, B. G. M.; Wijdeven, M. A.; Westermann, B.; Rutjes, F. P. J. T. *Eur. J. Org. Chem.* **2012**, 3543–3559. (c) Wilson, R. M.; Stockdill, J. L.; Wu, X.; Li, X.; Vadola, P. A.; Park, P. K.; Wang, P.; Danishefsky, S. J. *Angew. Chem. Int. Ed.* **2012**, *51*, 2834–2848. (d) Dömling, A.; Wang, W.; Wang, K. *Chem. Rev.* **2012**, *112*, 3083–3125. (e) *Isocyanide Chemistry*; Nenajdenko, V. G., Ed.; Wiley-VCH: Weinheim, 2012. (f) Gulevich, A. V.; Zhdanko, A. G.; Orru, R. V. A.; Nenajdenko, V. G. *Chem. Rev.* **2010**, *110*, 5235–5331. (g) *Multicomponent Reactions*; Zhu, J., Bienayme, H., Eds.; Wiley-VCH: Weinheim, 2005.

[43] Rohr, K.; Mahrwald, R. *Adv. Synth. Cat.* **2008**, *350*, 2877–2880.

[44] Richter, C.; Voigt, B.; Mahrwald, R. *RSC Adv.* **2015**, *5*, 45571–45574.

3

REDUCTIVE RING-OPENING IN DOMINO REACTIONS OF CARBOHYDRATES

Raquel G. Soengas, Sara M. Tomé, and Artur M. S. Silva
Department of Chemistry & QOPNA, University of Aveiro, 3810-193 Aveiro, Portugal

3.1 INTRODUCTION

The quest for synthetic efficiency in terms of minimization of synthetic steps, together with maximization of molecular complexity, is one of the main aims in modern organic synthesis [1]. In regard to this, the interest in the development of domino reactions for the synthesis of organic compounds has grown exponentially [2–4]. In domino or cascade conversions, several bond-forming transformations take place under the same reaction conditions and subsequent reactions are due to the functionality formed in the previous step. Domino reactions allow the highly efficient synthesis of complex organic compounds and, as these processes take place without intermediate recovery steps, their use drastically reduces operating times and costs as well as the consumption of chemicals and the use of energy [5].

Synthetic efficiency is particularly relevant in the area of carbohydrate chemistry. Owing to the remarkable biological relevance of carbohydrates [6], the synthesis of sugar derivatives has long been the goal pursued by researchers from various areas [7]. The complexity of carbohydrates, however, makes the synthesis of carbohydrates a challenging task. Carbohydrates are highly functionalized molecules, and as a result their use as building blocks in synthesis often requires many reaction steps and manipulation of several different protecting groups [8]. From this perspective, the application of domino reactions in carbohydrate chemistry has recently attracted

Domino and Intramolecular Rearrangement Reactions as Advanced Synthetic Methods in Glycoscience, First Edition. Edited by Zbigniew J. Witczak and Roman Bielski.

significant attention. Among the domino reactions that have been used in carbohydrate chemistry, we will focus on those that share as a common feature in the sequence a ring opening induced by metal-promoted reductive elimination.

Metal-promoted reductive β-elimination reactions in 1,2-difunctionalized substrates have been one of the most important means of generating C=C bonds [9]. In the past few years, a number of reductive elimination reactions of 1,2-difunctionalized compounds have been reported under various organometallic reagents, such as SmI_2 [10], Zn [11], or In [12]. In addition, those low-valent organometallic species exhibit excellent properties for sequential organic reactions and several examples have been described in which a metal-promoted elimination reaction occurs as a step of a sequential process [13].

In the past few years, the application of low-valent organometallic reagents to carbohydrate chemistry has attracted much interest largely due to the mild reaction conditions required and the excellent diastereoselectivities obtained due to the control of the process through chelation to proximal hydroxyl, amino, and keto substituents [14].

Our aim in this chapter is to present the most important advances in the domino reactions of carbohydrates based on the metal-promoted ring opening of cyclic sugars. These procedures, promoted by different low-valent organometallic reagents, start with the addition of the metal to an electrophilic site (i.e., an alkyl halide, a carbonyl, an alkene, or an alkyne), followed by reductive elimination to the open-chain sugar. From this open-chain sugar intermediate, the next reaction takes place under the same reaction conditions.

3.2 BERNET–VASELLA REACTION

In 1979, Bernet and Vasella reported the preparation of aldehydes via a zinc-mediated reductive elimination of haloglycosides [15]. In this pioneering work, activated zinc and acetic acid were added to a solution of methyl bromoglucoside **1** in ethanol and the solution was stirred at reflux to give aldehyde **2** (Scheme 3.1).

Under these conditions, several 5-halofuranosides and 6-halopyranosides ring-opened to the corresponding pent-4-enals and hex-5-enals, respectively [16, 17]. Mechanistically, the Bernet–Vasella reaction proceeds via oxidative insertion of zinc into the carbon–bromide bond, followed by reductive elimination with the loss of methoxide to give the aldehyde product (Scheme 3.2) [18].

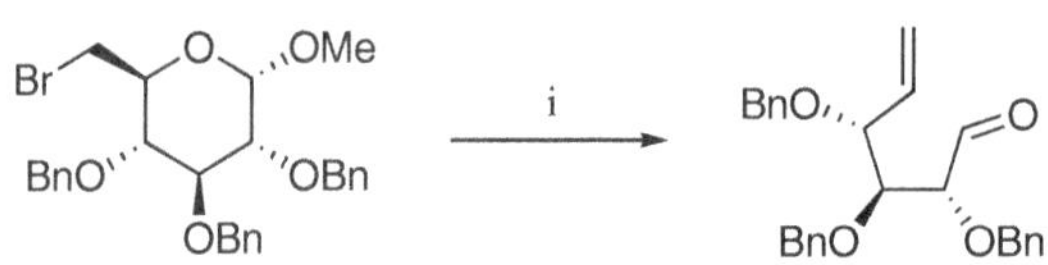

1 2

SCHEME 3.1 Reagents and conditions: (i) Zn, AcOH/EtOH, reflux.

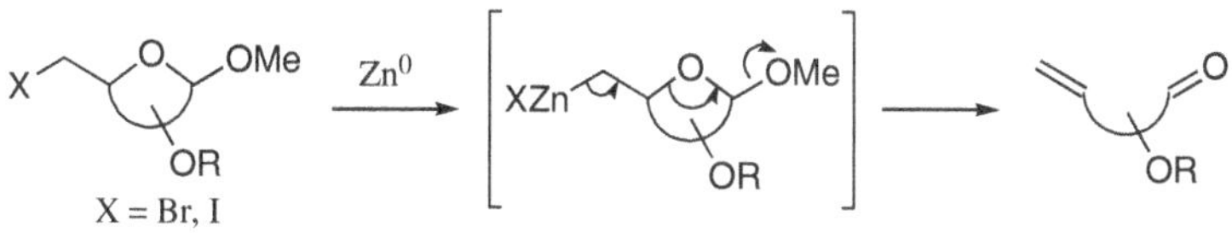

X = Br, I

SCHEME 3.2 Proposed mechanism for the Bernet–Vasella ring-opening.

The original procedure using acid-washed zinc powder in aqueous alcohols at reflux is still the most used protocol, but some other metallic sources were also employed in the past few years [19, 20]. For example, the use of more reactive zinc sources, as zinc–Rieke [21] or zinc–silver graphite [22,23], allowed the fragmentation to be carried out in tetrahydrofuran (THF) at room temperature. On the contrary, the fragmentation can also be effected by treatment of the halosugar with *n*-BuLi in THF at −80°C, as depicted in Scheme 3.3 for iodofuranoside **3** [24].

The Bernet–Vasella reductive ring opening has found many applications in carbohydrate chemistry and is a key step in several natural product synthesis [25, 26]. However, the instability of the liberated aldehyde can be a problem, leading to side reactions and decomposition [27, 28]. To overcome this limitation, the aldehyde has been trapped *in situ* by nitrogen or carbon nucleophiles. As a result, a series of domino reactions based on the Bernet–Vasella reaction has been described and applied to the total synthesis of chemically and biologically relevant products.

3.2.1 Domino Reductive Fragmentation/Reductive Amination

Bernotas et al. introduced a modification in the reductive fragmentation reaction to incorporate an *in situ* reductive amination of the alkenyl aldehyde [29]. Treatment of bromoglycoside **5** with an excess of zinc in the presence of benzylamine and sodium cyanoborohydride at reflux in a 19:1 mixture of PrOH/H_2O afforded aminoalkene **6** in a 70% yield. This is the first example of a domino Bernet–Vasella reaction/reductive amination and also of a reductive opening of an unprotected glycoside. After formation of the trifluoroacetate salt to prevent oxidation of the benzylamine, ozonolysis followed by hydrogenolysis finally gave iminosugar **7**, a potent inhibitor of β-glucosidase (Scheme 3.4).

More recently, primary amines were prepared from methyl glycosides in a single synthetic step by means of a domino Bernet–Vasella reaction/reductive amination using ammonia as the nitrogen source [30]. The protocol involved the overnight

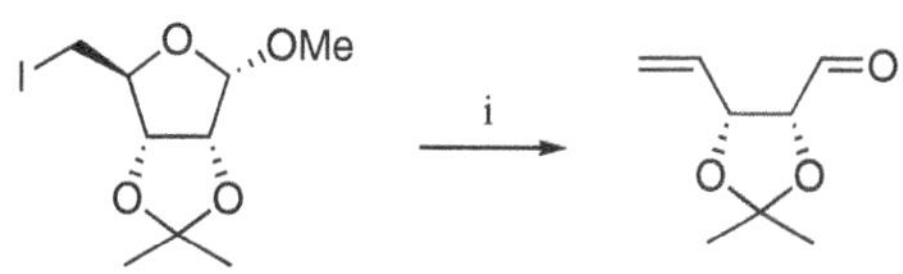

3 4

SCHEME 3.3 Reagents and conditions: (i) *n*-BuLi, THF, −80°C, quant.

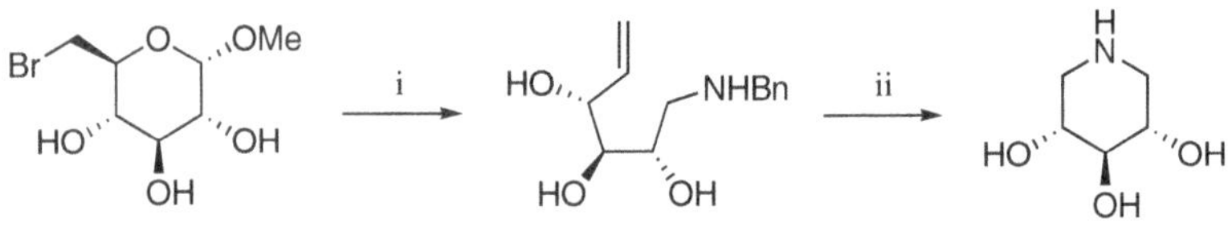

5 6 7

SCHEME 3.4 Reagents and conditions: (i) Zn, $BnNH_2$, $NaBH_3CN$, PrOH/H_2O, reflux, 70%; (ii) 1. TFA, CH_2Cl_2. 2. O_3 then SMe_2; (iii) H_2, Pd/C, 57% for the three steps.

reflux of a suspension of Zn, NH_4OAc, NH_3, $NaCNBH_3$, and the halosugar in ethanol. Under these conditions, the primary alkyl amines were produced as the major products in good yields, along with variable amounts of the dimeric secondary amine. For example, treatment of iodoglycoside **8** with Zn, excess NH_4OAc, NH_3, and $NaCNBH_3$ in ethanol at reflux for 18 h afforded primary amine **9** (a 95% yield) and the dimeric secondary amine **10** (ratio **9/10** >20:1) (Scheme 3.5).

3.2.2 Domino Reductive Fragmentation/Barbier-Type Allylation

In 1999, Hyldtoft et al. reported a zinc-mediated reductive elimination of ω-iodoglycosides in which the intermediate aldehyde was trapped *in situ* by a Barbier-type alkylation [31]. In this methodology, the reaction of various iodofuranosides and allyl bromide in the presence of activated zinc [32] afforded directly densely functionalized dienes in good yields. Zinc plays a dual role, first mediating the reductive fragmentation of the iodoglycoside, and second by activating the alkylating reagent [33,34]. Thus, a mixture of methyl iodofuranoside **3** and allyl bromide in THF/H_2O 4:1 was treated with activated zinc dust at 40°C under sonication, producing, after removal of the hydroxyl protecting groups with ion-exchange resin in the workup, a 4:1 mixture of diastereomeric dienes **11**α and **11**β in a quantitative yield (Scheme 3.6).

The combination of a reductive fragmentation with the reductive formation of a C—C bond incremented enormously the synthetic value of the Bernet–Vasella reaction [35]. For example, the resulting functionalized dienes can be subsequently converted into carbocycles by ring-closing metathesis (RCM) [36]. Thus, when the major diene isomer **11**α, obtained from the domino reductive fragmentation/allylation of iodoglycoside **12**, was treated with commercially available Grubbs' catalyst I in CH_2Cl_2, cyclohexene **13** was obtained in a 95% yield (Scheme 3.7) [37,38].

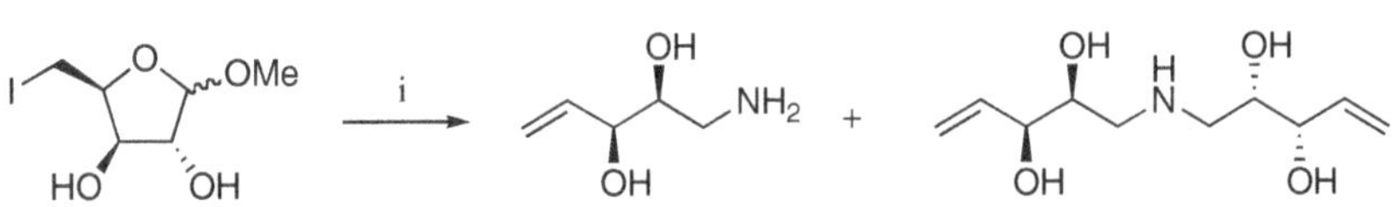

8 9 10

SCHEME 3.5 Reagents and conditions: (i) Zn, NH_3 (aq), $NaBH_3CN$, NaOAc, EtOH, reflux, 95%, ratio **9/10** >20:1.

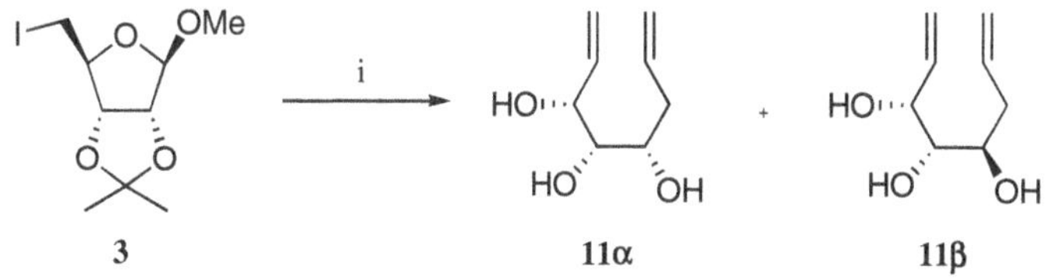

SCHEME 3.6 Reagents and conditions: (i) Zn, THF/H_2O 4:1, allyl bromide, 40°C, sonication, DOWEX 50W, quant.

This strategy based on the zinc-promoted domino reductive fragmentation/Barbier-type allylation followed by RCM has been used to prepare several carbocyclic natural products.

For example, Madsen et al. reported a short synthetic strategy for preparation of conduritols, using as a key step a domino zinc-mediated fragmentation/allylation of protected methyl 5-deoxy-5-iodo-D-pentofuranosides followed by ring-closing olefin metathesis [39]. Thus, methyl furanoside **3** and 3-bromopropenyl acetate were treated with zinc in a THF/H_2O mixture under sonication at 40°C. This resulted in one major product **14**, which was isolated in a 72% yield as a 4:1 mixture of two isomers. Conduritol D (**16**) was obtained from diene **14** when this compound was subjected to RCM followed by deprotection (Scheme 3.8).

Madsen and coworkers also described a concise synthetic route for the preparation of gabosine N (**22**), using as the key step a zinc-mediated domino fragmentation/allylation of methyl 5-deoxy-5-iodo-2,3-*O*-isopropylidene-α-D-ribofuranoside (**3**) with 3-benzoyloxy-2-methylallyl bromide (**17**) [40]. Treatment of furanoside **3** with zinc for 2 h followed by addition of bromide **17** gave an 85% yield of a 2:1 mixture of dienes from which major isomer **18** was separated by silica gel chromatography. The next step involved ring-closing olefin metathesis and was carried out with Grubbs' second-generation catalyst in refluxing dichloromethane. Under these conditions, the major diastereomer **18** cyclized cleanly to give cyclohexene **19** in a 97% yield. Treatment of alcohol **19** with dihydropyran gave fully protected **20** in a good yield, which was followed by removal of the benzoate to give allylic alcohol **21**. The oxidation of **21** with pyridinium dichromate (PDC), followed by deprotection under acidic conditions, finally afforded gabosine N (**22**) (Scheme 3.9).

Another interesting example is the efficient synthesis of the antitumor agent pancratistatin (**28**) from methyl 5-iodoribofuranoside, derived from xylose **23** [41].

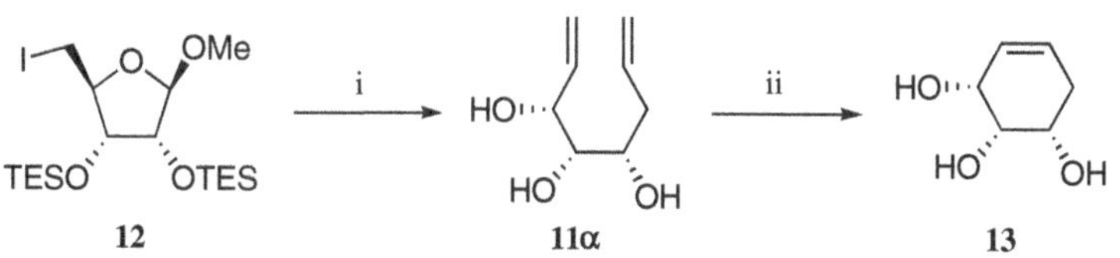

SCHEME 3.7 Reagents and conditions: (i) Zn, THF/H_2O 9:1, allyl bromide, 40°C, sonication, DOWEX 50W, 85%; (ii) Grubbs' catalyst I, CH_2Cl_2, 95%.

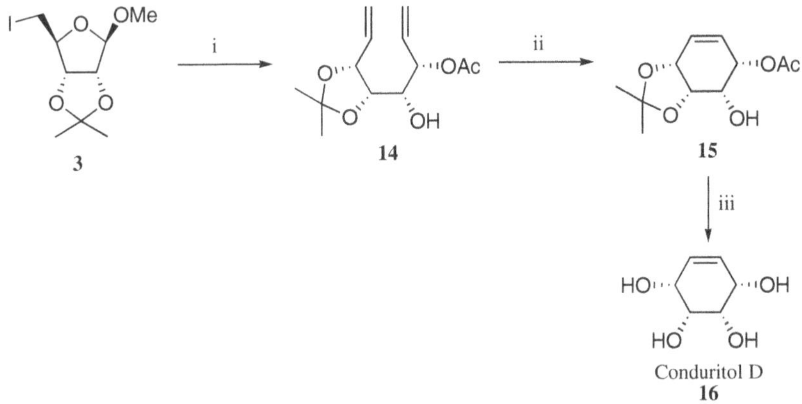

SCHEME 3.8 Reagents and conditions: (i) Zn, 3-bromopropenyl acetate, THF, H_2O, ultrasound, 40°C; (ii) Grubbs' catalyst II, CH_2Cl_2, 40°C; and (iii) AcOH, H_2O, 80°C, then MeOH, H_2O, NaOH, rt.

Cinnamyl bromide **24** and the iodocarbohydrate **23** are combined in a zinc-mediated fragmentation/allylation reaction to afford a highly functionalized 1,7-diene, which is then treated with Hoveyda–Grubbs' second-generation catalyst to produce a mixture of diastereomeric cyclohexenes **25** (35% yield) and **26** (32% yield). From the major isomer **25**, subsequent Overman rearrangement, dihydroxylation, and deprotection afford the natural product pancrastistatin (**28**) (Scheme 3.10).

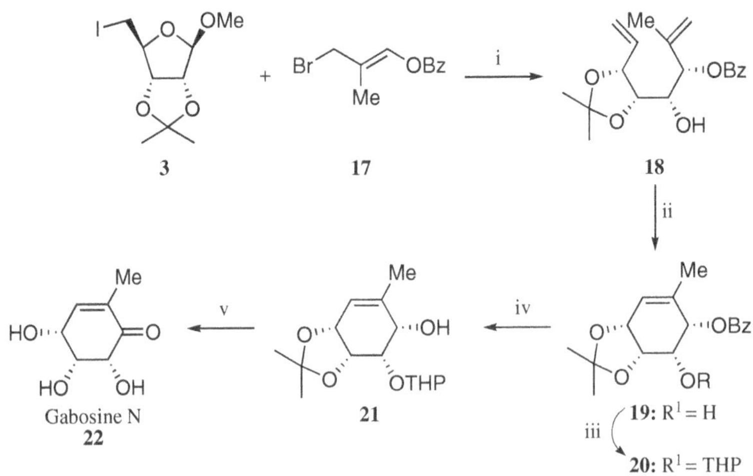

SCHEME 3.9 Reagents and conditions: (i) Zn, THF/H_2O, 40°C, sonication 85%; (ii) 10% Grubbs' catalyst I, CH_2Cl_2, 40°C, 97%; (iii) dihydropyran, pyridinium *p*-toluenesulfonate, CH_2Cl_2, rt, 75%; (v) NaOMe, MeOH, rt, 83%; and (vi) 1. PDC, CH_2Cl_2, rt, 71%. 2. AcOH, H_2O, rt to 40°C, 88%.

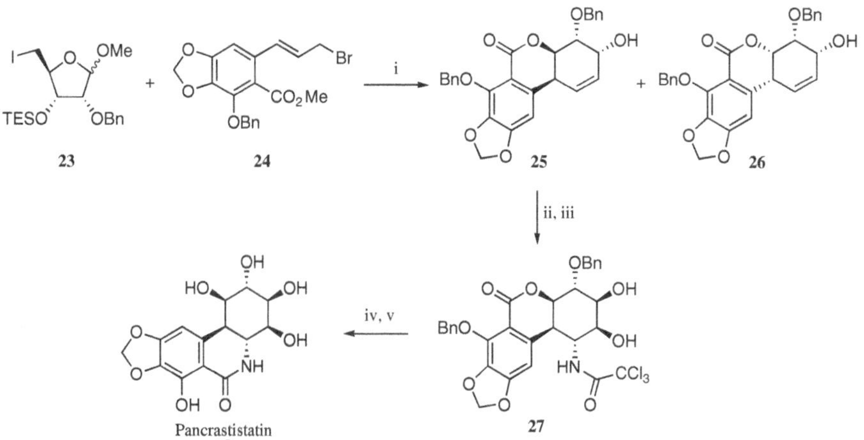

SCHEME 3.10 Reagents and conditions: (i) 1. Zn, THF/H_2O; MeOH, H^+. 2. K_2CO_3, MeCN. 3. Grubbs' catalyst II, CH_2Cl_2, 35% of 26 and 32% of 27; (ii) Cl_3CCN, 1,8-diazabicyclo[5.4.0]undec-7-ene and then 135°C, 64%; (iii) OsO_4, *N*-methylmorpholine oxide, 94%; (iv) K_2CO_3, MeOH, *N,N'*-dicyclohexylcarbodiimide (DCC), HOBt, 81%; and (v) H_2, $Pd(OH)_2/C$, 99%.

The total synthesis of alkaloid (+)-lycoricidine reported by Yadav et al. represents a further application of the sequence domino reductive fragmentation/allylation followed by ring-closing metathesis in the synthesis of bioactive natural products [42]. Thus, reaction of ω-iodoglycoside **29** with zinc/allyl bromide in THF/H_2O afforded diene **30** (87%, dr 85:15). Ring-closing metathesis of diene **30** followed by acetylation furnished cyclohexene **31**. Treatment of the acetate **31** with PhINTs in the presence of $Cu(acac)_2$, followed by sodium naphthalenide, afforded the aziridine **32** (67%), from which (+)-lycoricidine (**33**) can be easily obtained (Scheme 3.11).

The sequence metal-mediated domino fragmentation/allylation and ring-closing olefin metathesis of the resultant functionalized diene were also employed for the construction of the Tamiflu™ skeleton [43]. Thus, the Bernet–Vasella domino reaction of iodoribose derivative **34** in the presence of zinc dust and ethyl 2-(bromomethyl)acrylate in THF/H_2O (2:1) under sonication gave full conversion and a very high yield of the desired diene **35**. Using the second-generation Grubbs' catalyst, the diene **35** underwent RCM to provide 5-*epi*-shikimic acid **36** in a 60% yield. Shikimic acid derivative **36** was then efficiently transformed into oseltamivir phosphate (Tamiflu™) **37** (Scheme 3.12).

On the contrary, the intermediate aldehyde from the reductive ring-opening can be intercepted with an amine prior to the allylation [37]. The allylation takes then place on the formed imine and results in the introduction of an amino group. Thus, treatment of iodofuranose **3** with zinc, benzylamine, and allyl bromide succeeded in giving a 6:1 mixture of the amino dienes **38** and **39** in an 81% yield (Scheme 3.13).

Given that introduction of imino functionalities is of great importance for natural product synthesis [44], there are several examples of the use of this methodology in

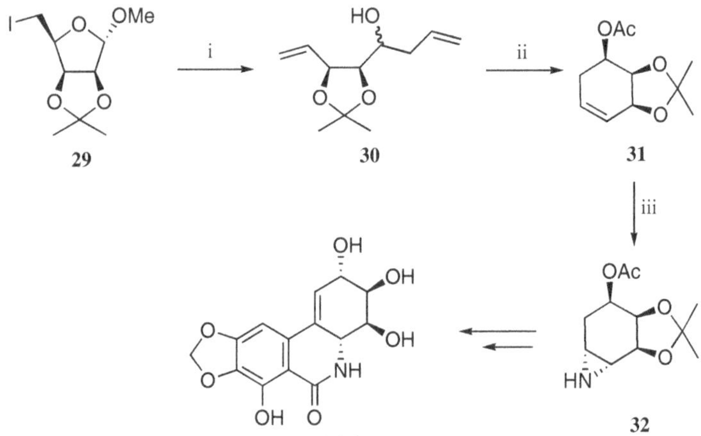

(+)-Lycoricidine
33

SCHEME 3.11 Reagents and conditions: (i) Zn, allyl bromide, THF/H_2O, 87%; (ii) 1. Grubbs' catalyst I, CH_2Cl_2, 84%. 2. Ac_2O, Et_3N, 4-dimethylaminopyridine, CH_2Cl_2, 92%; (iii) PhINTs, CH_3CN, Cu(acac)$_2$, 52%; and (iv) Na naphthalenide, dimethoxyethane, −70°C, 67%.

total synthesis. For example, this synthetic sequence was employed for the conversion of iodo glycopyranosides into enantiopure calystegines [45–48]. Thus, this procedure was used with iodopyranose **40** to develop the synthesis of diene precursor **41**, which was easily transformed by RCM into the 7-membered ring carbasugar **42**, main intermediate in the synthesis of calystegine A_3 (**43**) (Scheme 3.14) [45].

Using a similar strategy, Boyer et al. described a rapid and efficient route to (+)-calystegine B2, involving as the key step a triple domino reaction: zinc-mediated

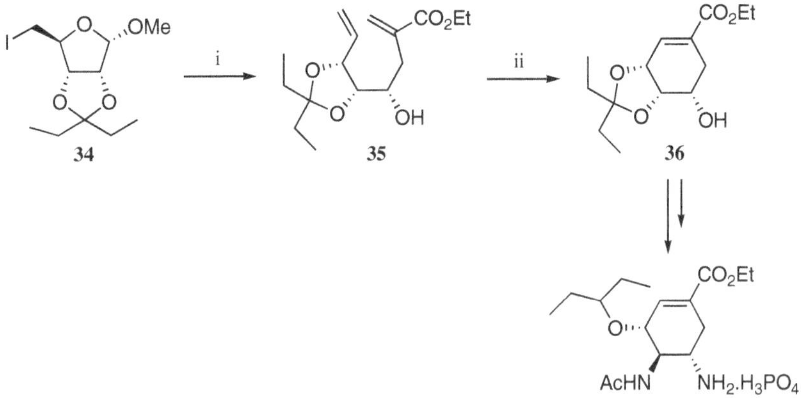

Tamiflu™
37

SCHEME 3.12 Reagents and conditions: (i) Zn, 2-(bromomethyl)acrylate, THF/H_2O 2:1, 40°C, sonication, 71%; and (ii) 10% Grubbs' catalyst II, CH_2Cl_2, 40°C, 60%.

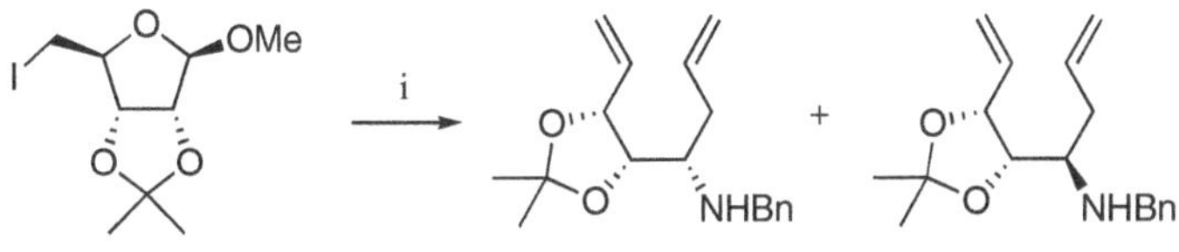

3 38 39

SCHEME 3.13 Reagents and conditions: (i) Zn, allyl bromide, benzylamine, THF, sonication, 40°C, 81%.

reductive ring-opening, imine formation, and allylation of 6-iodoglucopyranose derivative [46]. The synthesis started from 6-iodoglucopyranose **44**, which was subjected to zinc dust in the presence of benzylamine followed by the addition of allyl bromide in anhydrous THF under sonication, affording amino diene **45** as an 85:15 mixture of diastereomers. The major isomer was then protected as the benzyl carbamate and submitted to an RCM reaction using Grubbs' catalyst to yield **46**. Hydroboration of olefin **46** with borane–dimethyl sulfide complex followed by oxidative treatment gave a mixture of the regioisomeric alcohols **47** and **48** (2.6:1). Oxidation of the major regioisomer **47** with pyridinium chlorochromate (PCC), followed by hydrogenolysis of the resulting ketone with 10% Pd/C in aqueous acetic acid, afforded the desired (+)-calystegine B_2 (**50**) (Scheme 3.15).

3.2.3 Domino Reductive Fragmentation/Barbier-Type Propargylation

After establishing general reaction conditions for the zinc-mediated reductive ring-opening-allylation of iodoglycosides, Madsen et al. developed a similar procedure for the domino reaction of methyl 5-deoxy-5-iodo-pentofuranosides with propargyl

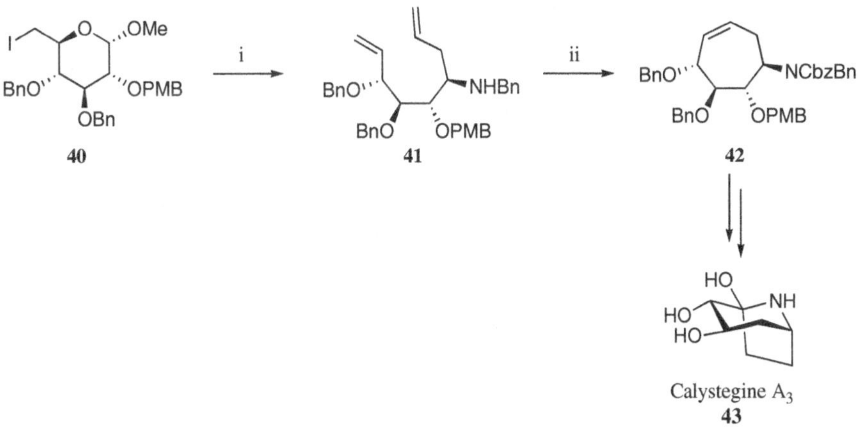

40 41 42

Calystegine A_3
43

SCHEME 3.14 Reagents and conditions: (i) Zn, TMSCl, THF, sonication, then $BnNH_2$, and then allyl bromide, sonication; and (ii) (a) CbzCl, $KHCO_3$, H_2O, CH_2Cl_2, (b) Grubbs' catalyst II, CH_2Cl_2.

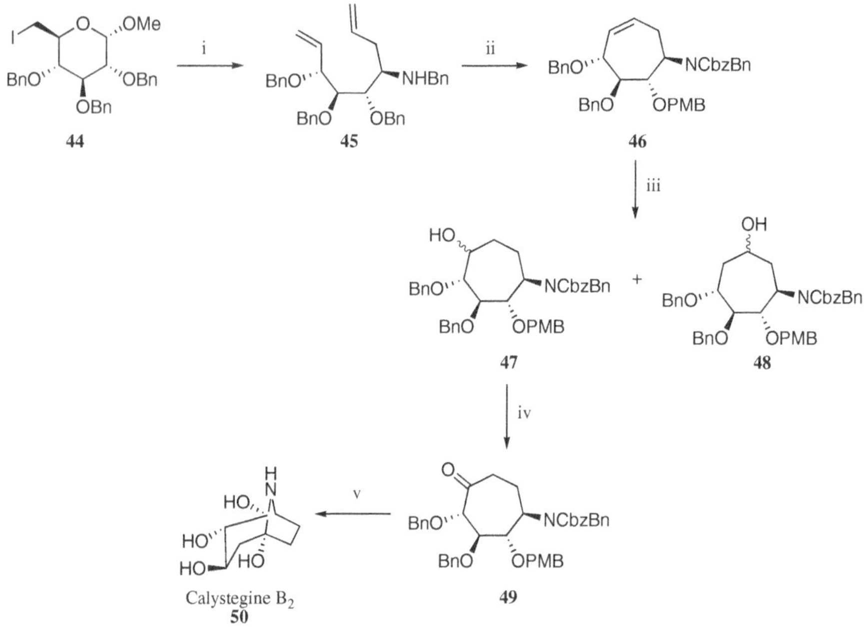

SCHEME 3.15 Reagents and conditions: (i) Zn, THF, $BnNH_2$, then allyl bromide, sonication, 73%; (ii) 1. CbzCl, $NaHCO_3$, AcOEt, 94%. 2. Grubbs' catalyst II, CH_2Cl_2, 97%; (iii) 1. BH_3-dimethyl sulfide, Et_2O, −30 to 0°C; 2. 30% H_2O_2, 2N NaOH, 61% of 47 and 24% of 48; (iv) PCC, CH_2Cl_2, 93%; and (v) 1. H_2, Pd/C, $AcOH/H_2O$; 2. NH_4OH, 79%.

bromide [49]. For this procedure, the mode of addition of propargyl bromide plays a major role. Carrying out the reaction on 5-iodo-ribofuranoside **3** with 5 equivalents of propargyl bromide all added from the start of the reaction gave a modest 37% yield of the 1,7-enyne **51**, together with the intermediate unsaturated aldehyde. However, shifting to dropwise addition of the propargyl bromide over 5 h by syringe pump led to the formation of desired enyne **51** in an improved 73% yield and in an excellent 9:1 diastereomeric ratio (Scheme 3.16).

The 1,7-enynes obtained by zinc-mediated ring opening and propargylation of iodofuranosides are suitable substrates for ring-closing enyne metathesis [50].

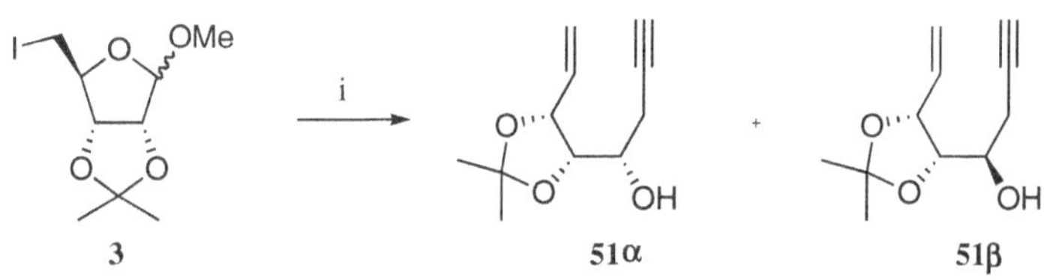

SCHEME 3.16 Reagents and conditions: (i) Zn, THF/H_2O 4:1, propargyl bromide, 40°C, sonication, 73%.

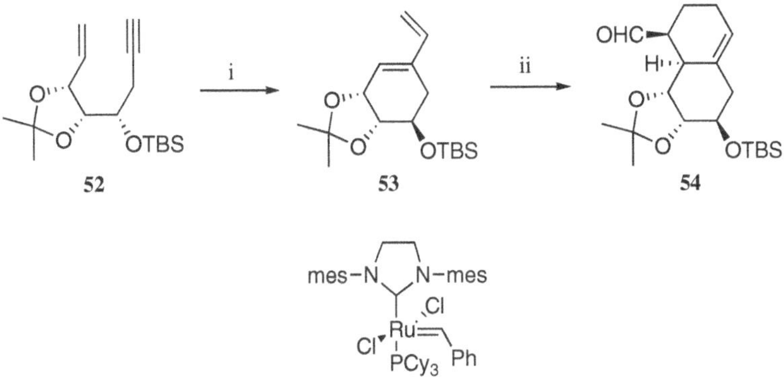

52 53 54

A

SCHEME 3.17 Reagents and conditions: (i) Catalyst A, CH_2Cl_2, ethylene atmosphere, 66%; and (ii) Acrolein, 60°C, 69%.

This methodology complements the carbocyclization method using reductive fragmentation and allylation followed by ring-closing olefin metathesis (Section 3.2.2). Furthermore, annulation of the product 1,3-dienes can be achieved through a Diels–Alder reaction. Thus, reaction of protected enyne **52** with 8% of ruthenium carbene catalyst **A** at room temperature in CH_2Cl_2 under an ethylene atmosphere afforded a 66% yield (72% based on recovered **52**) of 1-vinyl cyclohexene **53**. Warming 1,3-diene **53** in an acrolein solution at 60°C for 24 h gave mainly the Diels–Alder adduct **54**, which was isolated in a 69% yield (Scheme 3.17).

3.2.4 Domino Reductive Fragmentation/Vinylation

Following the success of the elimination–allylation and propargylation sequences, Madsen and coworkers investigated the corresponding elimination–vinylation reaction [37]. The vinylation, however, cannot be carried out directly under Barbier conditions because vinyl bromide does not insert zinc under the conditions for the reductive elimination; preformed divinyl zinc was used instead [51, 52]. Treatment of iodo furanoside **55** with zinc dust and divinyl zinc in THF gave diene **56** in a good yield (85%) and excellent diastereoselectivity (10:1) (Scheme 3.18).

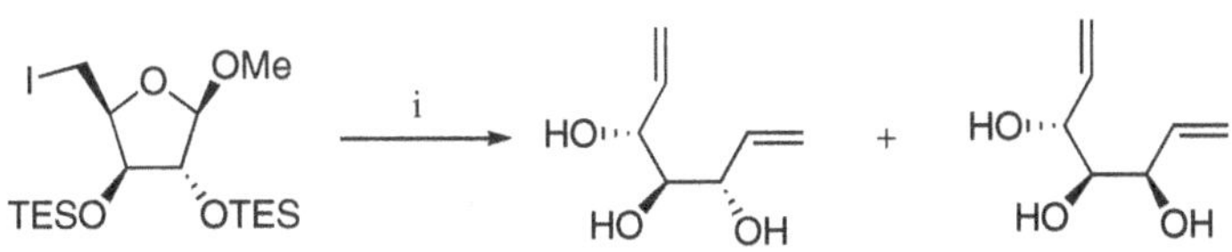

55 56α 56β

SCHEME 3.18 Reagents and conditions: (i) Zn, $Zn(CH{=}CH_2)_2$, THF, 40°C, sonication, DOWEX 50W, 85%.

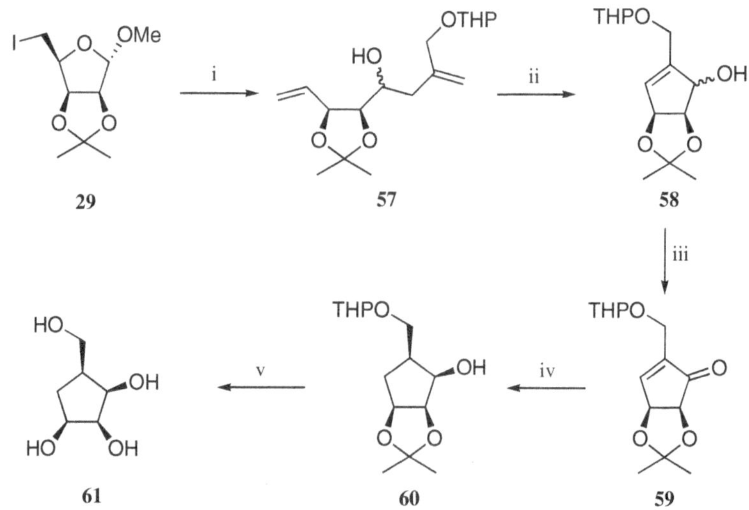

SCHEME 3.19 Reagents and conditions: (i) Mn/$CrCl_3$ (20:1), THF/ *N,N*-dimethylformamide (DMF) (1:1), 8 h then $NiCl_2$ (cation), 2-bromoprop-1-ene, TMSCl, 50°C, 7 h, TBAF, rt, 2 h, 55%; (ii) Grubbs' catalyst II, CH_2Cl_2, reflux, 6 h, 81%; (iii) PDC, CH_2Cl_2, 4 Å, molecular sieves, rt, 5 h, 98%; (iv) $CeCl_3 \cdot 7H_2O$, $NaBH_4$, CH_2Cl_2, 6 h, −78°C 93%; and (v) 1. Pd-C/H_2, MeOH, 2 h, 98%; 2. 2 M HCl in MeOH, 0°C, 15 min, 95%.

Rao and coworkers reported an efficient method for the synthesis of cyclopentitols from D-ribose, using a domino sequence of reductive-elimination and vinylation under Nozaki–Hiyama–Kishi conditions [53,54]. Thus, iodofuranoside **29** was treated with 2-bromoprop-1-ene in the presence of Mn/$CrCl_3$ and a catalytic amount of anhydrous $NiCl_2$, affording diene **57** in a 55% yield as an inseparable 1:1 mixture of diastereomers. Treatment of **57** with Grubbs' second-generation catalyst gave cyclic compound **58**, which was subjected to PDC oxidation to give enone **59** in a 98% yield. Luche reduction of **59**, followed by reduction of the olefinic double bond using Pd/C under H_2, gave cyclopentitol **60** in a 98% yield as the only product. Global deprotection of the THP and acetonide functionality in **60** gave the required carbasugar **61** in a 95% yield (Scheme 3.19).

3.2.5 Domino Reductive Fragmentation/Alkylation

Bercier et al. reported the domino reactions of iodofuranoses and various organometallic reagents [55]. The sequence involves an initial ring-opening by metal–halogen exchange and β-elimination, followed by *in situ* nucleophilic addition to the intermediate aldehyde. Thus, upon treatment with 3.5 equivalents of alkyl lithium in THF at 0°C, iodofuranose **62** was converted into diols **63** in excellent yields (Scheme 3.20).

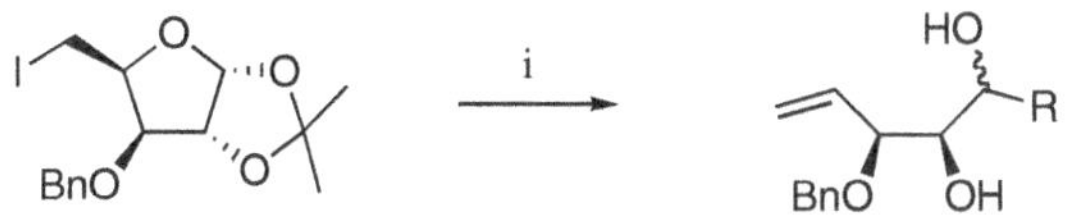

62

63a: R = *n*-Bu (88%, *syn/trans* 7:1)
63b: R = Me (90%, *syn/trans* 4:1)
63c: R = Ph (90%, *syn/trans* 6:1)

SCHEME 3.20 Reagents and conditions: (i) RLi, THF, 0°C, 8 h, 88–90%.

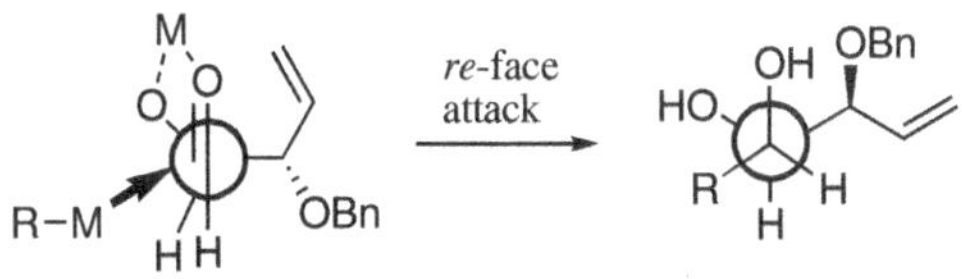

FIGURE 3.1 Major diastereomer according to α-chelated model.

The observed *syn*-selectivity (*R* for the new stereogenic center) is probably due to a α-chelation of lithium by the oxygen atoms of the generated alcoholate and of the carbonyl group (Figure 3.1) [56].

Taking into account that a wide range of organomagnesium reagents is commercially available, the scope of the domino reaction was extended to Grignard reagents. In this case, a larger excess of organometallic reagents was required for complete conversion. Thus, reaction of iodoglycoside **62** and 5.5 equivalents of Grignard reagents in diethyl ether at 0°C afforded the corresponding diols **63** in a good yield and moderate *syn*-selectivity (Scheme 3.21).

3.2.6 Domino Reductive Fragmentation/Olefination

As another possibility for introducing a double bond in the alkylation step, the intermediate aldehyde can be olefinated in the presence of carbon diiodide [57]. When iodo mannopyranoside **64** was treated with zinc, CH_2I_2, TMSCl, and $PbCl_2$, diene **56α** was obtained (Scheme 3.22) [37]. Since there are no stereocenters generated in this transformation, it is a better method for preparing diene **56α** than the elimination–vinylation reaction (Section 1.2, Scheme 3.18).

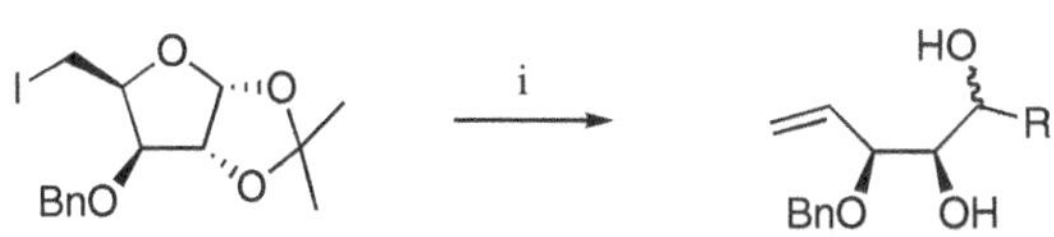

62

63b: R = Me (88%, *syn/trans* 4:1)
63c: R = Ph (87%, *syn/trans* 6:1)
63a: R = *i*-Pr (92%, *syn/trans* 4:1)
63a: R = All (91%, *syn/trans* 6:1)

SCHEME 3.21 Reagents and conditions: (i) RMgX, Et_2O, 0°C, 8 h, 87–92%.

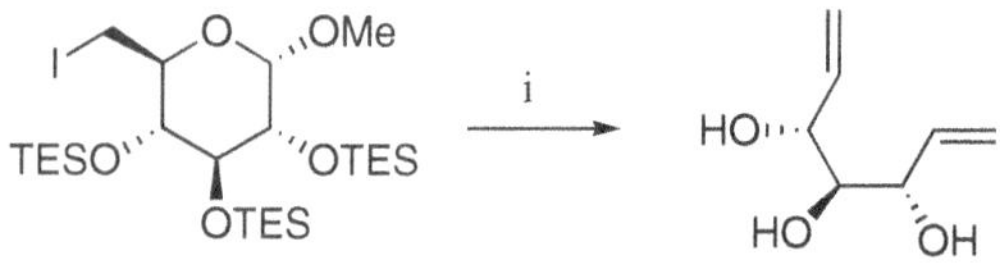

SCHEME 3.22 Reagents and conditions: (i) Zn, CH_2I_2, TMSCl, THF, 40°C, sonication, DOWEX 50W, 80%.

In 2010, Davies and coworkers reported a butyl lithium-promoted tandem reductive ring-opening/Wadsworth–Emmons olefination sequence for the preparation of a dienoate, intermediate in the synthesis of polyhydroxylated alkaloid (−)-7a-*epi*-hyacinthacine A1 [58]. Thus, a solution of D-ribose-derived iodoglycoside **3** and *tert*-butyl diethylphosphonoacetate in THF was treated with *n*-BuLi, to give an 88:12 (*E*)/(*Z*) mixture of olefin isomers, from which the desired (*E*)-isomer **65** was isolated by chromatographic separation. The doubly diastereoselective "matched" conjugate addition of lithium (*R*)-*N*-but-3-enyl-*N*-(*R*-methyl-*p*-methoxybenzyl)amide **66** [59] followed by *in situ* enolate oxidation with (−)-camphorsulfonyloxaziridine **67** produced derivative **68**, which on RCM afforded hexahydroazocine scaffold **69**. Subsequent treatment with I_2 resulted in transannular iodoamination accompanied by the loss of the *R*-methyl-*p*-methoxybenzyl group to give the corresponding pyrrolizidine scaffold **70** as a single diastereomer. Further functional group manipulations enabled the preparation of (−)-7a-*epi*-hyacinthacine A1 (**71**) (Scheme 3.23).

The reaction of 5-deoxy-5-iodopentofuranosides with α-silyl organolithium reagents affords 1,5-dienes in a domino reaction including a Peterson olefination step after addition on the intermediate aldehyde [55]. When iodofuranose **62** was treated with a (trimethylsilyl)methyl lithium solution (2 equivalents) in THF at 0°C, the domino sequence stopped at the nucleophilic addition stage, giving the vicinal diol **72** in a 69% yield and a high diastereomeric excess. However, when the transformation was conducted with 2.4 equivalents of the lithium reagent, diene **73** was obtained in a 78% yield (Scheme 3.24).

Treatment of **62** with α-lithio-α-trimethylsilyl dithioacetals (2.4 equivalents) led directly to the expected γ-unsaturated ketene dithioacetals **74** and **75** in 52 and 86% yield, respectively, as a result of a four-step domino reaction (Scheme 3.25).

3.2.7 Domino Reductive Fragmentation/Nitromethylation

In our group, we have described the synthesis of nitroalkenes starting from iodoglycosides via a novel tandem metallation–elimination–Henry reaction [60]. The reaction can be promoted either by the metal system zinc–indium or by *n*-BuLi, leading to different results in both cases. Thus, reaction of 5-iodo-ribofuranoside **3** with bromonitromethane in the presence of zinc (10 equivalents) and indium (1 equivalent) in THF/H_2O 4:1 under sonication at 40°C, afforded a 5:4 mixture of nitrohexene **76** and bromonitrohexene **77**, from which nitrohexene **76** was isolated in a 39% yield

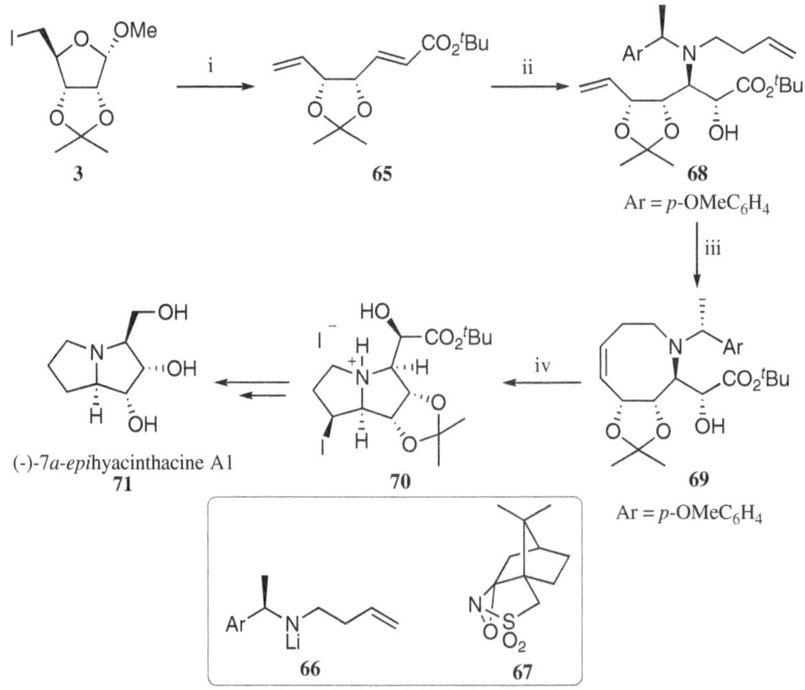

SCHEME 3.23 Reagents and conditions: (i) $(EtO_2)POCH_2CO_2{}^tBu$, *n*-BuLi, THF, −78°C to rt, 4 h, 55%; (ii) **66**, THF, −78°C, 2 h and then 67, −78°C to rt, 12 h, 50%; (iii) Grubbs' catalyst I, CH_2Cl_2, 30°C, 12 h, 73%; and (iv) I_2, $NaHCO_3$, rt, 12 h, 79%.

(dr α/β 5:2) and bromonitrohexene **77** was isolated in a 31% yield (dr 1*R*/1*S* 4:1) (Scheme 3.26).

The formation of the bromonitroalkene **77**, along with the expected nitroalkene **76**, can be explained on the basis of the two reaction pathways that the intermediate aldehyde can undergo, due to the presence of two different reactive species that are derived from bromonitromethane: nitronate **78** and bromonitronate **79**. The reaction of bromonitromethane and indium(0) give rise to the corresponding organoindium species **78**, upon which the reaction with the intermediate aldehyde afforded the nitroalkene **76** [61]. On the contrary, taking into account the high acidity of bromonitromethane, the trace of metal alkoxide released in the fragmentation reaction could

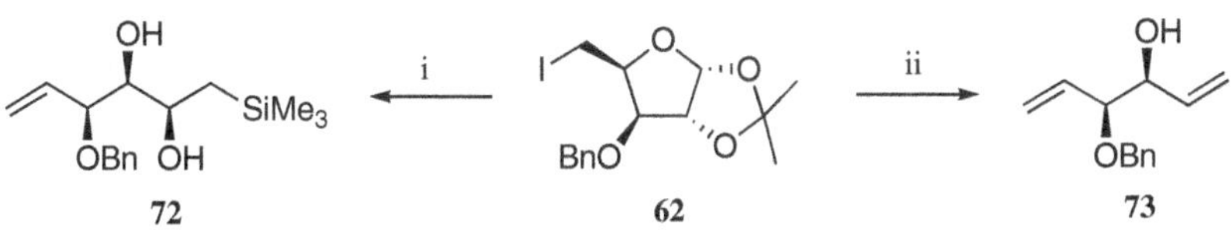

SCHEME 3.24 Reagents and conditions: (i) Me_3SiCH_2Li (2 equivalents), THF, 0°C, 69%; and (ii) 1. Me_3SiCH_2Li (2.4 equivalents), THF, 0°C; 2. aq H_2SO_4, 78%.

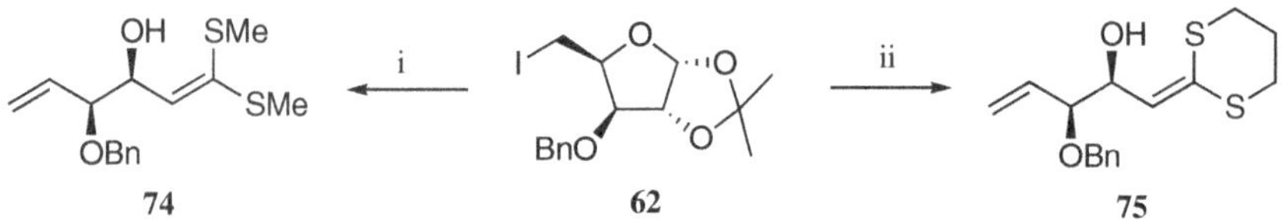

SCHEME 3.25 Reagents and conditions: (i) $Me_3SiC(SMe)_2Li$ (2.4 equivalents), THF, 0°C, 52%; and (ii) $Me_3SiC[S(CH_2)_2S]Li$ (2.4 equivalents), THF, 0°C, 86%.

be sufficient to abstract an acidic proton from bromonitromethane forming the corresponding bromonitronate **79**, which on reaction with the intermediate aldehyde afforded the bromonitrohexene **77** (Scheme 3.27).

On the contrary, treatment of a 1:1 mixture of methyl 5-iodoribofuranoside **3** and nitromethane in THF with 2.6 equivalents of *n*-BuLi promoted the domino Vasella reaction/Henry addition to give 1-nitro-6-hexene **76** in an 85% overall yield as a 1:3 α/β mixture of isomers (Scheme 3.28). The generality of this procedure was demonstrated when the methodology was successfully extended to iodopyranosides with different configurations and protections in the hydroxyl groups.

3.3 REDUCTIVE RING CONTRACTION

In contrast to the Bernet–Vasella reaction, which affords olefinic acyclic carbohydrates, a series of reductive rearrangements of cyclic sugars allows their direct conversion into carbocycles. The conversion of carbohydrate derivatives into functionalized and enantiomerically pure cycloalkanes is an extremely important objective and has attracted a great deal of attention [62,63]. In this regard, reductive rearrangements have proven to be highly efficient for the construction of the cycloalkane moiety. This straightforward approach to carbasugars relies on the structure of sugar enol-ethers, which inherently carry both masked nucleophilic and electrophilic functionalities [64].

Thus, after an initial reductive ring-opening step, the cycloalkane ring is constructed on subsequent *in situ* annulation of the electrophilic carbonyl and a newly formed nucleophilic organometalic site.

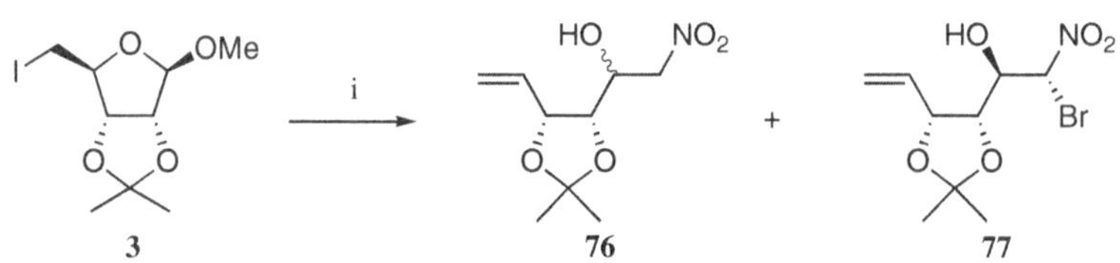

SCHEME 3.26 Reagents and conditions: (i) Zn, In, $BrCH_2NO_2$, THF/H_2O 4:1, 40°C, sonication.

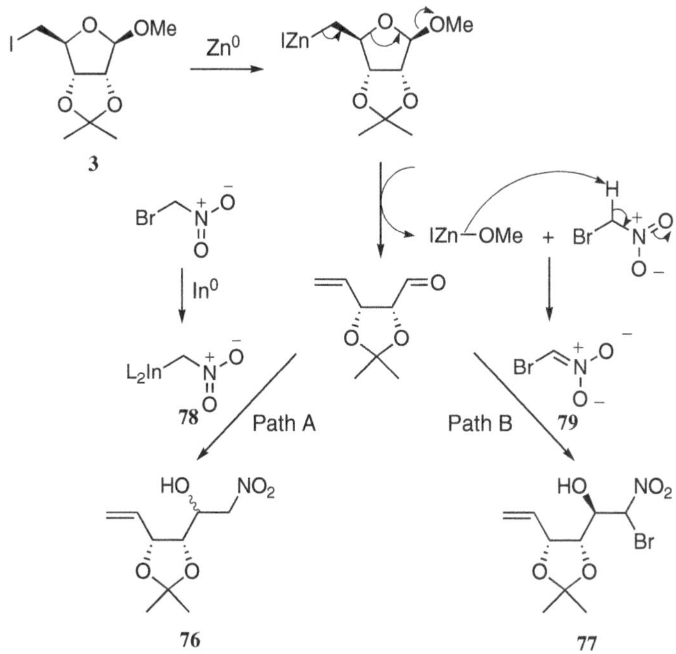

SCHEME 3.27 Mechanism for the formation of 1-nitro-6-hexenes **76** and 1-bromo-1-nitro-6-hexene **77** from 5-iodo-ribofuranoside **3**.

3.3.1 Ring Opening/Ketyl-Olefin Annulation

As explained in the previous section, the reductive dealkoxyhalogenation of iodoglycosides (Bernet–Vasella reaction) gives rise to ring-opened alken-5-enals, which are suitable substrates for intramolecular ketyl-olefin reductive coupling. Thus, 5-hexenals obtained by treatment of the corresponding methyl 6-deoxy-6-iodoglycosides with powdered zinc have been employed as substrates for an SmI_2-mediated radical cyclization, allowing the overall transformation of pyranoses into cyclopentanes [65]. A modified and more efficient domino procedure in which reductive ring-opening was executed under the action of samarium diiodide was successfully applied to the direct preparation of cyclopentanic derivatives from iodoglycosides [66]. Thus, treatment of iodoglucoside **80** with excess

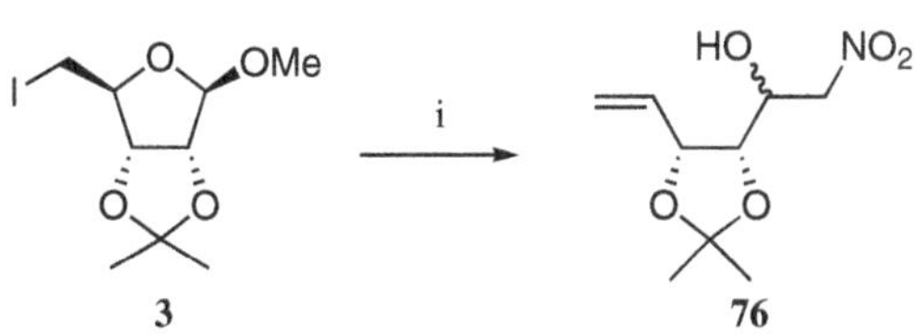

SCHEME 3.28 Reagents and conditions: (i) $BrCH_2NO_2$, *n*-BuLi, THF, −78°C to rt, 85%.

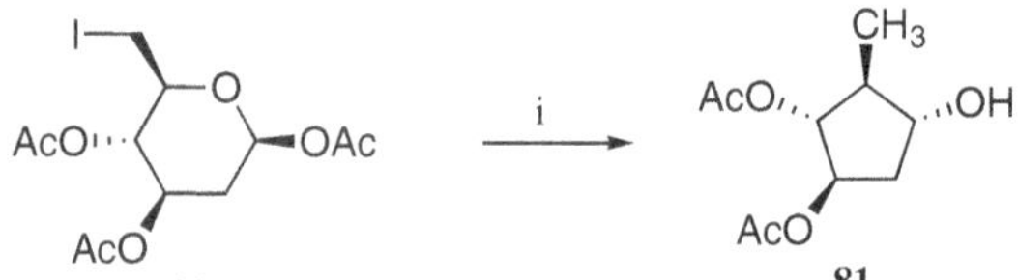

SCHEME 3.29 Reagents and conditions: (i) SmI_2, HMPA, THF, reflux, 70%.

SmI_2-THF/hexamethylphosphoramide (HMPA) under reflux afforded stereodefined cyclopentanol **81** (Scheme 3.29).

In order to effect the overall ring contraction, two complete reaction sequences are necessary. These include a reductive dealkoxyhalogenation to give the ring-opened hex-5-enal followed by an intramolecular ketyl-olefin reductive coupling, to afford the ring contracted organosamarium intermediate. A complete mechanism for the two individual subsequent steps is depicted in Scheme 3.30.

Intermolecular trapping of this organosamarium with an appropriate electrophile produces branched cyclopentanol derivatives. In addition, the presence of intramolecular radical traps allows the rapid assembly of bicyclic ring systems with stereodefined double bonds, via an appropriate intramolecular cyclization. For example, reaction of iodopyranoside **82** with SmI_2 in the presence of HMPA afforded bicyclic derivative **83** (Scheme 3.31) [67].

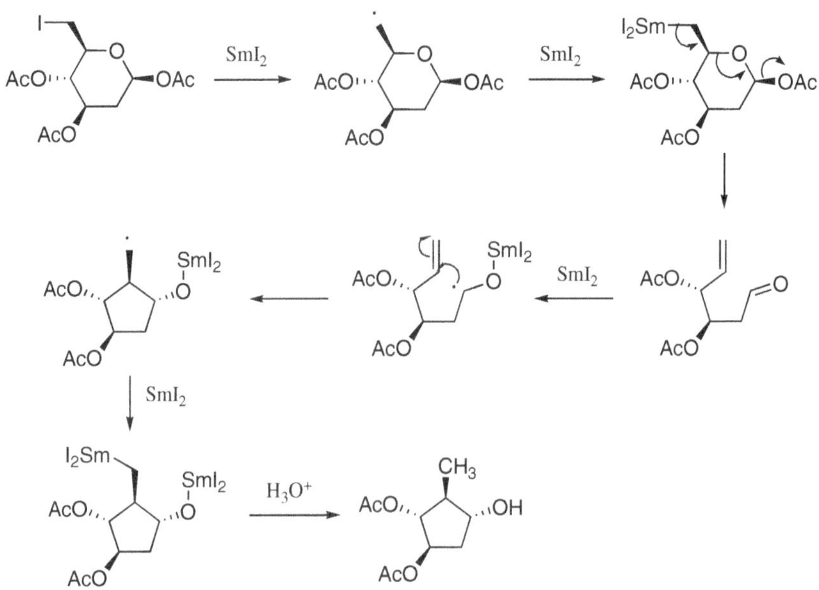

SCHEME 3.30 Proposed mechanism for the formation of cyclopentanol **81**.

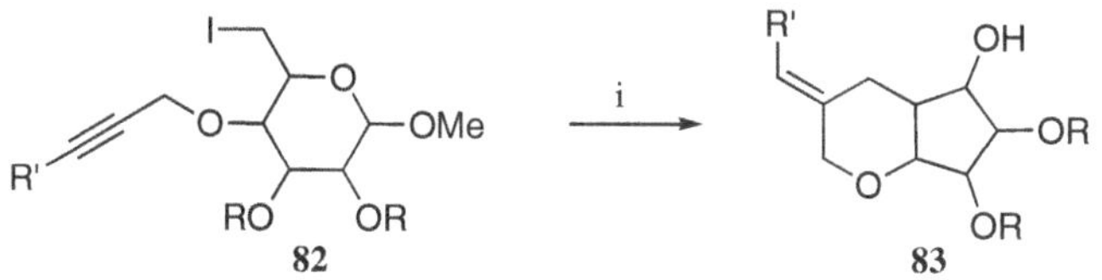

SCHEME 3.31 Reagents and conditions: (i) SmI_2, HMPA, THF.

After the dealkoxyhalogenation, the mechanism of the SmI_2-mediated cyclization of the intermediate aldehyde is shown in Scheme 3.32. The process is probably initiated by a single electron reduction of the aldehyde to form a ketyl radical, followed by an *exo*-cyclization onto the double bond. The resulting primary radical undergoes cyclization onto the tethered alkyne and the highly reactive vinyl radical that results is quenched by abstraction of a hydrogen atom from the solvent [68].

Similarly, Chiara et al. reported a cascade reaction of 6-deoxy-6-iodohexopyranosides promoted by samarium diiodide [69]. A series of 6-deoxy-6-iodohexopyranosides with different configurations and substitution patterns was treated with SmI_2 in THF/HMPA to give the desired cyclopentanols. For example,

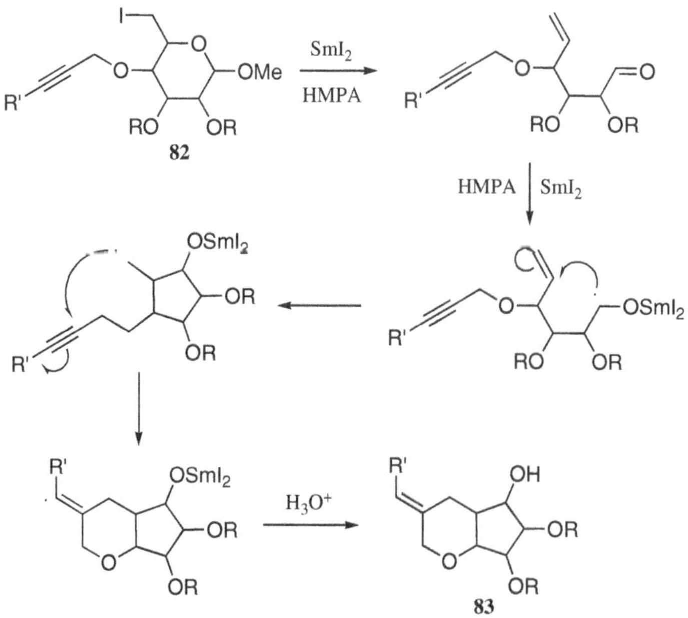

SCHEME 3.32 Proposed mechanism for the reductive ring opening and intramolecular radical trapping to produce **83**.

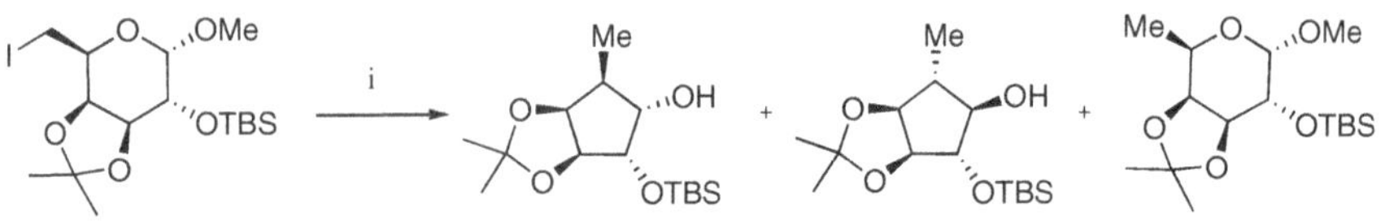

84 85 86 87

SCHEME 3.33 Reagents and conditions: (i) SmI_2, HMPA, THF, 47% of **85**, 17% of **86**, and 11% of **87**.

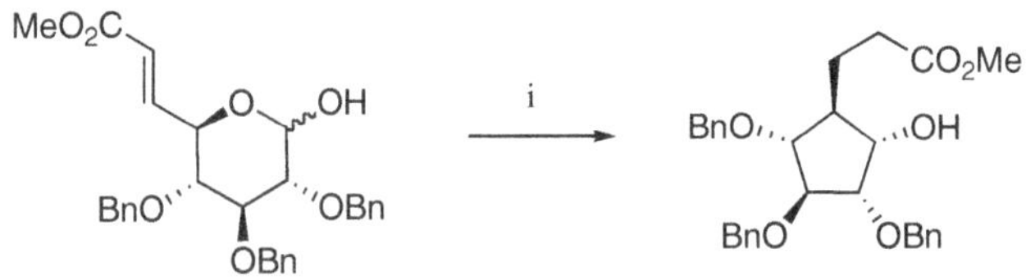

88 89

SCHEME 3.34 Reagents and conditions: (i) SmI_2, THF/MeOH 15:1, 0°C, 91%.

the reaction of pyranoside **84** with SmI_2 in THF and HMPA afforded cyclopentanols **85** and **86**, together with some of the dehalogenated product **87** (Scheme 3.33).

Using an approach that is conceptually similar to those described in Scheme 3.33, an α,β-unsaturated ester derivative of glucose has also been subjected to an SmI_2-mediated ring contraction [70]. Treatment of **88** with SmI_2 (5.0 equivalents) at 0°C in THF/MeOH (15/1) provided the cyclopentanol **89** as the sole product in an excellent yield (Scheme 3.34).

In order to gain more insight into the mechanism of this domino transformation, ester **88** was treated with 2.5 equivalents of SmI_2 in THF/MeOH (15/1), to afford the β,γ-unsaturated methyl ester **90**. Furthermore, coupling cyclization of **90** with 2.5 equivalents of SmI_2 in THF/MeOH (15/1), produced **89** as a single stereoisomer in an 87% yield. These results indicate that the reaction may involve sequential reductive ring opening and ketyl-olefin annulation. Thus, the ring-cleavage reaction of tetrahydropyran with SmI_2 is initiated by the single-electron reduction of the α,β-unsaturated ester moiety of **88**, forming the intermediate β,γ-unsaturated methyl ester **90** [71, 72]. The SmI_2-promoted 5-*exo-trig* ketyl-olefin cyclization of **90** generated *in situ* affords the observed cyclopentanol **89** (Scheme 3.35).

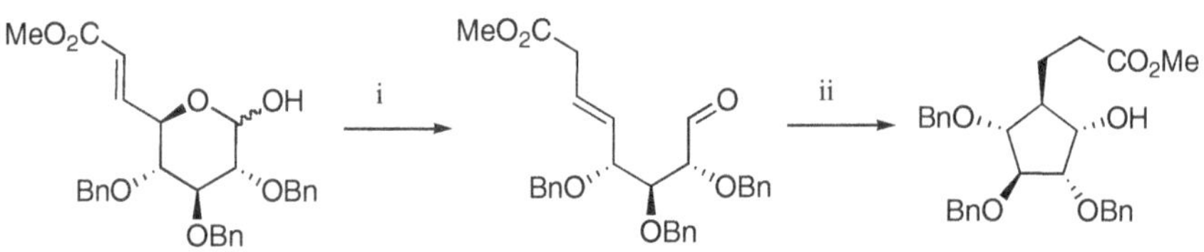

88 90 89

SCHEME 3.35 Reagents and conditions: (i) SmI_2, THF/MeOH 15:1, −40°C, 91%; and (ii) SmI_2, THF/MeOH 15:1, 0°C, 87% two steps.

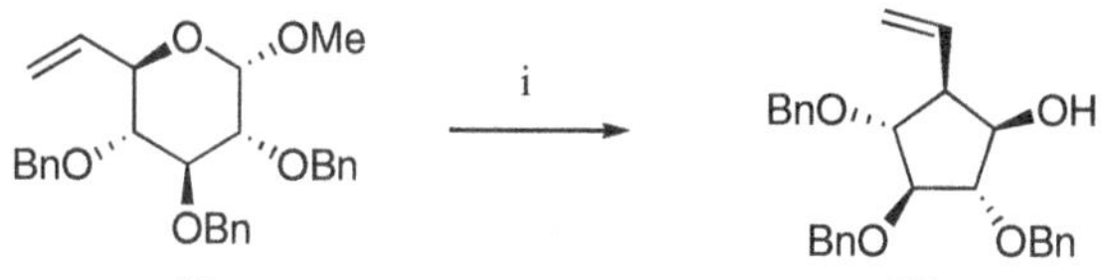

SCHEME 3.36 Reagents and conditions: (i) "Cp_2Zr"/THF then BF_3 Et_2O, −78°C to rt, 75%.

3.3.2 Ring Opening/Intramolecular Carbonyl Alkylation

Modified carbohydrates, where the terminal carbon has been elaborated into a vinyl unit, are useful precursors of cyclopentanes via formation and intramolecular reaction between allylmetal species and the aldehyde generated *in situ* [73].

In this regard, zirconium-butene complex ("Cp_2Zr") [74] was widely used to promote the carbocyclization of vinyl glycosides [75]. Thus, sequential treatment of 5-vinyl pyranose **91** with Cp_2Zr-(n-Bu)$_2$ (prepared *in situ* from zirconocene dichloride and 2 equivalents of butyl lithium) and boron trifluoride etherate in THF afforded vinyl cyclopentanol **92** as the major product in a 75% yield (Scheme 3.36) [76].

The reaction was extended to other 5-vinyl pyranoses and the major product has always a *cis*-configuration between the newly formed stereocenters and a *trans*-configuration between the vinyl and the adjacent substituent. Considering these results, the proposed mechanism for the "CpZr"–mediated carbocyclization is depicted in Scheme 3.37. Ligand exchange with the vinyl followed by reductive elimination of the alkoxy group produces a Z-allylic zirconacycle, which immediately cyclizes to the cyclopentane **92** (Scheme 3.37).

The excellent diastereoselectivity can be explained by considering the two possible chair-like transition states **B** and **C**. Reaction through transition state **C** is unfavorable due to steric repulsion between Cp and the benzyloxy substituents (Figure 3.2).

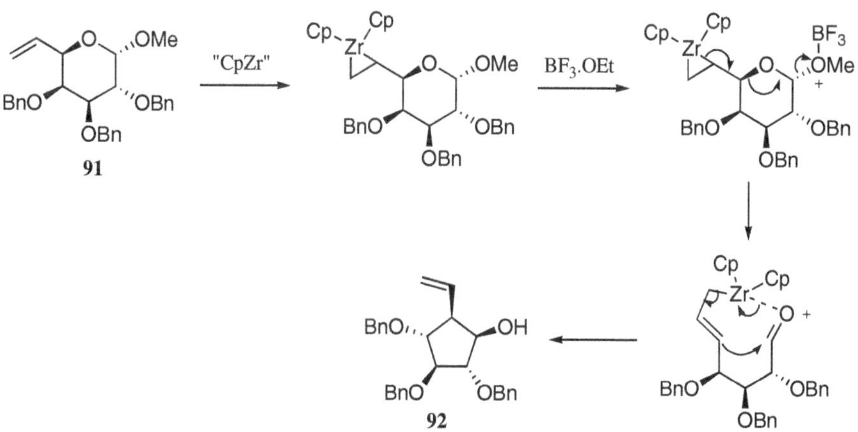

SCHEME 3.37 Proposed mechanism for the zirconium-promoted ring contraction of **91**.

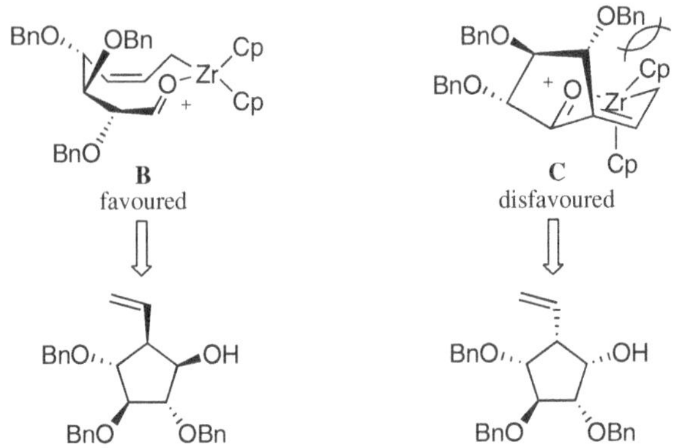

FIGURE 3.2 Two possible transition states for the ring contraction of 91.

As in the case of pyranosides, treatment of 4-vinylfuranosides with "Cp_2Zr" afforded vinylcyclobutanols in the same stereochemical outcome with that of the ring contraction of 5-vinylpyranosides [77]. For example, treatment of vinylfuranoside **93** with *in situ* generated Cp_2Zr-(n-Bu)$_2$ and boron trifluoride etherate in THF afforded vinyl cyclobutanol **94** in a 77% yield (Scheme 3.38).

As an example of the use of this methodology in total synthesis, the zirconocene-promoted ring contraction was the key step in the route to (+)-epiafricanol (**100**) reported by Paquette et al. [78]. Vinyl pyranose **95**, obtained from D-glucose in 12 steps, was treated with "Cp_2Zr" in toluene at −78°C and then with boron trifluoride etherate at room temperature, affording vinyl cyclopentane **96** in a 63% yield. Conversion of **96** to ketone **97** with periodinane as the oxidant in the presence of sodium bicarbonate and addition of the lithium derivative to 4,4-dimethyl-5-iodo-1-pentene **98** gave rise to **99**. Cyclization of **99** by RCM followed by cyclopropanation finally afforded the desired product **100** (Scheme 3.39).

Jenkins et al. described a route to ring-contracted analogues of the second messenger D-*myo*-inositol 1,4,5-trisphosphate, using a zirconium-mediated carbocyclization [79]. On sequential treatment with $Cp_2Zr(n\text{-}Bu)_2$ followed by $BF_3{\cdot}Et_2O$, glucopyranoside **101** afforded 5-vinylcyclopentane **102** as the major product. Removal of the *p*-methoxybenzyl groups of **102** and subsequent phosphorylation and deprotection afforded the first target 1,2,4-tris(phosphonooxy)-5-vinylcyclopentane **105** (Scheme 3.40).

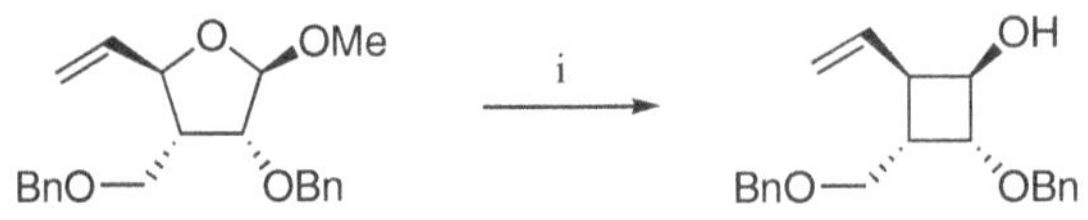

93 **94**

SCHEME 3.38 Reagents and conditions: (i) "Cp_2Zr"/THF then BF_3 Et_2O, −78°C to rt, 77%.

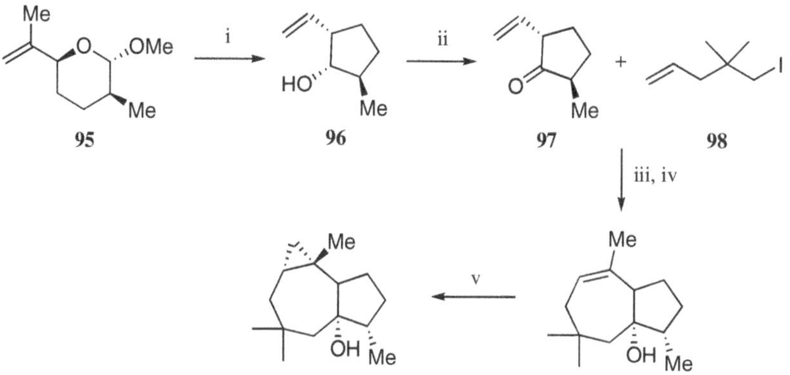

95 96 97 98

(+)-epiafricanol
100

99

SCHEME 3.39 Reagents and conditions: (i) "Cp_2Zr"/Toluene and then BF_3 Et_2O, −78°C to rt, 63%; (ii) Dess-Martin, $NaHCO_3$, CH_2Cl_2; (iii) *t*-BuLi, THF/pentane, −50°C, 45% for two steps; (iv) Grubbs' catalyst, CH_2Cl_2, 93%; (v) Et_2Zn, CH_2I_2, CH_2Cl_2, 87%.

Samarium(II) iodide in the presence of Pd(0) can also promote this transformation. An example is the transformation of 5-alkenylpyranoside **106** into homoallyl cyclopentanols **107** (82% yield) and **108** (12% yield) [80]. The first step is the formation of an allenylpalladium complex, followed by reduction by SmI_2 and subsequent intramolecular allylation of the originated carbonyl-tethered allenylsamarium (Scheme 3.41).

Comparing both processes, the difference lays on the stereochemical outcomes in the final cyclopentanes, depending on the particular metal employed. Thus, allylzirconium intermediates lead to a *syn*-relationship between vinyl and hydroxyl groups

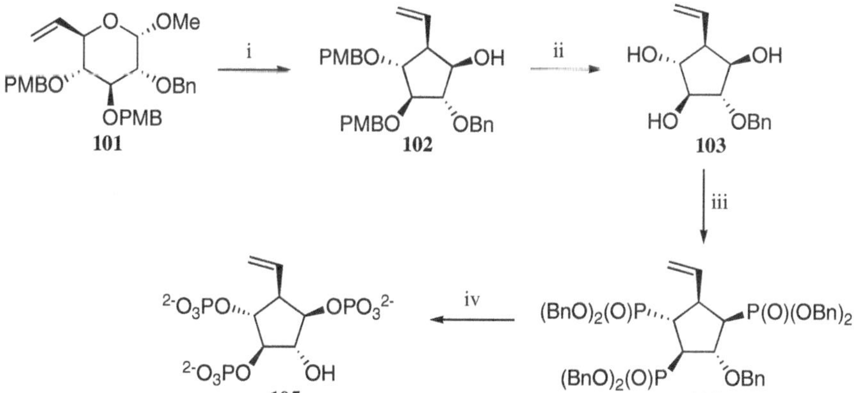

101 102 103

105 104

SCHEME 3.40 Reagents and conditions: (i) "Cp_2Zr"/THF then $BF_3{\cdot}Et_2O$, −78°C to rt; (ii) 1 M HCl/EtOH 1:2, reflux; (iii) 1. *i*-$Pr_2NP(OBn)_2$, 1*H*-tetrazole, CH_2Cl_2; 2. *m*-Chloroperbenzoic acid, −78°C; and (iv) Na/NH_3, −78°C.

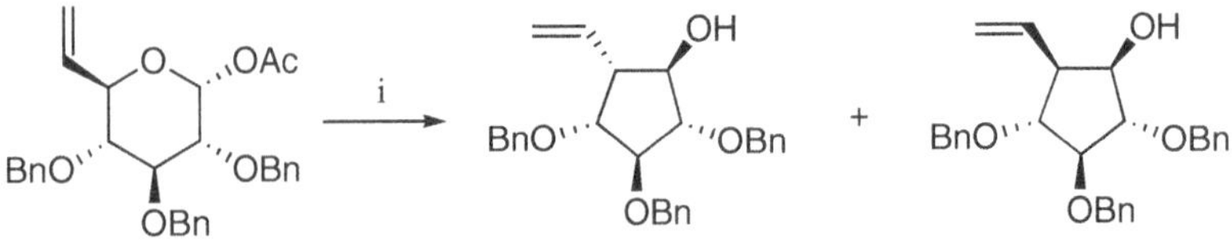

106 **107** **108**

SCHEME 3.41 Reagents and conditions: (i) SmI_2, $Pd(PPh_3)_4$, THF, 80°C.

in the cyclized products, whereas allylsamarium derivatives yield preferentially the corresponding *anti*-products [81].

Recently, the possibility of using the Pd(0)/Et_2Zn system in this transformation was also investigated [82]. Upon treatment of vinyl carbohydrate **109** with $Et_2Zn/Pd(PPh_3)_4/ZnCl_2$, the cyclopentanes **110**, **111**, **112**, and **113** were isolated (6% for **110**, 6% for **111**, 75% for **112**, and 13% for **113**) (Scheme 3.42).

This new carbohydrate ring-contraction reaction does not appear to follow the same stereochemical trends as either SmI_2 or "Cp_2Zr"-promoted analogous cyclizations. In general terms, the Et_2Zn-mediated reaction displays an overall preference for an *anti*-relationship between vinyl and adjacent OR groups at C(4). This effect has been previously observed in the carbocyclizations of related allylzirconiums and it was attributed to steric factors. However, in contrast with allylzirconium chemistry where a general preference for *anti*-C(4)/C(5) and *syn*-C(1)/C(5) relationships is observed, in these new Et_2Zn-promoted cyclizations it is clear that small structural changes are enough to break those tendencies. While diastereoselectivities are only moderate and substrate-dependent, the interest in this methodology lays on the possibility of having access to preparatively useful yields of isomers that are not available by using the alternative methodologies.

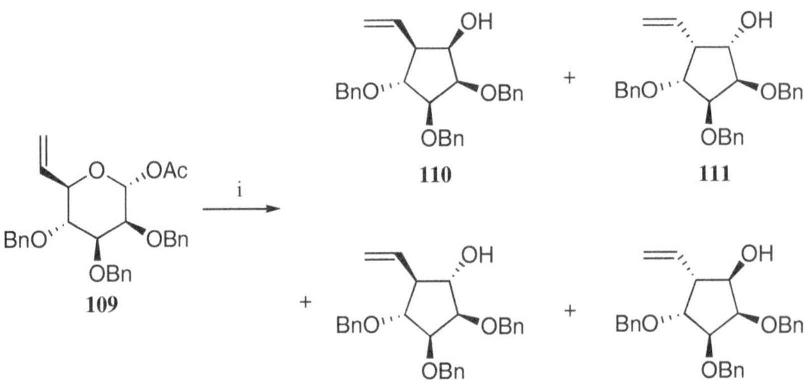

109 **110** **111** **112** **113**

SCHEME 3.42 Reagents and conditions: (i) Et_2Zn, $Pd(PPh_3)_4$, $ZnCl_2$, THF, rt.

3.4 CONCLUSIONS

The past several years have witnessed explosive developments in the application of low-valent organometallic reagents to carbohydrate chemistry. In addition to this, low-valent organometallic species exhibits excellent properties for sequential organic reactions. By combining both strategies, researchers found that reductive organometallic reagents could efficiently promote domino reactions on sugar derivatives. Most of the reported procedures, promoted by different low-valent organometallic reagents, start with the reductive elimination to an open-chain sugar intermediate, which immediately reacts with an organometallic species formed *in situ*.

In this review, we have discussed the most important advances in the domino reactions of carbohydrates based on the metal-promoted ring opening of cyclic sugars. We believe that we have painted an accurate picture on the advances made in the use of reducing metals in domino reactions of carbohydrate derivatives and the need to develop new processes. Further advances are expected to allow the facile and selective construction of highly functionalized small organic molecules from easily available sugar starting materials.

REFERENCES

[1] Trost, B. M. *Angew. Chem., Int. Ed. Engl.* **1995**, *34*, 259.
[2] Denmark, S. E.; Thorarensen, A. *Chem. Rev.* **1996**, *96*, 137.
[3] Parsons, P. J.; Penkett, C. S.; Shell, A. J. *Chem. Rev.* **1996**, *96*, 195.
[4] Eilbracht, P.; Bärfacker, L.; Buss, C.; Hollmann, C.; Kitsos-Rzychon, B. E.; Kranemann, C. L.; Rische, T.; Roggenbuck, R.; Schmidt, A. *Chem. Rev.* **1999**, *99*, 3329.
[5] Bruggink, A.; Schoevaart, R.; Kieboom, T. *Org. Proc. Res. Dev.* **2003**, *7*, 622.
[6] Gabius, H. J. *The Sugar Code: Fundamentals of Glycoscience*; Wiley-Blackwell, **2009**.
[7] Cipolla, L.; Araújo, A. C.; Bini, D.; Gabrielli, L.; Russo, L.; Shaikh, N. *Expert Opin. Drug Discov.* **2010**, *5*, 721.
[8] Hanessian, S. *Preparative Carbohydrate Chemistry*; Marcel Dekker Inc.: New York, **1997**.
[9] Concellón, J. M.; Rodríguez-Solla, H. *Chem. Soc. Rev.* **2004**, *33*, 599.
[10] Williams, D. B. G.; Caddy, J.; Blann, K. *Org. Prep. Proc. Int.* **2003**, *35*, 307.
[11] Classon, B.; Garegg, P. J.; Samuelsson, B. *Acta Chem. Scand.* **1982**, *36*, 251.
[12] Cho, S.; Kang, S.; Keum, G.; Kang, S. B.; Han, S.-Y.; Kim, Y. *J. Org. Chem.* **2003**, *68*, 180.
[13] Molander, G. A.; Harris, C. R. *Chem. Rev.* **1996**, *96*, 307.
[14] Soengas, R. G.; Estévez, A. M. *Curr. Org. Synth.* **2013**, *10*, 183.
[15] Bernet, B.; Vasella, A. *Helv. Chim. Acta* **1979**, *62*, 1990.
[16] Bernet, B.; Vasella, A. *Helv. Chim. Acta* **1979**, *62*, 2400.
[17] Bernet, B.; Vasella, A. *Helv. Chim. Acta* **1979**, *62*, 2411.
[18] Cintas, P. *Activated Metals in Organic Synthesis*, CRC Press, Taylor & Francis Group, **1993**.

[19] Kleban, M.; Kautz, U.; Greul, J.; Hilgers, P.; Kugler, R.; Dong, H.-Q.; Jäger, V. *Synthesis* **1997**, 634.

[20] Ovaa, H.; Codée, J. D. C.; Lastdrager, B.; Overkleeft, H. S.; van der Marel, G. A.; van Boom, J. H. *Tetrahedron Lett.* **1999**, *40*, 5063.

[21] Skaanderup, P. R.; Hyldtoff, L.; Madsen, R. *Monatsh. Chem.* **2002**, *133*, 467.

[22] Fürster, A.; Jumbam, D.; Teslic, J.; Weidmann, H. *J. Org. Chem.* **1991**, *56*, 2213.

[23] Fürster, A.; Baumgartner, J.; Jumbam, D. *J. Chem. Soc. Perkin Trans. 1* **1993**, 131.

[24] Palmer, A. M.; Volker, J. *Eur. J. Org. Chem.* **2001**, 1293.

[25] Désiré, J.; Prandi, J. *Tetrahedron Lett.* **1997**, *38*, 6189.

[26] Hümmer, W.; Dubois, E.; Gracza, T.; Jäger, V. *Synthesis* **1997**, 634.

[27] Ferrier, R. J.; Furneaux, R. H.; Prasit, P.; Tyler, P. C.; Brown, K. L.; Gainsford, G. J.; Diehl, J. W. *J. Chem. Soc. Perkin Trans. 1* **1983**, 1621.

[28] Dransfield, P. J.; Moutel, S.; Shipman, M.; Sik, V. *J. Chem. Soc. Perkin Trans. 1* **1999**, 3349.

[29] Bernotas, R. C.; Papandreou, G.; Urbach, J.; Ganem, B. *Tetrahedron Lett.* **1990**, *31*, 3393.

[30] Dangerfield, E. M.; Plunkett, C. H.; Win-Mason, A. L.; Stocker, B. L.; Timmer, M. S. M. *J. Org. Chem.* **2010**, *75*, 5470.

[31] Hyldtoft, L.; Poulsen, C. S.; Madsen, R. *Chem. Commun.* **1999**, 2101.

[32] Smith, C. R. *Synlett* **2009**, *9*, 1522.

[33] Einhorn, C.; Luche, J. L. *J. Organometallic Chem.* **1987**, *322*, 177.

[34] Wilson, S. R.; Giazzarpmo, M. E. *J. Org. Chem.* **1989**, *54*, 3087.

[35] Jorgensen, M.; Hadwiger, P.; Madsen, R.; Stutz, A. E.; Wrodnigg, T. M. *Curr. Org. Chem.* **2000**, *4*, 565.

[36] Prunet, J. *Eur. J. Org. Chem.* **2011**, 3634.

[37] Hyldtoft, L.; Madsen, R. *J. Am. Chem. Soc.* **2000**, *122*, 8444.

[38] Skaanderup, P. R.; Madsen, R. *J. Org. Chem.* **2003**, *68*, 2115.

[39] Keinicke, L.; Madsen, R. *Org. Biomol. Chem.* **2005**, *3*, 4124.

[40] Monrad, R. N.; Fanefjord, M.; Hansen, F. G.; Jensen, N. M. E.; Madsen, R. *Eur. J. Org. Chem.* **2009**, 396.

[41] Dam, J. H.; Madsen, R. *Eur. J. Org. Chem.* **2009**, 4666.

[42] Yadav, J. S.; Satheesh, G.; Murthy, C. V. S. R. *Org. Lett.* **2010**, *12*, 2544.

[43] Wichienukul, P.; Akkarasamiyo, S.; Kongkathip, N.; Kongkathip, B. *Tetrahedron Lett.* **2010**, *51*, 3208.

[44] Nakagawa, N. *J. Chem. Soc. Jpn.* **2003**, *123*, 225.

[45] Skaanderup, P. R.; Madsen, R. *Chem. Commun.* **2001**, 1106.

[46] Boyer, F.-D.; Hanna, I. *Tetrahedron Lett.* **2001**, *42*, 1275.

[47] Skaanderup, P. R.; Madsen, R. *J. Org. Chem.* **2003**, *68*, 2115.

[48] Monrad, R. N.; Pipper, C. B.; Madsen, R. *Eur. J. Org. Chem.* **2009**, 3387.

[49] Poulsen, C. S.; Madsen, R. *J. Org. Chem.* **2002**, *67*, 4441.

[50] Aubert, C.; Buisine, O.; Malacria, M. *Chem. Rev.* **2002**, *102*, 813.

[51] Fürstner, A.; Weidmann, H. *J. Org. Chem.* **1989**, *54*, 2307.

[52] Stout, E. I.; Doane, W. M.; Trinkus, V. C. *Carbohydr. Res.* **1976**, *50*, 282.

[53] Mishra, G. P.; Rao, B. V. *Tetrahedron Asymmetry* **2011**, *22*, 812.

[54] Mishra, G. P.; Kumar, B. S.; Rao, B. V. *Tetrahedron Asymmetry* **2012**, *23*, 1161.

[55] Bercier, A.; Plantier-Royon, R.; Portella, C. *Tetrahedron* **2010**, *66*, 4109.

[56] Luchetti, G.; Ding, K.; d'Alarcao, M.; Kornienko, A. *Synthesis* **2008**, 3142.

[57] Lombardo, L. *Org. Synth.* **1987**, *65*, 81.

[58] Brock, E. A.; Davies, S. G.; Lee, J. A.; Roberts, P. M.; Thomson, J. E. *Org. Lett.* **2010**, *13*, 1594.

[59] Davies, S. G.; Fletcher, A. M.; Roberts, P. M.; Smith, A. D. *Tetrahedron* **2009**, *65*, 10192.

[60] Soengas, R. G.; Silva, A. M. S. *Eur. J. Org. Chem.* **2013**, 5022.

[61] Soengas, R. G.; Estévez, A. M. *Eur. J. Org. Chem.* **2010**, 5190.

[62] Ferrier, R. J.; Middleton, S. *Chem. Rev.* **1993**, *93*, 2779.

[63] Redlich, H. *Angew. Chem., Int. Ed. Engl.* **1994**, *33*, 1345.

[64] Dalko, P. I.; Sinaÿ, P. *Angew. Chem. Int. Ed.* **1999**, *38*, 773.

[65] Grové, J. J. C.; Holzapfel, C. W.; Williams, D. B. G. *Tetrahedron Lett.* **1996**, *37*, 1305.

[66] Grové, J. J. C.; Holzapfel, C. W.; Williams, D. B. G. *Tetrahedron Lett.* **1996**, *37*, 5817.

[67] Grové, J. J. C.; Holzapfel, C. W. *Carbohydr. Lett.* **1997**, *2*, 329.

[68] Fevig, T. L; Elliott, R. L.; Curran, D. P. *J. Am. Chem. Soc.* **1988**, *110*, 5064.

[69] Chiara, J. L.; Martínez, S.; Bernabe, M. *J. Org. Chem.* **1996**, *61*, 6488.

[70] Kan, T.; Nara, S.; Ozawa, T.; Shirahama, H.; Matsuda, F. *Angew. Chem. Int. Ed.* **2000**, *39*, 355.

[71] Molander, G. A.; La Belle, B. E.; Hahn, G. *J. Org. Chem.* **1986**, *51*, 5259.

[72] Enholm, E. J.; Schreier, J. A. *J. Org. Chem.* **1995**, *60*, 1110.

[73] Zhou, J.; Wang, G.; Zhang, L.-H.; Ye, X.-S. *Curr. Org. Chem.* **2006**, *10*, 625.

[74] Negishi, E.; Cederbaum, F. E.; Takahashi, T. *Tetrahedron Lett.* **1986**, *27*, 2829.

[75] Hanzawa, Y.; Ito, H.; Taguchi, T. *Synlett* **1995**, 299.

[76] Ito, H.; Motoki, Y.; Taguchi, T.; Hanzawa, Y. *J. Am. Chem. Soc.* **1993**, *115*, 8835.

[77] Redlich, H. *Angew. Chem. Int. Ed. Engl.* **1994**, *33*, 1345.

[78] Paquette, L. A.; Arbit, R. M.; Funel, J. A.; Bolshakov, S. *Synthesis* **2002**, 2105.

[79] Jenkins, D. J.; Riley, A. M.; Potter, B. V. L. *J. Org. Chem.* **1996**, *61*, 7719.

[80] Aurrecoechea, J. M.; Lopez, B.; Arrate, M. *J. Org. Chem.* **2000**, *65*, 6493.

[81] Aurrecoechea, J. M.; Gil, J. H.; López, B. *Tetrahedron* **2003**, *59*, 7111.

[82] Aurrecoechea, J. M.; Arrate, M.; Gil, J. H.; López, B. *Tetrahedron* **2003**, *59*, 5515.

4

DOMINO REACTIONS TOWARD CARBOHYDRATE FRAMEWORKS FOR APPLICATIONS ACROSS BIOLOGY AND MEDICINE

Vasco Cachatra and Amélia P. Rauter
Universidade de Lisboa, Faculdade de Ciências, Departamento de Química e Bioquímica, Centro de Química e Bioquímica, Carbohydrate Chemistry Group, Campo Grande, Lisboa, Portugal

4.1 INTRODUCTION

A domino reaction was defined by Tietze [1] as a chemical process involving two or more bond-forming transformations that take place under the same reaction conditions without adding additional reagents and catalysts, in which the subsequent reactions result as a consequence of the functionality formed in the previous step. Although this definition is widely used by synthetic chemists, a few examples of domino reactions where bond cleavage occurs, like in a deprotection of a determined functionality, are here considered included in this concept. Domino reactions benefit from the fact that more than one transformation can be accomplished without the need for purification of the intermediates, which turns out to be time-consuming in a synthetic strategy that involves various steps.

In this chapter, domino reactions are reviewed, in particular those that afford synthetic scaffolds from carbohydrate frameworks, used as building blocks for a wide variety of molecules with biological interest.

Domino and Intramolecular Rearrangement Reactions as Advanced Synthetic Methods in Glycoscience, First Edition. Edited by Zbigniew J. Witczak and Roman Bielski.

4.2 DOMINO REACTIONS TOWARD BUTENOLIDES FUSED TO SIX-MEMBERED RING SUGARS AND THIO SUGARS

Sugars fused to butenolides are bicyclic structures with potential as scaffolds for carbohydrate transformations and as biologically active molecular entities. The five-membered ring butenolides embody an α,β-unsaturated lactone that is a structural feature known to act as a Michael acceptor, and responsible for a number of bioactivities such as antibiotic, antihelmintic, antifungal, insecticidal, antitumor, antiviral, anti-inflammatory, cytostatic, antiallergenic, and anti-human immunodeficiency virus (HIV) activities [2]. It occurs in nature linked to a diversity of families, namely to steroid glycosides such as digitoxin and digoxin, both of them cardiac glycosides, (Figure 4.1) and the latter is considered an important drug in a basic health system by the World Health Organization [3]. Also, patulin is a natural product produced by fungi species of the genera *Byssochlamys*, *Penicillium*, and *Aspergillus* [4]. This compound posseses antibiotic properties and antibacterial activity although it also demonstrated acute toxicity and genotoxicity, and was teratogenic and immunotoxic to animals, while no conclusive studies have been conducted in humans. Patulin is now under strict control by the authorities, due to its formation in rotten apples and in processed apple products [4].

When fused to sugars, butenolide reactivity may change because of stereochemical hindrance issues and sugars also become conformationally restricted, possibly altering the biological properties, for example, by enhancing selectivity or activity. Hence, synthetic approaches to such structures became a challenge for glycochemists. The synthesis of such molecular entities succeeded via an efficient and high-yield domino reaction, starting from Wittig products type **2**, easily synthesized from diacetoneglucose **1**, by oxidation with pyridinium chlorochromate (PCC) and olefination with [(ethoxycarbonyl)methylene]triphenylphosphorane [5]. Under acid-promoted hydrolysis of the isopropylidene groups in the presence of a strong acid resin and methanol, sugar ring opening also takes place to give **4**, followed by nucleophilic attack by the hydroxy group at position 5 to the aldehyde function affording the

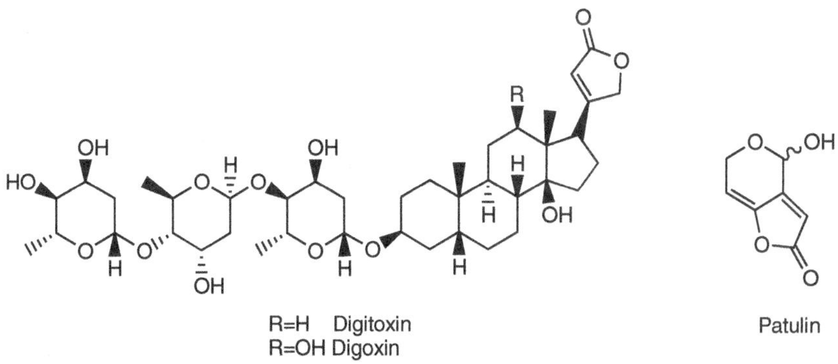

FIGURE 4.1 Structure of the cardiac glycosides digitoxin and digoxin and of patulin as examples of biologically active natural products containing a butenolide in their structure.

pyranose system as in **5** and **6**. Under such acidic conditions, transesterification occurs by addition–elimination reaction of the hydroxy group at positions 4 or 2, depending on the alkene configuration, to the ester functionality, to give compounds **7** or **9**, respectively. This domino reaction affords, in one single step, the bicyclic products isolated as their fully protected acetates in a good yield [5]. This acid-promoted mechanism is supported by the results obtained when this methodology was applied to 5-*O*-pivaloyl protected hexofuranoses. The pivaloyl group is not cleaved under these conditions, and bicyclic products could not be formed with these substrates, as illustrated in Scheme 4.1 [5].

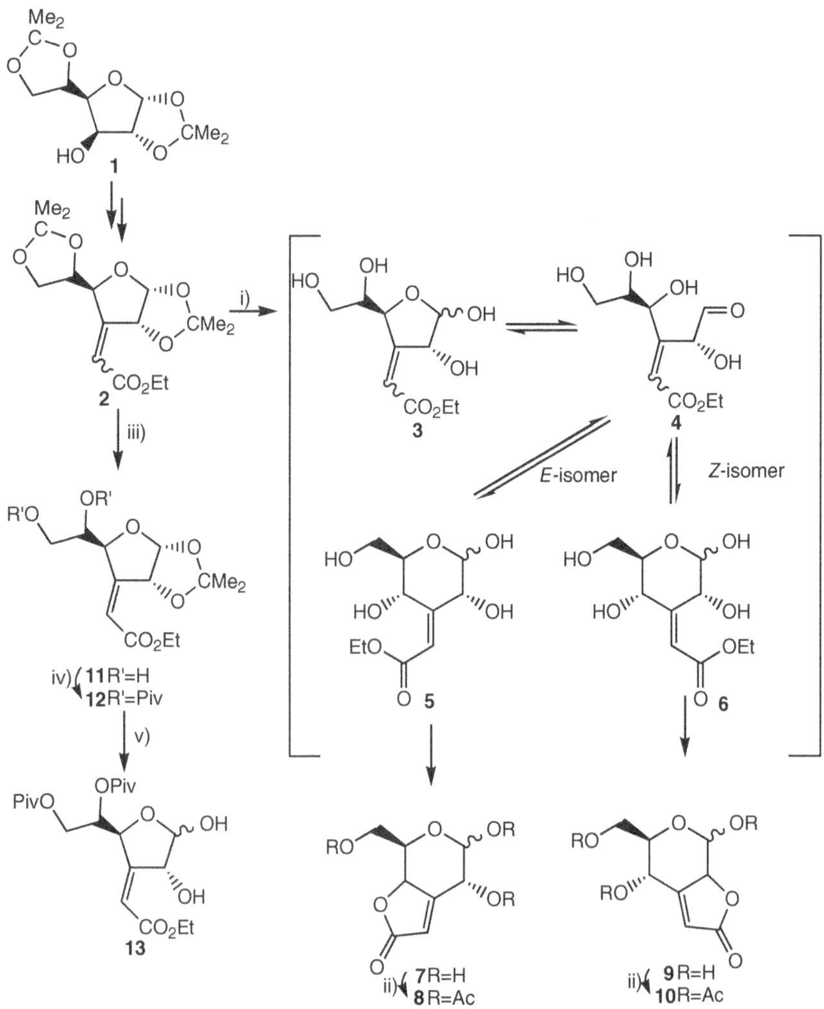

SCHEME 4.1 Reagents and conditions: (i) IR-120 H^+ resin, MeOH, reflux, overnight; (ii) Ac_2O, py, rt, 5 min, 63% for **8** and 90% for **10**, over two steps; (iii) AcOH 60% aq., rt, 20 h, 95%; (iv) PivCl, py/DCM, 0°C, 1 h, 27%; and (v) AcOH 70% aq., reflux, 5 h, 62%.

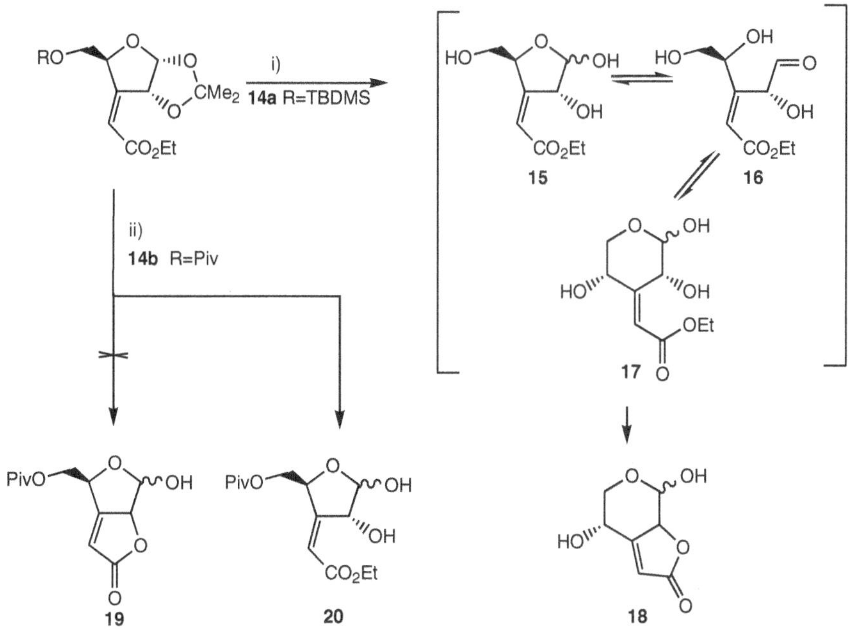

SCHEME 4.2 Reagents and conditions: (i) IR-120 H^+ resin, MeOH, reflux, overnight, 79%; and (ii) AcOH 70% aq., reflux, 5 h, 70%.

This synthetic strategy was applied to pentofuranoses, namely to compounds type **14** with the primary hydroxy group protected with *tert*-butyldimethylsilyl (TBDMS) (**14a**). Both the isopropylidene and TBDMS groups are acid labile, and furanose ring opening is followed by pentopyranose formation and lactonization to give the bicyclic compound **18** in a 79% yield in one single step from **14a**. When OH-5 was protected with a pivaloyl group, neither a six-membered ring nor a lactone was formed (Scheme 4.2).

This domino reaction, which transforms furanose sugars, in one single step, into butenolides fused to six-membered ring sugars, is easily carried out, not expensive, high yielding, and generates complex bicyclic molecules from sugar Wittig products, which are also accessed in a good yield.

This type of compounds showed weak or moderate antibacterial and antifungal activities, with selectivity toward *Bacillus subtilis* and *Candida albicans* [5]. These bicyclic structures may offer a sterical hindrance to the nucleophilic attack to the electrophilic double bond conjugated to the carbonyl group, thus rendering it less susceptible to a Michael addition. Nevertheless, these molecules are unique scaffolds for further transformations into analogues of biologically active compounds, for example, patulin, aiming at new, active, and nontoxic products.

The introduction of an endocyclic sulfur atom replacing the sugar oxygen usually changes molecule chemical and biological properties. Differences in electronegativity and polarizability, as well as in the conformation and flexibility of the pyranose ring,

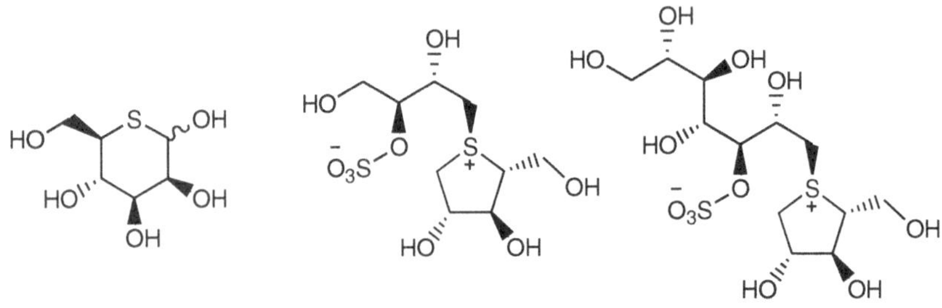

5-Thiomannose Salacinol Kotalanol

FIGURE 4.2 Structure of naturally occurring thio sugars.

together with the distinct chemical reactivity of the sulfide when compared to the corresponding ether, are most likely key factors that influence the biological profile of thio sugars, some of which exhibiting potent activity as glycosidase inhibitors, as anticancer agents and are effective against infectious diseases [6]. The first naturally occurring 5-thiosugar was 5-thiomannose, isolated from the marine sponge *Clathria pyramida* (*Lendenfeld*) [7]. Among the naturally occurring thio sugars, salacinol and kotalanol (Figure 4.2) containing a sulfate moiety should be highlighted for their potent activity as glycosidase inhibitors [8]. In addition, thio sugars are very important sugar mimetics to study the cell surface recognition, and all the mechanisms are involved in the process [9]. However, most of the thio sugars are of synthetic origin and the methodology developed to prepare sugar-fused butenolides succeeded to access thio sugar-fused butenolides in a good yield, by installing a sulfhydryl group in position 5 as shown in compound **22**, which is then transformed by domino reaction to give the expected bicyclic thio sugar **25** (Scheme 4.3) [10].

Another particularly relevant issue to be highlighted is that D- or L- configuration of the sugar fused to the butenolide is determined by the absolute configuration of hexofuranose starting material carbon 5. In the presence of (5*R*)-configuration, the pyranosyl moiety belongs to D-series (see Scheme 4.1) while inversion of configuration at position 5 leads to a bicycle possessing an L-sugar, as illustrated with compound **25** (Scheme 4.3).

4.3 EXPLORATORY CHEMISTRY FOR AMINO SUGARS' DOMINO REACTIONS

In an attempt to access the imino sugar analogues bearing the butenolide fused to the pyranose ring, the amino sugar **27** was used as a starting material but the single compound obtained (**28**) resulted from isopropylidene hydrolysis (Scheme 4.4) [11]. Neutralization with either aqueous NaOH or Et_3N to avoid formation of ammonium salt led to the formation of compound **32**, embodying a butenolide C–C linked to a functionalized chain with an alcohol and an amide. The proposed mechanism relies on the transformation of the intermediate sugar-fused butenolide **29**, which

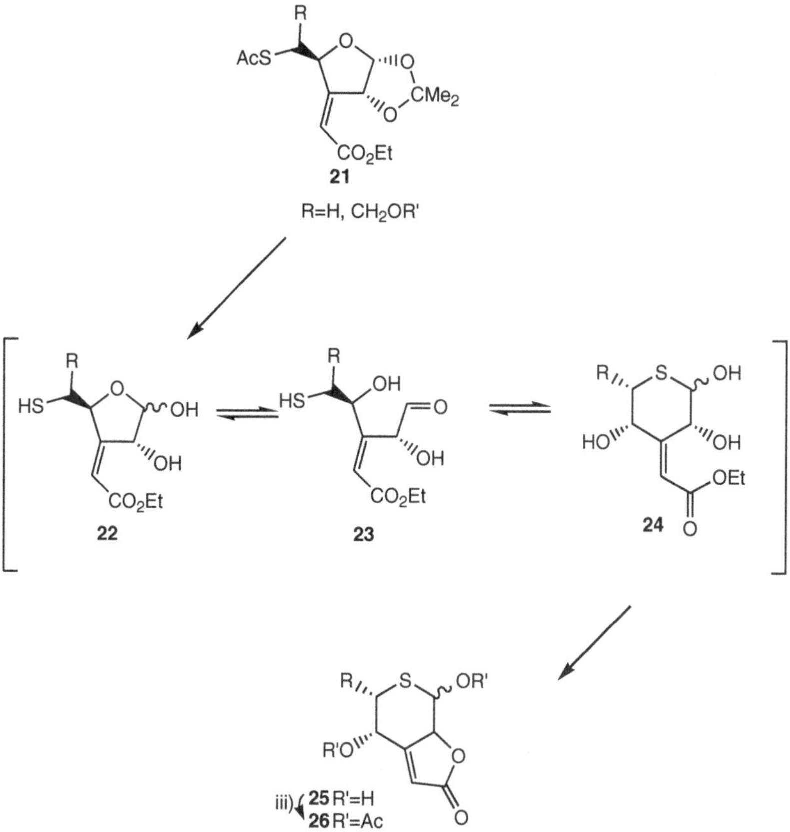

SCHEME 4.3 Reagents and conditions: (i) NaOH, MeOH/H_2O, rt, 15 min; (ii) AcOH 70% aq., reflux, 2 h, 35%, over two steps; and (iii) Ac_2O, py, rt, 5 min, 80%.

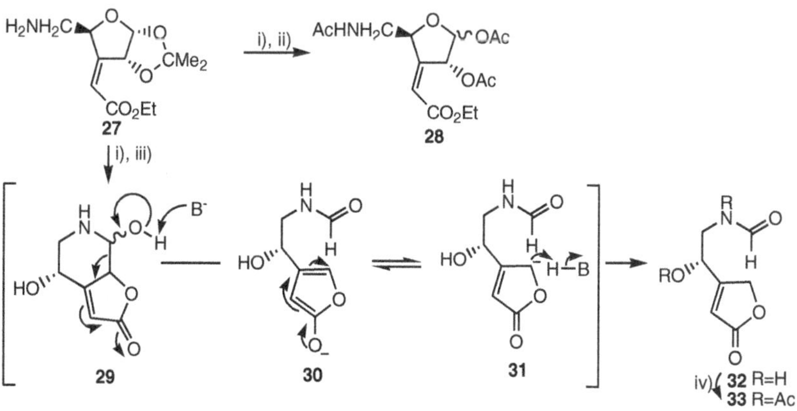

SCHEME 4.4 Reagents and conditions: (i) TFA 60% aq., rt, 10 min; (ii) Ac_2O, py, rt, 5 min, 80% over two steps; (iii) NaOH 1M, EtOH/H_2O, until pH 8, 45%, over two steps; or NEt_3, EtOH/H_2O until pH 10, 65%, over two steps; and (iv) Ac_2O, catalytic DMAP, py, rt, 1 h, 84%.

undergoes deprotonation of the anomeric hydroxy group and cleavage of the C—C bond between position 1 and 2 with consequent formation of the carbanion ion stabilized by resonance, that, in the presence of the *in situ* formed acid leads to compound **32** in a 65% yield. Butenolides type **32**, bearing an amide functionality, are building blocks, for example, for pyrrolizidine alkaloids [12] and analogues of compounds type **34**, reported to have antilarval settlement activity, an important feature to avoid marine organisms to deposit in marine installations (Figure 4.3) [13].

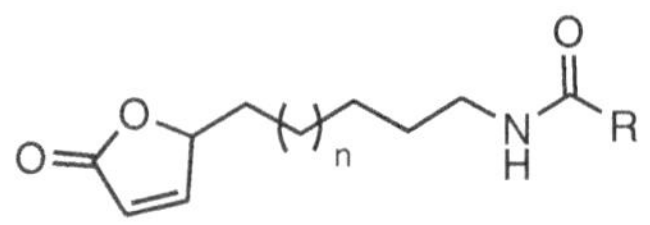

34

FIGURE 4.3 Butenolide functionalized amides with antilarval settlement activity.

The access to products type **37** (Scheme 4.5), bearing an α,β-unsaturated δ-lactam fused to a furanose, succeeded via domino reaction of a Wittig product type **35**, embodying an azide functionality at C-6, Staudinger reduction to afford an amine was followed by intramolecular cyclization resulting from nucleophilic addition of the formed amine to the ester carbonyl group with subsequent elimination of the ethoxy group to give **37**, isolated in an 89% yield, also characterized as its *N*-acetyl derivative **38** (Scheme 4.5) [11]. This domino reaction is easy to run and high yielding and gives access to compounds exhibiting the lactam functionality, which is a structural feature installed in precursor molecules for the synthesis of biologically active alkaloids [14], and pharmacological drugs, namely the δ-amino acid homobaclofen (**39**) [15] (Figure 4.4), a homologue of the γ-aminobutyric acid B ($GABA_B$) receptor, that demonstrated a remarkable pharmacological profile [16].

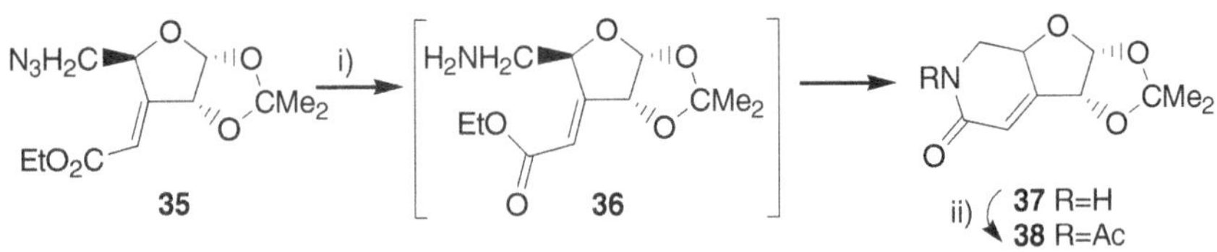

SCHEME 4.5 Reagents and conditions: (i) PPh_3, H_2O/THF, rt, 16 h, then DCM, rt, 16 h, 89%; and (ii) Ac_2O, catalytic DMAP, py, rt, 16 h, 85%.

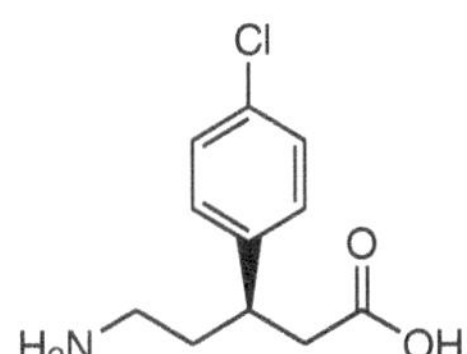

39

FIGURE 4.4 Structure of homobaclofen **39**.

6-Azido sugars type **40** are unique substrates to afford 1,2-dihydropyridin-3-ones type **50** via domino reaction. After hydrolysis of the isopropylidene group of **40**, domino reaction starts with hydrogenation of the azide to the amine functionality, leading to the formation of a six-membered ring hemiaminal intermediate, which suffers water elimination to give an imine, isomerization to enol and tautomerization to the keto form. Boc protection and acetylation of the free hydroxy group were followed by elimination of acetic acid to afford the α,β-unsaturated ketone **50** starting from **40** in 42% overall yield (Scheme 4.6) [11].

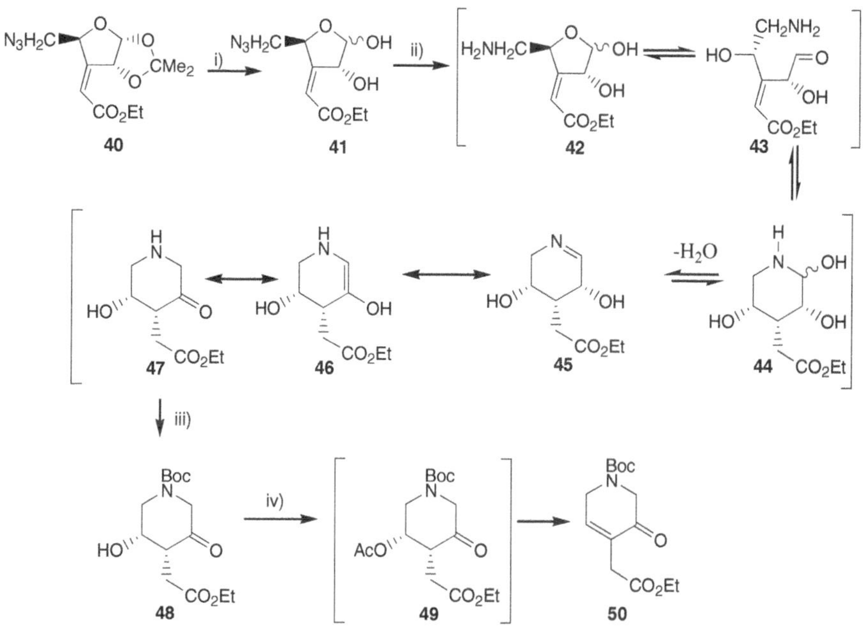

SCHEME 4.6 Reagents and conditions: (i) TFA 60% aq., rt, 5 min, 94%; (ii) H_2, 10%Pd/C, EtOH, rt, 40 min; (iii) Boc_2O, rt, 1 h 30 min; and (iv) Ac_2O, py, 30 min, 42% overall yield.

1,2-Dihydropyridin-3-ones are important building blocks for the synthesis of biologically active compounds, namely of a diversity of imino sugar glycosidase inhibitors. A variety of bioactive molecules also embody this structural feature [17], as illustrated by the potent γ-secretase inhibitor **51** [18] (Figure 4.5), one of the target enzymes for the treatment of Alzheimer's disease (AD). Dihydropyridinones also offer the structural requirements to act as valuable synthons for development of new neuronal nicotinic acetylcholine receptor (nAChR) ligands that require a π-system, namely a heteroaromatic ring or a carbonyl group as a hydrogen bond acceptor, and a cyclic amine group [19]. Dihydropyridinones type **52** have been described as memory enhancers in relation to their nicotinic acetylcholine receptor activity [19]. Receptor nAChR became a therapeutic target for a wide range of neurological disorders namely attention-deficit hyperactivity disorder, Tourette's syndrome, schizophrenia,

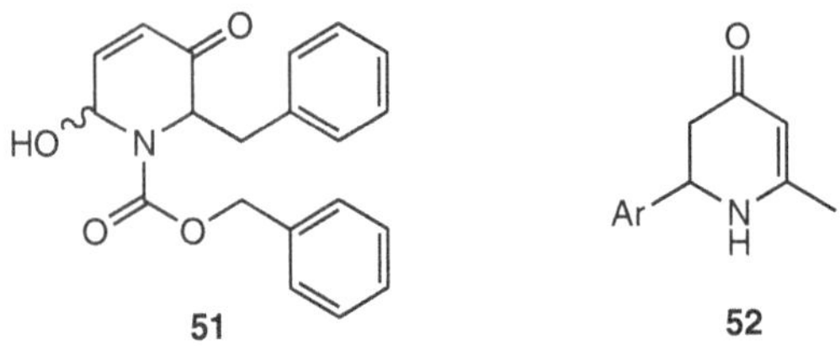

51 52

FIGURE 4.5 Examples of biologically active compounds embodying a dihydropyridinone.

AD, and Parkinson's disease. Hence, organic and glycochemists face the challenge to find easy and low-cost synthetic approaches to dihydropyridinones. The efficient protocol illustrated in this chapter (Scheme 4.6) that results from domino reactions gives an important contribution to this area of research.

4.4 DOMINO REACTIONS TOWARD SUGAR RING CONTRACTION

4.4.1 Pyrano–Furano Ring Contraction

Construction of structures embodying two five-membered rings fused is rather difficult and such compounds may be prepared via multistep syntheses in low overall yield. However, they are easily accessed in Brønsted or Lewis acid medium, starting from bicyclic structures embodying a five-membered ring fused to a six-membered ring. This bicycle is a structural feature of the miharamycins, with a core structure embodying the sugar bicycle elongated with an amino acid functionality with its anomeric position N^9-linked to a 2-acetamidopurine, whose first synthesis dates from 2008 [20, 21]. These antimicrobial agents are natural fungicides quite potent against *Pyricularia oryzae*, which causes the rice blast disease and is considered a bioterrorism agent. When structures type **53** [20] (Figure 4.6) were treated with concentrated sulfuric acid in acetic anhydride, aiming at the replacement of the anomeric methoxy group by acetoxy, a better leaving group for the coupling with the purine base, a domino reaction took place leading to a ring contraction and formation of compound **55** isolated in 85% yield (Scheme 4.7). When acetolysis was carried out

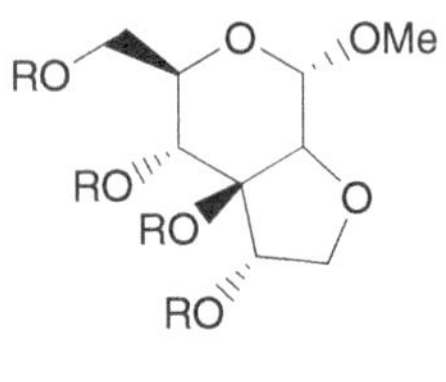

53 R=Ac
54 R=Bn

FIGURE 4.6 Bicyclic sugar core of the miharamycins.

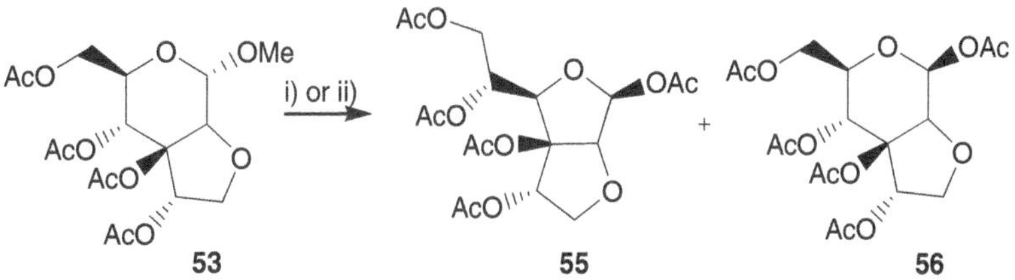

SCHEME 4.7 Reagents and conditions: (i) H_2SO_4, Ac_2O, 0°C, 45 min, 85% for **55** and 10% for **56**; and (ii) H_2SO_4 5% in AcOH, Ac_2O, 0°C, 45 min, 70% for **56** and 20% for **55**.

with compound **53** in 5% sulfuric acid in acetic acid, the major product was the expected compound **56** and the ring contraction product **55** was a minor reaction product isolated in only 20% yield.

This domino reaction starts with protonation of the endocyclic oxygen, ring opening and formation of the oxonium ion **59** (Scheme 4.8), migration of the acetyl group

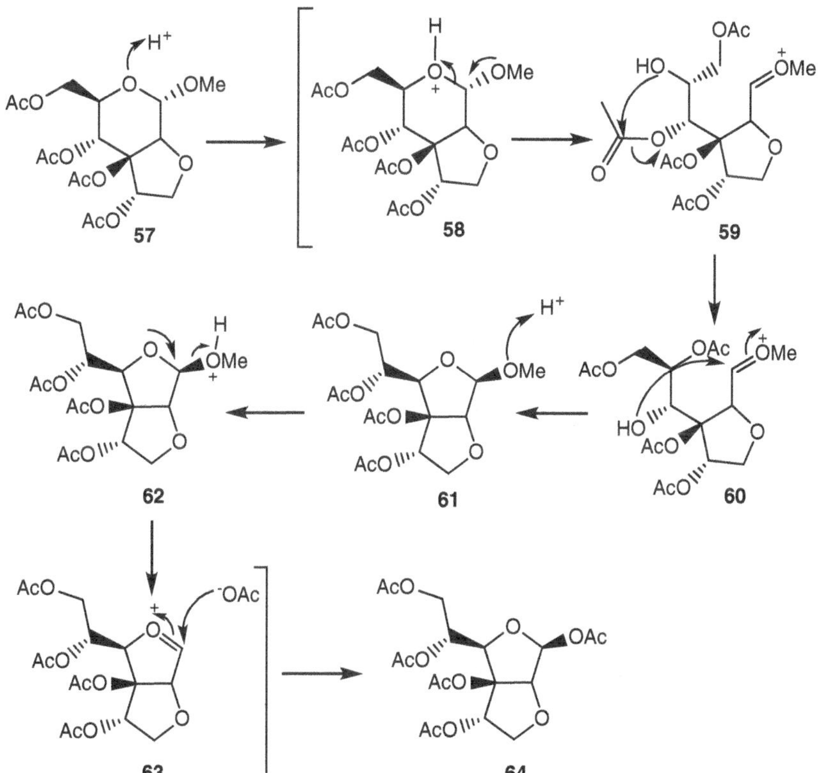

SCHEME 4.8 Domino reaction catalyzed by Brønsted acid comprising ring opening by acetal cleavage, transesterification, ring closure with subsequent acetal formation, and nucleophilic attach to the oxycarbenium ion to give compound **64**.

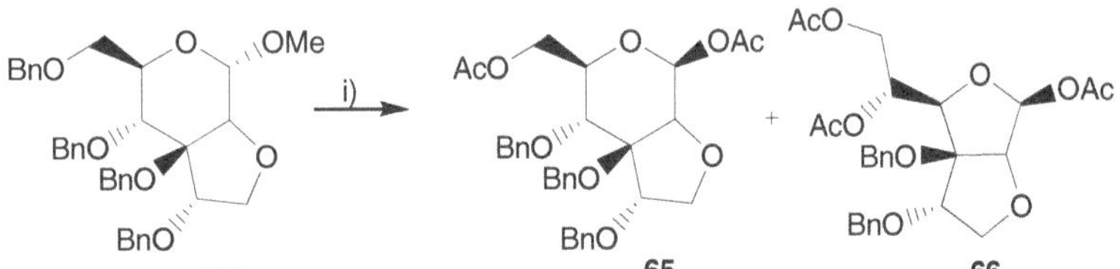

54 65 66

SCHEME 4.9 Reagents and conditions: (i) H_2SO_4 99% in AcOH, Ac_2O, 0°C → 20°C, 25 min, 48% for **65** and 32% for **66**.

from position 4 to position 5, and finally ring closure by nucleophilic attack of the hydroxy group at position 4 to the oxonium ion **60**, resulting in the formation of **61** that embodies a five-membered ring fused to the furanoside. Protonation of this intermediate methyl furanoside, cleavage of methanol with concomitant formation of the oxycarbenium ion **63**, and attack of the acetate ion give the fully acetyl protected bicycle **64**.

While for the per-*O*-benzylated analogue of **53**, the results were similar to those obtained with the per-*O*-acylated compound, reaction with the per-*O*-benzylated compound **54** gave an opposite ratio for the formed bicycles, exhibiting the five-membered ring fused to a five-membered ring, or the six-membered ring fused to a five-membered ring, being the major compound the one embodying the six-membered ring (Scheme 4.9). This result seems to be due to the low migration ability of the benzyl group, when compared with that of the acetyl group, minimizing the formation of the intermediate with OH-4 that is essential to the preparation of compounds type **66** [20].

A similar ring contraction was obtained under Lewis acid catalysis, when the elongated bicycles type **67,** with 6*R* or 6*S* configuration, were debenzylated in the presence of boron trichloride (Scheme 4.10). The cleavage of the endocyclic O—C1 bond of the pyranose ring results in the oxonium ion formation that suffers then the nucleophilic attach of OH-4 to generate the β-furanoside **70** in a good yield [21].

This ring contraction was promoted by the Lewis acid BCl_3. When the removal of the benzyl group was performed by hydrogenation (Scheme 4.11), no ring contraction took place and the expected azide reduction proceeded in high yield.

Compound **71** has a masked amino acid, while **72** already has an amino acid moiety, which makes both of them suitable building blocks for the synthesis of

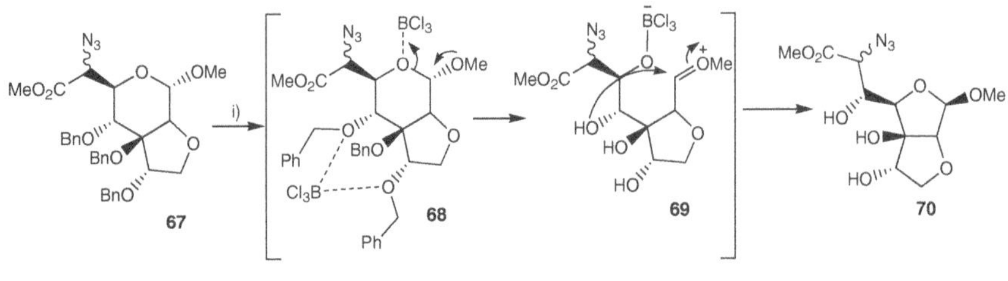

67 68 69 70

SCHEME 4.10 Reagents and conditions: (i) BCl_3 (1M in DCM), DCM, −78°C → 0°C, 7 h, 65%.

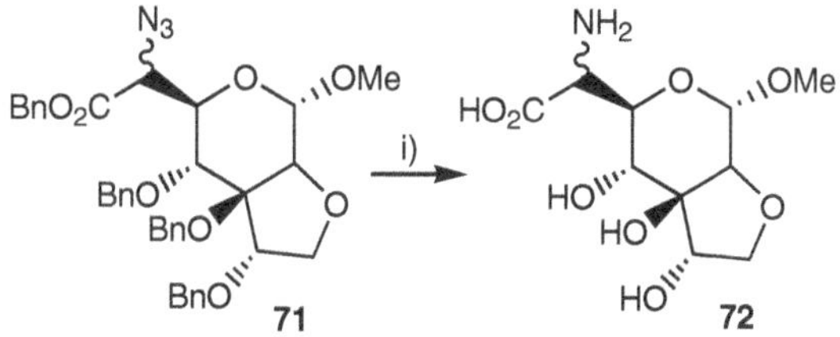

SCHEME 4.11 Reagents and conditions: (a) H_2, 10% Pd/C, EtOAc/MeOH/HCl 1M, overnight, quant.

peptide mimetics [22]. These compounds may be a valuable asset for understanding the folding of proteins and its role in compound biological activity.

The bis(tetrahydrofuran)-fused structure is found in synthetic nucleosides that are used as mimetics to study the structural constriction of the sugar moiety [23] and is also present in a variety of bioactive molecules, for example, goniofufurone [24] and in two potent anti-HIV drugs, brecanavir and darunavir (Figure 4.7), the latter approved by the Food and Drug Administration for the treatment of HIV infection as a viable option to patients with drug-resistant HIV infection [25,26].

4.4.2 Ring Contraction of Furans to Oxetanes

Oxetanes are four-membered cyclic ethers and have a wide range of applications. They are used as monomers in materials science, as building blocks in synthetic chemistry for countless reactions, and are present in natural products and biologically active compounds [27]. Oxetin was the first natural product reported to contain an oxetane ring, isolated from a fermentation broth of *Streptomyces* sp. (Figure 4.8). This compound exhibited herbicidal effect and inhibited *Bacillus subtilis* and *Pyricularia oryzae* [28]. Oxetanocin A is another example of a natural product containing an oxetane ring, which was isolated from the fermentation broth of *Bacillus megaterium* [29]. This adenine nucleoside showed a broad spectrum of antiviral activities, namely against herpes simplex virus 1 and 2, varicella zoster virus, human cytomegalovirus, and HIV [30].

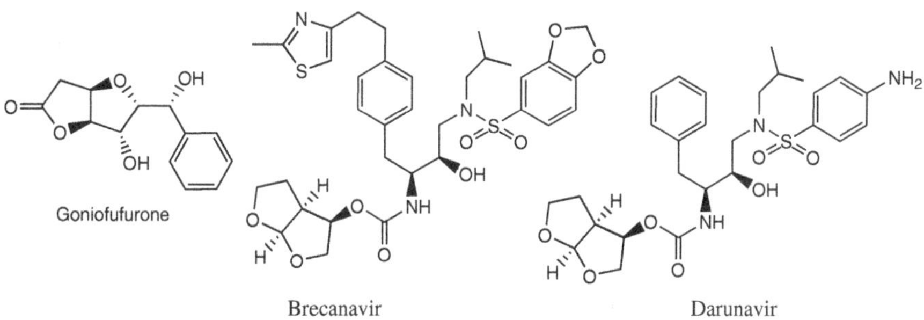

FIGURE 4.7 Structure of goniofufurone, brecanavir, and darunavir.

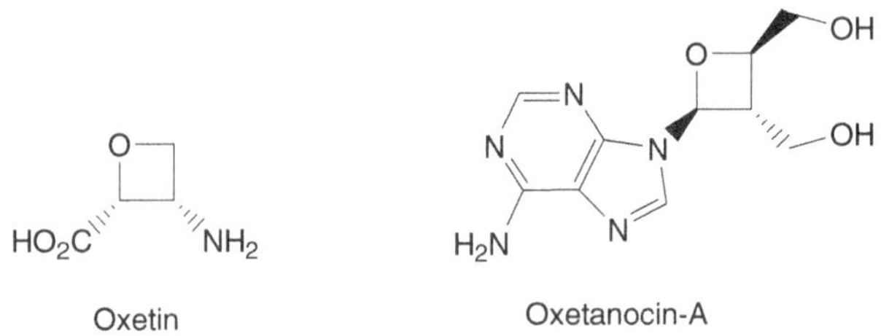

FIGURE 4.8 Examples of bioactive natural products with an oxetane ring in their structure.

Oxetanes have been used in medicinal chemistry to improve physicochemical properties and as structurally restricted mimetics to study different biological processes [27]. Furthermore, oxetane amino acids are commonly used as monomers in peptide synthesis to investigate how their rigid structure affects the biological role of proteins. Their low flexibility affords a well-defined spatial distribution of its substituents leading to a more efficient binding to the target receptor.

Ring contraction of 2-hydroxy gamma-lactones bearing a triflyl substituent in position 2 to oxetane carboxylic esters was first described by Fleet and coworkers already in 1990 [31]. Starting from a five-membered aldonolactone bearing a good leaving group at position 2, for example, a triflate, oxetane formation is achieved in basic medium in methanol. The methoxy anion formed promotes an addition–elimination reaction to the lactone with the subsequent formation of an alkoxy ion (Scheme 4.12), a strong nucleophile that promotes nucleophilic substitution at position 2, with rejection of the triflate group and formation of the oxetane ring.

This reaction has been applied to the synthesis of oxetane β-amino acids mainly by the Fleet's group and oxetane δ-amino acids by Rauter's group, in collaboration with Wessel's group, in order to study the potential of oxetane amino acids as monomers for β-/δ-peptide foldamers, a class of unnatural oligomers with propensities to adopt specific and compact conformations for potential clinical applications.

Fleet's group applied this procedure to prepare oxetin analogues from aldonolactones prepared from either L-rhamnose (**77**) [32, 33] or D-xylose (**82**) [32, 34], affording β-amino acid precursors type **78** and **83**, respectively. Triflation of the free OH followed by nucleophilic displacement led to the preparation of oxetanes with different configuration and substitution at the β-position (Scheme 4.13). The final steps involve hydrolysis of the ester and reduction of the azide to the corresponding amine, giving monomeric structures for the synthesis of small peptides.

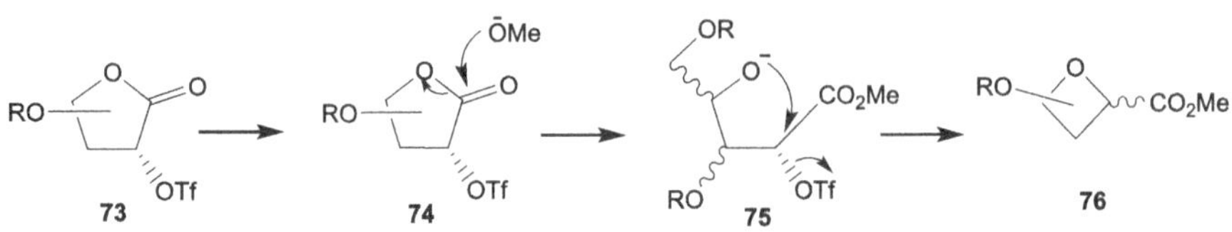

SCHEME 4.12 Oxetane formation by ring contraction of aldonolactones.

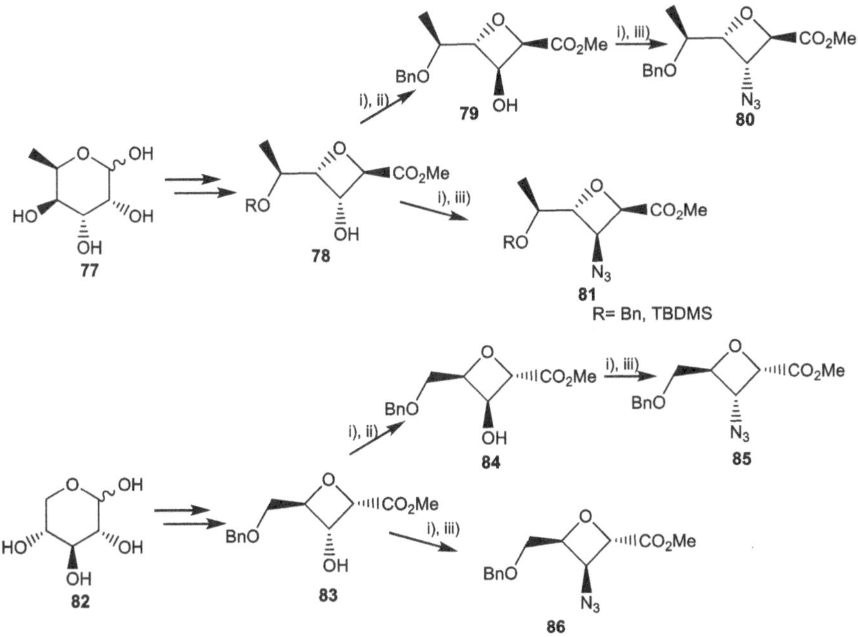

SCHEME 4.13 Reagents and conditions: (i) Tf_2O, py; (ii) CF_3COOCs, MeCOEt; and (iii) NaN_3, DMF, 82% for **80**, 73% for **81** (R=TBDMS), 93% for **81** (R=Bn), 80% for **85**, and 90% for **86**.

Synthesis of oxetane δ-amino acids was also accomplished by ring contraction of aldonolactones, namely xylonolactone (**87**) [35] and ribonolactone (**92**) [36], as shown in Scheme 4.14. Installation of the amine group at the primary alcohol succeeded via debenzylation of oxetane carboxylic esters, triflation of the primary alcohol, and nucleophilic displacement with an azide. Reduction of the azide and enzymatic hydrolysis of the ester gave compound **90** bearing a free secondary OH in a high yield. Fluorinated analogues were also synthesized by reaction with DAST of the oxetane carboxylic esters with the Boc protected amine, followed by ester hydrolysis with LiOH to afford compounds **91** and **94**, differing in the stereochemistry of the oxetane ring.

Xylonolactone **95**, embodying an azide functionality, was submitted to the ring contraction conditions to determine whether it was possible to access 5-azidooxetanes in a smaller number of steps 5-azido oxetanes (Scheme 4.15). While **96** was obtained in a 67% yield, ring contraction of its epimer **98** afforded a mixture of the α and β anomers isolated in a 57% yield for **99** and a 7% yield for **101**. Azide reduction and ester hydrolysis afforded Boc protected δ-amino acids in a good yield [37].

Oxetane δ-amino acids and their 5-azido ester precursors have been submitted to a number of transformations, showing the synthetic potential of these structures. 1,2,4-Oxadiazole, commonly used as bioisosteric replacement for amides, esters, and isothioureas [38] was introduced into structures type **103** [39] (Scheme 4.16).

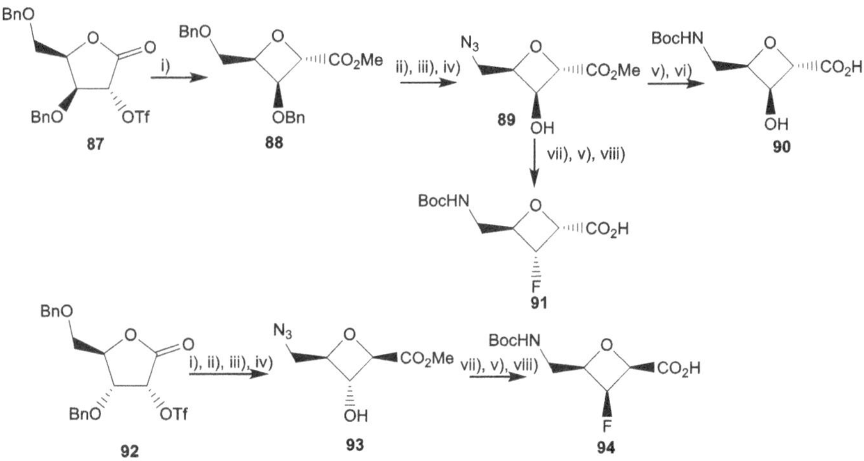

SCHEME 4.14 Reagents and conditions: (i) K_2CO_3, MeOH, −12°C, 30 min, 90% for **88**; (ii) H_2, Pd/C, MeOH/dioxane 1:1, rt, 40 min; (iii) Tf_2O, Et_2O/DCM 5:1, 4 Å molecular sieves, −15°C, 50 min; (iv) LiN_3, acetone, rt, 30 min, 55% for **89**, 53% for **93**; (v) H_2, Pd/C, EtOAc, Boc_2O, rt, 2 h; (vi) L2- *Candida antarctica*, TBME/H_2O, 45°C, 3 d, 88% for **90**; (vii) DAST, MeCN, −20°C to reflux, 1 h; (viii) LiOH 1N, HCl 1N, 0–5°C, 1 h, 62% for **91**, 50% for **94**.

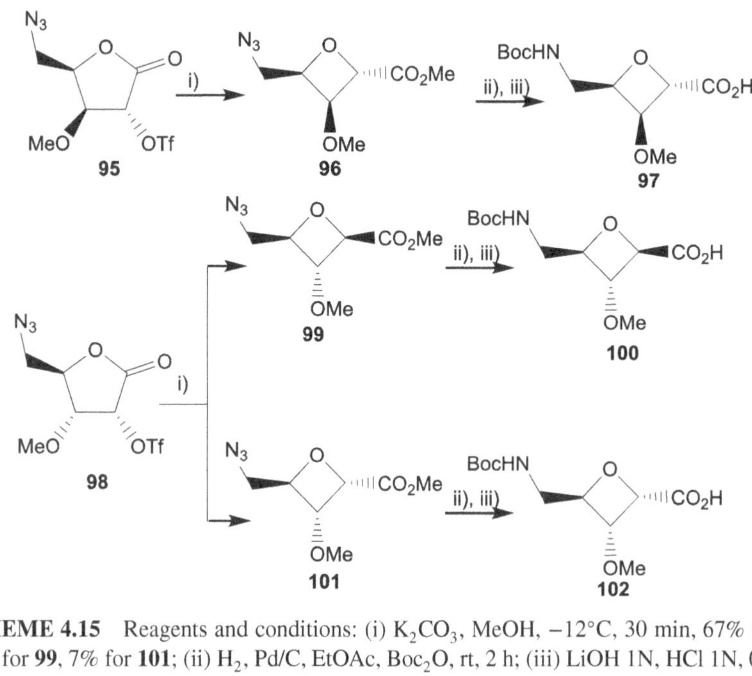

SCHEME 4.15 Reagents and conditions: (i) K_2CO_3, MeOH, −12°C, 30 min, 67% for **96**, 53% for **99**, 7% for **101**; (ii) H_2, Pd/C, EtOAc, Boc_2O, rt, 2 h; (iii) LiOH 1N, HCl 1N, 0–5°C, 1 h, 74% for **97**, 72% for **100**, and 78% for **102**.

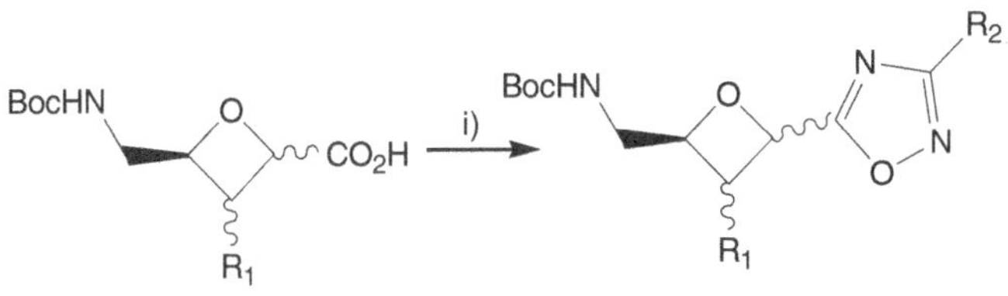

103 **104**

R_1=OH, OPMB, OMe, F
R_2=Ph, *p*-OMePh, *p*-ClPh, *p*-MePh, Py

SCHEME 4.16 Reagents and conditions: (i) (a) HATU, DIPEA, rt, DMF; (b) $R_2C(NH)$=NHOH; and (c) 80°C.

Using click chemistry 5-azido esters type **105** were transformed into compounds type **106** bearing a 1,2,3-triazole ring (Scheme 4.17) [39], also promising as amide bond isoster [40].

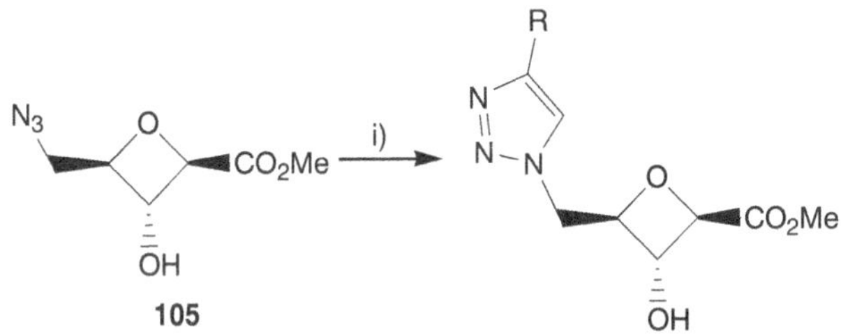

105 **106**

R=Aryl substituents

SCHEME 4.17 Reagents and conditions: (i) alkyne, sodium ascorbate, $CuSO_4{\cdot}5H_2O$, DMF.

4.5 MACROCYCLIC BISLACTONE SYNTHESIS VIA DOMINO REACTION

Bislactones are found in nature in a variety of structures and compounds, with this framework often presenting very interesting biological activities. We highlight some examples, namely paracaseolide A (Figure 4.9), isolated from the stem bark of the Chinese mangrove *Sonneratia paracaseolaris*, that is a phosphatase (CDC25B) inhibitor at micromolar concentrations [41]. This enzyme is important in the cell-cycle progression and its inhibition has proven effective against tumor cells. Isoavenaciolide is another bislactone isolated from the fermentation broth of *Aspergillus* and *Penicillium* species, presenting a broad spectrum of antibacterial and antifungal activities and inhibiting vaccinia 1-related (VHR) phosphatase activity [42]. Ardimerin is a dimeric bislactone isolated from *Ardisia japonica* that showed radical scavenging activity. The digallate derivative of ardimerin showed inhibitory activity against HIV-1 and HIV-2 RNase H *in vitro* with micromolar values of IC_{50} [43].

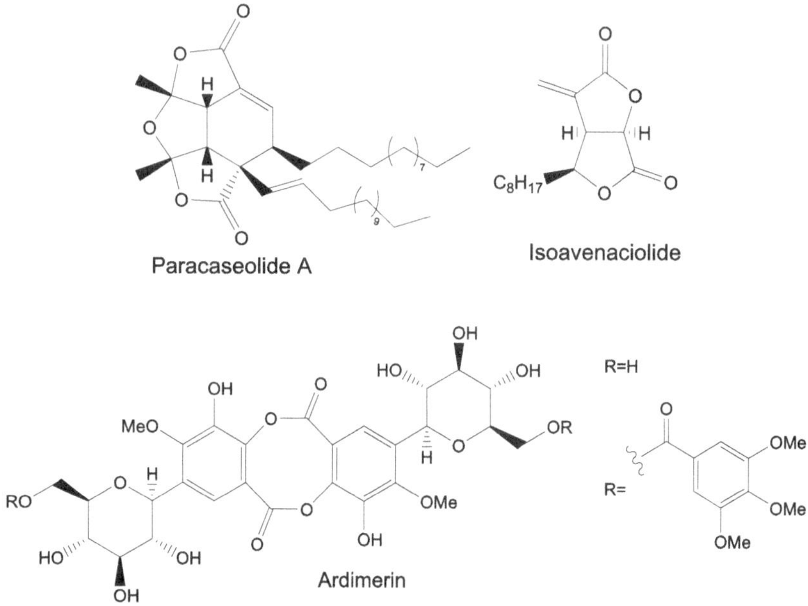

FIGURE 4.9 Examples of bioactive natural products embodying bislactones in their structure.

The synthesis of bislactones requires usually several steps, which is a drawback for the access to such structures. Our approach is based on a domino reaction to afford a sugar bislactone, starting from partially protected sugars bearing two free vicinal hydroxy groups [44]. Selective pivaloylation of positions 2 and 6 of methyl glucopyranoside or benzylidene acetal formation in positions 4 and 6 afforded the scaffolds for the PCC oxidation, illustrated in Scheme 4.18 with the dipivaloyl starting material. The diulose formed by oxidation of both the free hydroxy groups reacts with the monomer to give a bis(hemiacetal) formed by a nucleophilic addition of the free OH to a carbonyl group to give a hemiacetal. The oxidative cleavage of the C–C bond affords the macrocyclic bislactone in a moderate yield. Both bislactones showed moderate antifungal activity against *Candida albicans* and weak activity against *Bacillus subtilis*.

4.6 SUGAR DEOXYGENATION BY DOMINO REACTION

Deoxy sugars are an important class of carbohydrates present in a wide variety of biologically active natural products and synthetic compounds [45]. To cite some examples, the natural cardiac glycosides digoxin and digitoxin have in their structure 2-deoxy sugar units. Among the synthetic structures, 2-deoxy alkyl glycosides exhibit unique antibacterial properties against *Bacillus* species, especially *Bacillus anthracis* [46]. While the 2-deoxy sugars are the most studied class of deoxy compounds, deoxygenation at other positions also produces bioactive compounds (Figure 4.10).

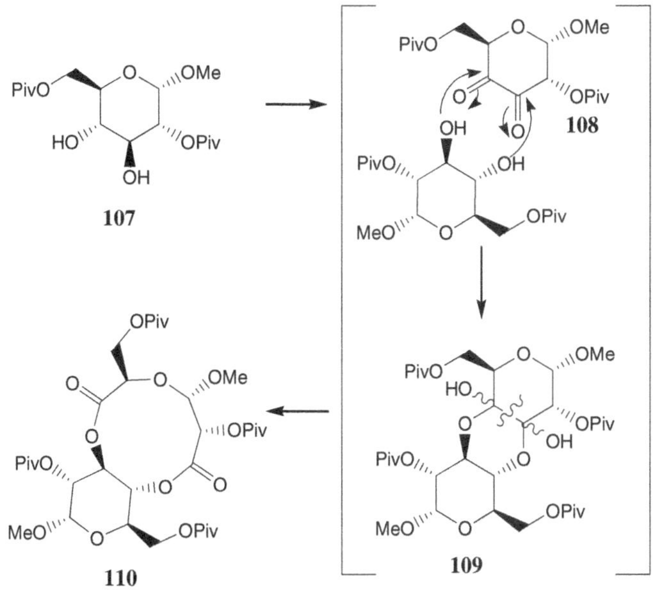

SCHEME 4.18 Domino reaction to access sugar bislactones in the presence of PCC.

4-Deoxy sugars showed an inhibitory effect on the biosynthesis of heparan sulphate [47], and a 4-deoxy branched-chain sugar is a structural feature of amipurimycin, isolated from *Streptomyces novoguineensis*, that presented a strong antifungal activity against *Pyricularia oryzae* [48].

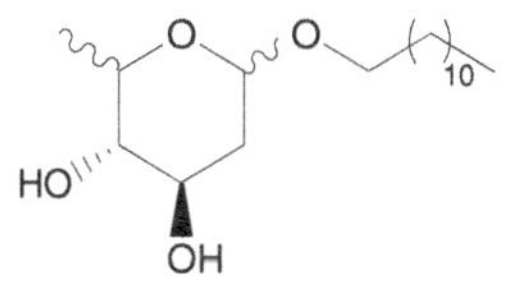

Antimicrobial surface active deoxy glycosides

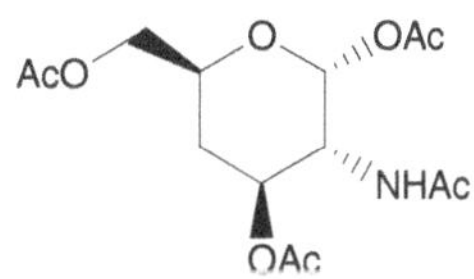

Inhibitor of heparan sulfate biosynthesis

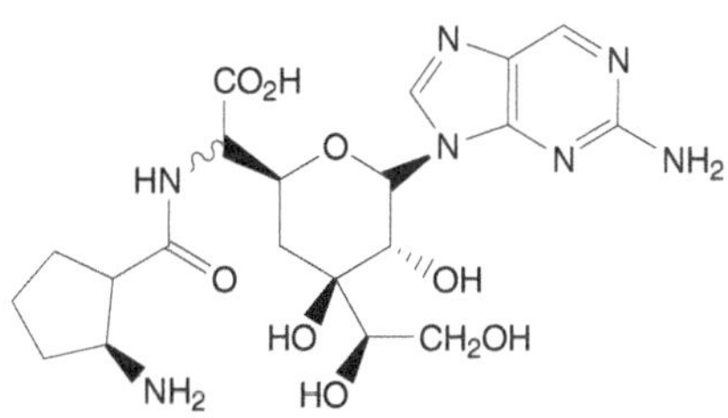

Amipurimycin

FIGURE 4.10 Examples of bioactive compounds exhibiting deoxy sugars in their structure.

While deoxygenation of position 2 can be readily accomplished via electrophilic addition to glycals, deoxygenation of other positions is accomplished by a variety of methods, involving more than one reaction step. However, we have developed a domino reaction that succeeds to afford α-deoxy keto sugars in one step, starting from their readily available α-hydroxy keto sugar precursors, in the presence of bromine or iodine/imidazole and triphenylphosphine [46]. Triphenylphosphine activates the hydroxy group and a nucleophilic substitution type 2 affords the halogenide, which reacts *in situ* to give an enolate, from which the 2-deoxy keto sugar is formed (Scheme 4.19). This methodology was applied to several carbohydrate scaffolds bearing α-hydroxy lactones, α-hydroxy esters, and α-hydroxy amides, and has proven a versatile methodology to afford α-deoxygenation of carbonyl functionalities in good yields [49,50].

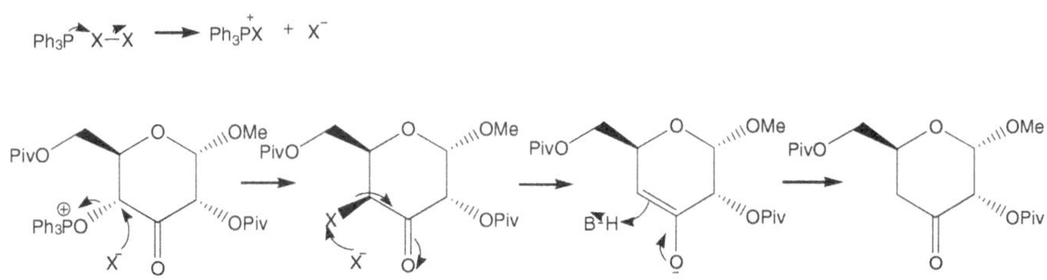

SCHEME 4.19 Domino reaction affording a 4-deoxy keto sugar directly from the hydroxyl keto sugar precursor.

4.7 CONCLUSIONS

With this critical review, we have highlighted the versatility of domino reactions with carbohydrate scaffolds. This was not a comprehensive review, but we have shown that a diversity of structures and transformations, namely bicyclic fused sugars, macrocyclic bislactones, ring expansion and contraction, sugar deoxygenation, are easily carried out through domino reactions, alternatively to the usual multistep syntheses known for such types of compounds and synthetic pathways. We hope to encourage the scientific community to further develop domino reactions aiming to find the answer to other challenges facing organic chemists today.

REFERENCES

[1] Tietze, L. F. *Chem. Rev.* **1996**, *96*, 115–136.

[2] Patel, R. M.; Puranik, V. G.; Argade, N. P. *Org. Biomol. Chem.* **2011**, *9*, 6312–6322.

[3] *Who Model List of Essential Medicines*, 18th Edition, 2013, as cited in http://www.who.int/medicines/publications/essentialmedicines/en>/index.html

[4] Cunha, S. C.; Faria, M. A.; Pereira, V. L.; Oliveira, T. M.; Lima, A. C.; Pinto, E. *Food Control* **214**, *44*, 185–190.

[5] (a) Xavier, N. M., Rauter, A. P. *Org. Lett.* **2007**, *9*, 3339–3341. (b) Xavier, N. M.; Silva, S.; Madeira, P. J. A.; Florêncio, M. H.; Silva, F. V. M.; Justino, J.; Thiem, J.; Rauter, A. P. *Eur. J. Org. Chem.* **2008**, *2008*, 6134–6143.

[6] Witczak, Z. J.; Culhane, J. M. *Appl. Microbiol. Biotechnol.* **2005**, *69*, 237–244.

[7] Capon, R. J.; MacLeod, J. K. J. Chem. Soc., Chem. Commun. **1987**, 1200–1201.

[8] Riedner, J.; Robina, I.; Fernández-Bolaños, J. G.; Gómez-Bujedo, S.; Fuentes, J. Tetrahedron Asymmetry. **1999**, *10*, 3391–3401.

[9] Jayakanthan, K.; Mohan, S.; Pinto, B. M. *J. Am. Chem. Soc.* **2009**, *131*, 5621–5626.

[10] Xavier, N. M.; Madeira, P. J. A.; Florêncio, M. H.; Rauter, A. P. *Eur. J. Org. Chem.* **2009**, *2009*, 4983–4991.

[11] Xavier, N. M.; Queneau, Y.; Rauter, A. P. *Eur. J. Org. Chem.* **2011**, *2011*, 713–720.

[12] Dai, M.; Zhang, X.; Khim, S.-K.; Schultz, A. G. *J. Org. Chem.* **2005**, *70*, 384–387.

[13] Li, Y.; Zhang, F.; Xu, Y.; Matsumura, K.; Han, Z.; Liu, L.; Lin, W.; Jia, Y.; Qian, P.-Y. *Biofouling* **2012**, *28*, 857–864.

[14] Chen, B.-F.; Tasi, M.-R.; Yang, C.-Y.; Chang, J.-K.; Chang, N.-C. *Tetrahedron* **2004**, *60*, 10223–10231.

[15] He, Z.-T.; Wei, Y.-B.; Yu, H.-J.; Sun, C.-Y.; Feng, C.-G.; Tian, P.; Lin, G.-Q. *Tetrahedron* **2012**, *68*, 9186–9191.

[16] Jensen, A. A.; Madsen, B. E.; Krogsgaard-Larsen, P.; Bräuner-Osborne, H. Eur. J. Pharmacol. **2001**, 177–180.

[17] Jida, M.; Ollivier, J. *Eur. J. Org. Chem.* **2008**, *2008*, 4041–4019.

[18] Takahashi, Y.; Fuwa, H.; Kaneko, A.; Sasaki, M.; Yokoshima, S.; Koizumi, H.; Takebe, T.; Kan, T.; Iwatsubo, T.; Tomita, T.; Natsugari, H.; Fukuyama, T. *Bioorg. Med. Chem. Lett.* **2006**, *16*, 3813–3816.

[19] Leflemme, N.; Dallemagne, P.; Rault, S. *Tetrahedron* **2004**, *60*, 4861–4865.

[20] Marcelo, F. *Total Synthesis and Stereochemical Assignment of Miharamycins*. PhD Thesis, University of Lisbon, University Pierre and Marie Curie–Paris VI, 2008.

[21] Marcelo, F.; Jiménez-Barbero, J.; Marrot, J.; Rauter, A. P.; Sinaÿ, P.; Blériot, Y. *Chem. Eur J.* **2008**, *14*, 10066–10073.

[22] Sharma, G. V. M.; Sridhar, T.; Reddy, P. P.; Kunwar, A. C. *Eur. J. Org. Chem.* **2013**, *2013*, 3543–3554.

[23] Mathé, C.; Périgaud, C. *Eur. J. Org. Chem.* **2008**, *2008*, 1489–1505.

[24] Yadav, V.; Agrawal, D. Chem. Commun. **2007**, 5232–5234.

[25] Velthuisen, E. J.; Baughman, T. M.; Johns, B. A.; Temelkoff, D. P.; Weatherhead, J. G. *Eur. J. Med. Chem.* **2013**, *63*, 202–212.

[26] Kulkarni, M. G.; Shaikh, Y. B.; Borhade, A. S.; Dhondge, A. P.; Chavhan, S. W.; Desai, M. P.; Birhade, D. R.; Dhatrak, N. R.; Gannimani, R. *Tetrahedron Asymmetry.* **2010**, *21*, 2394–2398.

[27] Davis, O. A.; Bull, J. A. *Angew. Chem.* **2014**, *53*(51), 14230–14234.

[28] Ōmura, S.; Murata, M.; Imamura, N.; Iwai, Y.; Tanaka, H. *J. Antibiot.* **1984**, *37*, 1324–1332.

[29] (a) Shimada, N.; Hasegawa, S.; Harada, T.; Tomisawa, T.; Fujii, A.; Takita, T. *J. Antibiot* **1986**, 1623–1625; (b) Nakamura, H.; Hasegawa, S.; Shimada, N.; Fujii, A.; Takita, T.; Iitaka, Y. *J. Antibiot.* **1986**, 1626–1629.

[30] Hoshino, H.; Shimizu, N.; Shimada, N.; Takita, T.; Takeuchi, T. *J. Antibiot.* **1987**, 1077–1078.

[31] Witty, D. R.; Fleet, G. W. J.; Vogt, K.; Wilson, F. X.; Wang, Y.; Storer, R.; Myers, P. L.; Wallis, C. J. *Tetrahedron Lett.* **1990**, *31*, 4787–4790.

[32] Barker, S. F.; Angus, D.; Taillefumier, C.; Probert, M. R.; Watkin, D. J.; Watterson, M. P.; Claridge, T. D. W.; Hungerford, N. L.; Fleet, G. W. J. *Tetrahedron Lett.* **2001**, *42*, 4247–4250.

[33] Johnson, S. W.; Jenkinson, S. F.; Angus, D.; Jones, J. H.; Fleet, G. W. J.; Taillefumier, C. *Tetrahedron Asymmetry* **2004**, *15*, 2681–2686.

[34] Jenkinson, S. F.; Harris, T.; Fleet, G. W. J. *Tetrahedron Asymmetry* **2004**, *15*, 2667–2679.

[35] Lucas, S. D.; Iding, H.; Alker, A.; Wessel H. P.; Rauter, A. P. *J. Carbohydr. Chem.* **2006**, *25*, 187–196.

[36] Lucas, S. D.; Rauter, A. P.; Schneider, J.; Wessel, H. P. *J. Carbohydr. Chem.* **2009**, *28*, 431–446.

[37] Lucas, S. D.; Rauter, A. P.; Wessel, H. P. *J. Carbohydr. Chem.* **2008**, 172–187.

[38] Lucas, S. D.; Fischer, H.; Alker, A.; Rauter, A. P.; Wessel, H. P. *J. Carbohydr. Chem.* **2011**, *30*, 498–548.

[39] Yerande, S. G.; Ghaisas, A. B.; Newase, K. M.; Wang, W.; Wang, K.; Dömling, A. *J. Heterocyclic Chem.* **2014**, *51*, 1752–1756.

[40] Monceaux, C. J.; Hirata-Fukae, C.; Lam, P. C.-H.; Totrov, M. M.; Matsuoka, Y.; Carlier, P. R. *Bioorg. Med. Chem. Lett.* **2011**, *21*, 3992–3996.

[41] Guney, T.; Kraus, G. A. *Org. Lett.* **2013**, *15*, 613–615.

[42] Santos D.; Ariza, X.; Garcia, J.; Lloyd-Williams, P.; Martínez-Laporta, A.; Sánchez, C. *J. Org. Chem.* **2013**, *78*, 1519–1524.

[43] Mavlan, M.; Ng, K.; Pansear, H.; Yepremyan, A.; Minehan, T. G. *Org. Lett.* **2014**, *16*, 2212–2215.

[44] Rauter, A. P.; Piedade, F.; Almeida, T.; Ramalho, R.; Ferreira, M. J.; Resende, R.; Amado, J.; Pereira, H.; Justino, J.; Neves, A.; Silva, F. V. M.; Canda, T. *Carbohydr. Res.* **2004**, *339*, 1889–1897.

[45] Issa, J. P.; Bennett, C. S. *J. Am. Chem. Soc.* **2014**, *136*, 5740–5744.

[46] Martins, A.; Santos, M. S.; Dias, C.; Serra, P.; Cachatra, V.; Pais, J.; Caio, J.; Teixeira, V. T.; Machuqueiro, M.; Silva, M. S.; Pelerito, A.; Justino, J.; Goulart, M.; Silva, F. V.; Rauter, A. P. *Eur. J. Org. Chem.* **2013**, *2013*, 1448–1459.

[47] Kisilevsky, R.; Szarek, W. A.; Ancsin, J. B.; Elimova, E.; Marone, S.; Bhat, S.; Berkin, A. *Am. J. Pathol.* **2004**, *164*(6), 2127–2137.

[48] (a) Harada, S.; Kichi, T. *J. Antibiot.* **1977**, *30*, 11–16. (b) Iwasa, T.; Kishi, T.; Matsura, K.; Wakae, O. *J. Antibiot.* **1977**, *30*, 1–10.

[49] Rauter, A. P.; Fernandes, A. C.; Czernecki, S.; Valery, J.-M. *J. Org. Chem.* **1996**, *61*, 3594–3598.

[50] Rauter, A. P.; Fernandes, A. C.; Figueiredo, J. A. *J. Carbohydr. Chem.* **1998**, *17*(7), 1037–1045.

5

MULTISTEP TRANSFORMATIONS OF *BIS*-THIOENOL ETHER-CONTAINING CHIRAL BUILDING BLOCKS: NEW AVENUES IN GLYCOCHEMISTRY

DANIELE D'ALONZO, GIOVANNI PALUMBO, AND
ANNALISA GUARAGNA
Department of Chemical Sciences, University of Napoli Federico II, Napoli, Italy

5.1 INTRODUCTION

The involvement of carbohydrates in a wide variety of biological processes and thereby their large use as biochemical probes and therapeutic agents have justified the exceptional synthetic interest toward this class of biomolecules and their analogues over the last decades [1, 2]. However, despite their wide commercial availability, the use of naturally occurring sugars as a source of starting building blocks does not always represent a convenient synthetic strategy. Especially with regard to the synthesis of glycostructures bearing rare or unnatural sugars, common carbohydrate-based routes usually require time-consuming, lengthy synthetic sequences involving protective group-based strategies of carbohydrate core manipulation. Alternatively, the de novo synthetic methodologies—relying on the assembly of carbohydrate precursors with the desired functionality and stereochemistry from structurally simpler starting materials—have demonstrated a well-established synthetic utility [3, 4]. Expeditious, efficient, and stereoselective procedures, including asymmetric epoxidations [5], Achmatowicz rearrangements [6], Tsuji–Trost rearrangements [7], ring-closing metatheses [8], organocatalyzed aldol additions [9], or chemoenzymatic transformations [10] as the key steps have enabled rapid access

Domino and Intramolecular Rearrangement Reactions as Advanced Synthetic Methods in Glycoscience, First Edition. Edited by Zbigniew J. Witczak and Roman Bielski.

to a wide variety of mono- [11] and oligosaccharides [3], glycomimetics [12, 13], biomimetics [14], and carbohydrate-containing complex natural products and their unnatural congeners [15]. In some cases, the superiority of the de novo methodologies over the common carbohydrate-based approaches has also been demonstrated [16]. Even more importantly, given the high degree of stereocontrol typically available in the construction of contiguous oxygen-bearing stereocenters, flexible strategies leading to the target glycostructures, their epimers, or even their enantiomers [17], have been developed for various medicinal chemistry applications [3, 15]. Overall, a wealth of untapped synthetic opportunities in carbohydrate chemistry is being explored by de novo synthetic methods, aiming at the identification of novel biologically important molecules via the preparation of carbohydrate-based building blocks of whatever structure and complexity.

5.2 (5,6-DIHYDRO-1,4-DITHIIN-2-YL)METHANOL: NOT SIMPLY A HOMOLOGATING AGENT

The arsenal of de novo approaches for carbohydrate synthesis currently includes many elegant solutions based on the (asymmetric) carbon chain elongation of various electrophiles [18–20]. In this area, our synthetically available [21, 22] (5,6-dihydro-1,4-dithiin-2-yl)methanol **1** (Scheme 5.1) has represented a useful and versatile building block in a plethora of synthetic programs. In its protected form [**2a**: R = 4-methoxybenzyl (PMB); **2b**: R = allyl], the 3-*C*-lithiated version of **1** (i.e., **3a-b**; Scheme 5.1) has been ascertained to act as a synthetic equivalent of an allylic alcohol (**I**) or a β-acyl vinyl anion (**II**), thus displaying either regular or umpolung reactivity features. The synthetic potential of **3** as a homologating agent of a wide variety of (a)chiral electrophiles including halides [23], epoxides [(*R*)- and (*S*)-**4**] [24], aldehydes **6** [25, 26], esters **7** [27, 28], and lactones **5** [29, 30] has been extensively recognized over the last decade (Scheme 5.2). In addition, depending on the specific synthetic aims, the use of **1** and the corresponding coupling products **8-11** as sugar bioisosteres [31], spacers [30] or as carbohydrate components within various mono- [27, 32] and oligomeric structures [33, 34] has been reported.

A peculiarity of the dithiin-containing systems (**2**, **12**) lies in the opportunity to remove the sulfur bridge moiety in a highly stereoselective manner leading,

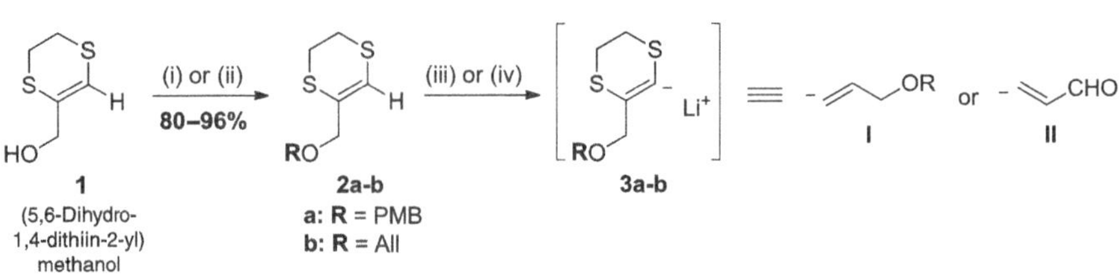

SCHEME 5.1 Reagents and conditions: (i) NaH, PMBCl, dimethylformamide (DMF), rt (room temperature); (ii) NaH, AllBr, DMF, rt; (iii) BuLi, tetrahydrofuran (THF), –78°C; and (iv) lithium diisopropylamide *in situ*, THF, –78°C.

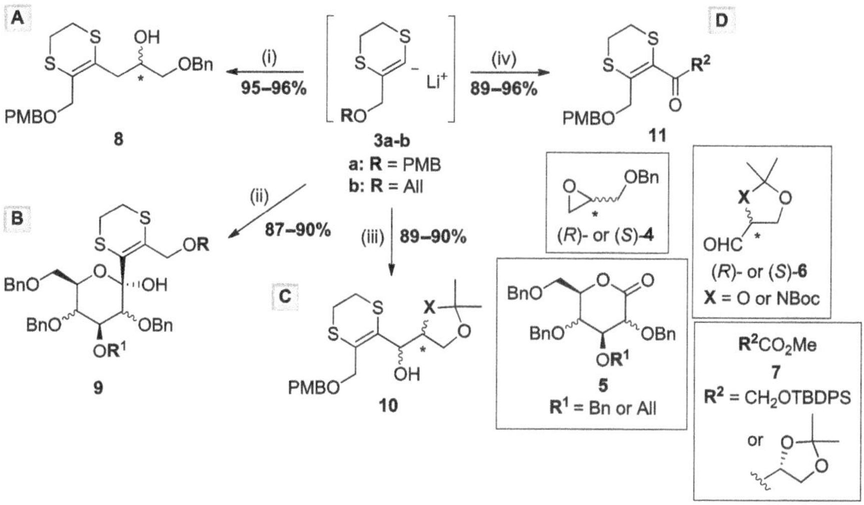

SCHEME 5.2 Reagents and conditions: (i) Ti(*O*-*i*-Pr)$_4$, (*R*)- or (*S*)-**4**, THF, –78°C, and then –40°C; (ii) **5**, THF, –78°C; (iii) (*R*)- or (*S*)-**6**, THF or Et_2O, –78°C; and (iv) **7**, THF, –78°C.

depending on the synthetic conditions, only to cis (Raney-Ni) [35] or trans (LAH/Ti(*O*-*i*-Pr)$_4$/quinoline) [36] olefins **13** and **14** (Scheme 5.3).

However, the most relevant synthetic trait of this molecule is its capacity to activate neighboring functional groups embedded in its structural framework (Schemes 5.4–5.6, 5.10–5.13, and 5.15). In one of the earliest examples devised to this end, cleavage of the PMB ethers **2** and **15** with 2,3-dihydro-5,6-dicyano-1,4-benzoquinone (DDQ) in dichloromethane/H_2O emulsions selectively afforded the expected primary alcohol **16** [25,26] or a formyl function **17** [23, 37] (Scheme 5.4). The latter transformation is not typically observed starting from the corresponding sulfur-free allyl PMB ethers [38].

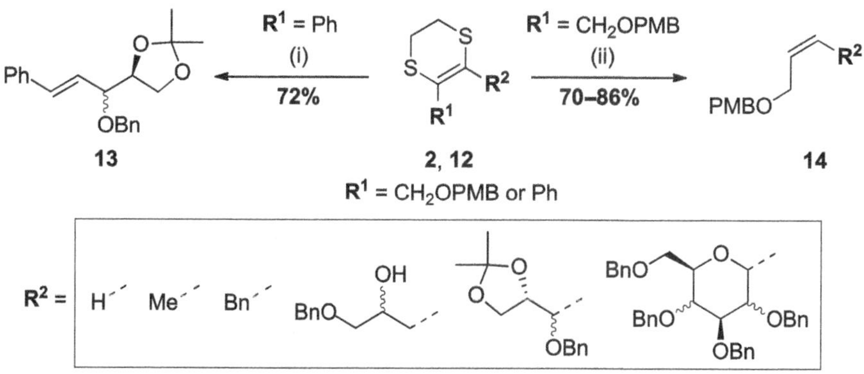

SCHEME 5.3 Reagents and conditions: (i) $LiAlH_4$, Ti(*O*-*i*-Pr)$_4$, quinoline, ethanol, 50°C; and (ii) Raney-Ni, THF, 0°C or rt.

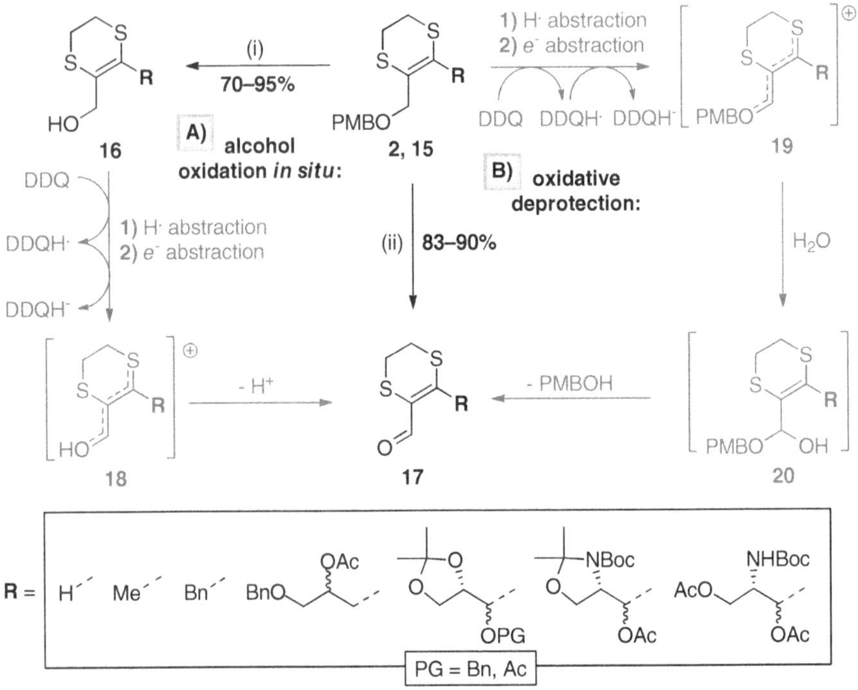

SCHEME 5.4 Reagents and conditions: (i) DDQ, CH_2Cl_2:H_2O = 9:1, rt; and (ii) DDQ, CH_2Cl_2:H_2O = 18:1, rt.

The conversion of **2** and **15** into the corresponding α,β-unsaturated aldehydes **17** has been judged to rely on one or both of the two competing mechanisms depicted in Scheme 5.4. The detection, on some substrates [27, 37], of the transient allylic alcohol **16** clearly suggests that an early PMB group removal followed by a subsequent oxidation *in situ* of **16** under single electron transfer (SET) conditions [39] may take place (Scheme 5.4A). The latter transformation is supposed to occur via formation of the stabilized cation intermediate **18**, overall obtained by H· and e^- abstractions from **16** by DDQ. Similarly, when no allylic alcohol intermediate, even at low temperature, could be detected, a direct oxidative deprotection mechanism has been postulated, involving formation of the hemiacetal intermediate **20** via cation **19** (Scheme 5.4B).

Starting from this early finding, extensive studies aimed at exploring the activation conditions of various functional groups, including ethers [23], aldehydes [24], alcohols [25], ketones [28], and acetals [28, 32] by our heterocyclic system have then been undertaken. Exploiting the reactivity of various *bis*-thioenol ether-containing substrates when exposed to either acid, alkaline or radical conditions, novel multistep processes involving domino, one-pot, or domino-one pot transformations have been devised, enabling the rapid construction of carbohydrate precursors for various synthetic applications.

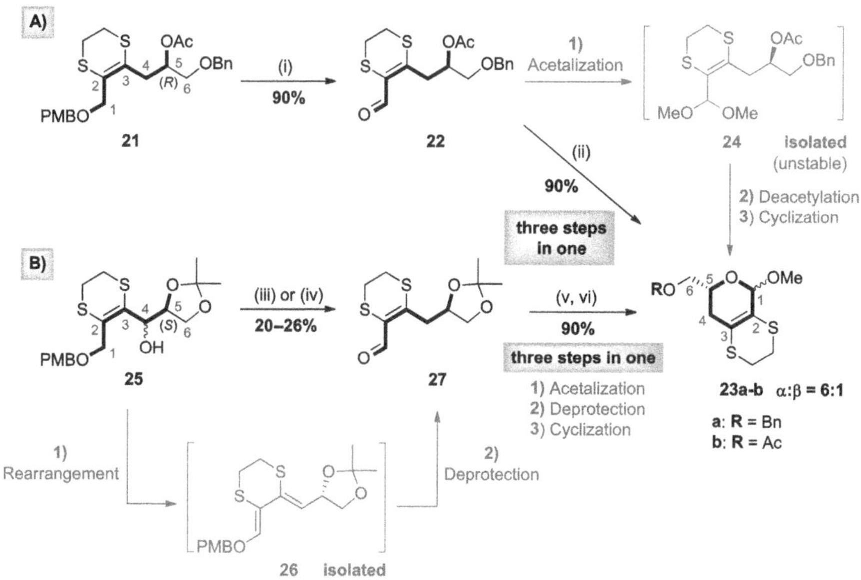

SCHEME 5.5 Reagents and conditions: (i) DDQ, CH_2Cl_2:H_2O = 18:1, rt; (ii) TMSOTf, TEA, MeOH, rt; (iii) (a) NaH, DMF, rt; (b) column chromatography, (iv) (a) $(CO)_2Cl_2$, DMSO (dimethyl sulfoxide), CH_2Cl_2, –78°C; (b) Et_3N, –78 to 0°C; (c) column chromatography, (v) Amberlyst 15, MeOH, rt, and (vi) Ac_2O, pyridine, rt.

5.3 SULFUR-ASSISTED MULTISTEP PROCESSES AND THEIR USE IN THE DE NOVO SYNTHESIS OF GLYCOSTRUCTURES

5.3.1 Three Steps in One Process: Double Approach to 4-Deoxy L-(and D-)-Hexoses

A first look at the synthetic potential of our heterocyclic system was given during the synthesis of the rare 4-deoxy L- (and D-)-hexoses [24,25]. In this context, two different routes have been reported to enable access to the key hexopyranoside precursor **23** (Scheme 5.5). In early investigations, synthesis of **23** was carried out starting from acetate **21** (in turn obtained from glycidyl ether (*R*)-**4**; Scheme 5.2A). After the DDQ-mediated conversion of **21** into aldehyde **22** was achieved, treatment of the latter with trimethylsilyltrifluoromethanesulfonate (TMSOTf)/triethylamine (TEA) in MeOH promoted three sequential transformations in one process, involving (1) aldehyde acetalization, to give an isolable (albeit unstable) di-*O*-methyl acetal **24**; (2) acetyl group cleavage of **24**; and (3) intramolecular transacetalization, leading to methyl acetal **23a** in a high yield (90%) and moderate stereoselectivity (α:β = 6:1; Scheme 5.5A). An alternative approach to the synthesis of **23**, even if in lower overall yields (o.y.), was developed starting from alcohol **25** [in turn obtained from protected glyceraldehyde (Scheme 5.2C) or methyl glycerate (Scheme 5.2D)]. Treatment of **25** under either acid [$(CO)_2Cl_2$] [27b] or alkaline (NaH) [25] conditions unexpectedly

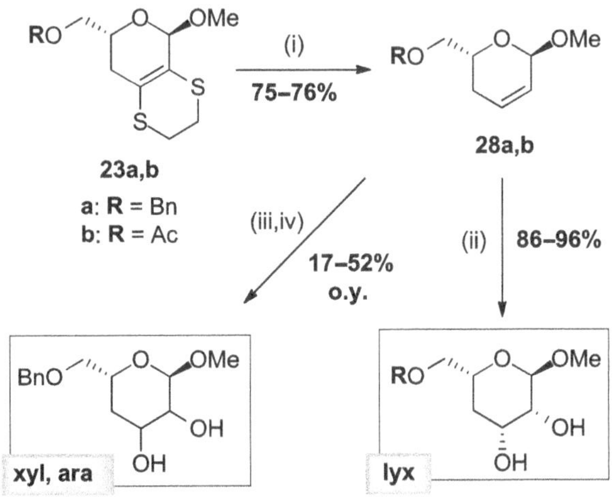

SCHEME 5.6 Reagents and conditions: (i) from **23a**: Raney-Ni, THF, rt; from **23b**: Raney-Ni, THF, 0°C; (ii) OsO_4, *N*-methylmorpholine-*N*-oxide (NMO), *tert*-Butyl-OH/acetone, rt; (iii) *m*CPBA, CH_2Cl_2, rt; and (iv) 1N NaOH, 95°C.

gave diene **26**, as a result of a sulfur-assisted dehydration reaction (Scheme 5.5B). A subsequent deprotection step was then observed during the chromatographic purification of **26**, quantitatively providing aldehyde **27**. The latter was converted into bicycle **23b** by a multistep reaction, involving (1) dimethoxyacetalization, (2) isopropylidene ring cleavage, and (3) cyclization (95% after acetylation of the crude residue, α:β = 6:1).

With α-pyranosides **23a** and **23b** in hand, removal of dithiodimethylene bridge (Raney-Ni) led to the corresponding olefins **28a-b** (75–76%). Hence, *syn*-dihydroxylation (OsO_4) of **28a-b** afforded, as the only detected stereoisomer, the *lyxo*-configured 4-deoxy-L-hexopyranoside (96% from **28a**; 86% from **28b**). Conversely, *anti*-dihydroxylation (*meta*-chloroperoxybenzoic acid [*m*CPBA] and then NaOH) of **28a** gave both the *arabino*- (52% o.y.) and *xylo*- (17% o.y.) configured 4-deoxy-L-hexoses (Scheme 5.6). It is worthy to mention that, exploiting a mirrored methodology, the enantiomeric acetate *ent*-**21** was obtained from glycidyl ether (*S*)-**4** (data not shown), opening the way to the synthesis of 4-deoxy-D-hexoses [24].

5.3.2 Five Steps in One Process: The Domino Way to L-Hexoses (and Their Derivatives)

Most of the multistep reactions triggered by our heterocyclic system relied on its combined use with DDQ [27, 28, 32]. In early studies, the versatile reactivity of DDQ—a result of its combined electron-transfer, oxidative, and acidic properties [39]—was exploited to enable domino cyclizations starting from protected *sec*-alcohols **29a-b** on the way to the de novo synthesis of the whole series of the rare L-hexoses (Scheme 5.7) [27]. Treatment of **29a-b** [in turn obtained from protected glyceraldehyde (Scheme 5.2C) or methyl glycerate (Scheme 5.2D)] with DDQ in a 3/1 CH_2Cl_2/MeOH

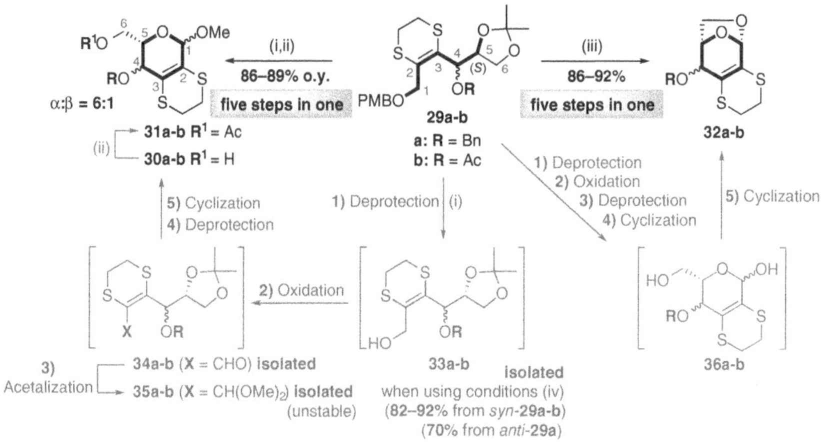

SCHEME 5.7 Reagents and conditions: (i) DDQ, CH_2Cl_2:MeOH = 3:1, rt; (ii) Ac_2O, Py, rt; (iii) DDQ, CH_2Cl_2:H_2O = 18:1, rt; and (iv) DDQ, CH_2Cl_2:H_2O = 9:1, rt.

solution provided methyl hexopyranosides **31a-b** after acetylation of the crude residues (86–89%). This DDQ-mediated conversion was the result of five sequential transformations carried out in a single process, comprising (1) PMB group removal of **29** to give alcohols **33**, (2) oxidation of the free primary hydroxyl function of **33** to the corresponding aldehydes **34**, (3) dimethoxyacetalization of **34**, (4) isopropylidene ring cleavage, and (5) ring closure. The reaction mechanism was ascertained by detection of most intermediates, including the rather unstable di-*O*-methyl acetals **35a-b**, isolated by a previously reported stepwise route [25, 27b]. Similarly, 1,6-anhydrosugar precursors **32a-b** were obtained when treating **29a-b** with DDQ in an 18/1 CH_2Cl_2/H_2O emulsion (86–92%). In this case, the process is reasonably the result of five sequential conversions [(1) PMB group removal, (2) primary alcohol oxidation, (3) isopropylidene ring cleavage, (4) 1,5-cyclization, and (5) 1,6-cyclization], as deduced by the thin layer chromatography detection of transient lower-R_f species (likely ascribable to **36a-b**) rapidly converting into the higher R_f tricyclic compounds **32a-b**.

After dithioethylene bridge removal of bi- and tricyclic intermediates **31** and **32**, access to the whole series of L-hexoses (Scheme 5.8) with a high degree of stereoselectivity was then achieved exploiting the selectivity of the *syn*- or *anti*-dihydroxylation reactions. As an example, treatment of (1*R*,4*S*,5*S*)-**31a** with Raney-Ni in THF at 0°C followed by the *syn*-dihydroxylation under Upjohn conditions (OsO_4/NMO) provided the methyl hexopyranoside with L-*manno* configuration as the only detected epimer [25]; similarly, *cis*-2,3-diols of L-hexoses with *allo*, *gulo*, and *talo* configuration were selectively synthesized (dr = 19:1 or higher) from (1*S*,4*S*,5*S*)-**32a**, (1*R*,4*R*,5*S*)-**31a**, and (1*R*,4*R*,5*S*)-**31b**, respectively [27]. An analogous synthetic path, but involving *anti*-dihydroxylation reactions (methyl(trifluoromethyl)dioxirane [TFDO]/KOH or TFDO/$HClO_4$), gave the corresponding *trans*-2,3-diols of L-hexoses with *altro*,

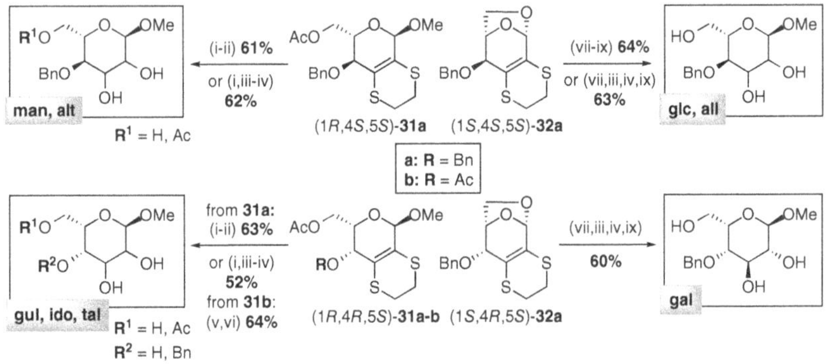

SCHEME 5.8 Reagents and conditions: (i) Raney-Ni, THF, 0°C; (ii) OsO_4, NMO, *tert*-Butyl-OH-acetone-H_2O, rt; (iii) Oxone, Na_2EDTA_{aq}, $NaHCO_3$, CF_3COCH_3, CH_3CN, 0°C; (iv) KOH_{aq}, reflux; (v) MeONa, MeOH, rt; (vi) OsO_4, tetramethylethylenediamine, CH_2Cl_2, –78°C; (vii) Raney-Ni, acetone, 0°C to rt; (viii) OsO_4, pyridine, rt; and (ix) TMSOTf, MeOH, 50°C.

ido, *gluco*, and *galacto* configuration (dr = 9:1 or higher) from (1*R*,4*S*,5*S*)-**31a**, (1*R*,4*R*,5*S*)-**31a**, (1*S*,4*S*,5*S*)-**32a**, and (1*S*,4*R*,5*S*)-**32a**, respectively (Scheme 5.8) [27].

Monosaccharide precursors (1*R*,4*S*,5*S*)-**31b** and (1*S*,4*S*,5*S*)-**32b** also represented useful intermediates en route to sugar-modified nucleosides **38** and **40** (Scheme 5.9). On one side, synthesis of L-hexopyranosyl nucleosides **38a-b** was accomplished via stereoselective *N*-glycosidation [40] (TMSOTf, **B**, bis(trimethylsilyl) acetamide

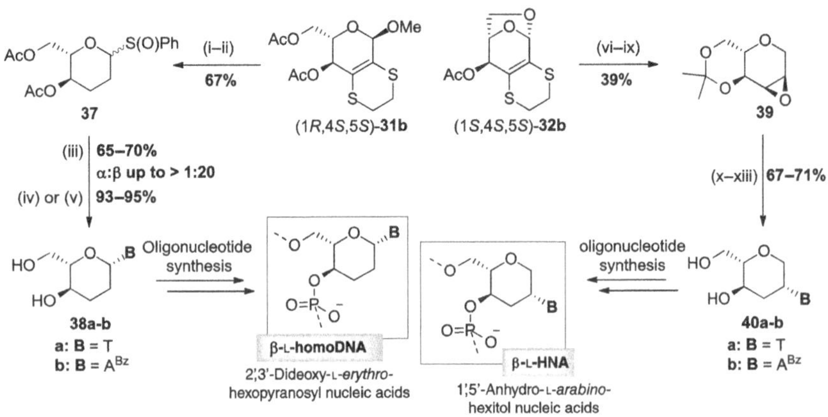

SCHEME 5.9 Reagents and conditions: (i) Raney-Ni, THF, rt; (ii) (a) $BF_3{\cdot}OEt_2$, PhSH, CH_2Cl_2, –20°C, and (b) *m*CPBA, CH_2Cl_2, –78°C; (iii) TMSOTf, **B**, BSA, CH_3CN 60°C; (iv) NH_3/MeOH, rt; (v) MeONa, MeOH, 0°C; (vi) Raney-Ni, acetone, 0°C; (g) TMSOTf, Et_3SiH, CH_2Cl_2, 0°C; (viii) (a) MeONa, MeOH, rt, and (b) *m*CPBA, CH_2Cl_2, 0°C to rt; (ix) dimethoxypropane, pyridinium *p*-toluenesulfonate, acetone, rt; (x) **B**, 1,8-diazabicyclo[5.4.0]undec-7-ene, DMF, 90°C; (xi) **B** = thymine: $NaOH_{aq}$, CS_2, $BrCH_2CH_2CN$, dimethyl sulfoxide, 0°C, and then Bu_3SnH, AIBN, toluene, reflux; **B** = adenine: $NaOH_{aq}$, CS_2, $BrCH_2CH_3$, DMF, 0°C, and then Bu_3SnH, AIBN, toluene, reflux; (xii) 80% $AcOH_{aq}$, 60°C; and (xiii) **B** = adenine: TMSCl, BzCl, Py, 0°C to rt, and then NH_4OH, 0°C.

[BSA], α:β up to > 1:20) of sulfoxide intermediate **37**, in turn obtained by the manipulation of **36**. On the other side, sugar ring elaboration of (1*S*,4*S*,5*S*)-**32b** provided oxirane **39**, from which L-hexitol nucleosides **40a-b** were obtained after nucleobase insertion (**B**, 1,8-diazabicyclo[5.4.0]undec-7-ene), Barton-McCombie deoxygenation (NaOH, CS_2, $BrCH_2CH_2CN$ or $BrCH_2CH_3$, and then Bu_3SnH, azobisisobutyronitrile [AIBN]) and acetonide removal (AcOH) [41]. Nucleosides **38** and **40** were employed for the synthesis of two novel oligonucleotide systems, namely the 2′,3′-dideoxy-β-L-*erythro*hexopyranosyl nucleic acids (β-L-homoDNA) and 1′,5′-anhydro-L-*arabino*-hexitol nucleic acids (β-L-HNA), whose base-pairing properties were studied in the frame of a project aimed at revisiting the role of chirality in the hybridization processes [33, 34, 42].

5.3.3 Up to Six Steps in One Process: 4′-Substituted Nucleoside Synthesis

Further synthetic opportunities deriving from the combined use of our heterocyclic system with DDQ were provided by the study of the reactions of ketones **41a-b** (Scheme 5.10) and alcohols **43a-b** (Scheme 5.11) [28, 32]. Treatment of

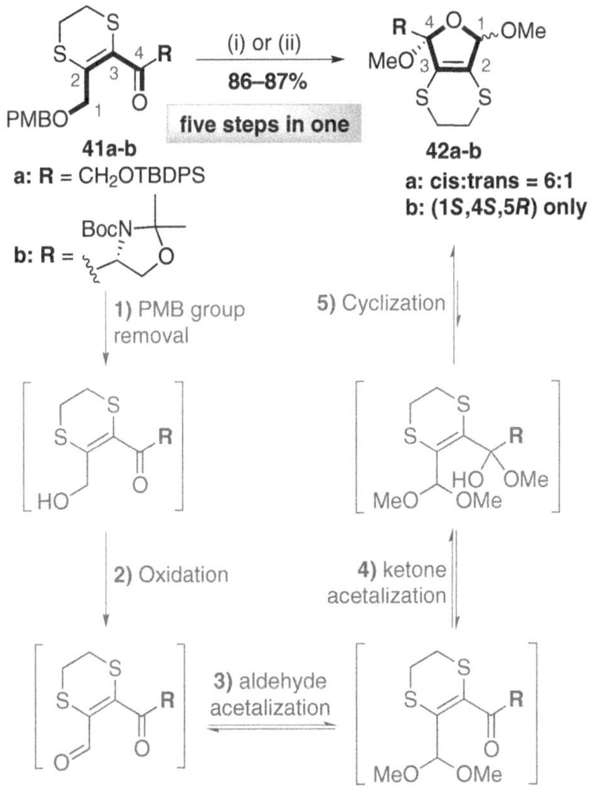

SCHEME 5.10 Reagents and conditions: (i) from **41a**: DDQ (1.2 equivalents), CH_2Cl_2:MeOH = 18:1, rt; and (ii) from **41b**: DDQ (1.8 equivalents), C_6H_6:MeOH = 3:1, rt.

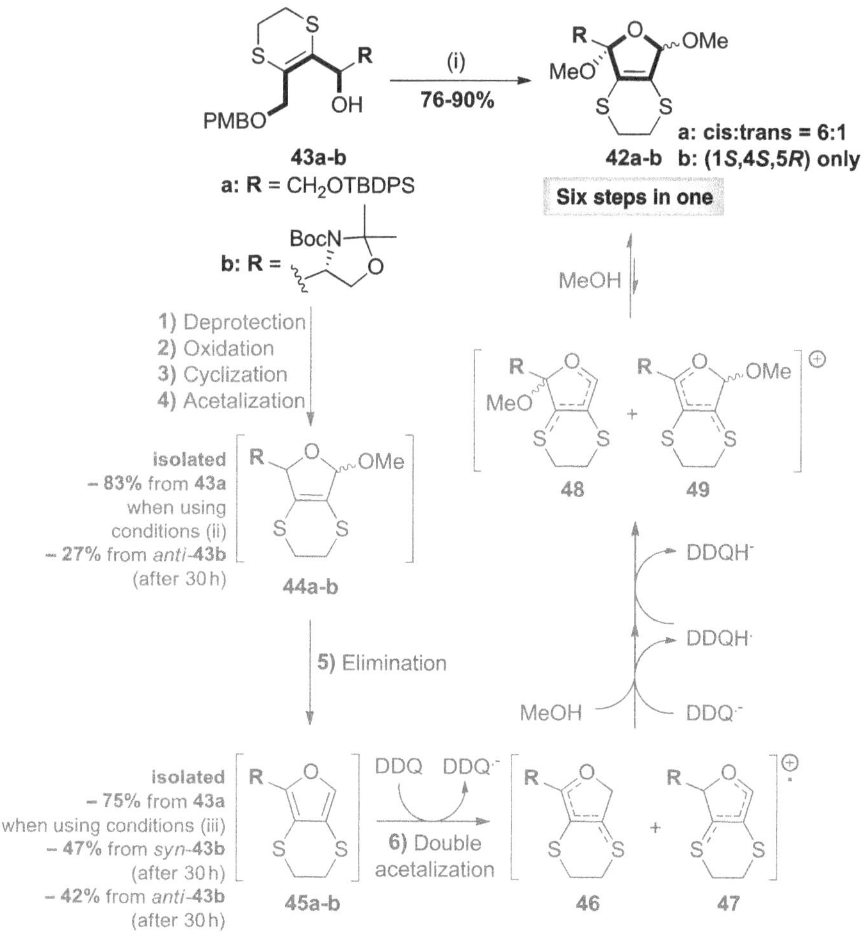

SCHEME 5.11 Reagents and conditions: (i) DDQ (1.8 equivalents), C_6H_6:MeOH = 3:1, rt, 3–48 h; (ii) DDQ (1.2 equivalents), CH_2Cl_2:MeOH = 18:1, rt, 0.5 h; and (iii) DDQ (1.2 equivalents), C_6H_6:MeOH = 18:1, rt, 3 h.

either ketone **41a** (obtained starting from methyl glycolate **7**; Scheme 5.2D) or **41b** (obtained from the protected D-serinal; Scheme 5.2C) with DDQ (1.8 equivalents) in MeOH-containing binary mixtures gave, after 3–24 h at rt, the *bis*-methyl acetals **42a-b** in very good 86–87% yields (Scheme 5.10). Although no intermediate could be isolated in this case, based on previous data the process is conceivably the result of five sequential transformations, involving (1) PMB group removal, (2) oxidation of the resulting primary alcohol, (3) aldehyde acetalization, (4) ketone acetalization, and (5) cyclization. When the achiral model system **41a** was employed as a starting material, a moderately high diastereoselectivity was found (cis–trans = 6:1) [28, 43]. On the contrary, starting from **41b** (bearing the chiral and bulky oxazolidine ring), the formation of a single isomer **42b** having 1*S*,4*S*,5*R* configuration (out of four theoretically achievable stereoisomers) was detected (Scheme 5.10) [32]. Similarly, treatment of alcohol **43a** with DDQ (1.2 equivalents) in an 18/1 CH_2Cl_2/MeOH solution directly provided the bicyclic dihydrofuran **44a** in a good 83% yield (cis–trans = 3:1) after four sequential transformations: (1–2) as in Scheme 5.10, (3) cyclization, and (4) acetalization (Scheme 5.11). However, when the reaction was carried out using a larger amount of DDQ (1.8 equivalents), methyl acetal **44a** rapidly disappeared in favor of the formation of furan **45a** at first, and of *bis*-acetal **42a** then. Furan **45a** could also be smoothly isolated as the main product (75%) using slightly different conditions (DDQ: 1.2 equivalents; C_6H_6:MeOH = 18:1, 3 h) (Scheme 5.11) [28]. The same path was observed starting from alcohols *syn*- and *anti*-**43b**, although in this case only small amounts of the weakly stable **44b** (27%, isolated only starting from *anti*-**43b**) and of furan **45b** (47% from *syn*-**43b**; 42% from *anti*-**43b**) were recovered if the reaction was stopped after 30 h. At this stage, formation of a certain amount of *bis*-acetal **42b** was already rather apparent by thin layer chromatography; after 48 h, the complete conversion of **44b** and **45b** into **42b** for either *syn*- or *anti*-**43b** was then accomplished (76%) [32]. Overall, the process **43a-b**→**42a-b** comprises six sequential transformations in one step: (1–4) as aforementioned, (5) elimination, and (6) double acetalization (Scheme 5.11). Despite the efficiency of the conversions (76–90%), during the reaction **43b**→**42b**, the appearance of other products was also detected, suggesting the occurrence of additional synthetic transformations (see Section 5.3.4).

Surprisingly, the stereoselectivity of the reaction starting from alcohols **43a-b** [**42a**: cis–trans = 6:1; **42b**: the only (1*S*,4*S*,5*R*) isomer recovered] was identical to that previously observed from ketones **41a-b** (Scheme 5.10); in addition, the same process, carried out from starting materials not bearing the *bis*-thioenol ether bridge, runs far less efficiently and without any stereoselection [28]. A role of the 1,4-dithiinyl moiety influencing rate and stereoselectivity of this process was, therefore, rather apparent. We hypothesized that the observed stereoselectivity could be the result of a sulfur-assisted equilibrium between the (up to four) diastereomeric species of **42** toward the isomer having the methoxy groups in a cis relationship. This process would be driven by the formation of ions **46–47** and **48–49** (Scheme 5.11), whose stability, owing to the wide sulfur-assisted delocalization of the positive charge, would significantly contribute to accelerate the rate of a process otherwise hardly achievable. In support of this assumption, the conversion **43a**→**42a** was performed in an NMR tube (Scheme 5.12) [28]. First, we appreciated the disappearance of the ^1H signals of **43a** [e.g., H4 (red) at 5.17 ppm; Scheme 5.12A] and concurrently the gradual appearance of those belonging to acetal **50** [H1(cis) and H1(trans) (purple) at 5.66 and 5.82 ppm, respectively; Scheme 5.12B], furan **45a** [e.g., H1 (green) at 6.89 ppm; Scheme 5.12B], and the *bis*-acetal **51** [H1(cis) and H1(trans) (blue) at 5.62 and 5.67 ppm, respectively; Scheme 5.12B–C]. In addition, the time-dependent changes in the cis–trans ratio (e.g., $t = 2$ h, cis–trans ~ 2:1; $t = 4$ h, cis–trans ~6:1; Scheme 5.12B–C) witnessed the existence of a thermodynamic equilibrium between *cis*-**51** and *trans*-**51**.

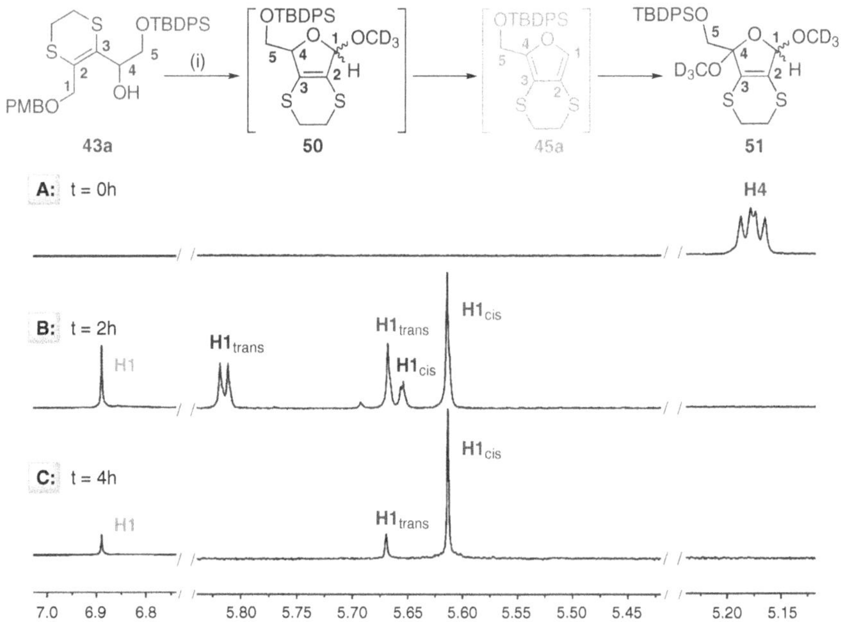

SCHEME 5.12 Reagents and conditions: (i) DDQ (1.8 equivalents), C_6D_6:CD_3OD = 3:1, rt, 5 h.

The process reported in Scheme 5.11 represents an example of diversity-oriented synthesis, since structurally different compounds (**42**, **44**, and **45**) can be obtained from a single starting material by suitably varying the reaction conditions. Likewise, because the same reaction product **42** is synthesized from stereochemically different starting materials (*syn*- and *anti*-**43**), this reaction is amenable to be also regarded as a stereoconvergent process. In addition, the term "chemoconvergent" was in this case used for the first time [32] to describe a chemical transformation where reagents bearing different functional groups (such as **41** and **43**) provided a same reaction product (**42**).

The capacity of the 1,4-dithiinyl ring to activate the neighboring acetal functions was further witnessed by some simple model reactions (Scheme 5.13). On one side, DDQ-promoted double transacetalizations of **42a** with readily available alcohols (Scheme 5.13A) gave the corresponding *bis*-alk(yn)yl acetals **51** and **52a-c** in high yields (71–90%) and good to excellent stereoselectivities (cis–trans up to >20:1). On the other side, we found that **42a** reacted under common Vorbrüggen conditions (N^4-AcC, BSA, TfOTMS) much faster (0.5 h) and slightly more stereoselectively (cis–trans = 4:1) than the corresponding desulfurized **53** (48 h, cis–trans = 1:1) (Scheme 5.13B). Starting from either nucleosides **53** or **54**, access to biologically interesting α- and β-4′-methoxy-2′,3′-dideoxy-2′,3′-didehydronucleosides (and, potentially, a number of other 4′-alkoxynucleosides) was then easily provided [28].

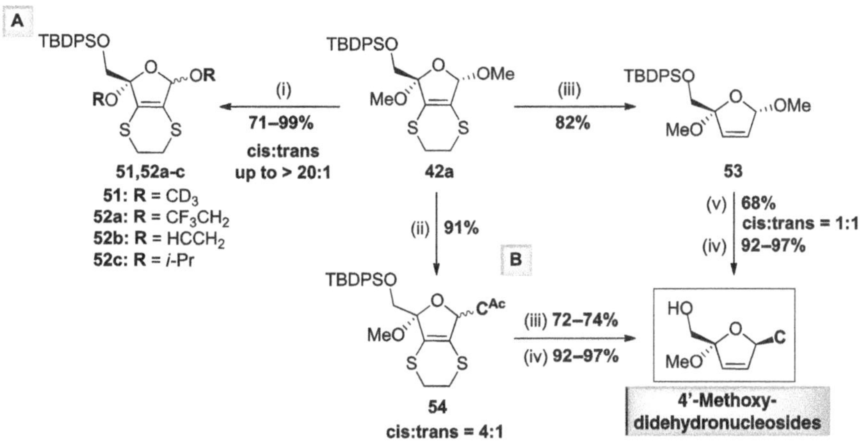

SCHEME 5.13 Reagents and conditions: (i) DDQ (1.8 equivalents), C_6H_6/ROH, rt, 24–48 h; (ii) N^4-AcC, BSA, TMSOTf, dichloroethane (DCE), DMF (6 equivalents), rt, 0.5 h; (iii) Raney-Ni, acetone, 0°C; (iv) (a) tetrabutylammonium fluoride, THF, rt and (b) MeONa, MeOH, rt; and (v) N^4-AcC, BSA, TMSOTf, DCE, DMF (6 equivalents), rt, 48 h.

5.3.4 Eight Steps in One Process: Beyond Achmatowicz Rearrangement

The 2,5-dialkyloxy-2,5-dihydrofuran motif contained in compounds **42a-b** (Schemes 5.10–5.13) has been recognized as a key intermediate in the oxidative ring expansion of furfuryl alcohols (**X** = O) or amines (**X** = NPG) **55** into the corresponding 3-pyrones (X = O) or dihydropyridines (**X** = NPG) **56** (the well-known Achmatowicz rearrangement; Scheme 5.14). Accordingly, we thought to utilize **42b** as an intermediate for additional DDQ-mediated transformations, conveniently leading to Achmatowicz-type products in one step from acyclic starting materials, which is a synthetic target reached only rarely so far [44].

Compared to the reaction conditions reported in Scheme 5.11, larger amounts of DDQ (2.5 equivalents) in a 3/1 DCE/MeOH solution were used in this case, paying attention to warm the reaction mixture (60°C) as soon as conversion of starting alcohols *syn*/*anti*-**43b** into *bis*-acetal **42b** was achieved (48 h). Under these conditions, oxazolidine ring cleavage was sluggish (8 h) oxazolidine ring cleavage was observed, providing **57** in an excellent 91% yield (Scheme 5.15). Surprisingly, the simple

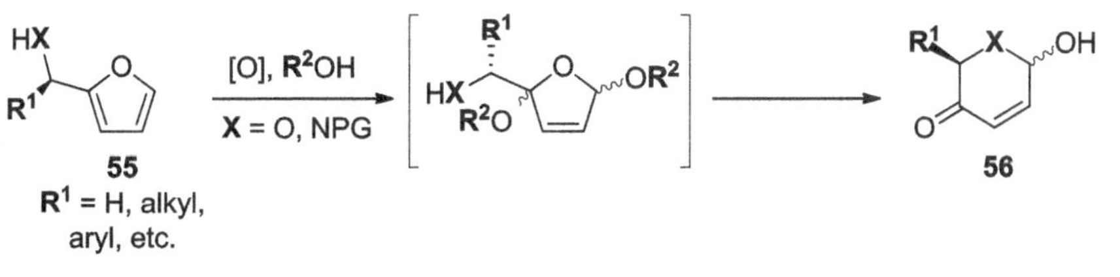

SCHEME 5.14 The Achmatowicz rearrangement.

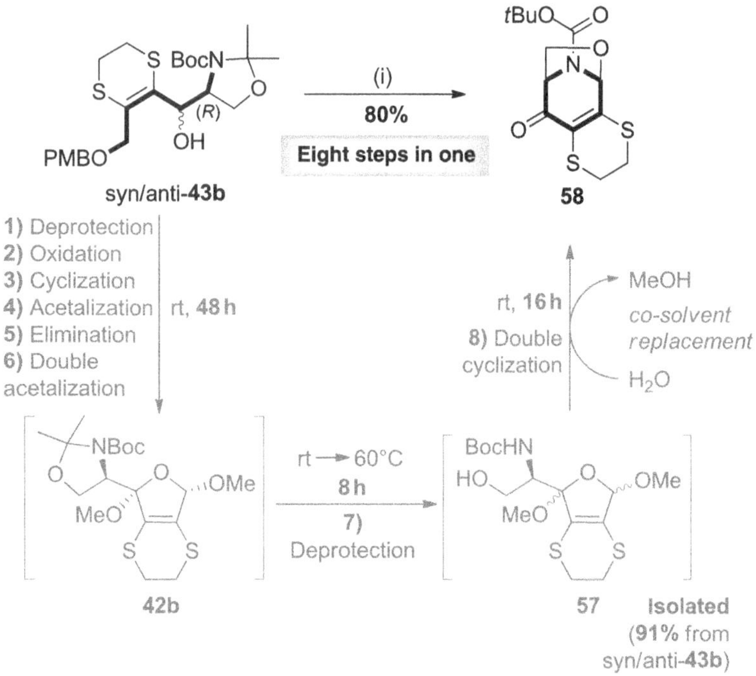

SCHEME 5.15 Reagents and conditions: (i) (a) DDQ (2.5 equivalents), DCE:MeOH = 18:1, rt to 60°C, 56 h, and (b) co-solvent replacement (from MeOH to H_2O), rt, 16 h.

oxazolidine ring hydrolysis of *bis*-acetal **42b** acted as a stereodivergent process, since a mixture of 4 chromatographically separable stereoisomers **57** was obtained from the single species (1*S*,4*S*,5*R*)-**42b**. In line with earlier remarks, this unusual loss in the stereochemical integrity of **42b** during oxazolidine hydrolysis was actually judged as a further proof of the thermodynamically driven equilibrium affecting the acetal functions of dihydrofurans **42a-b**: while the bulky and rigid oxazolidine group of **42b** conceivably gave one main, energetically more stable stereoisomer, the presence of the smaller and conformationally more flexible (2-hydroxyethyl)carbamate moiety in **57** was not supposed to provide any stereoselective effect to the adjacent stereocenters.

With alcohol **57** in hand, one-pot co-solvent change (MeOH was replaced with H_2O) then enabled the conversion (16 h) of the latter into the tricyclic compound **58** (Scheme 5.15). Taken overall, the process is the result of at least eight sequential transformations carried out under domino-one pot conditions: (1-6) as in Scheme 5.11; (7) oxazolidine hydrolysis; and (8) double cyclization [32]. Given the excellent degree of the conversion (80% from *syn*/*anti*-**43b**), this method can be reasonably considered an advantageous alternative to the existing aza-Achmatowicz protocols on the way to various nitrogen-containing heterocycles, including enantiomerically pure iminosugars [45] or other piperidine-based natural products [46].

5.4 CONCLUDING REMARKS

Among the de novo methodologies aiming at the synthesis of carbohydrates and their structurally related analogues, only a few of them can claim to make use of such a powerful and versatile synthon like the (5,6-dihydro-1,4-dithiin-2-yl)methanol (**1**). Besides its primary application as three-carbon homologating agent, the synthetic utility of this heterocyclic system lies in the capacity to activate neighboring functional groups—whether alcohols, ethers, aldehydes, ketones, and acetals—promoting multistep reactions (domino, one-pot or domino-one pot processes) finalized to the rapid assembly of five- and six-membered carbohydrate precursors. The most fruitful results were provided by the combined use of **1** (and its derivatives) with DDQ, exploiting, on the one hand, the synthetic versatility of the latter—enabling various synthetic transformations through SET mechanisms—and, on the other hand, the capacity by the dithiodimethylene moiety of **1** to promote these transformations stabilizing the intermediates generated by such SET reactions. As a result of this synergy, new synthetic routes in glycochemistry have been opened up, such as those leading to rare monosaccharides, nitrogen-containing heterocycles, sugar-modified nucleosides, or nucleic acid analogues. The synthetic potential of **1** carries likewise the attendant expectation of further creative chemistry in this area for the near future.

5.5 ACKNOWLEDGMENTS

The work described in this book chapter has been the result of intense synthetic efforts by undergraduate, postgraduate, PhD students, and postdoctoral researchers carried out over the last 10 years. The authors thank present and past laboratory members for their invaluable contribution to the project. Particularly, the authors convey special appreciation to Dr. Carmela Napolitano, Dr. Stefano D'Errico, Dr. Concetta Paolella, and Dr. Antonio Dell'Isola, whose passion and dedication has allowed to disclose the unexpected chemistry behind this project.

REFERENCES

[1] (a) *Glycoscience: Chemistry and Chemical Biology*; Fraser-Reid, B. O., Tatsuta, K., Thiem, J., Coté, G. L., Flitsch, S., Ito,Y., Kondo, H., Nishimura, S.-I., Yu, B., Eds.; Springer: Berlin, **2008**. (b) *Carbohydrate Chemistry: Chemical and Biological Approaches*, Vol. 38; Rauter, A. P., Lindhorst, T., Eds.; Royal Society of Chemistry: Cambridge, 2012.

[2] (a) Stütz, A. E. *Iminosugars as Glycosidase Inhibitors: Nojirimycin and Beyond*; Wiley-VCH: Weinheim, **1999**. (b) *Iminosugars: From Synthesis to Therapeutic Applications*; Compain, P., Martin, O. R., Eds.; John Wiley & Sons Ltd: West Sussex, 2007.

[3] Cuccarese, M. F.; Li, J. J.; O'Doherty, G. A. In *Modern Synthetic Methods in Carbohydrate Chemistry: From Monosaccharides to Complex Glycoconjugates*; Werz, D. B., Vidal, S., Eds.; Wiley-VCH Verlag GmbH & Co.: KGaA Weinheim, **2014**; pp 1–28.

[4] D'Alonzo, D.; Guaragna, A.; Palumbo, G. *Curr. Org. Chem.* **2009**, *13*, 71–98.

[5] Johnson, R. A.; Sharpless, K. B. In *Catalytic Asymmetric Synthesis, 2nd Edition*; Ojima, I., Ed.; Wiley-VCH: New York, **2000**; pp 357–398.

[6] Palframan, M. J.; Pattenden, G. *Chem. Commun.* **2014**, *50*, 7223–7242.

[7] Trost, B. M.; Crawley, M. L. *Chem. Rev.* **2003**, *103*, 2921–2944.

[8] Grubbs, R. *Tetrahedron* **2004**, *60*, 7117–7140.

[9] Trost, B. M.; Brindle, C. S. *Chem. Soc. Rev.* **2010**, *39*, 1600–1632.

[10] Koeller, K. M.; Wong, C.-H. *Nature* **2001**, *409*, 232–240.

[11] Mlynarski, J.; Gut, B. *Chem. Soc. Rev.* **2012**, *41*, 587—596.

[12] Dragutan, I.; Dragutan, V.; Demonceau, A. *RSC Adv.* **2012**, *2*, 719–736.

[13] Whalem, L. J.; Wong, C.-H. *Aldrichim. Acta* **2006**, *39*, 63–71.

[14] Trost, B. M. *J. Org. Chem.* **2004**, *69*, 5813–5837.

[15] Zhou, M.; O'Doherty, G. *Curr. Top. Med. Chem.* **2008**, *8*, 114–125.

[16] Guo, H.; O'Doherty, G. A. *Angew. Chem. Int. Ed.* **2007**, *46*, 5206–5208.

[17] Wang, H.-Y. L.; Xin, W.; Zhou, M.; Stueckle, T. A.; Rojanasakul, Y.; O'Doherty, G. A. *ACS Med. Chem. Lett.* **2011**, *2*, 73–78.

[18] Dondoni, A.; Marra, A. *Chem. Rev.* **2004**, *104*, 2557–2599.

[19] Enders, D.; Voith, M.; Lenzen, A. *Angew. Chem. Int. Ed.* **2005**, *44*, 1304–1325.

[20] Casiraghi, G.; Battistini, L.; Curti, C.; Rassu, G.; Zanardi, F. *Chem. Rev.* **2011**, *111*, 3076–3154.

[21] Guaragna, A.; Pedatella, S.; Palumbo, G. In *e-Encyclopedia of Reagents for Organic Synthesis*; Paquette, L. A., Ed.; John Wiley & Sons, Inc.: New York, **2008**.

[22] **1** can be obtained in a multigram scale starting from commercial methyl pyruvate through a three-step procedure (86% overall yield) (see Reference 21).

[23] Caputo, R.; Guaragna, A.; Palumbo, G.; Pedatella, S. *J. Org. Chem.* **1997**, *62*, 9369–9371.

[24] Caputo, R.; De Nisco, M.; Festa, P.; Guaragna, A.; Palumbo, G.; Pedatella, S. *J. Org. Chem.* **2004**, *69*, 7033–7037.

[25] Guaragna, A.; Napolitano, C.; D'Alonzo, D.; Pedatella, S.; Palumbo, G. *Org. Lett.* **2006**, *8*, 4863–4866.

[26] Guaragna, A.; D'Errico, S.; D'Alonzo, D.; Pedatella, S.; Palumbo, G. *Org. Lett.* **2007**, *9*, 3473–3476.

[27] (a) D'Alonzo, D.; Guaragna, A.; Napolitano, C.; Palumbo, G. *J. Org. Chem.* **2008**, *73*, 5636–5639. (b) Guaragna, A.; D'Alonzo, D.; Paolella, C.; Napolitano, C.; Palumbo, G. *J. Org. Chem.* **2010**, *75*, 3558–3568.

[28] Paolella, C.; D'Alonzo, D.; Palumbo, G.; Guaragna, A. *Org. Biomol. Chem.* **2013**, *11*, 7825–7829.

[29] Caputo, R.; Guaragna, A.; Palumbo, G.; Pedatella, S.; Solla, F. *Eur. J. Org. Chem.* **2002**, 534–536.

[30] Caputo, R.; Festa, P.; Guaragna, A.; Palumbo, G.; Pedatella, S. *Carbohydr. Res.* **2003**, *338*, 1877–1880.

[31] Paolella, C.; D'Alonzo, D.; Guaragna, A.; Cermola, F.; Palumbo, G. *Tetrahedron Lett.* **2010**, *51*, 6060–6063.

[32] Guaragna, A.; Dell'Isola, A.; D'Errico, S.; Palumbo, G.; D'Alonzo, D. *Tetrahedron Lett.* **2014**, *55*, 7007–7010.

[33] D'Alonzo, D.; Van Aerschot, A.; Guaragna, A.; Palumbo, G.; Schepers, G.; Capone, S.; Rozenski, J.; Herdewijn, P. *Chem. Eur. J.* **2009**, *15*, 10121–10131.

[34] D'Alonzo, D.; Amato, J.; Schepers, G.; Froeyen, M.; Van Aerschot, A.; Herdewijn, P.; Guaragna, A. *Angew. Chem. Int. Ed.* **2013**, *52*, 6662–6665.

[35] Caputo, R.; Palumbo, G.; Pedatella, S. *Tetrahedron* **1994**, *50*, 7265–7268.

[36] Caputo, R.; Longobardo, L.; Palumbo, G.; Pedatella, S.; Giordano, F. *Tetrahedron* **1996**, *52*, 11857–11866.

[37] Guaragna, A.; D'Alonzo, D.; De Nisco, M.; Pedatella, S.; Palumbo, G. *Nucleos. Nucleot. Nucl.* **2007**, *26*, 959–962.

[38] Horita, K.; Tadao, Y.; Tanaka, T.; Oikawa, Y.; Yonemitsu, O. *Tetrahedron* **1986**, *42*, 3021–3028.

[39] Buckle, D. R.; Collier, S. J.; McLaws, M. D. In *e-Encyclopedia of Reagents for Organic Synthesis*; Paquette, L. A., Ed.; John Wiley & Sons, Inc.: New York, **2005**.

[40] D'Alonzo, D.; Guaragna, A.; Van Aerschot, A.; Herdewijn, P.; Palumbo, G. *J. Org. Chem.* **2010**, *75*, 6402–6410.

[41] D'Alonzo, D.; Guaragna, A.; Van Aerschot, A.; Herdewijn, P.; Palumbo, G. *Tetrahedron Lett.* **2008**, *49*, 6068–6070.

[42] D'Alonzo, D.; Guaragna, A.; Palumbo, G. *Chem. Biodivers.* **2011**, *8*, 373–413.

[43] Differently from what commonly reported for the nomenclature of cyclic compounds bearing doubly substituted stereocenters [Cross, L. C.; Klyne, W. *Pure Appl. Chem.* **1976**, *45*, 11–30], for our convenience the "cis/trans" notation will be hereafter employed to indicate the relative configuration of the methoxy groups in **41a** and its achiral congeners.

[44] Ransborg, L. K.; Lykke, L.; Hammer, N.; Næsborg, L.; Jørgensen, K. A. *Chem. Commun.* **2014**, *50*, 7604–7606.

[45] D'Alonzo, D.; Guaragna, A.; Palumbo, G. *Curr. Med. Chem.* **2009**, *16*, 473–505.

[46] Ciufolini, M. A.; Hermann, C. Y. W.; Dong, Q.; Shimizu, T.; Swaminathan, S.; Xi, N. *Synlett* **1998**, 105–114.

6

THIO-CLICK AND DOMINO APPROACH TO CARBOHYDRATE HETEROCYCLES

ZBIGNIEW J. WITCZAK AND ROMAN BIELSKI

Department of Pharmaceutical Sciences, Wilkes University, Wilkes-Barre, PA, USA

6.1 INTRODUCTION

The classical domino reaction proceeds via the formation of the thermodynamically stable intermediate, which is immediately converted into the new, energy-richer cyclization or rearrangement product with an aromatic or heterocyclic structure.

Examples of the formation of the heterocyclic systems from carbohydrate precursors with concurrent cyclization or intermolecular rearrangement often include those in which the core fragment is functionalized with a specific reactive moiety capable of reacting with other highly reactive molecules, such as aldehydes or ketones. The syntheses of fully diversified *N*-, *O*-, heterocyclic systems from carbohydrate precursors and functionalized amino acids templates have been reviewed during the last decade [1–5]. The excellent accounts by Lichtenthaler [1] deal with the synthesis of carbohydrate *N*-heterocycles and by Dondoni [4] with heterocyclic *C*-glycoconjugates synthesized via asymmetric multicomponent reaction using Bignelli and Hantzsch cyclocondensation approaches.

6.2 CLASSIFICATION AND REACTION MECHANISM

The domino reactions can be divided into several categories. The most often reported category is a stereoselective domino-oxa-Michael aldol reaction of α,β-unsaturated ketones with aldehydes.

Domino and Intramolecular Rearrangement Reactions as Advanced Synthetic Methods in Glycoscience, First Edition. Edited by Zbigniew J. Witczak and Roman Bielski.

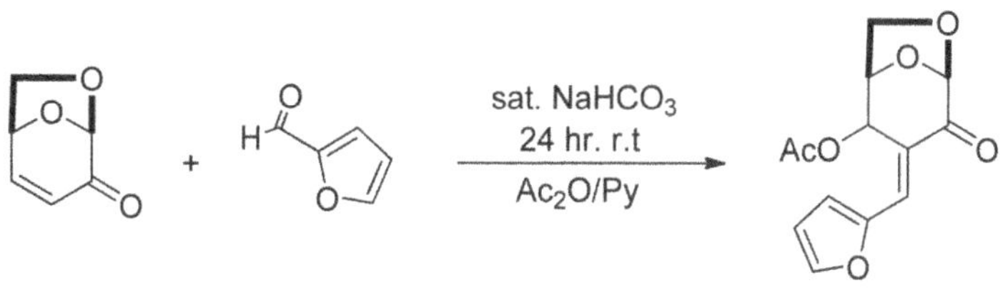

SCHEME 6.1 Levoglucosenone reaction with furfural.

In this category, chiral α,β-unsaturated ketone, levoglucosenone, undergoes a stereoselective base-catalyzed domino reaction with furfural. The reaction was performed in saturated aqueous solution of sodium bicarbonate at room temperature for 24 h. This particular process was serendipitously discovered during the production of levoglucosenone via pyrolysis of cellulose (Scheme 6.1) [6].

Under these conditions, furfural forms a carbon–carbon bond at C-3 position of levoglucosenone. Interestingly, it was also reported that the reaction of benzaldehyde with levoglucosenone under the same conditions was not successful in producing a similar adduct.

In contrast, 2-hydroxybenzaldehydes react with levoglucosenone via stereoselective domino oxa-Michael-aldol reaction as reported by Samet and coworkers [7]. The phenolate-anion attacks from the site opposite to the anhydro-ring of levoglucosenone. The reaction is depicted in Scheme 6.2.

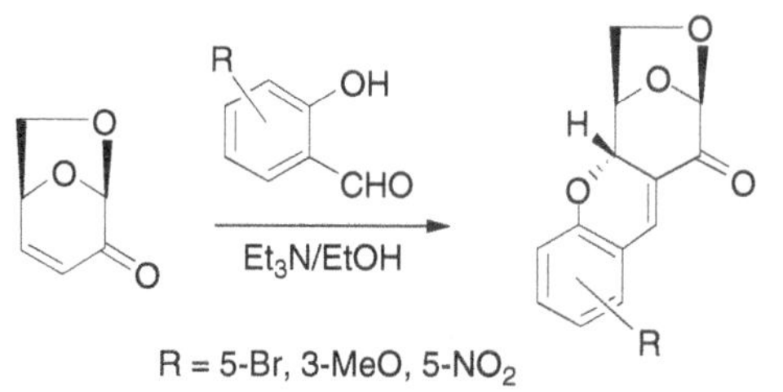

SCHEME 6.2 Stereoselective domino reaction of levoglucosenone with o-hydroxybenzaldehyde.

Our laboratory has developed analogous domino thio-click Michael-aldol approach to heterocyclic thio-chromene systems [8], using 2-thio salicylic aldehyde. The base-catalyzed thio-click reaction proceeds via a conventional stereoselective Michael addition reaction of thiol to C-4 position of levoglucosenone and a concomitant domino cyclization of intermediate thio adduct to form C—C bond at C-3 of functionalized levoglucosenone as illustrated in Scheme 6.3.

The functionalization of C-2 position of dihydrolevoglucosenone via stereoselective addition of Grignard reagents produces ethenyl and ethynyl derivatives, which undergo 3+2 dipolar cycloaddition with the formation of functionalized isoxazoles and 1,2,3-triazole derivatives [9] as depicted on Scheme 6.4.

In a similar domino reaction, acetoxy pyranone and 1,3-dicarbonyls such as acetylacetone undergo Michael addition followed by a concomitant cycloacetalization

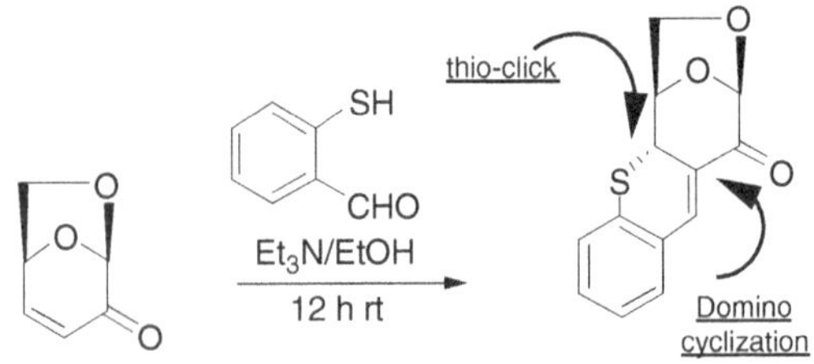

SCHEME 6.3 Stereoselective domino reaction of levoglucosenone with o-thiohydroxybenzaldehyde.

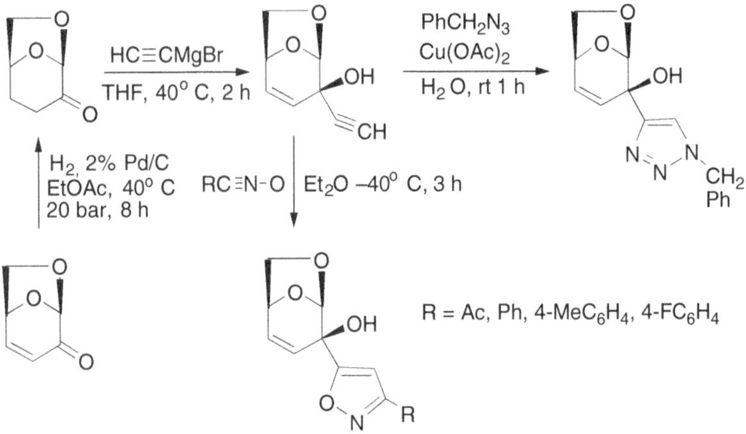

SCHEME 6.4 Formation of functionalized isoxazoles and 1,2,3-triazole derivatives starting from levoglucosenone.

[10]. This new process as reported by Ramasastry [10] proceeds via a classical base-catalyzed cascade Michael addition and cyclization of formed enols with *in situ*-generated oxonium ions. Interestingly, pyranones having a quaternary carbon produced the bicyclic furopyranones in good yields (75–80%). The reaction is depicted in Scheme 6.5.

The intramolecular azide-olefin [3+2] cycloaddition of functionalized carbohydrate scaffolds to benzooxazocines and benzooxazocines via a domino Cu-catalyzed aryl azidation with subsequent stereoselective aziridine formation was recently reported by Chattopadhyay and coworkers [11]. This novel strategy is illustrated in Scheme 6.6.

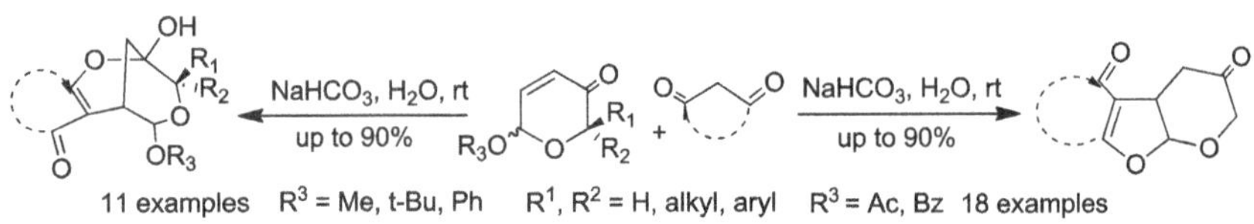

SCHEME 6.5 Synthesis of bicyclic furopyranones starting from carbohydrate enones.

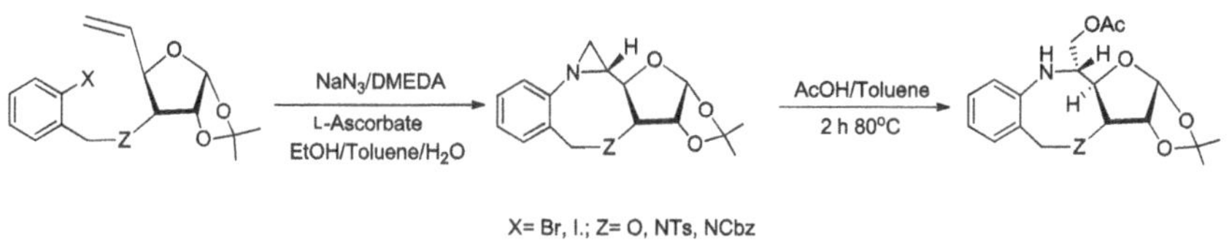

SCHEME 6.6 Itramolecular azide-olefin [3+2] cycloaddition of functionalized carbohydrate scaffolds to benzooxazocines and benzodiazocines.

The use of *N,N*-dimethylethylenediamine (DMEDA) as a base catalyst in an aqueous solution of ethanol/toluene for the stereoselective azidation under click reaction conditions (copper (II) salts and L-ascorbate) is the very efficient way to accomplish this [3+2] cycloaddition. The intermediate addition product formed under the reaction conditions is an unstable triazoline, which immediately decomposes (with the loss of nitrogen) with the exclusive formation of aziridines-fused furobenzoxazocine or benzodiazocine as depicted in Scheme 6.7.

Bauman and Axendale [12] reported a new approach to synthesized functionalized thioimidazoles via three-component reaction using unprotected aldoses and ketoses. This sequential approach is based on the Marckwald reaction as illustrated in Scheme 6.8.

An interesting approach to synthesizing benzopyrans starting form 3,4,6,-tri-*O*-acetyl-D-galactal was reported by Williams and coworkers [13]. The reaction proceeds via aluminum triflate-catalyzed *C*-coupling of various functionalized phenols with unsaturated D-galactal system. The synthetic route to this new class of *C*-functionalized benzopyrans is illustrated in Scheme 6.9.

Mahrwald and coworkers [14] reported a new synthetic approach to carbohydrate heterocycles via three-component reactions starting from unprotected maltose, L-proline, and ethyl isocyanoacetate. The process is catalyzed by *N,N*-diisopropylethylamine (Huning's base) and produces glycopeptide mimetic as functionalized lactone, which was isolated in a 36% yield as syn/anti mixture 83/17. The methodology is depicted in Scheme 6.10.

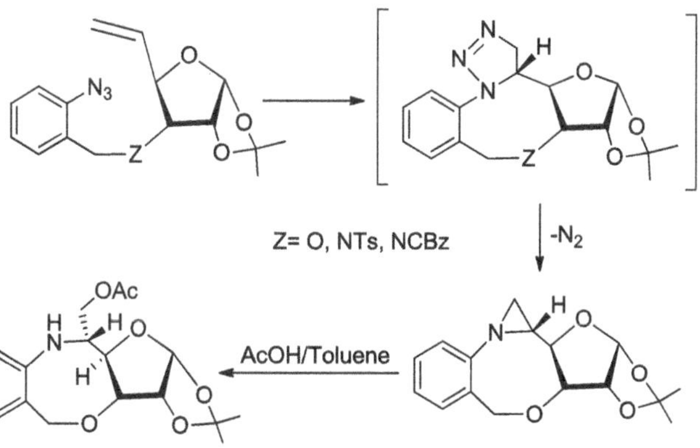

SCHEME 6.7 Formation of aziridines-fused furobenzoxazocine or benzodiazocine.

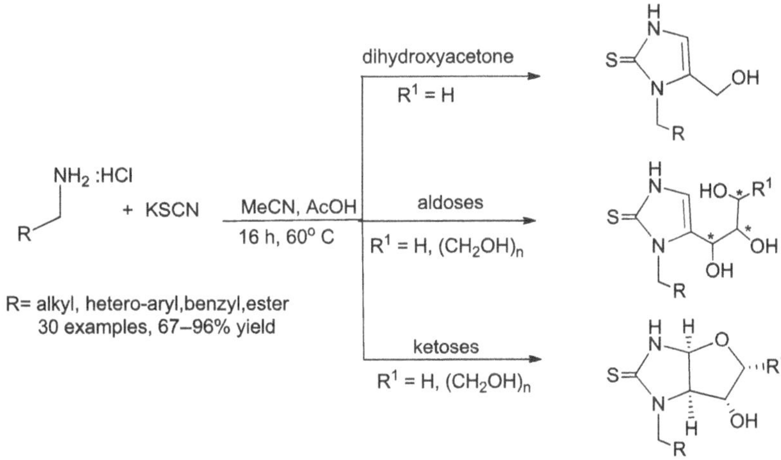

SCHEME 6.8 Synthesis of functionalized thioimidazoles via three-component reaction using unprotected aldoses and ketoses.

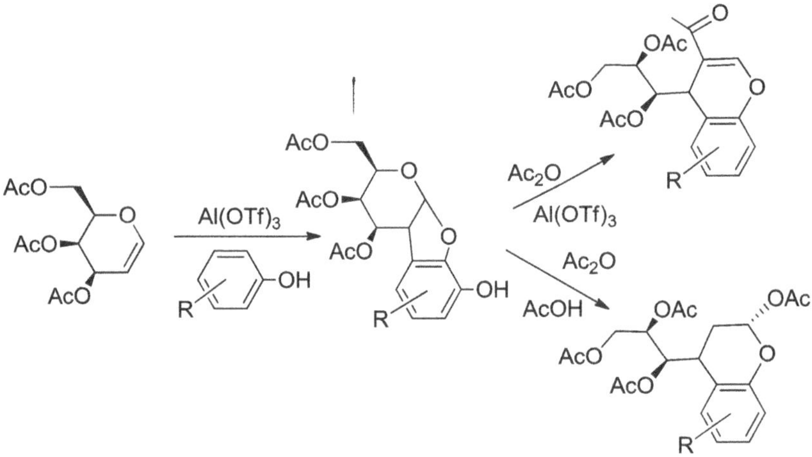

SCHEME 6.9 Williams's synthesis of benzopyrans starting form 3,4,6,-tri-Oacetyl-D-galactal.

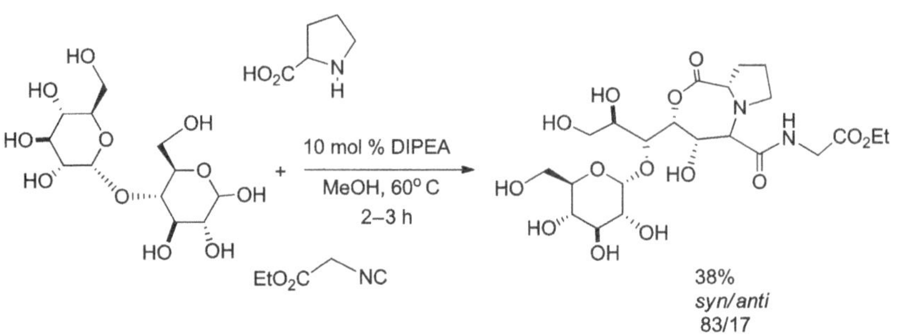

SCHEME 6.10 Mahrwald's synthesis of carbohydrate heterocycles starting from maltose, L-proline, and ethyl isocyanoacetate.

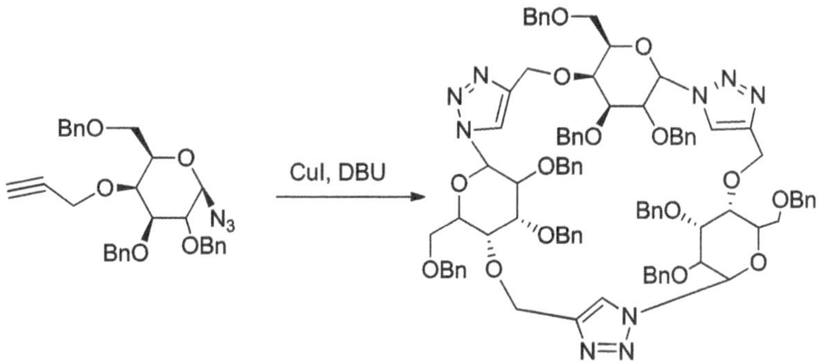

SCHEME 6.11 Synthesis of C_3- and C_2-symmetric oligosaccharide macrocycles starting from anomeric azide with 4-propargyl ether at the opposing terminus.

Gin and coworkers [15] described a highly convergent strategy for the synthesis of C_3- and C_2-symmetric oligosaccharide macrocycles starting from anomeric azide with 4-propargyl ether at the opposing terminus. The reaction proceeds via triple CuI-catalyzed Huisgen cycloaddition with the formation of a macrocycle as symmetrical cyclotrimer is illustrated in Scheme 6.11.

Analogously, functionalized α-(1,4)-linked mannose unit as anomeric azide with 4′-propargyl ether at opposite end undergoes a similar cycloaddition with the formation of cyclodimer in an 89% yield. Interestingly, molecular modeling of these macrocycles clearly indicates their sterically congested cavities. That would be important for the installation of orthogonal protecting groups handles allowing for the site-selective manipulation and their practical use for supramolecular chemistry.

A very interesting new concept of using domino reactions to access highly substituted chromans and isochromans from carbohydrates was published by Werz and coworkers [16] (see also Chapter 8). The synthetic methodology is depicted in Scheme 6.12.

The concept was further summarized and considered as exploring a new field of chemical biology and highlighted in "Hot News & Views" section of *Nature Chemical Biology* [17].

6.3 CONCLUSIONS

These recent applications of diverse domino reactions in the synthesis of many functionalized carbohydrates are in accordance with many great developments in

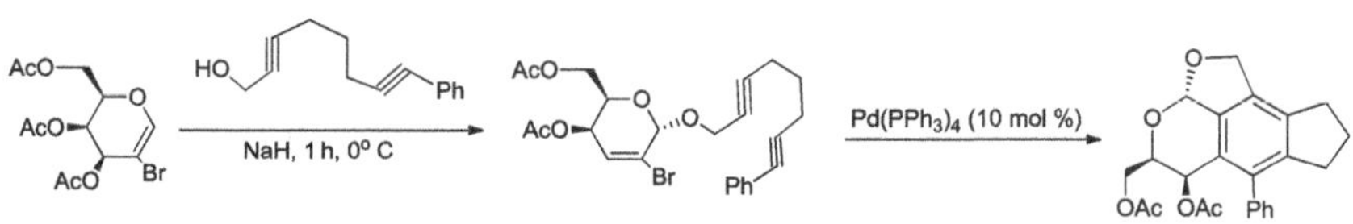

SCHEME 6.12 Werz's synthesis of highly substituted chromans and isochromans from carbohydrates.

organic synthesis of the last decade. Generally, rearrangement/domino approaches provide mild reaction conditions for the formation of various new bonds such as C—C, C—N, C—S, and this type of new bond formation is exploited in many syntheses of carbohydrate heterocycles and *C*- and *N*-glycoconjugates chemistry. The novel developments in synthetic organic chemistry enabling applications to specific synthetic targets, including many natural carbohydrate heterocycles, will be steadily growing in the next decade. All of the new strategies currently available or under development will constitute a significant milestone in the area of glycoscience and carbohydrate therapeutics. Therefore, the development of these new carbohydrate synthetic methodologies is of utmost importance and will be closely followed as many new biological targets will constitute promising prospects for the future of glycoscience.

REFERENCES

[1] Lichtenthaler, F. W. *Acc. Chem. Res.* **2002**, *35*, 728–737.

[2] *Synthesis of Naturally Occurring Nitrogen Heterocycles From Carbohydrates*; El Sayed, H., ElAshry, A., El, Nemr, Eds.; Blackwell Publishing: Oxford, **2005**.

[3] Heterocycles from carbohydrate precursors. In *Topics in Heterocyclic Chemistry, 7*; Springer: Heidelberg, **2007**.

[4] Dondoni, A.; Massi A. *Acc. Chem. Res.* **2006**, *39*, 451–463.

[5] Osborn, H. M. I.; Gemmell, N.; Harwood, L. M. *J. Chem. Soc., Perkin Trans.* **2002**, *1*, 2419–2438.

[6] Nishikawa, T.; Araki, H.; Isobe, M. *Biosci. Biotechnol. Biochem.* **1998**, *62*, 190–192.

[7] Samet, A. V.; Lutov, D. N.; Firgang, S. I.; Lyssenko, K. A.; Semenov, V. V. *Tetrahedron Lett.* **2011**, *52*, 3026–3028.

[8] Witczak, Z. J. Thio-click and domino reaction tandem approach to carbohydrate heterocycles, abstracts of 248th ACS National Meeting, August 12–17, 2014, San Francisco, CA, Abstract CARB-23.

[9] Brel, V. K.; Samet, A. V.; Konyushkin, L. D.; Stash, A. I.; Belsky, V. K.; Semenov, V. V. *Mendeleev Commun.* **2015**, *25*, 44–46.

[10] Kasare, S.; Bankar, S. K.; Ramasastry, S. V. *Org. Lett.* **2014**, *16*, 4284–4287.

[11] Bhattacharya, D.; Ghorai, A. U.; Pal, N.; Maiti, C.; Chattopadhyay, P. *RSC Adv.* **2014**, *4*, 4155–4162.

[12] Baumann, M.; Baxendale, I. R. *Org. Lett.*, **2014**, *16*, 6076–6079.

[13] Simelane, S. B.; Kinfe, H. H.; Muller, A.; Williams, B. G. *Org. Lett.*, **2014**, *16*, 4543–4545.

[14] Voigt, B.; Linke, M.; Mahrwald, R. *Org. Lett.* **2015**, *17*, 2606–2609.

[15] Bodine, K. D.; Gin, D. Y.; Gin, M. S. *Org. Lett.* **2005**, *7*, 4479–4482.

[16] Leibeling, M.; Koester, D. C.; Pawliczek, M.; Schild, S. C.; Werz, D. B. *Nat. Chem. Biol.* **2010**, *6*, 199–201.

[17] Young, D. W. *Nat. Chem. Biol.* **2010**, *6*, 174–175.

7

CONVERTIBLE ISOCYANIDES: APPLICATION IN SMALL MOLECULE SYNTHESIS, CARBOHYDRATE SYNTHESIS, AND DRUG DISCOVERY

SOUMAVA SANTRA, TONJA ANDREANA, JEAN-PAUL BOURGAULT, AND PETER R. ANDREANA

Department of Chemistry and Biochemistry and School of Green Chemistry and Engineering, University of Toledo, Toledo, OH, USA

7.1 INTRODUCTION

Isocyanides, also known as isonitriles or carbylamines, are characterized by a primary isocyanide (R-NC) functional group and pungent odor. According to the Hückel, MNDO, and *ab initio* molecular orbital calculations, the electronic configuration of isocyanide is represented by an iminocarbene resonance hybrid (Eq. 7.1) [1]. Experimental studies have confirmed that the resonance hybrid is closely represented by the triple-bonded resonance structure [2]. Electronically, isocyanides resemble that of carbon monoxide (Eq. 7.2) and recently, isocyanides have been used as substitutes for carbon monoxide in organometallic transformations [3].

$$\underset{\mathbf{1}}{\text{R–NC}} \;\Bigg|\; \underset{\mathbf{1a}}{\text{R–}\ddot{\text{N}}\text{=C:}} \longleftrightarrow \underset{\mathbf{1b}}{\text{R–}\overset{\oplus}{\text{N}}\equiv\overset{\ominus}{\text{C}}\text{:}} \tag{7.1}$$

$$\underset{\mathbf{2}}{\text{CO}} \;\Bigg|\; \underset{\mathbf{2a}}{\text{:}\ddot{\text{O}}\text{=C:}} \longleftrightarrow \underset{\mathbf{2b}}{\overset{\oplus}{\text{O}}\equiv\overset{\ominus}{\text{C}}\text{:}} \tag{7.2}$$

Domino and Intramolecular Rearrangement Reactions as Advanced Synthetic Methods in Glycoscience, First Edition. Edited by Zbigniew J. Witczak and Roman Bielski.

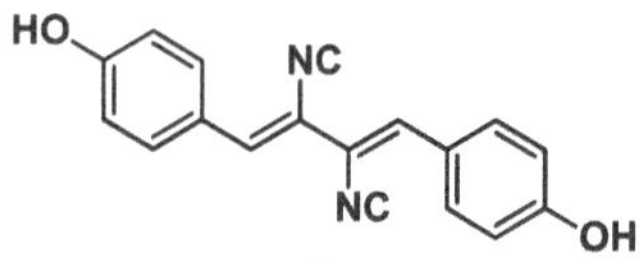

3

FIGURE 7.1 Xanthocillin.

Isocyanides are versatile functional groups due to the presence of a divalent carbon atom—the site of protonation [4]. A limited number of compounds belong to the same class of proton acceptors as do isocyanides, which make prevalent their uniqueness.

Isocyanides are quite stable to basic conditions but do not bode well in acidic media; they are known to hydrolyze, oligomerize, or even polymerize to helical polymers [5]. Isocyanides are also characterized by their ease of radical formation and undergo isomerization under thermal conditions to the corresponding cyanide, thus the prefix *iso* [6, 7].

The first reported isocyanide synthesized was by Lieke in 1859, when he tried to make allyl cyanide using allyl iodide and silver cyanide [1a]. What he synthesized instead was a foul smelling compound that he first believed to be a nitrile. It would be years before the structures of isocyanides were formally identified [1c, 2]. The first naturally occurring isocyanide was discovered by Rothe in 1950 and later used as the antibiotic xanthocillin (**3**, Figure 7.1) [8].

Gautier and Hoffman developed the first syntheses of isocyanides in 1867 (Scheme 7.1). Gautier's work with isocyanides pioneered the discovery of the isomeric relationship between isocyanides and nitriles [9]. Hoffman's synthesis is currently used as a chemical test for primary amines.

Other common syntheses include the reduction of isocyanates [10] and nucleophilic displacement of alcohols [11]. Yet the most common method to synthesize isocyanides is via the dehydration of formamides (Scheme 7.2). This route (Method A), though more efficient than the previous syntheses, is undesirable due to the use

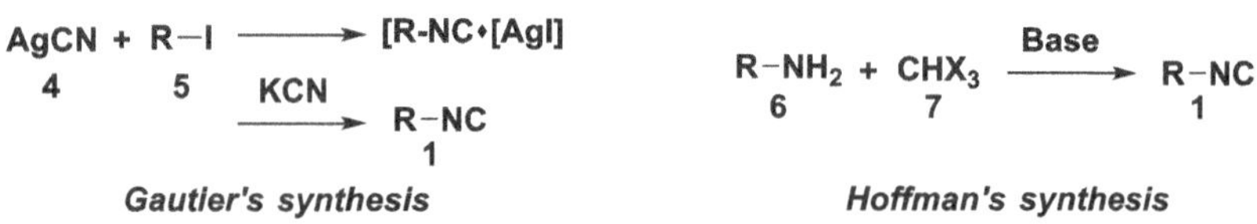

SCHEME 7.1 Syntheses of isocyanides by Gautier and Hoffman.

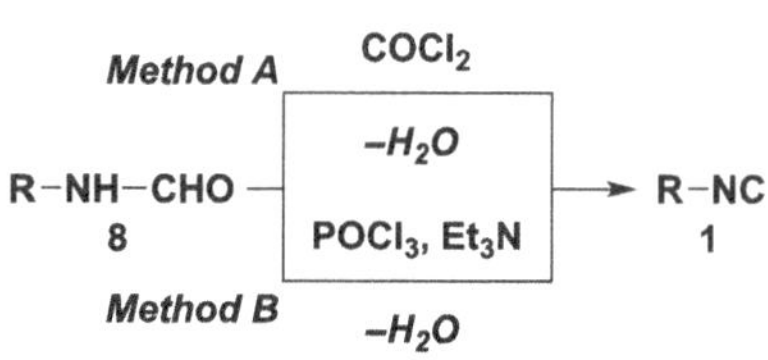

SCHEME 7.2 Dehydration of formamides for the synthesis of isocyanides.

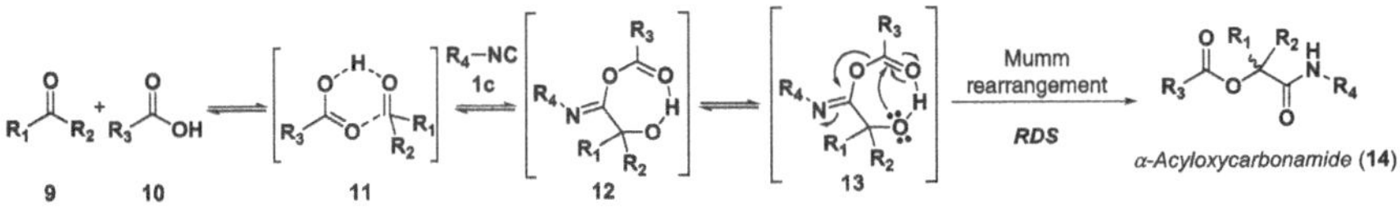

SCHEME 7.3 The Passerini reaction.

of phosgene [12]. van Leusen and coworkers later developed an alternative method (Method B) of dehydration using phosphorous oxychloride [13].

The synthesis of isocyanides was slow to evolve, consisting of only 12 that were documented before 1960. Their offensive odor made them an unappealing target to investigate even though researchers were intrigued by their reactivity. Several synthetic methodologies have been reported in the last half century [14]. With the emergence of increased efficient methodologies, more than two dozen isocyanides are now commercially available and hundreds more have been reported in the literature.

A peak in the interest of isocyanides began with the commencement of their use in multicomponent reactions (MCRs). MCRs incorporate at least three starting materials, which react to form a product in such a fashion that a majority of the atoms of the starting materials can be found in the product. MCRs have been frequently used by synthetic chemists as a facile means to generate molecular diversity from multifunctional substrates that react subsequently in an intramolecular fashion [15]. MCRs are perhaps the most promising, efficient, and powerful method for covering maximum chemical space by generating varying skeletal diversity. As processes that avoid time-consuming and costly purification procedures, as well as protection-deprotection steps, MCRs are inherently more environmentally benign and atom economical [16].

Passerini introduced the first isocyanide-based MCR, utilizing *p*-isocyanide-azobenzene, in 1921. His work was centered around three-component reactions (3CRs), converting carbonyl compounds **9**, carboxylic acids **10**, and isocyanides **1c** into α-acyloxycarbonamides **14** (Scheme 7.3) [17]. More recently, Maeda and coworkers have proposed that the P-3CR is a four-component reaction (4CR) via an artificial force-induced reaction method [18].

As more isocyanides became available, Ivor Ugi began working on 4CR s with isocyanides, introducing his first isocyanide multicomponent coupling reaction (IMCR) in 1959 [19]. The four components of Ugi's reaction (Scheme 7.4) are aldehydes **9a**,

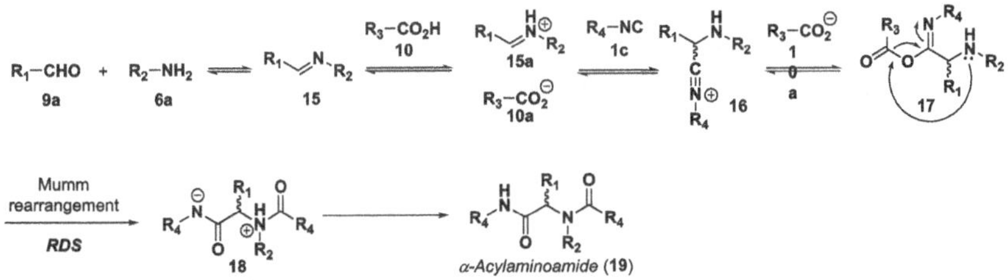

SCHEME 7.4 The Ugi reaction.

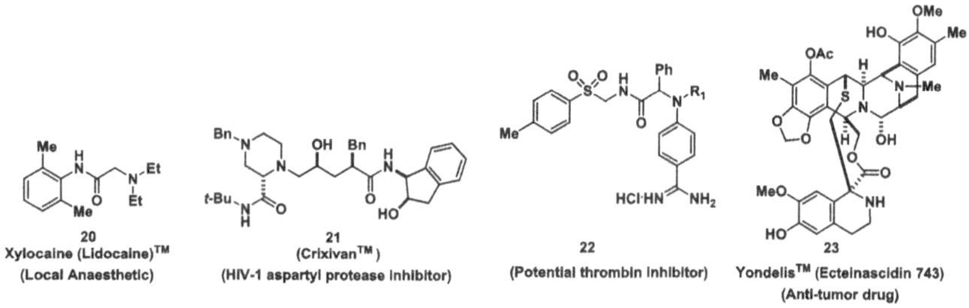

FIGURE 7.2 U-4CR in drug discovery and synthesis.

amines **6a**, acids **10**, and isocyanides **1c**. This "aza-version" of the P-3CR provides α-acylaminoamide **19** as the final product and it is known as the U-4CR.

As work progressed on the Ugi reaction (U-4CR), the usefulness of the variability of the components and products was quickly realized. Today, MCRs have emerged as powerful tools for the synthesis of natural products, small molecules, combinatorial syntheses, and diversity-oriented synthesis [20]. More importantly, the U-4CR has found ample application in the drug discovery process in recent years (Figure 7.2) [21]. Ironically, the U-4CR has provided improved yields of intermediates where traditional synthetic methods have failed. With the recent advancement of high throughput screening [22], MCRs have become a rapid platform for the drug development and discovery process among scientists in academia as well as industry [23].

The limited commercial availability, often challenging synthesis, volatility, and offensive odors of isocyanides had hindered the use of the Ugi reaction despite enormous appeal. Ugi admits that "the development of the chemistry of isonitriles has probably suffered through the characteristic odor of volatile isonitriles, which has been described by Hofmann and Gautier as "'highly specific, almost overpowering', 'horrible', and 'extremely distressing'" [24]. People who have inhaled volatile isocyanides, such as allyl, benzyl, methyl, or *tert*-butyl isocyanide (TBIC), over a long period of time report the sensory perception of the smell of hay. Long-term inhalation has also been said to increase the intensity of dreams at night [15a].

Other liquid isocyanides, such as the positional isomeric picolyl isocyanides, on the contrary, do not smell at all. The isocyanide derived from L-phenylglycine methyl ester smells of rhubarb. Toluenesulfonylmethyl isocyanide (TosMIC) and adamantyl isocyanide are not as offensive as others. To avoid the distressing smell of isocyanides, there is a continuous effort among the scientific community to develop fragrant smelling isocyanides.

The Ugi reaction is also limited in its postcondensation modification. Commercially available isocyanides introduce a secondary amide functionality into the reaction that can be difficult to hydrolyze in the absence of harsh conditions, which represents a problem for highly functionalized compounds. Thus, in recent years, there has been a growing interest among researchers to develop "convertible isocyanides" (CICs), or isocyanides with an easily removable protecting group (PG). These CICs (at times referred to as "universal isocyanides") eliminate the need to synthesize and store unstable isocyanides and allow for the expanded scope of Ugi reactions with the ability to "convert" in a single step to a variety of easily obtained products and enhanced diastereoselectivity, when applicable. CICs have been found to be useful for the synthesis of small molecules as well as complex natural products having biological activities related to human health.

7.2 CONVERTIBLE ISOCYANIDES

CICs are isocyanides containing a PG that allows for selective cleavage of the *N*-terminal amide Ugi postcondensation product without interfering with the other amide moiety contained within the α-acylaminoamide (**19**, Scheme 7.4). However, this strategy has been less successful for post-Passerini modification due to the presence of more reactive ester group (**14**, Scheme 7.3). The attractiveness of the CIC lies within its ability to use a variety of conditions for the amide cleavage. CICs allow for the introduction of relatively unavailable input in library synthesis by providing it in a convertible form.

7.2.1 CIC Employed in the Ugi Reaction

7.2.1.1 p-Tolylsulfonylmethyl Isocyanide TosMIC (**1e**) was purportedly synthesized in 1967 and published for the first time by van Leusen and Strating in 1970 [25]. Preparation from methyl isocyanide (**1d**) and *p*-tolylsulfonyl chloride (**24**) in the presence of ^{n}BuLi (Scheme 7.5) provides the compound in an excellent yield. TosMIC is a stable, odorless, white solid and can be stored at room temperature for extended periods of time.

TosMIC is characterized by the α-acidity of the methyl group and α-anion stabilizing effect of tosyl and isocyanide groups and the electrophilicity of the isocyanide group. Since the time it was discovered, the use of TosMIC has been expanding rapidly and it has found ample application in organic synthesis as a versatile building block. α-Metallated TosMIC possesses a nucleophilic center that can attack polar bonds and the isocyanide group that allows subsequent heterocyclization process. Thus, TosMIC has found widespread application in the synthesis of heterocycles, especially imidazoles (**26**), tosyl substituted thiazoles (**27**), pyrroles (**30**), oxazoles (**32**), and α-hydroxyaldehydes (**34**) (Scheme 7.6) [26].

The majority of these reactions take place in the presence of a suitable base (e.g., ^{t}BuOK, K_2CO_3, NaH). The shortcoming of the methodology is that 4-ethoxy-2-oxazolines (e.g., **33**, Scheme 7.6), which are important intermediates for the synthesis

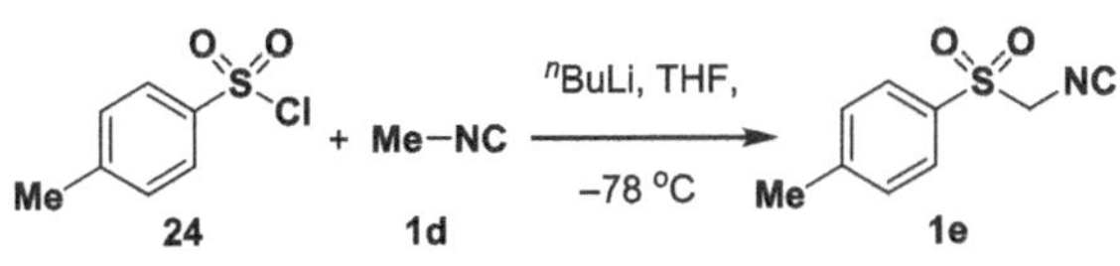

SCHEME 7.5 Synthesis of TosMIC.

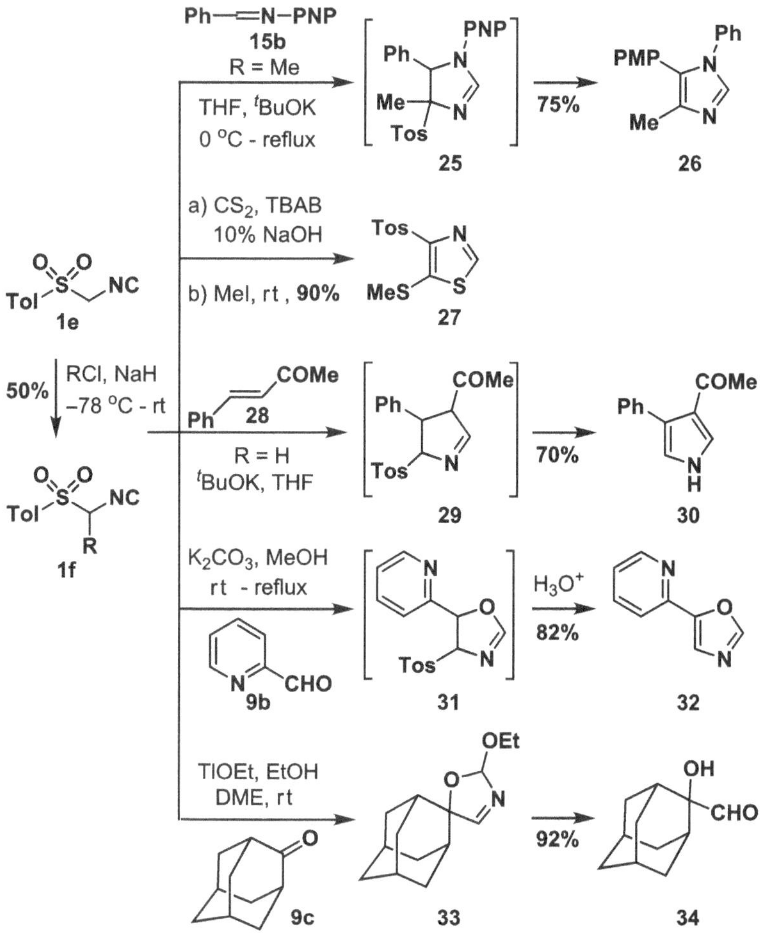

SCHEME 7.6 Application of TosMIC (**1e**) in heterocycle synthesis.

of α-hydroxyaldehydes (e.g., **34**, Scheme 7.6), can be synthesized only in the presence of highly toxic thallium ethoxide.

van Leusen and coworkers have extended the use of TosMIC to modify side chains of 17-oxosteroids (**35**) to synthesize pharmacologically useful 17-(hydroxyacetyl)-Δ^{16}-steroids (**37**) by employing a phase transfer catalyst (Scheme 7.7) [27].

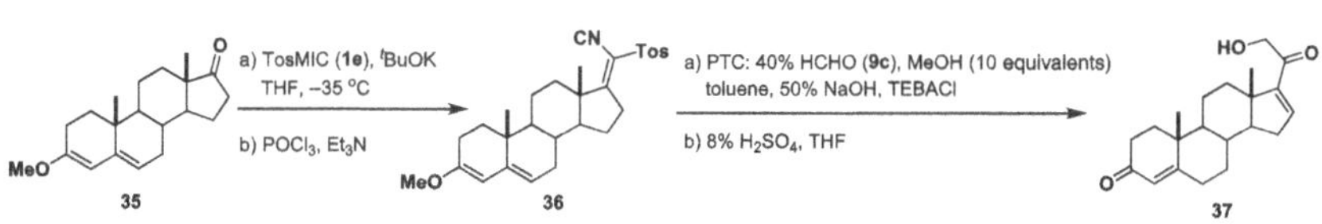

SCHEME 7.7 Application of TosMIC (**1e**) in steroid modifications.

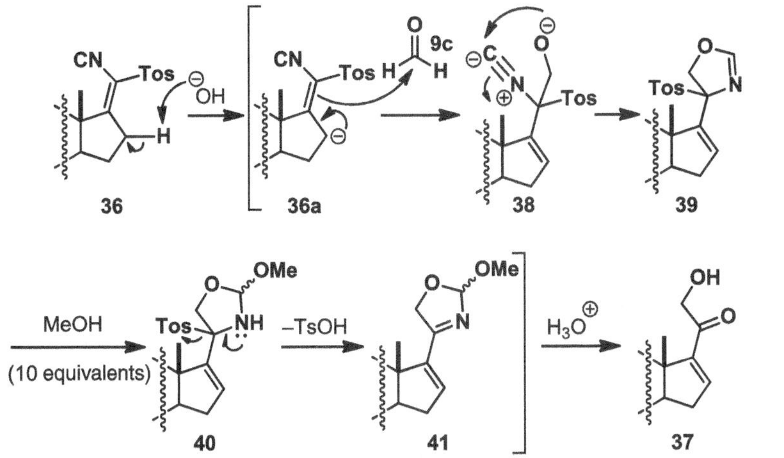

SCHEME 7.8 Mechanism of formation of **37**.

The first step of this method involves a Knoevenagel-type condensation of TosMIC to provide intermediate **36a** [28]. It is noteworthy that the conditions used helped (a) cope with the lower activity of the 17-oxo group of **35** and (b) avoid elimination of sulfinic acid. The second step involves the condensation of **36a** with formaldehyde (**9c**) to form intermediate **38**, which then undergoes an internal nucleophilic attack onto the isocyanide carbon to generate the oxazoline intermediate **39**. Addition of 10 equivalents of MeOH to the phase transfer catalyst directly generates oxazoline **41** via the elimination of tosyl sulfonic acid from **40**. Acid-catalyzed hydrolysis provides the target compound **37** (Scheme 7.8).

TosMIC has also found application in natural product synthesis. Yadav et al. have utilized TosMIC in their synthesis of Tricos-9Z-ene (muscalure, **47**), which is a sex pheromone of common house fly [29]. The key step (Scheme 7.9) in the synthesis is the reduction of the dialkylated TosMIC (**46**) in the presence of Li/liquid NH_3 to obtain muscalure (**47**).

A similar approach was employed by Yadav and coworkers [29] in the synthesis of (+)-1*R*,5*S*,7*R*-*exo*-Brevicomis (**53**), which is the active pheromone in the Western pine beetle (Scheme 7.10).

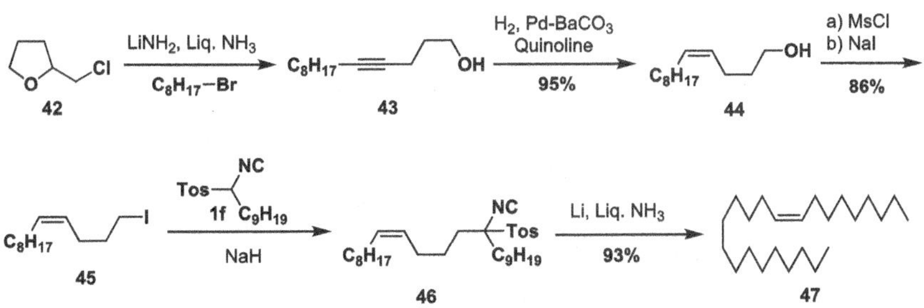

SCHEME 7.9 Modified TosMIC (**1f**) in natural product synthesis.

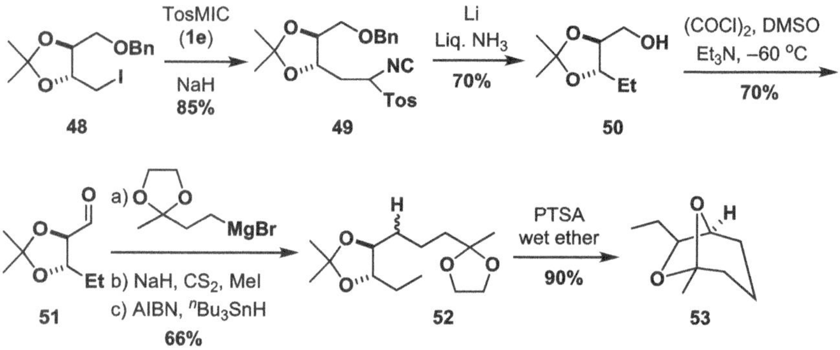

SCHEME 7.10 Synthesis of (+)-1*R*,5*S*,7*R*-*exo*-Brevicomis (**53**).

Recently, Varquero and coworkers employed TosMIC in their synthesis of the core structure of **54** (Scheme 7.11) for access into the Variolins [30] (**55–57**), a family of alkaloids isolated from the sponge *Kirtpatrickia variolosa* Kirkpatrick [31]. Variolins contain the rare pyrrolo[1,2-*c*]pyrimidine system and also claim to have antiviral and antiproliferative activities against P388 leukemia cells [31b].

The mechanism of this unusual cyclocondensation reaction involves the initial nucleophilic attack of the TosMIC anion to the bromomethylazole, followed by an intramolecular transfer of the *N*-protecting methoxycarbonyl to the TosMIC giving rise to **63**. Subsequent intramolecular cyclization of **63** followed by a 1,2-elimination of toluenesulfonate affords the azolopyrimidine derivative **60** (Scheme 7.12).

More recently, Di Santo et al. utilized TosMIC to synthesize 2*H*-pyrrolo[3,4-*c*]quinolines [32]. Michael addition of TosMIC onto **65** and subsequent cyclization yielded pyrrole derivative **66**. Reductive cyclization followed by reduction of the amide and oxidation by MnO_2 provided the tricyclic product **67** (Scheme 7.13).

7.2.1.2 1-Oxy-2-picolyl Isocyanide 1-Oxy-2-picolyl isocyanide (**1g**) was first synthesized by Mizuno and coworkers in 1980 [33]. Commercially available

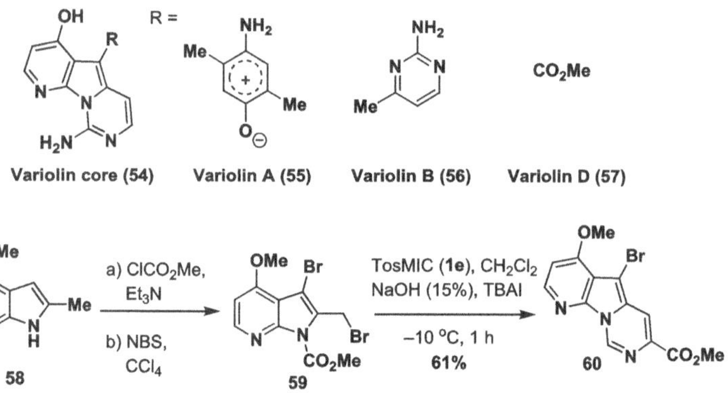

SCHEME 7.11 Synthesis of Variolin core **60**.

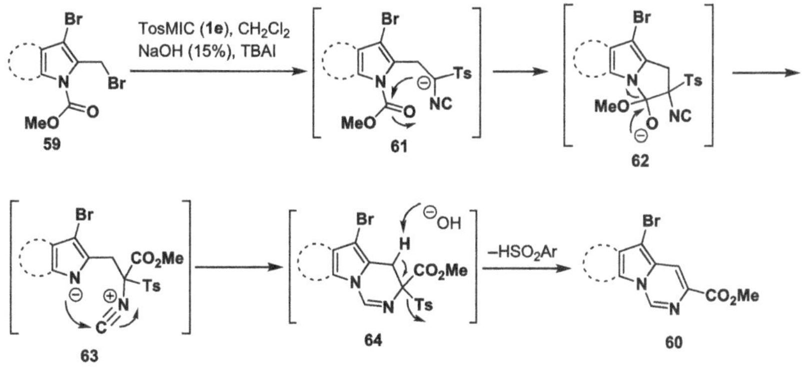

SCHEME 7.12 Mechanism of formation of **60**.

2-pyridylmethyl alcohol (**68**) can be converted to the corresponding 2-picolylamine 1-oxide (**6b**) via meta-chloroperoxybenzoic acid (*m*-CPBA) oxidation followed by treatment with saturated NH_4OH at 0°C and passing through DOWEX 1 (OH^- exchange resin). Refluxing **6b** in a solution of ethyl formate provides *N*-(2-picolyl 1-oxide) formamide (**8a**) exclusively. Refluxing the formamide **8a** in the presence of CCl_4, Et_3N, and triphenylphosgene at 60°C for 3 h provides the desired 1-oxy-2-picolyl isocyanide (OPIC) (**1g**) in a 71% yield (Scheme 7.14). It is noteworthy that usual dehydration methods such as the phosphoryl chloride-pyridine method, trialkylamine method, and the cyanuric chloride-K_2CO_3 method do not provide the desired product **1g**.

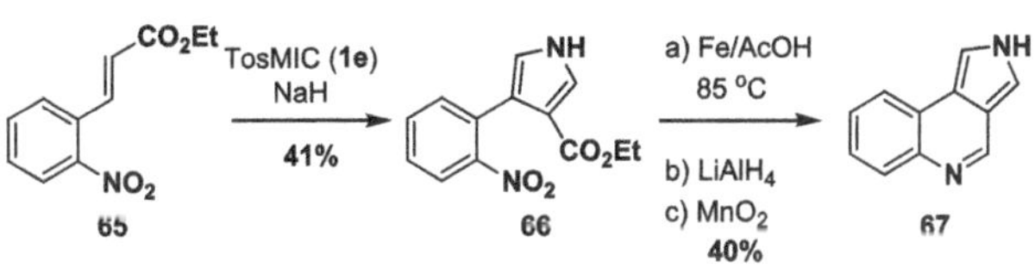

SCHEME 7.13 Application of **1e** to isoquinoline synthesis.

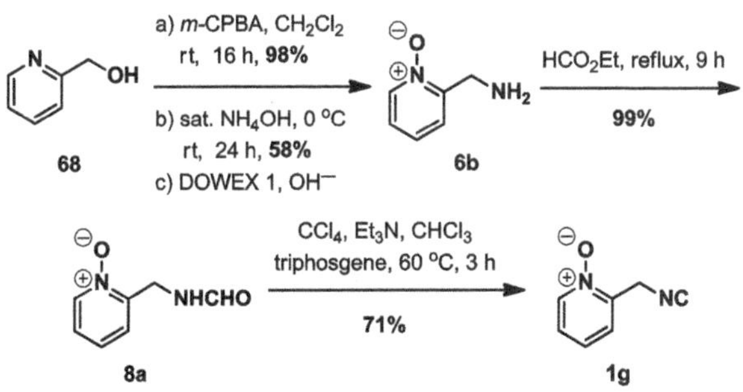

SCHEME 7.14 Synthesis of OPIC (**1g**).

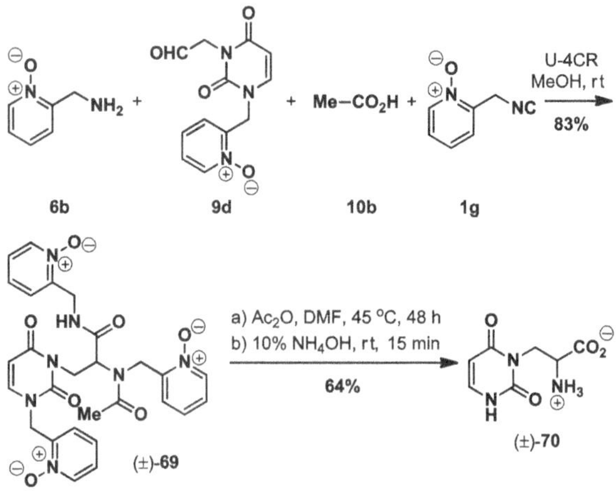

SCHEME 7.15 Synthesis of (±)-isowillardiine (**70**).

Mizuno and coworkers have applied OPIC (**1g**) in the one-pot synthesis of potential antimetabolite (±)-isowillardiine (**70**) via the U-4CR [33]. (±)-Isowillardiine is produced during the germination of pea seeds and its biosynthesis has drawn the attention of the scientific community. Since *N*-benzyl groups have proven to be difficult to remove and not compatible with other functionalities under hydrogenation conditions, Mizuno's strategy was to use 1-oxy-2-picolyl (OP) group in the U-4CR, which can be removed under mild conditions using Ac_2O and methanolic ammonia [34]. Thus, 2-picolylamine 1-oxide (**6b**), 1-(2-picolyl 1-oxide)-3-(formylmethyl)uracil (**9d**), acetic acid (**10b**), and OPIC (**1g**) provided the Ugi product **69** in an 83% yield. The OP groups were cleaved simultaneously by heating (±)-**69** in dimethylformamide (DMF) and Ac_2O followed by treatment with 10% NH_4OH solution (Scheme 7.15).

7.2.1.3 Diphenylmethyl Isocyanide Diphenylmethyl isocyanide or benzhydryl isocyanide (**1h**) is known as the Hofheinz (Isenring) CIC. This odorless crystalline isocyanide can be easily prepared from *N*-(diphenylmethyl)-formamide (**8b**) using thionyl chloride and sodium carbonate (Scheme 7.16) [35].

In 1981, Isenring and Hofheinz discovered that diphenylmethyl isocyanide can act as a CIC when the Ugi product was treated with N_2O_4 in $CHCl_3$ at 0°C and warmed to room temperature [36]. The Ugi product α-acylaminoamide **71** forms a nitrosoamide intermediate **72**, which converts slowly into the diazo species **73**

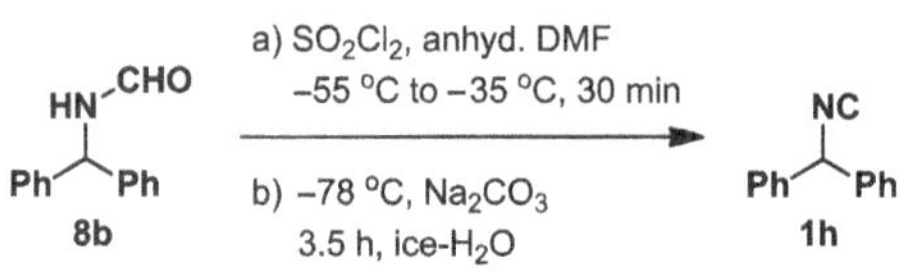

SCHEME 7.16 Synthesis of diphenylmethyl isocyanide (**1h**).

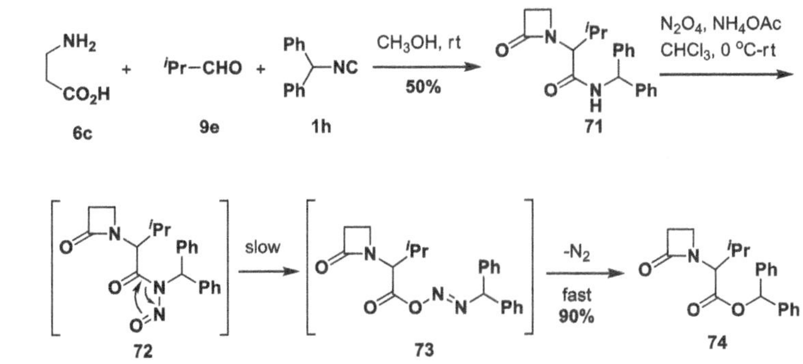

SCHEME 7.17 DPMIC (**1h**) as a CIC.

that decomposes very fast to provide the corresponding ester **74** in a high yield (Scheme 7.17). The limitation of this method is that an α-ketoaldehyde cannot be employed in the Ugi reaction as the ensuing diphenylmethyl-amides will contain enolizable β-keto groups that are easily *C*-nitrosated and therefore prevents clean conversion into corresponding esters.

They applied this methodology for the total synthesis of nocarcidins (see Scheme 7.20 first) [37]. Nocardicin A (**85**), isolated in 1975 by Fujisawa as a fermentation product of *Nocardia* species, is the first example of a monocyclic β-lactam with potential useful antibacterial activity [38].

All nocardicins contain 3-ANA (**79**) as a common nucleus and differ only in the structure of the *N*-acyl side chain [39]. Two key intermediates, 3-ANA (**79**) and the side chain **82**, were synthesized from amino acids and the key step involves the N_2O_4-mediated conversion of amides **77** and **80** to the corresponding ester **78** and alcohol **81**, respectively (Schemes 7.18 and 7.19).

7.2.1.4 ***O-Benzyloxy Phenyl Isocyanide*** *O*-(Benzyloxy)phenyl formamide (**1i**) was synthesized by dehydration of *O*-(benzyloxy)phenyl formamide (**8c**) in the presence of phosphoryl chloride and diisopropylamine (Scheme 7.21) [40]. The use of triethylamine decreases the overall yield of the desired product.

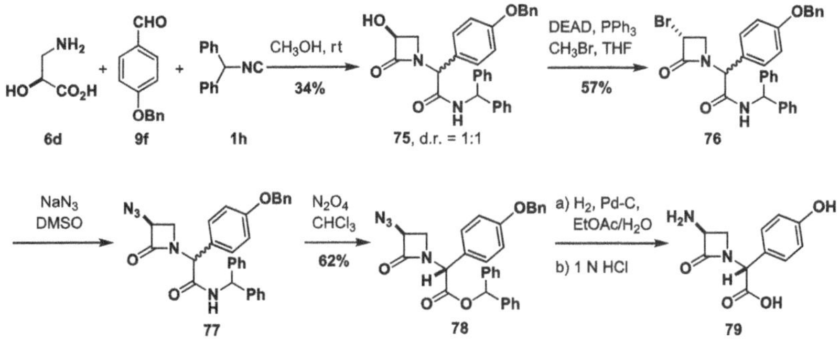

SCHEME 7.18 Synthesis of 3-aminonocardinic acid (3-ANA, **79**).

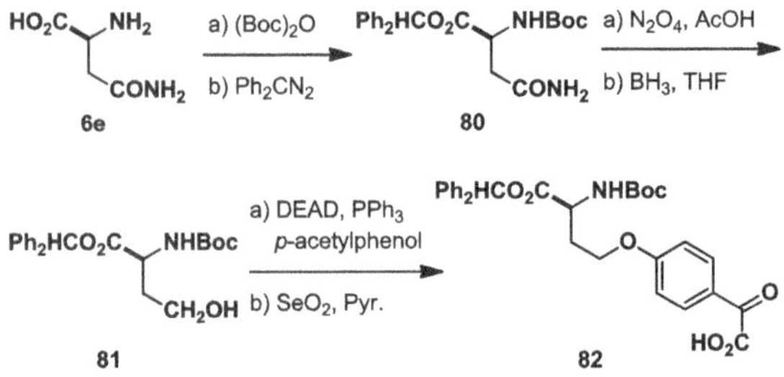

SCHEME 7.19 Synthesis of the side chain of norcacidin alkaloids.

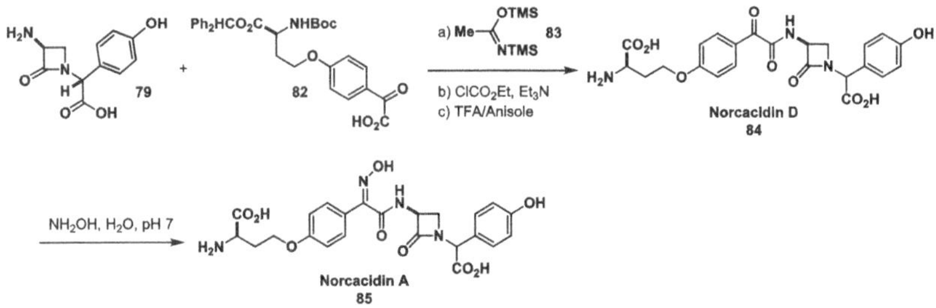

SCHEME 7.20 Synthesis of norcacidin alkaloids.

Borthwick and coworkers have used *O*-benzyloxy phenyl isocyanide (**1i**) in the U-4CR to synthesize 2,5-diketopiperazines (DKPs) and converted it into the cyclic carbamate **88** in the presence of 1,1′-carbonyldiimidazole [41]. The cyclic carbamate **88** is a highly activated amide and undergoes aminolysis, hydrolysis, and methanolysis to generate products like **89**, **90**, and **91**, respectively (Scheme 7.22). Compound **89** is shown to act as a potent, selective, and orally bioavailable oxytocin antagonist [41]. However, compounds of the type **89** were reported to have low solubility and relatively high log *D*. This resulted in replacement of the 3,4-difluorobenzaldehyde with 2-methyloxazole-4-aldehyde in the U-4CR. Compound **92** (GSK221149A) has been noted to have nanomolar affinity for the oxytocin receptor with >1400-fold selectivity over the closely related vasopressin receptors. Compound **92** (Figure 7.3) has also shown to have a good rat pharmacokinetic profile and low human microsomal clearance and inhibits oxytocin-induced contraction *in vivo* in the anesthetized rat [42].

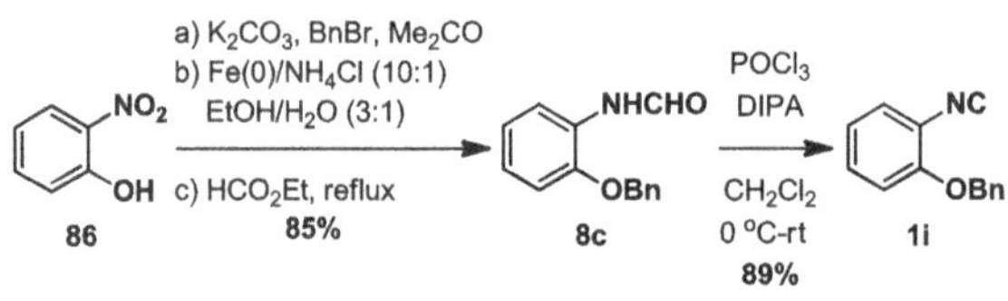

SCHEME 7.21 Synthesis of *O*-benzyloxy phenyl isocyanide (**1i**).

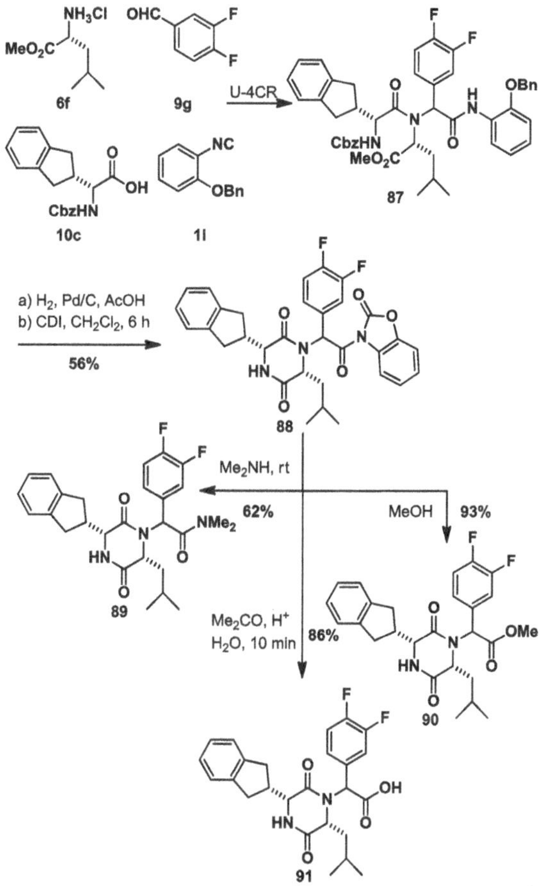

SCHEME 7.22 Synthesis of oxytocin receptor antagonists using *O*-benzyloxy phenyl isocyanide (**1i**) in the U-4CR.

7.2.1.5 Benzotriazol-1-yl-methyl Isocyanide The benzotriazol-1-yl-methyl isocyanide (BetMIC, **1j**), first synthesized by Katritzky et al., is known as Katritzky's CIC [43]. It can be easily synthesized in two steps from commercially available benzotriazole (Bet, **93**), paraformaldehyde (**9c**), and formamide (**9h**) to initially give

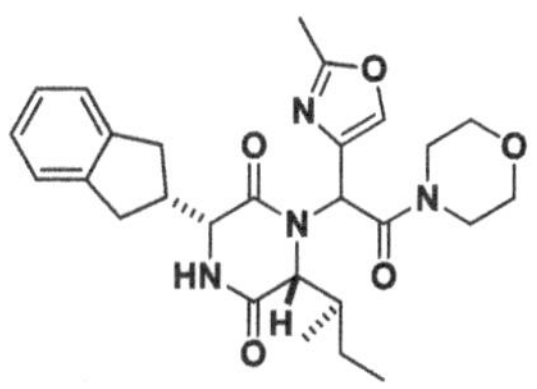

FIGURE 7.3 DKP **92** as a potent oxytocin antagonist.

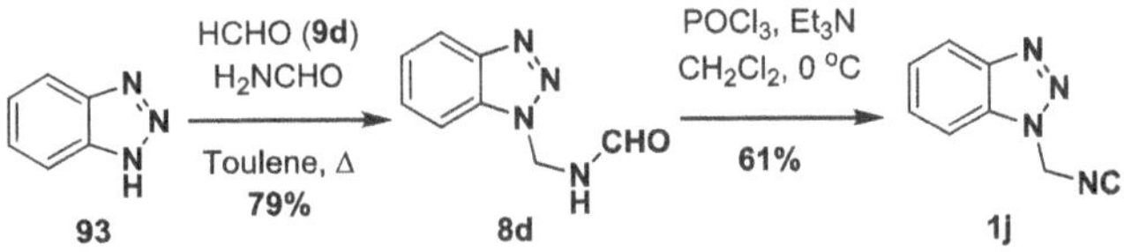

SCHEME 7.23 Synthesis of BetMIC (**1j**).

the Mannich product (**8d**), which is then converted to BetMIC (**1j**) with $POCl_3$ and triethylamine (Scheme 7.23). It is a stable, nonhygroscopic odorless golden yellow solid.

BetMIC is characterized by two key properties; first, the good leaving ability of the benzotriazole (Bet) moiety and, second, the high acidity of the methylene protons. BetMIC is reactive with ketones to afford 4-ethoxy-2-oxazolines (**94**), which can be hydrolyzed to synthetically useful α-hydroxyaldehydes (**95**). It must be noted that unlike TosMIC, BetMIC forms **94/96** without the presence of highly toxic thallium alkoxides. The reaction works well in the absence of ethanol and provides better yields when using aliphatic ketones. With aromatic ketones, for example, benzophenone, partial conversion to oxazoline occurs probably due to steric effects. Reaction of BetMIC with aldehydes typically generates oxazoles **97** (Scheme 7.24) [43,44].

In the presence of base, BetMIC undergoes cycloaddition with carbonyl group to form the oxazole intermediate **98** (Scheme 7.25). In the case of aldehydes, base-mediated elimination of Bet (or EtOH) provides the final product **99** and gives back Bet (**93**). On the contrary, in case of ketones, the elimination cannot occur and instead a substitution reaction takes place by a nucleophilic solvent (e.g., EtOH).

In a similar fashion, Katritzky and coworkers have shown that aldimines and alkenes also undergo cycloaddition with BetMIC (**1j**) in the presence of a suitable base under refluxing conditions to afford substituted imidazoles **101** and pyrroles **104**, respectively (Scheme 7.26) [45]. The mechanism of formation of **101** and **104** is similar to that shown in Scheme 7.25.

Recently, Dömling and coworkers employed BetMIC in the U-4CR [46] to synthesize α-aminomethyl tetrazoles, like **107**, which are isosteric to α-amino acids and thus have drawn major attention in medicinal chemistry to alter unfavorable ADME

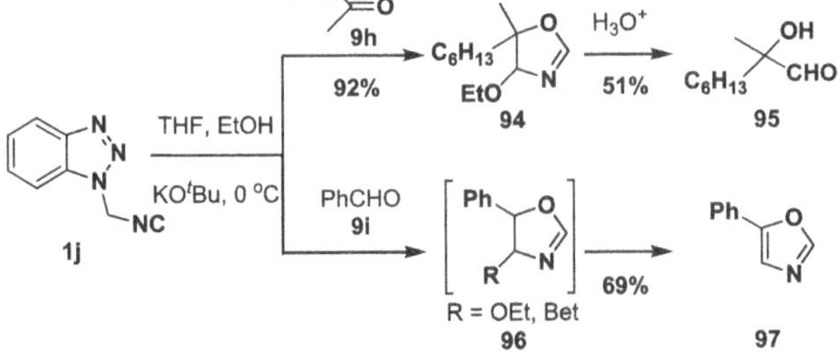

SCHEME 7.24 Synthesis of oxazoles and α-hydroxyaldehydes using BetMIC (**1j**).

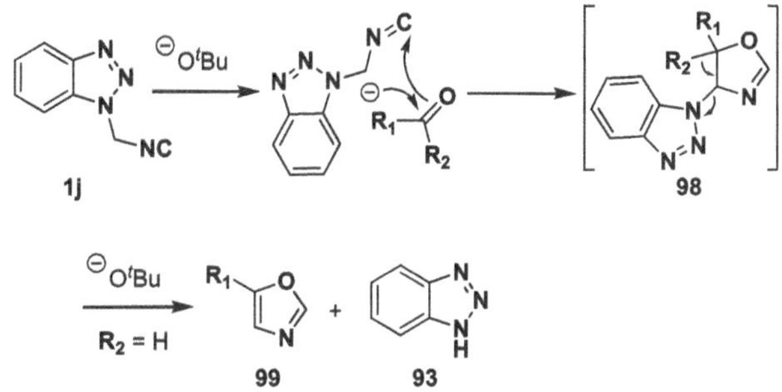

SCHEME 7.25 General mechanism of the reaction between BetMIC (**1j**) and carbonyl compounds.

(absorption, distribution, metabolism, and excretion) properties and/or to access free parent space [47]. In the tetrazole variation of the U-4CR, BetMIC and TMS-N_3 react with enamine **105** to form diheterocyclic *N*,*N*-aminal **106**, which can be hydrolyzed under mild acidic conditions to afford the desired tetrazole **107** (Scheme 7.27). The hydrolysis reaction can be performed after isolation of **106** or in one pot; however, the later gives a slightly lower yield.

7.2.1.6 2-Pyridylisocyanide 2-Pyridylisocyanide (PyIC, **1k**) was first synthesized by Ugi in 1965 [48]. It can be readily synthesized from 2-amino pyridine (**6g**) via a conventional method as shown in Scheme 7.28 [49].

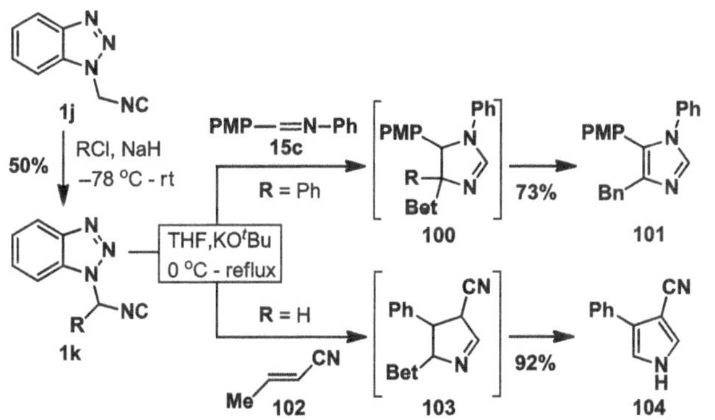

SCHEME 7.26 Synthesis of imidazoles (**101**) and pyrroles (**102**) using BetMIC (**1j**).

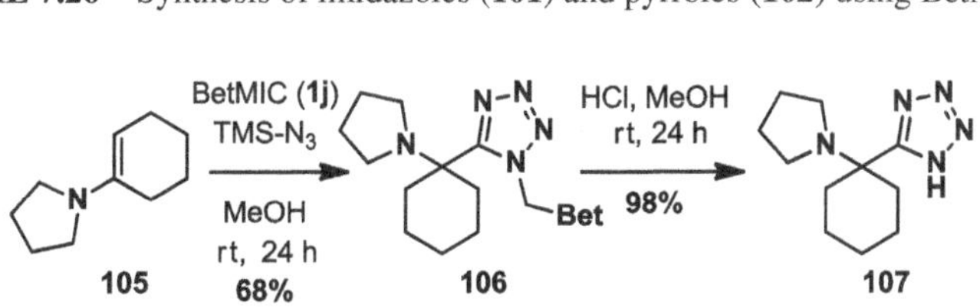

SCHEME 7.27 Synthesis of α-aminomethyl tetrazoles using BetMIC (**1j**).

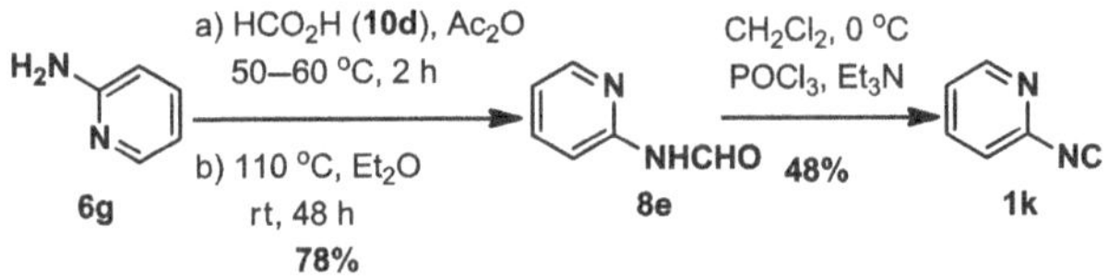

SCHEME 7.28 Synthesis of PyIC (**1k**).

The first synthesis of pentasubstituted pyrroles was achieved by using PyIC (**1k**) in the U-4CR on a solid phase [50]. According to Mjalli and coworkers, pyridine was necessary as a cosolvent in the U-4CR as it helped buffer the reaction mixture and stabilize the isocyanide. The acyclic Ugi product **108** can be directly converted to the münchnone **109** by heating (100°C) in acetic anhydride and trapped *in situ* with alkynes bearing electron-withdrawing groups via 1,3-diploar cycloaddition to afford pyrrole **110** (Scheme 7.29).

7.2.1.7 Phenyl Isocyanide In 1972, Ugi and coworkers first synthesized phenyl isocyanide (PhIC) (**1l**) via the Hoffmann carbylamine reaction [51]. Mjalli and coworkers [50] have also synthesized it following a procedure reported by Walborsky and Ronman [52] (Scheme 7.30). It can be purified by distillation to obtain a light yellow oil, which has a boiling point of 53°C.

Phenyl isocyanide (PhNC) has been utilized in the U-4CR to synthesize pentasubstituted pyrroles [50]. The U-4CR product **111** was converted to the corresponding *N*-acyl-*N*-alkyl-α-amino acid **112** with the treatment with Boc_2O-4-dimethylaminopyridine (DMAP) in tetrahydrofuran (THF) followed by 1 N LiOH

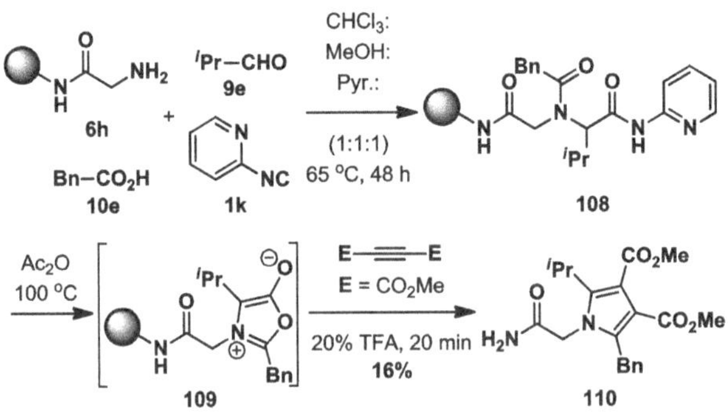

SCHEME 7.29 Synthesis of pentasubstituted pyrroles.

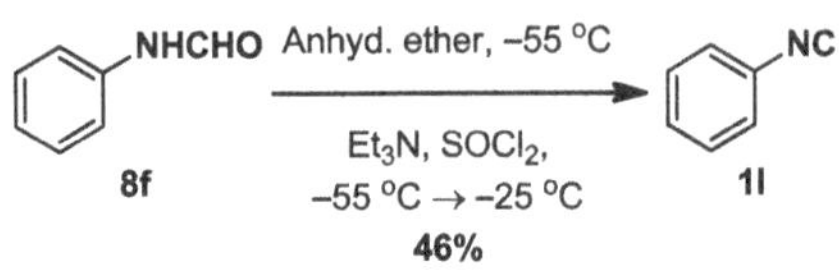

SCHEME 7.30 Synthesis of PhIC (**1l**).

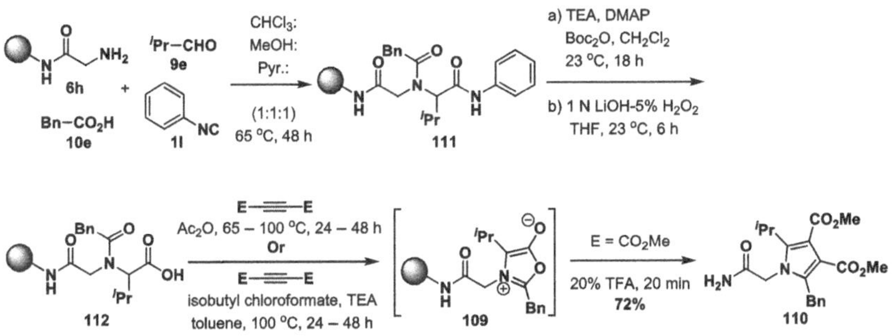

SCHEME 7.31 Application of PhNC (**1l**) in pyrrole synthesis.

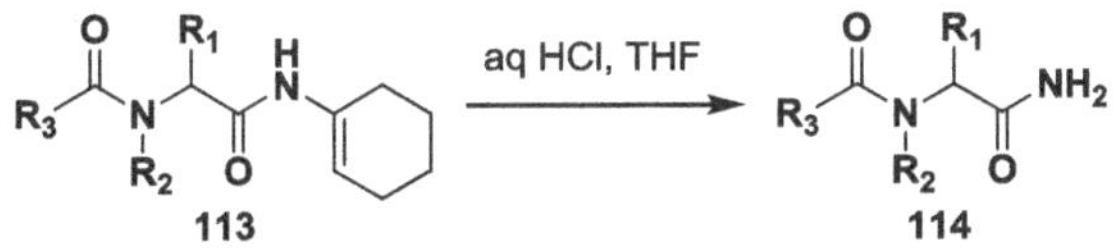

SCHEME 7.32 Conversion of enamide into primary amide using acidic medium.

and 5% H_2O_2 in THF at 23°C. It must be noted that when they used benzyl isocyanide, the reaction was very slow and provided a low yield. The acid can be converted to pyrrole **110** by treating either with neat acetic anhydride (65–100°C) or with isobutyl chloroformate-triethylamine (TEA)-toluene (65–100°C, 24–48 h) and subsequent removal from the solid support by 20% trifluoroacetic acid (TFA) in CH_2Cl_2 (Scheme 7.31).

7.2.1.8 Cyclohexenyl Isocyanide Cyclohexenyl isocyanide (**1m**, CyEIC) was first synthesized by Ugi and coworkers in 1963 [53]. They reported that the enamide product **113**, derived from the use of 1-isocyanocyclohexene in the U-4CR, can be converted into a primary amide **114** through acidic hydrolysis (Scheme 7.32).

CyEIC (**1m**) can be easily synthesized from commercially available cyclohexanone (**9j**). Later, Armstrong and coworkers modified the synthesis of **1m** [54]. A Strecker reaction using **9j** with aqueous sodium cyanide and ammonium chloride in diethyl ether forms α-aminonitrile, which was converted into the formamide **8g**. The formamide was purified and subjected to ^{t}BuOK followed by dehydration in the presence of triphosgene (Scheme 7.33). CyEIC can be stored indefinitely at –30°C; however, it is highly air-sensitive and will turn dark upon exposure.

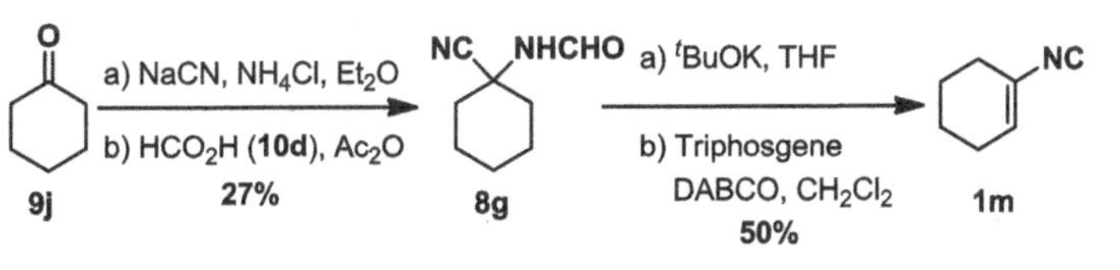

SCHEME 7.33 Synthesis of CyEIC (**1m**).

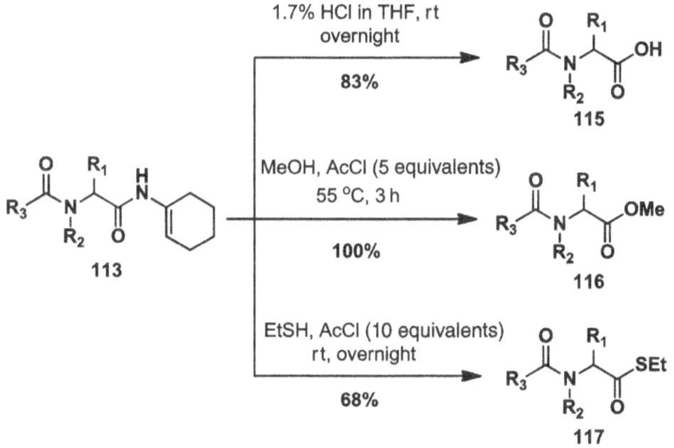

SCHEME 7.34 Cleavage conditions of the "universal convertible" cyclohexenamide group.

Reinvestigation of Ugi's observation on the use of **1m** in the U-4CR by Keating and Armstrong transformed it into the "universal CIC" [54]. Thus, U-4CR enamide product **113** can be converted into acids **115**, esters **116**, and thioesters **117** under various conditions (Scheme 7.34) in good to excellent yields [54].

Armstrong proposed a mechanism for conversion of the cyclohexenamide into various products (Scheme 7.35) [54]. In the presence of acid, protonation of the enamide **113** forms activated iminium intermediate **118**, which undergoes an internal nucleophilic attack by the *N*-acyl group to form oxazolinium-5-one (münchnone) intermediate **119** and eliminates cyclohexaimine. External nucleophilic attack (typically by the solvents used) onto the reactive münchnone provides product **120**. It must be noted that lack of an *N*-acyl group or its rapid loss prevents formation of münchnone **119**.

The münchnone **119** derived from U-4CR product can undergo resonance and form anhydro-5-hydroxyoxazolium hydroxide species **121** (Scheme 7.36), which

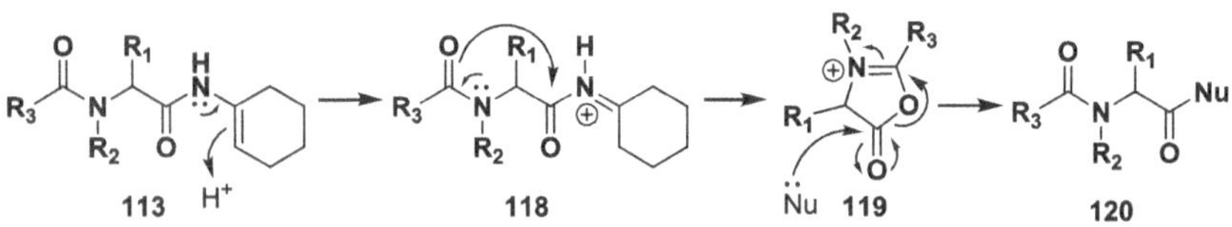

SCHEME 7.35 Mechanism of cyclohexenamide conversion.

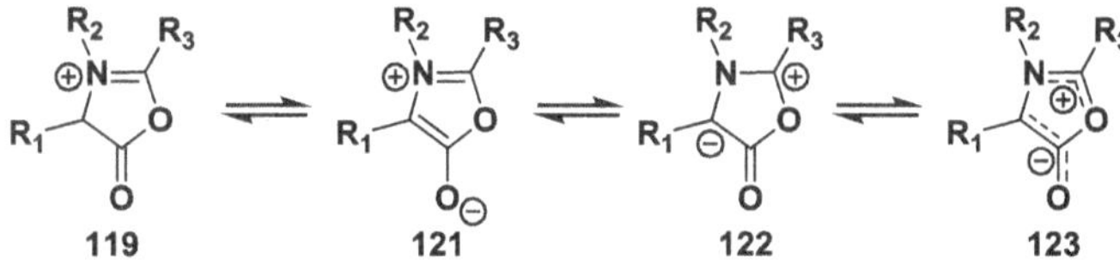

SCHEME 7.36 Possible resonance structures of münchnone.

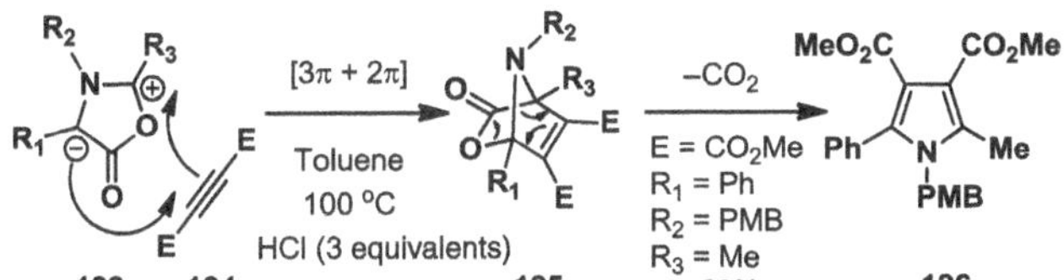

SCHEME 7.37 Synthesis of pyrroles via cyclohexenamide-derived münchnone **122**.

are well known to undergo cycloadditions with a wide variety of dipolarophiles [55].

Armstrong and coworkers capitalized the 1.3-dipole intermediate **122** to synthesize polysubstituted pyrroles via a [3π + 2π] cycloaddition reaction with mono and disubstituted acetylenes in the presence of an acid. As previously mentioned, protonation of the enamide followed by cyclization and loss of cyclohexanimine forms the active intermediate **119**, which loses a proton to form the 1,3-dipole **122**. A [3π + 2π] cycloaddition reaction of **122** with acetylene **124** forms intermediate **125**, which rapidly loses CO_2 to provide the aromatized product pyrrole **126** (Scheme 7.37). Some of the key features of this reaction are:

1. In contrary to other cyclodehydration reactions that are typically run at 50–60°C, higher temperature and toluene are necessary for better or observable yields.
2. Stronger electron-withdrawing groups on acetylenic dipolarophile provide better yields.
3. Steric effects on intermediate **125** influence the reaction yield.

As an extension of their methodology, Armstrong and coworkers have synthesized 2-acetamido-2-deoxyhexose via the use of tethered internal nucleophiles. Thus, the U-4CR of aldehyde **9k**, derived from D-arabinose, *p*-methoxybenzylamine (**6i**), acetic acid (**10b**), and **1m**, provided enamide **127** in a high yield as a 3.4:1 mixture of inseparable diastereomers at the α-carbon. In the presence of mild acidic conditions (HCl generated from AcCl), cleavage of the isopropylidene PG and protonation of the enamide followed by münchnone **128** formation takes place. Internal nucleophilic attack by the secondary alcohol (attack by the primary alcohol is unfavorable as it forms a seven-membered ring) forms the expected six-membered 2-acetamido-2-deoxy-D-mannono-δ-lactone **129** (Scheme 7.38) [56].

In an approach to avoid PGs, they have also shown that "off-the-shelf" anthranilic acid (**10f**) can be used directly in the U-4CCR to obtain biologically validated 1,4-benzodiazepine-2,5-dienones (1,4-BDZs) **130** in a high yield (Scheme 7.39) [56,57].

Hulme and coworkers applied Armstrong's isocyanide to synthesize another natural scaffold DKP **131** in a solution phase (Scheme 7.40) in good yield [58].

Further advancement has been made to employ Armstrong's CIC to perform U-4CR on a solid phase to synthesize pentasubstituted pyrroles [59] similar to that shown in Scheme 37.

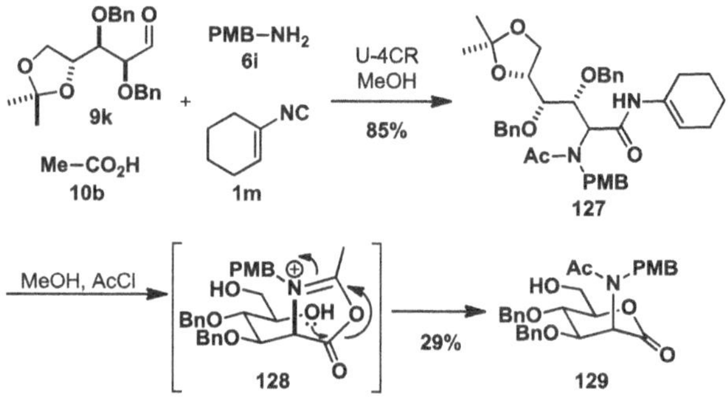

SCHEME 7.38 Synthesis of 2-acetamido-2-deoxy-D-mannono-δ-lactone (**129**).

Although Armstrong's CIC (**1m**) has proven to be useful in synthesizing a diverse-array of small molecule scaffolds, the common procedures for the preparation of **1m** are not practical for large-scale synthesis [52c, 53, 60]. The overall yield is generally low due to a large number of steps, decomposition of product during distillation, and difficulties arising from coevaporation of the product while the solvent is removed after chromatography. Furthermore, it has a quite unpleasant odor. To overcome these problems, Martens and coworkers developed a shorter route to synthesize **1n**. Commercially available cyclohexanone derivative **9m** was converted to the formamide **8f** and dehydrated in the presence of phosgene to obtain **1n** (Scheme 7.41) [61]. The synthesis can be performed on a large scale. Due to the less than unpleasant odor of **1m**, **1n** is much easier to handle and can be stored at 0°C under argon for several months.

They have utilized this new CIC (**1n**) based off Armstrong's CIC in the U-4CR to synthesize nucleic acid mimics called peptide nucleic acids (PNAs) monomers [62].

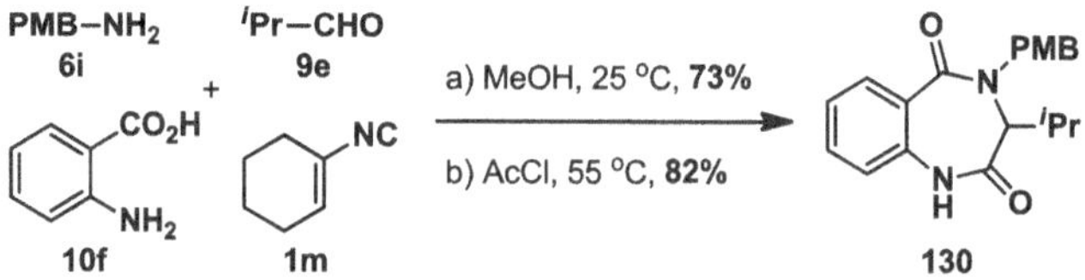

SCHEME 7.39 Synthesis of 1,4-BDZs **130**.

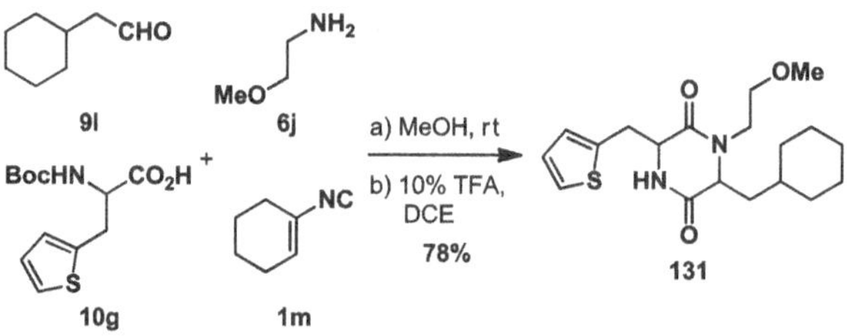

SCHEME 7.40 Synthesis of DKPs **131** using **1m**.

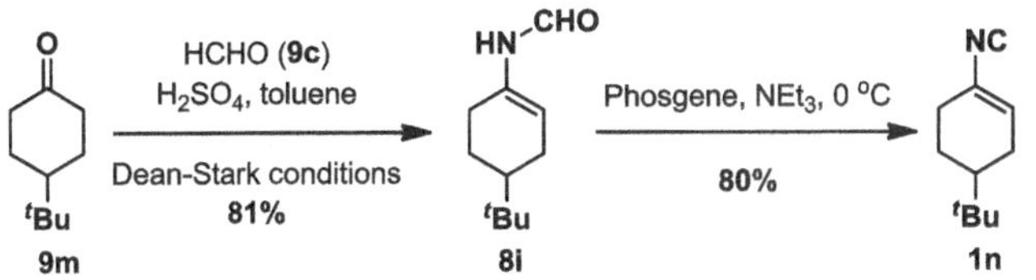

SCHEME 7.41 Synthesis of CIC **1n**.

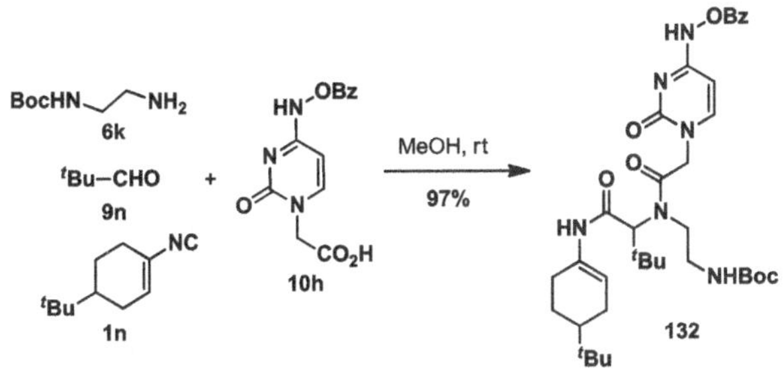

SCHEME 7.42 Synthesis of totally protected PNA-monomers **132**.

The one-pot U-4CR provides totally protected PNA monomer **132** in a good yield (Scheme 7.42). Other natural nucleobase acetic acids in the U-4CR also provide a moderate to excellent yield except guanine-based acetic acid due to its low solubility. The cyclohexenamide can be cleaved using THF/concentrated HCl, similar to the conditions used by Armstrong et al. to obtain the corresponding sterically crowded or unusual α-amino acid analogs [60].

Recently, Hulme and coworkers have reported a convenient two-step synthesis of quinoxaline-benzodiazepines (e.g., **133**) and bisbenzodiazepines (e.g., **134**) by utilizing **1n** from Boc-protected amines via the U-4CR under microwave irradiation (MWI) (Scheme 7.43) [63].

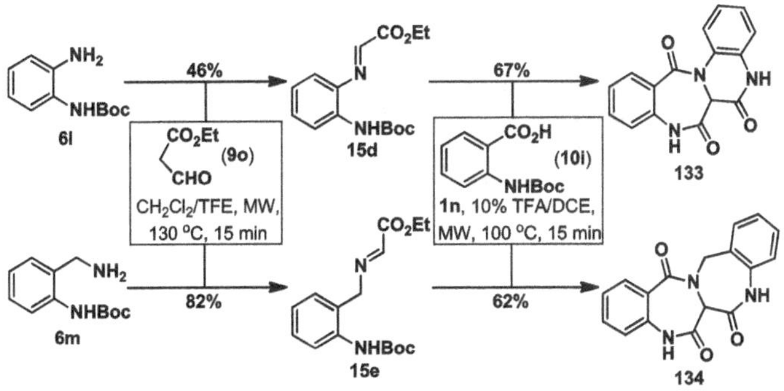

SCHEME 7.43 Synthesis of quinoxaline-benzodiazepines (**133**) and bisbenzodiazepines (**134**) using **1n**.

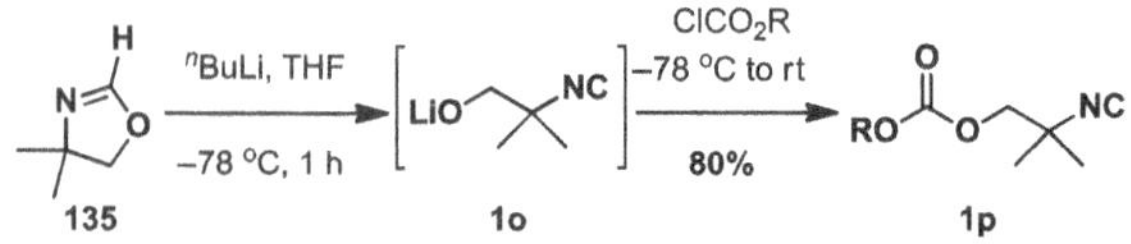

SCHEME 7.44 Synthesis of CIC **1p**.

7.2.1.9 [(2-Isocyano-2-methyl)-propyl-1-]-alkyl-carbonate [(2-Isocyano-2-methyl)-propyl-1-]-alkyl-carbonates (**1p**) were developed by Lindhorst in collaboration with Ugi and have been referred to as the Lindhorst isocyanides [64]. The synthesis of **1p** and its analog are very simple (Scheme 7.44). Commercially available 4,4-dimethyl-2-oxazoline **135** can be deprotonated by ^{n}BuLi in anhydrous THF at −78°C and the resulting lithium alcoholate **1o** is captured with alkyl chloroformate (R = Me, Et, allyl, Bn) to provide 80% of the desired product as colorless crystals or oil (Scheme 7.44). It can be further purified by vacuum distillation. This one-step method offers several advantages: (i) it does not involve phosgene or derivatives thereof, (ii) (β-isocyano-ethyl)-alkyl carbonates are stable and can be stored at room temperature, and (iii) it can be employed in a large-scale synthesis.

Lindhorst and coworkers have shown that after the U-4CR, **1q** can be cleaved under basic conditions (KO^{t}Bu) to obtain a variety of α-amino esters **137**, α-amino acids **138**, and BDZs **139** (Scheme 7.45). In contrast to Armstrong's CIC **1m**, basic conversion of U-4CR products derived from **1p** to α-amino esters and α-amino acids is not limited by the structure of the residual molecule.

The postulated mechanism of basic cleavage is shown in Scheme 7.46. In the first step, deprotonation of the secondary amide **136** by KO^{t}Bu forms the intermediate **140**, which then undergoes an intramolecular cyclization to afford the cyclic *N*-acyl

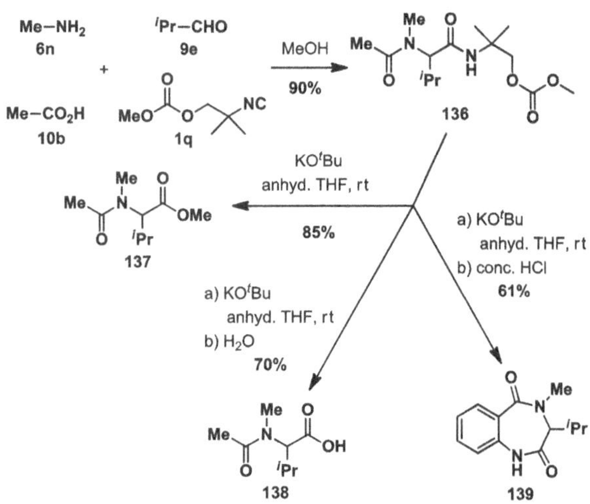

SCHEME 7.45 Application of Lindhorst/Ugi's CIC **1q** in synthesis.

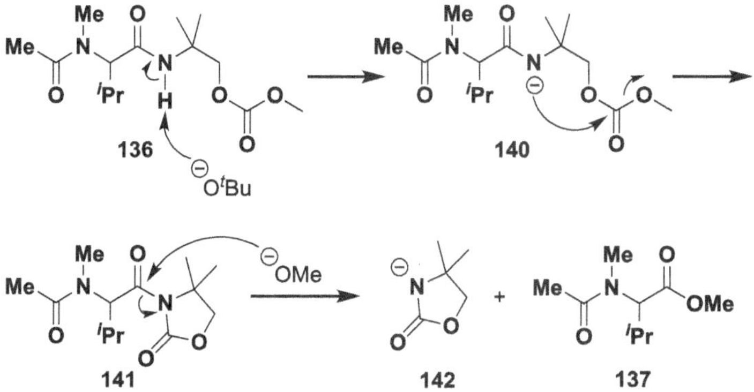

SCHEME 7.46 Mechanism of basic hydrolysis of Lindhorst's CIC **1q**.

urethane **141**. Liberated alcoholate converts **141** *in situ* to 4,4-dimethyl-2-oxazoline-2-one anion **142** and the corresponding products, for example, **137**.

Formation of the cyclic *N*-acyl urethane **141** requires very anhydrous conditions. In the presence of a small amount of water, potassium hydroxide (KOH) forms from KO^{t}Bu and leads to partial saponification of the carbamate moiety of **136** and liberates CO_2 prior to the intramolecular cyclization, ultimately resulting in the formation of alcohol **145** (Scheme 7.47).

Fukuyama and coworkers have synthesized [(2-isocyano-2-methyl)-propyl-1-]-phenyl-carbonate by a modification of the Lindhorst and Ugi protocols. Instead of using alkyl chloroformate to capture the lithium alcoholate **1o**, they used aryl chloroformate (phenyl chloroformate) to obtain **1r** (Scheme 7.48) [65]. The phenyl analog is odorless and stable under chromatographic separation.

Fukuyama and coworkers have utilized **1r** in the U-4CR to synthesize *N*-acyl thio esters. The U-4CR product **146** can be converted to the corresponding *N*-acyl oxazolidinone **147** by treating with ^{t}BuOK in anhydrous THF in the presence of 4Å molecular sieves. Unlike alkyl alkoxides (e.g., methoxide), phenoxide anion does not attack the *N*-acyl oxazolidinone **147** because of weak nucleophilicity (pK_a = 10). Molecular sieves play a critical role in preventing hydrolysis of the starting

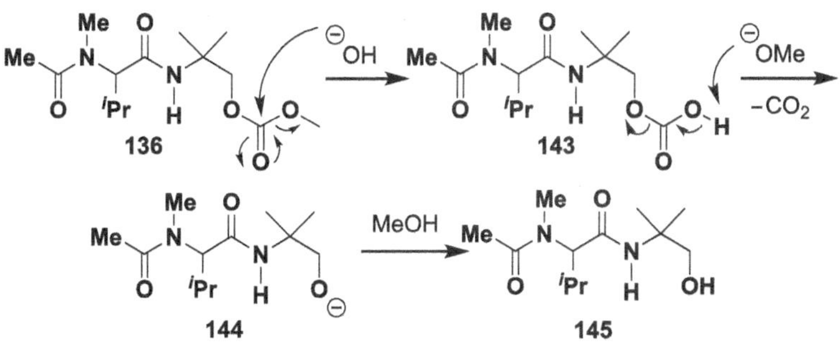

SCHEME 7.47 Formation of **145** from a carbamate under moist conditions.

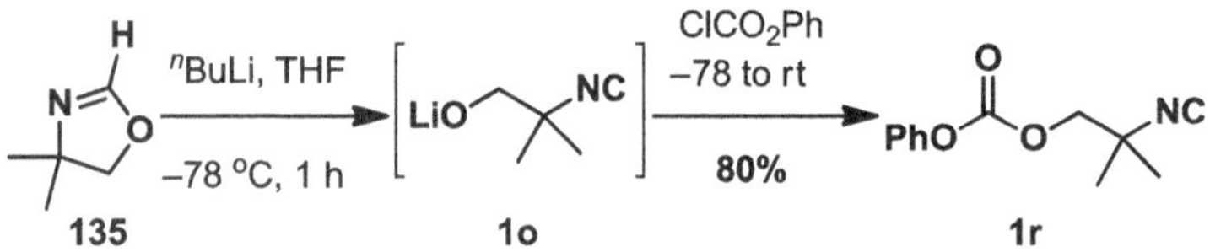

SCHEME 7.48 Synthesis of [(2-isocyano-2-methyl)-propyl-1-]-phenyl-carbonate (**1r**).

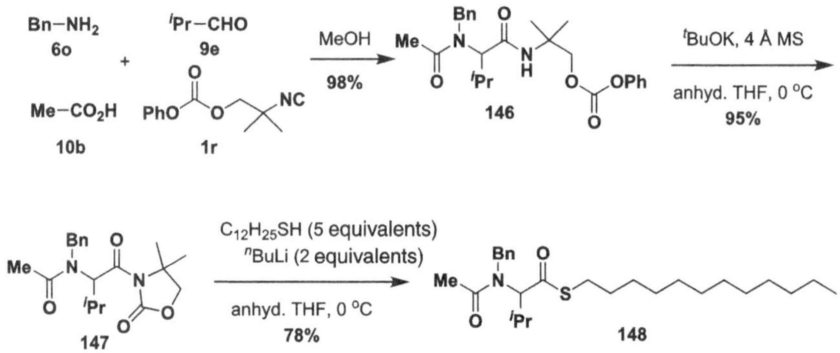

SCHEME 7.49 Synthesis of *N*-acyl thiol esters **148**.

material and/or *N*-acyl oxazolidinone to the corresponding acids and alcohols by moisture. The *N*-acyl oxazolidinones can be converted to corresponding thio esters **148** in the presence of ^{n}BuLi and *n*-dodecanethiol at 0°C (Scheme 7.49). At a higher temperature, addition to the carbonyl group of the oxazolidinone ring takes place. Thiol esters can be converted to the corresponding aldehydes or ketones in the presence of palladium catalyst [66]. Thus, the method provides an additional handle toward the synthesis of α-amino acid derivatives.

7.2.1.10 Tert-butyl(2-isocyanobenzyloxy)dimethyl Silane In 1999, Linderman and coworkers developed *tert*-butyl(2-isocyanobenzyloxy)dimethyl silane (**1s**) [67]. It can be synthesized from readily available 2-amino benzyl alcohol (**6p**). Chemoselective *O*-silylation of **6p** using sodium hydride and *tert*-butyldimethylsilyl chloride (TBDMSCl) provides **6q**, which is *N*-formylated using the mixed anhydride to obtain formamide **8j**. Dehydration of **8j** with 1,4-diazabicyclo[2.2.2]octane (DABCO) and triphosgene generates **1s** in a good yield (Scheme 7.50). It can be purified by chromatography on deactivated silica gel or by Kugelrohr distillation and is quite stable. Although **1s** has been known to be stable when stored at −18°C and can be prepared on a reasonable scale, it is best to prepare formamide **8j** in large quantities and convert to **1s** as needed.

They have utilized this new CIC (**1s**) in the U-4CR of benzylamine (**6o**), benzaldehyde (**9i**), and formic acid (**10d**) to obtain acyclic U-4CR product **149**. Acid-catalyzed deprotection of the silyl group followed by an intramolecular conversion of the amide to ester via a six-membered transition state provides ester **151** in a quantitative yield after basification with saturated $NaHCO_3$ (Scheme 7.51).

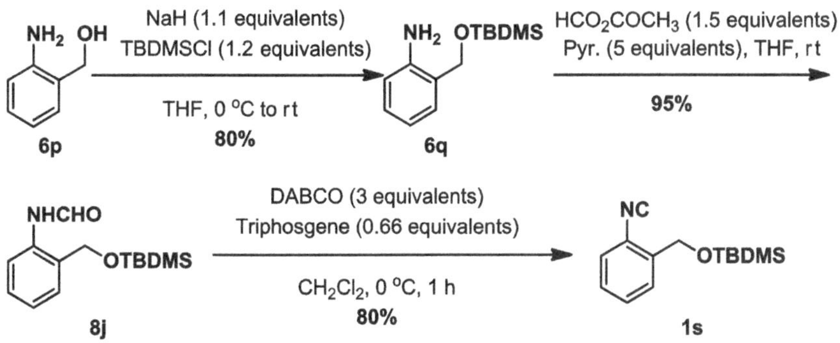

SCHEME 7.50 Synthesis of Linderman's CIC **1s**.

As an extension of their methodology, Linderman and coworkers have utilized CIC **1s** in the asymmetric U-4CR to synthesize amino acid derivatives, using D-galactosylamine auxiliary **6r** [67] under the conditions reported by Kunz and Pfrengle [68]. The U-4CR provides a good yield of (*R*)-valine derivative **152** with excellent diastereoselectivity (97:3). Other aldehydes also provide good to excellent yields and excellent diastereoselectivity. It is noteworthy that the diastereoselectivity of the reaction rapidly degrades with increasing temperature. The minor diastereomer can be easily removed by flash chromatography. The *N*-terminal amide of **152** can be converted to corresponding amino acid (*R*)-valine **154** via acid hydrolysis of D-galactosylamine auxiliary and formyl group simultaneously without isolating intermediate ester **153** (Scheme 7.52).

In a similar fashion, using D-arabinosylamine (**6s**) in the U-4CR, (*S*)-amino acids **158** can be obtained in excellent diastereoselectivity (Scheme 7.53). Remarkably, the diastereoselectivity was much higher than similar asymmetric U-4CR reported by Kunz and Pfrengle [68].

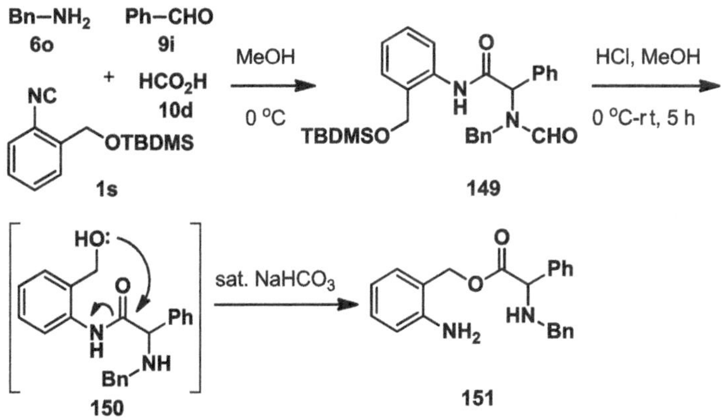

SCHEME 7.51 Use of CIC **1r** in the U-4CR.

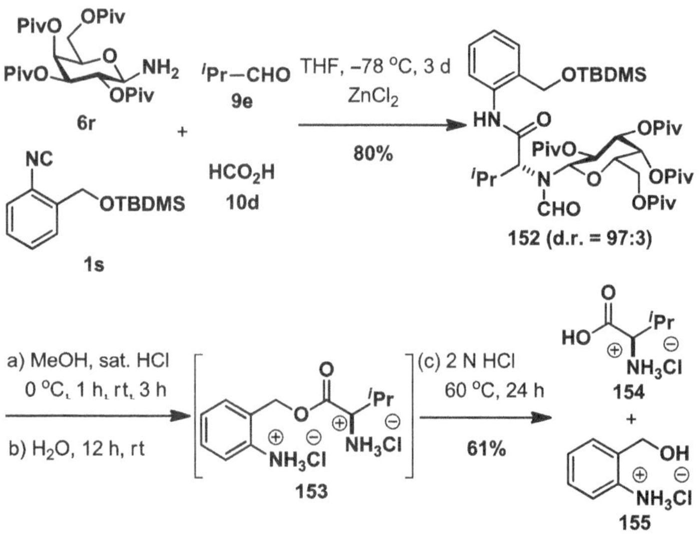

SCHEME 7.52 Synthesis of (*R*)-amino acids (**154**) using Linderman's CIC (**1s**) via a diastereoselective U-4CR.

The chiral auxiliary sugar can be isolated by pentane extraction of the crude hydrolysis reaction mixture. The amino acids can then be purified by ion exchange chromatography. The shortcoming of this methodology is that simple amino acids such as phenylalanine are difficult to separate from the hydrolyzed by-product 2-aminobenzyl alcohol (**155**). According to Linderman and coworkers, the size of the

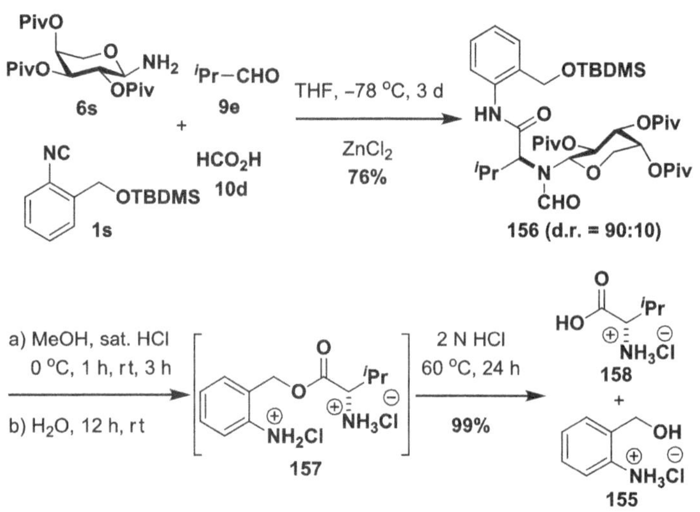

SCHEME 7.53 Synthesis of (*S*)-amino acids (**158**) using Linderman's CIC (**1s**) via a diastereoselective U-4CR.

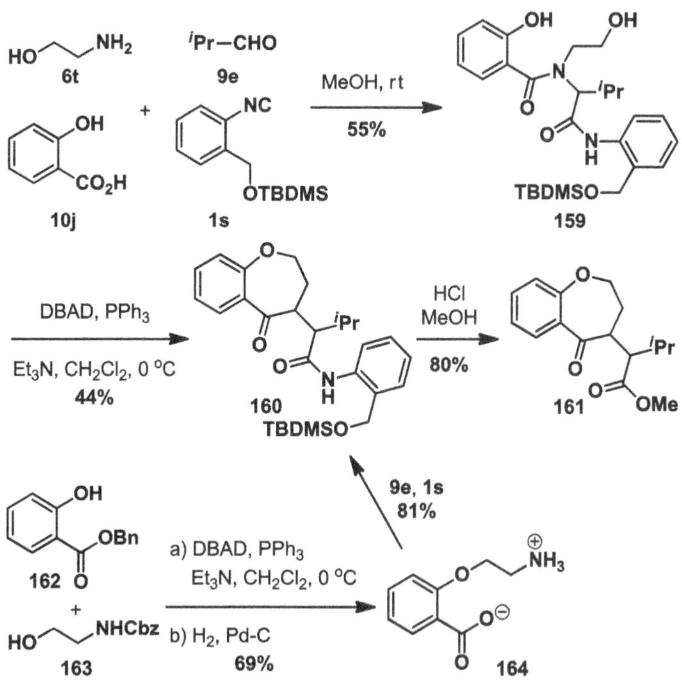

SCHEME 7.54 Synthesis of dihydrobenzoxapinones (**160**).

silyl ether of the isocyanide component also influences diastereoselectivity and bulky silyl ethers provided the highest diastereoselectivity with a compromised lower yield [67].

Recently, Banfi et al. have demonstrated that Linderman's CIC **1s** can be employed in the U-4CR to synthesize dihydrobenzoxapinones **160** and subsequently cleaved to obtain corresponding methyl ester **161** (Scheme 7.54)

Williams and coworkers have used Linderman's CIC **1s** to synthesize analogs of polyoxins and nikkomycins, which are nucleoside-peptide antibiotics that display biological activity against fungal chitin synthase derived from *Candida albicans* and/or *Saccharomyces cerevisiae* [69]. It is noteworthy that chitin is absent in mammals and plants, and thus CSs are attractive drug targets for inhibition in both fungi and insects. The U-4CR of 2,4-dimtheoxybenzylamine (**6u**), aldehyde **9p**, isooxazole-based acid **10k**, and CIC **1s** provides **165** in a 45% yield. Cleavage of the CIC in the presence of methanolic HCl provides the polyoxin/nikkomycin analog **166** (Scheme 7.55) [70].

7.2.1.11 ***n-Butyl Isocyanide*** Although *n*-butyl isocyanide (NBIC, **1t**) has been known for quite a while, the characteristic low odor and commercial availability make it very attractive for use in synthesis. Traditionally, **1t** has been used as a noncovertible isocyanide (NCIC) in MCR. Hulme and coworkers reported, for the first time, that NBIC (**1t**) can be used as a simple and inexpensive CIC and as an

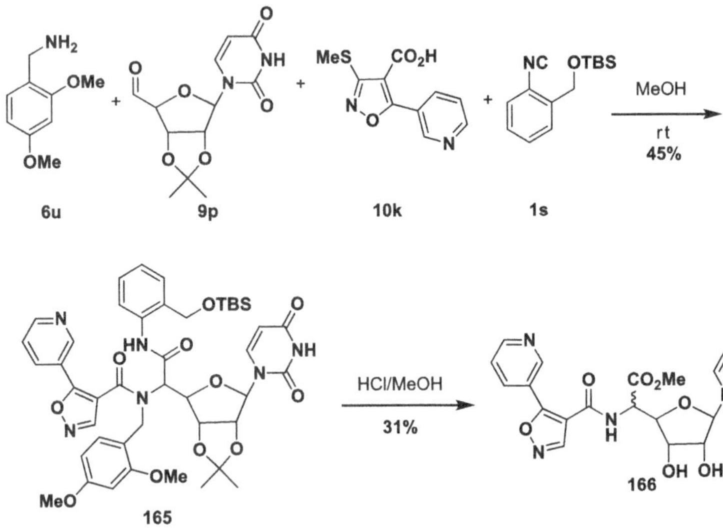

SCHEME 7.55 Synthesis uracil polyoxin C analog **166**.

alternative to synthetically designed CICs. They have shown that **1t** can be cleaved in the presence of 20% TFA [71] to synthesize biologically relevant dihydroquinoxalin-2-one **169** from the acyclic Ugi product **167** (Scheme 7.56).

Upon further modification, MWI was able to cleave the ^{n}Bu group in a shorter time period. They have employed this modification to synthesize DKP **170** and BDZ **171** via the U-4CR (Scheme 7.57).

Recently, Hulme and coworkers have reported a convenient two-step synthesis of quinoxaline-benzodiazepine (**133**) by utilizing **1t** from Boc-protected amines via the U-4CR under MWI (Scheme 7.58) [63].

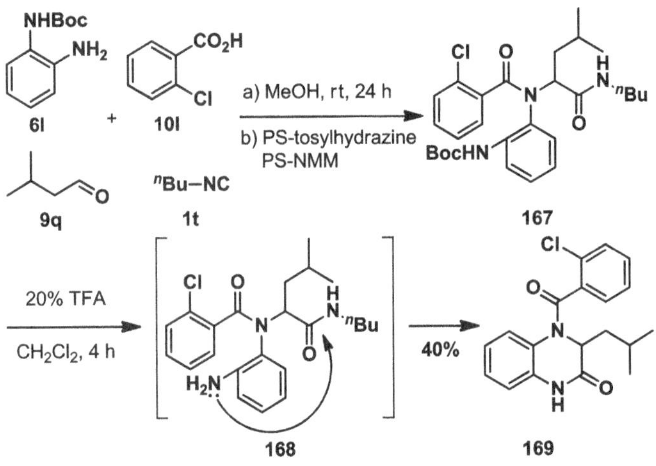

SCHEME 7.56 Synthesis of dihydroquinoxalin-2-one **169**.

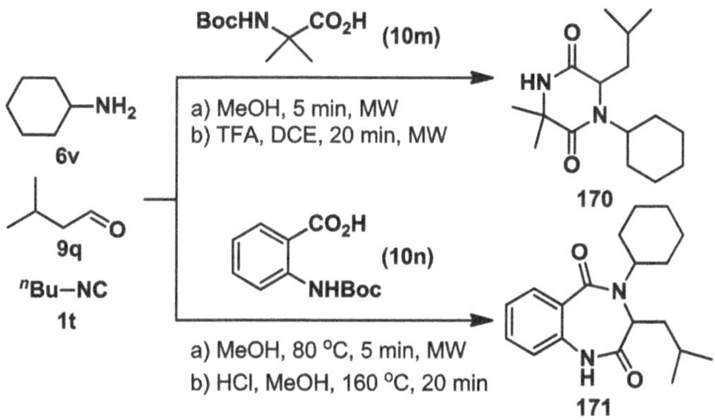

SCHEME 7.57 Synthesis of DKP **170** and BDZ **171**.

7.2.1.12 2-Isocyano Vinyl and Phenyl Esters Pirrung and coworkers took advantage of the metallation of oxazoline **172** at the 2-position (Scheme 44), as reported by Lindhorst et al. [64], and elegantly developed (*Z*)-isocyanovinyl (**1u**) and 2-isocyanophenyl (**1w**) esters [72]. They can be easily prepared in a good to excellent yield from commercially available oxazole (**172**) and benzoxazole (**174**) by treating with ^{n}BuLi in THF at −78°C and capturing the lithium alcholate with various acyl chlorides (Scheme 7.59). These new CICs are very unique and have a fragrant-like smell (Table 7.1). They are stable at room temperature when stored under argon and can be synthesized on a large scale (e.g., 5 g).

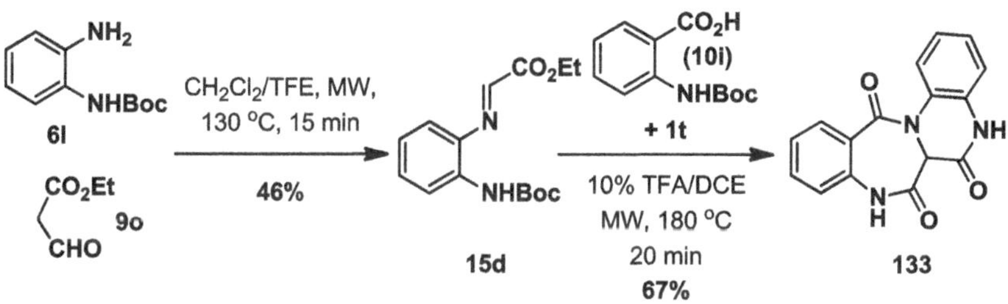

SCHEME 7.58 Synthesis of quinoxaline-benzodiazepines by using **1t**.

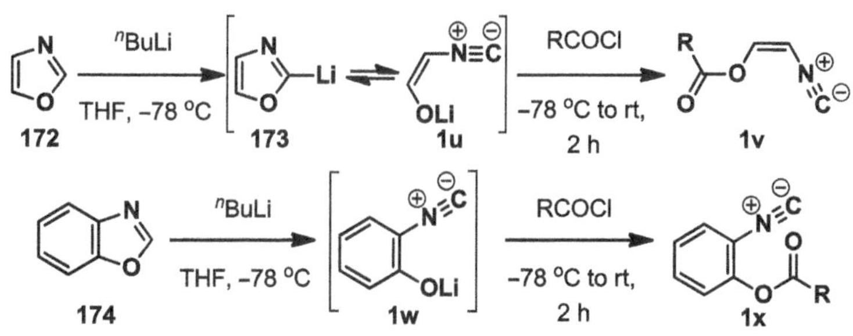

SCHEME 7.59 Synthesis of 2-isocyano vinyl (**1u**) and phenyl (**1w**) esters.

TABLE 7.1 Odor profile of 2-isocyano vinyl (1u) and phenyl (1w) esters

Entry	RCOCl	Yield %/Compound	Odor
1u	AcCl	75%/$\mathbf{1u_1}$	Mild isocyanide
1u	*O*-Formylmandeloyl-Cl	95%/$\mathbf{1u_2}$	Soy
1w	AcCl	85%/$\mathbf{1w_1}$	Malt
1w	PivCl	92%/$\mathbf{1w_2}$	Natural rubber
1w	MeC_6H_4COCl	96%/$\mathbf{1w_3}$	Cresote
1w	BocCl	83%/$\mathbf{1w_4}$	Taffy
1w	$MeOC_6H_4COCl$	96%/$\mathbf{1w_5}$	Mild cherry
1w	NCC_6H_4COCl	90%/$\mathbf{1w_6}$	Old wood
1w	C_6H_5COCl	93%/$\mathbf{1w_7}$	Mild petroleum

They have shown that these isocyano esters can be utilized in the U-4CCR and the resulting acyclic product **175** can be converted to the corresponding ester **176** quantitatively under Armstrong's conditions (Scheme 7.60).

The use of CIC $\mathbf{1w_3}$ derived from benzoxazole in the U-4CR provides the acyclic product **177**, which under previous conditions (Scheme 7.60) gave a quantitative yield of ester **176** along with benzoxazole derivative **178** (Scheme 7.61).

They were also able to synthesize pentasubstituted pyrroles **126** (Scheme 7.62) by refluxing the acyclic U-4CR product **179** in toluene at 100°C in the presence of HCl and dimethylacetylenedicarboxylate.

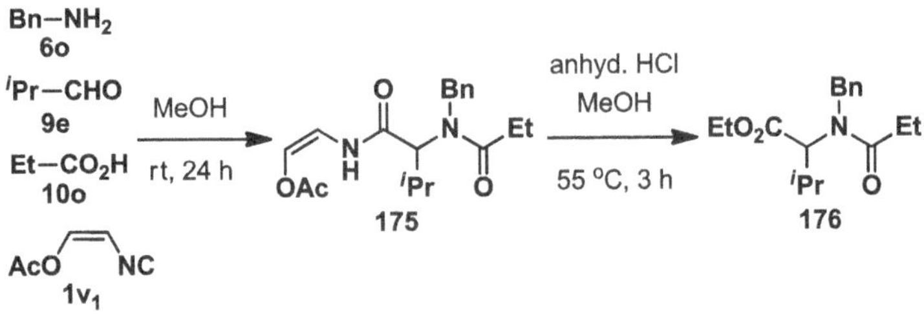

SCHEME 7.60 Cleavage of $\mathbf{1v_1}$ under Armstrong's conditions.

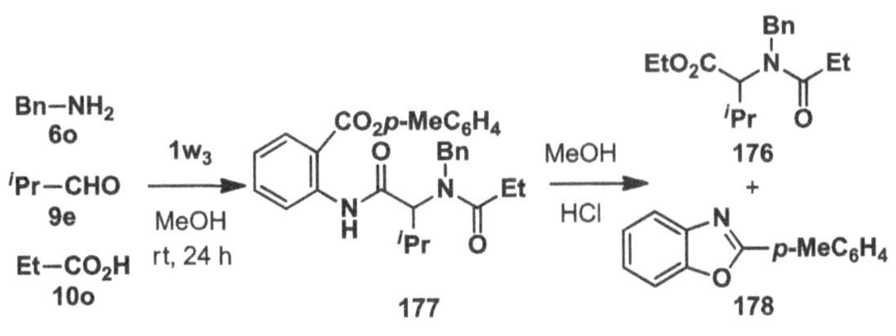

SCHEME 7.61 Benzoxaole formation from cleavage of CIC $\mathbf{1w_3}$.

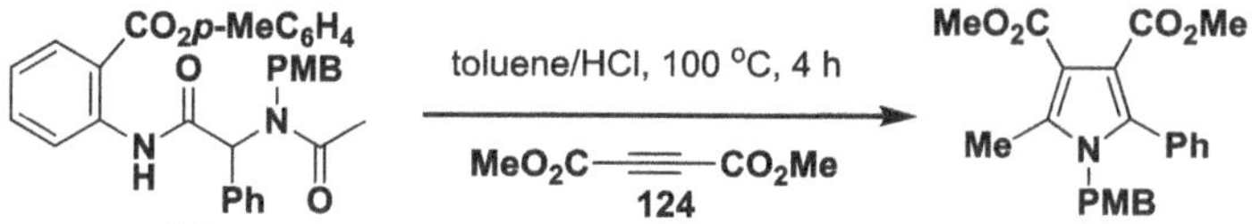

179 126

SCHEME 7.62 Synthesis of pentasubstituted pyrroles **126** using dimethylacetylenedicarboxylate.

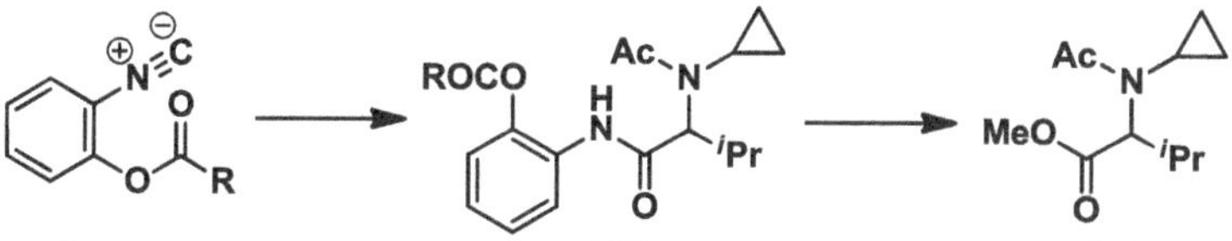

$1w_5$-w_7 180 181

SCHEME 7.63 Effect of substituents on rate of conversion of **180** to **181**.

Pirrung and coworkers have shown that the conversion rate of the U-4CR product **180** to the corresponding ester **181** is influenced by the electronic nature of *para*-substitutents (R) of the phenyl ring on the CICs $\mathbf{1w_5}$–$\mathbf{w_7}$ (Scheme 7.63). Electron-donating groups (e.g., OMe) accelerate the rate of conversion (0.5 h), whereas electron-withdrawing groups (e.g., CN) slow down the rate of conversion (2.5 h).

The postulated mechanism for the formation of esters from the Ugi products is depicted in Scheme 7.64. Internal nucleophilic attack of the amide nitrogen lone pair onto the protonated ester carbonyl of **183** (derived from isocyanide) forms the intermediate **184**, which equilibrates with **185**. Electron-donating groups on the phenyl ring accelerate the formation of intermediate **186** and help stabilize it by forming intermediate **187** after the loss of water. Nucleophilic attack by the solvent used at the amide carbonyl of **187** (derived from isocyanide) provides the desired esters **181**.

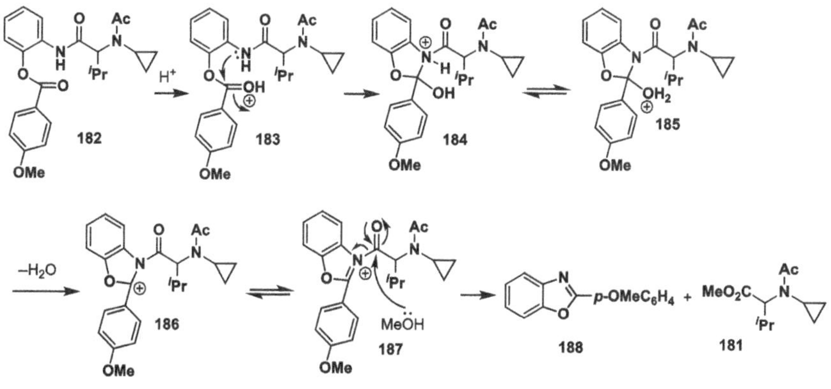

SCHEME 7.64 Mechanism of formation of esters from the U-4CR products.

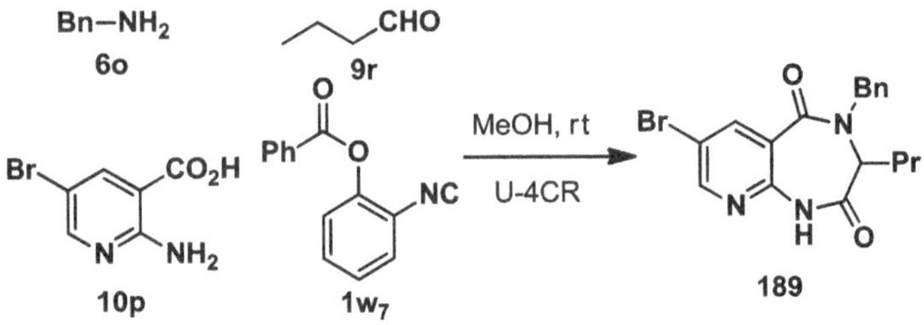

SCHEME 7.65 Synthesis of pyridodiazepinediones using CIC $\mathbf{1w_7}$.

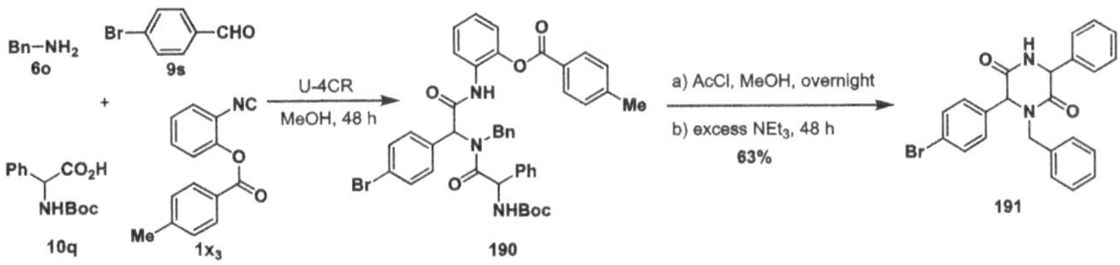

SCHEME 7.66 Synthesis of DKP 191 using CIC $\mathbf{1w_3}$.

Recently, De Borggraeve and coworkers have utilized Pirrung's CIC $\mathbf{1w_7}$ to synthesize pyridodiazepinediones **189** via the U-4CR (Scheme 7.65) [73].

More recently, Fei and coworkers have employed the Ugi/deprotection/activation/cyclization strategy for their synthesis of DKPs **191** using CIC $\mathbf{1w_3}$ (Scheme 7.66) [74]. Anhydrous HCl, which forms *in situ*, helps deprotect the Boc group and activates the amide carbonyl group derived from $\mathbf{1w_3}$.

7.2.1.13 1-Isocyano-2-(2,2-dimethoxyethyl)benzene 1-Isocyano-2-(2,2-dimethoxyethyl)benzene (DMEBIC, **1x**) was developed by Kobayashi and coworkers [75]. It is also known as the indole isocyanide. It can be conveniently synthesized from commercially available *o*-nitro-toulene as shown in Scheme 7.67. Leimgruber–Batcho enamine reaction provided **193**, which can be converted to

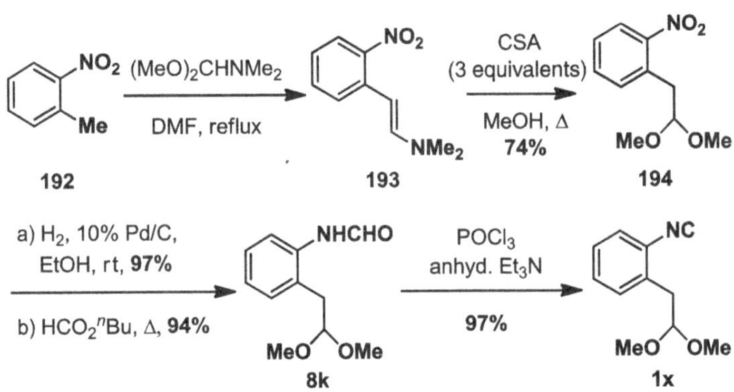

SCHEME 7.67 Synthesis of DMEBIC (**1x**).

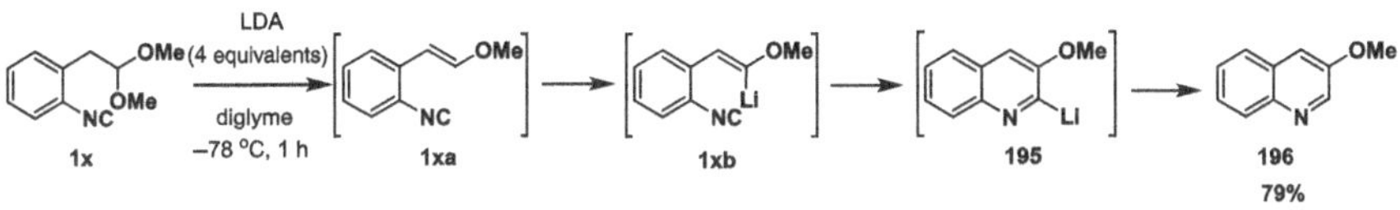

SCHEME 7.68 Synthesis of isoquinoline 196 from CIC **1y**.

2-(2-nitrophenyl)-acetaldehyde dimethyl acetal (**194**) by refluxing with camphor sulfonic acid (CSA) for 9 h [76]. Hydrogenation followed by formylation with butyl formate (or hexyl formate) under refluxing conditions provides the formamide **8k**. Dehydration of **8k** in the presence of $POCl_3$ and anhydrous triethylamine (Et_3N) provides **1x** as a pale yellow liquid.

Kobayashi and coworkers have developed an efficient method to synthesize isoquinolines by utilizing **1x** [75]. Treatment of **1x** with 4 equivalents of lithium diisopropyl aluminum forms **1xa** via the elimination of methoxide from benzyl anion. The lithiated intermediate **1xb** then undergoes an intramolecular cyclization to provide the desired isoquinolinone **196** in a 79% yield. It is noteworthy that the use of less than 4 equivalents of lithium diisopropyl aluminum in the reaction lowers the yield of the expected isoquinolines and **1x** can be recovered (Scheme 7.68).

Buller and coworkers [77] have demonstrated that **1x** can be used in the Ugi four-component three-centered reaction (U-4C-3CR) along with γ-keto acid **10r** to synthesize pyroglutamic acid derivatives. Condensation of levulinic acid with CIC **1x** and *p*-methoxybenzylamine (**6i**) forms the acyclic product **197**, which can be converted to *N*-acylindole **198** by refluxing in benzene in the presence of CSA. Treatment of **198** with cesium carbonate in a 1:1 mixture of DMF and water provides the acid **199**. The presence of other nucleophiles (e.g., alcohols, thiols, amines) provides corresponding esters, thioesters, and amides (e.g., **200**, **201**, and **202**), respectively. The *N*-acylindole **198** can also be converted to the corresponding aldehyde **203** by reduction with sodium borohydride (Scheme 7.69) [77]. Although the method provides good yields of esters, thioesters, and amides, the limitation of this method is that secondary alcohols (isopropanol) and amines (morpholine) do not react with the *N*-acylindole **198** due to steric hindrance. They have also been able to synthesize bicyclic pyroglutamic acid derivatives using this methodology [78].

Kobayashi and coworkers have applied this methodology to synthesize omuralide (**207**) (Scheme 7.70) [79], which is derived from the natural product lactacystin (**204**, Figure 7.4) and has been shown to be a selective inhibitor of the *20S* proteasome found in mammalian and bacterial cells [80]. Proteasome inhibition has become an attractive and novel alternative therapy to cure cancer in recent years [81].

Diastereoselective U-4CR of chiral γ-keto acid **10s** with **6i** and **1x** in 2,2,2-trifluoroethanol at room temperature provides γ-lactam **205** in a 78% yield as a single diastereomer. The diastereoselectivity is believed to arise from the exclusive axial approach of small nucleophiles onto the carbonyl group of 1,3-dioxan-4-ones [82]. Deprotection of the acetonide with catalytic amount of CSA in MeOH followed by protection of the diol to the corresponding diacetate and further treatment with catalytic amount of CSA in benzene provides the *N*-acylindole **206**. The *N*-acylindole

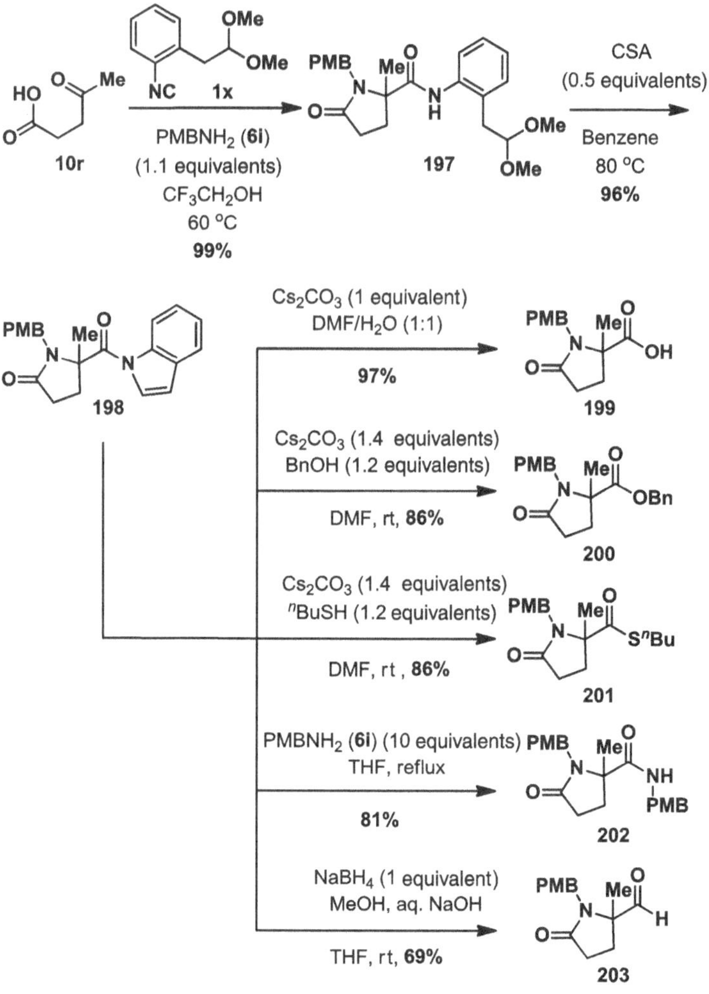

SCHEME 7.69 Synthetic application of *N*-acylindole **198**.

can be converted into omuralide (**207**) via a series of transformations (Scheme 7.70). The limitations of this approach are as follows: (i) the acidic conditions used to form the *N*-acylindole intermediate **206** from the sterically hindered *exo*-anilide **205** are not compatible with the adjacent unprotected alcohols and (ii) direct intramolecular conversion of the activated *N*-acylindole intermediate **206** to the β-lactone with simultaneous expulsion of indole does not occur readily.

In another application, Kobayashi and coworkers have employed CIC **1x** in the U-4C-3CR to synthesize (±)-dysibetaine (**212**) [83], which was isolated from Micronesian sponge *Dysidea herbacea* [84] and has shown potential neuroexcitotoxin activity. Condensation of γ-keto acid **10t** with ammonium acetate and CIC **1x** provides acyclic Ugi product **208**. It must be noted that other ammonia sources (e.g., ammonium halide

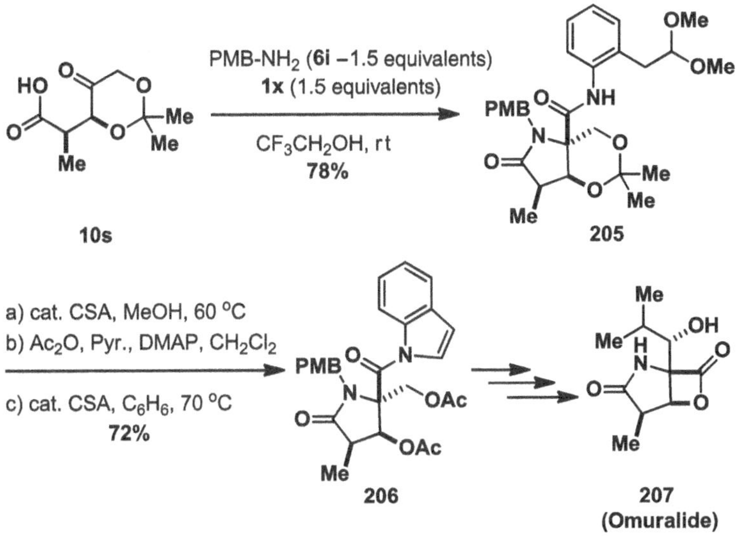

SCHEME 7.70 Application of CIC **1y** in the total synthesis of omuralide (**207**).

salts) except hexamethyldisilazane did not provide the expected U-4C-3CR product **208**. Silyl ether deprotection, mesylation, and *N*-acylindole formation followed by cleavage of *N*-acylindole provides ester **211**, which can be converted to the final product **212** after a series of transformations (Scheme 7.71). Following a similar strategy, they were able to synthesize enantiopure (−)-dysibetaine.

Kreye and coworkers have utilized CIC **1x** in their U-4CR for the synthesis of α,ω-diene monomer **213** (Scheme 7.72) [85]. Employing an acyclic diene metathesis polymerization strategy, the monomers were subsequently polymerized using the second-generation Hoveyda–Grubbs catalyst in the presence of 1,4-benzoquinone (1,4-BQ) as an additive to obtain compound **214**. Hydrogenation followed by the treatment with pyridinium *p*-toluenesulfonate provided the *N*-acylindole intermediate **215**. Under various conditions, **215** provided polymers containing amide and acid (**216**, **217**, and **218**) functionalities that can be useful for further modification of the side chain. This method clearly exemplifies that judicious application of IMCRs can lead to a new era in polymer synthesis.

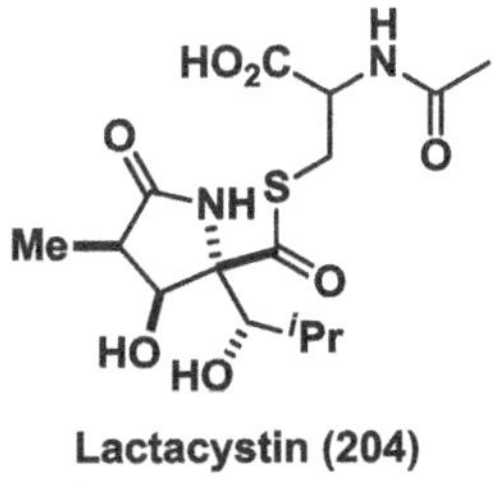

Lactacystin (204)

FIGURE 7.4 Lactacystin.

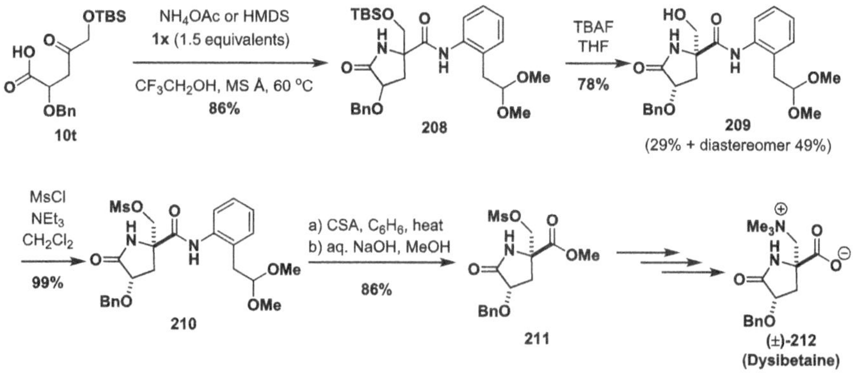

SCHEME 7.71 Application of CIC **1y** in the total synthesis of (±)-dysibetaine (**212**).

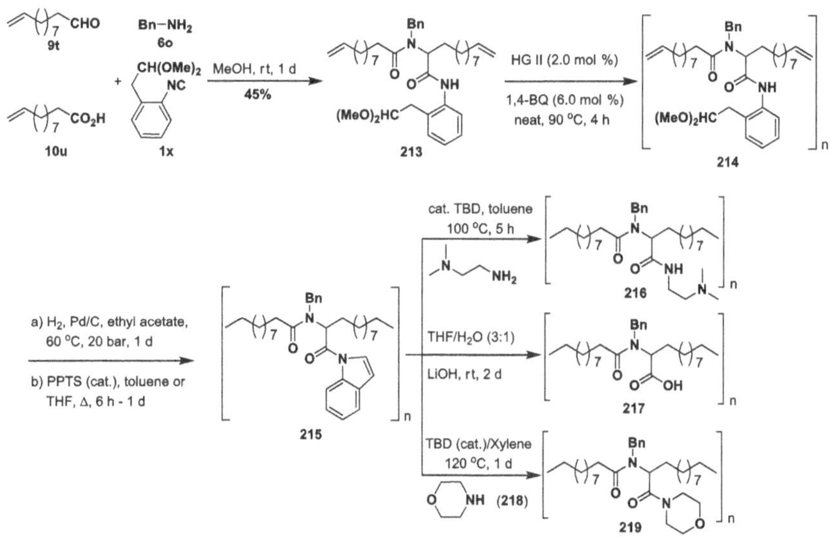

SCHEME 7.72 Application of CIC **1x** in a polymer synthesis.

7.2.1.14 3-Methoxy-2-nitrophenyl Isocyanide Martens and coworkers [61] have utilized 3-methoxy-2-nitrophenyl isocyanide (MNPIC) (**1y**) to synthesize fully protected PNA monomers [62] **220**. The anilide can be hydrolyzed by methanolic KOH to obtain the corresponding acid **221** (Scheme 7.73). Structurally, the PNAs are quite similar to the backbone of DNA or RNA. The oligomers of PNAs have become quite a useful tool in biomedical discoveries.

Recently, Baldoli and coworkers have employed **1y** in the U-4CR to synthesize chiral chromium tricarbonyl-labeled thymine PNA monomers **224** [86]. These new PNA monomers might be utilized to improve cell permeability (which is poor for PNAs) and resolve analytical problems by their unique spectroscopic and reactivity properties. Thus, the U-4CR of pre-formed geometric isomer benzaldimine **15f**

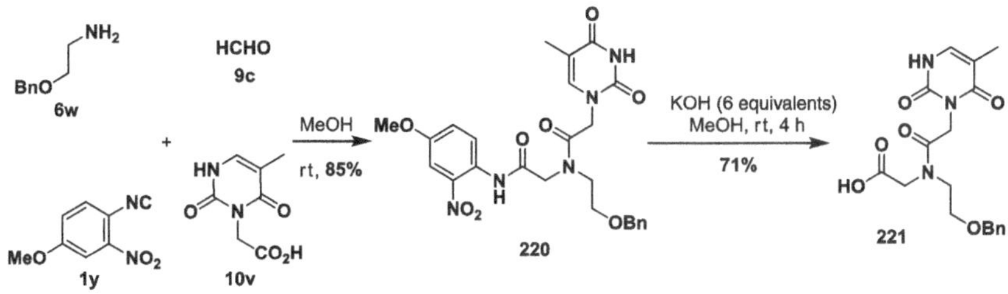

SCHEME 7.73 Application of 1y in the synthesis of PNA monomer **221**.

(from *o*-methylbenzaldehyde·$Cr(CO)_3$ (**9u**) and *N*-Cbz-ethylendiamine (**6x**)) with 1-carboxymethylthymine (**10v**) and MNPIC (**1y**) provides the acyclic anilide product **222** in a 52% yield as 1:1 mixture of diastereomers (Scheme 7.74). The two diastereomers can be separated by silica gel column chromatography to obtain enantiopure complexes. Although chiral benzaldehyde chromiumtricarbonyl complexes are known to provide high diastereoselectivity, the *ortho*-substitutions on benzaldehyde do not render any diastereochemical control [87]. The probable cause for the lack of diastereoselectivity is the heating necessary to promote the reaction. The anilide can be hydrolyzed to the corresponding potassium salt **223** using 1.5 equivalents of KOH in methanol at room temperature. In the presence of an electron-withdrawing nitro group at the *ortho*-position, a slightly weakened anilide amide bond makes it hydrolyzable under quite harsh conditions. It is noteworthy that the hydrolysis conditions are compatible with the use of the $Cr(CO)_3$ and Cbz groups.

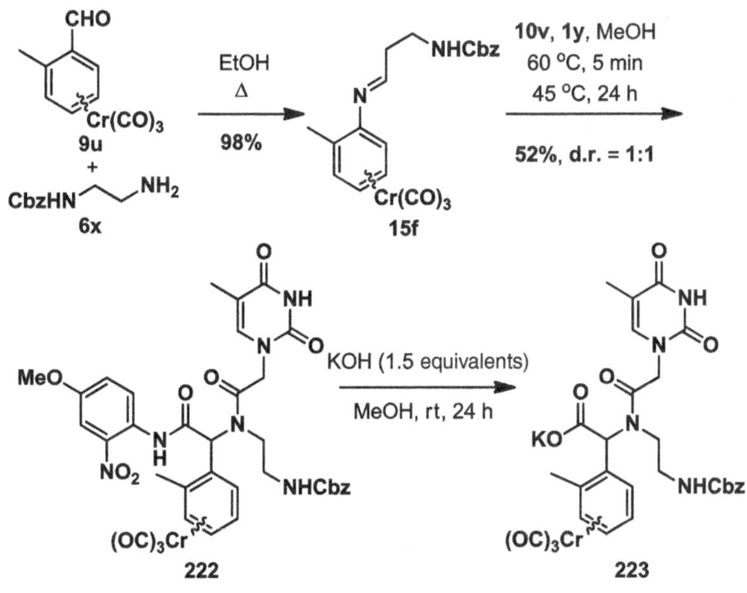

SCHEME 7.74 Synthesis of transition metal labeled PNA monomers (**223**).

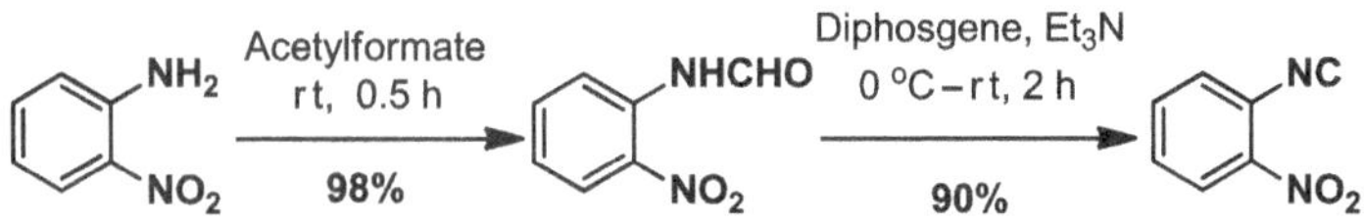

6y 8l 1z

SCHEME 7.75 Synthesis of CIC **1z**.

7.2.1.15 O-Nitrophenyl Isocyanide Although *o*-nitrophenyl isocyanide (ONPIC, **1z**) was known for quite a while, Hahn and coworkers have recently reported the first synthesis of ONPIC [88] as a ligand in metal complexation. It can be readily synthesized from commercially available *o*-nitroaniline **6x** via formamide **8l** formation followed by dehydration (Scheme 7.75).

In order to overcome some limitations over their previous methodology (i.e., the use of DMEBIC (**1x**)) employed in the synthesis of omuralide (**207**), Kobayashi and coworkers have developed an extraordinary methodology where they have utilized ONPIC (**1z**) in the U-4C-3CR [89]. Hydrogenation of the acyclic Ugi product **224** provides corresponding amine **225**. Diazotization under neutral condition by isoamyl nitrite in chloroform at room temperature forms the diazo intermediate **226**, which undergoes a *5-endo-dig* cyclization *in situ* to provide the *N*-acylbenzotriazole **227** (Scheme 7.76). Due to the presence of two additional electronegative nitrogen atoms, benztotriazoles (Bet's, (**93**)) are better leaving groups, which make the carbonyl group of the *N*-acylbenzotriazole highly activated. Furthermore, *N*-acylbenzotriazoles are known as the stable nonmoisture sensitive acid chloride equivalent. Thus, a variety of nucleophiles, for example, alcohols, thiols, amines, and Grignard reagents, react with **227** in the presence of 0.1 equivalent of triethylamine to provide corresponding esters (**228**, **229**), thiols (**230**), amides (**231**), and ketones (**233**), respectively. In addition, reduction of **227** by sodium borohydride provides corresponding alcohol **234** (Scheme 76). It is noteworthy to mention that all products can be obtained in a pure form without silica gel column chromatography because the by-product benzotriazole (**93**) can be removed by washing the crude reaction mixture with a 1 N NaOH solution. However, secondary and tertiary alcohols, as well as amines, do not react with the *N*-acylbenzotriazole **227** due to steric hindrance around the carbonyl carbon.

Next, they applied this improved methodology to synthesize the fused γ-lactam β-lactone core **239** of omuralide (**207**). Stereoselective U-4C-3CR reaction of γ-keto acid **10s**, *p*-methoxybenzylamine (**6i**), and **1z** provides acyclic anilide product **235**. Deprotection of acetonide group using CSA in MeOH followed by hydrogenation provides the aniline derivative **236**, which directly forms the γ-lactam β-lactone intermediate **238** in the presence of isoamyl alcohol and catalytic amount of triethylamine. Silyl protection of alcohol **238** without purification affords **239** in a 71% yield (Scheme 7.77).

Kobayashi and coworkers have further translated the concept of CIC in their synthetic approach toward the synthesis of hasubanan alkaloid (Figure 7.5) Hasubanonine (**240**) [90]. Although the biological activity of **240** has not been evaluated yet, structurally it is similar to morphine-related alkaloids and therefore may

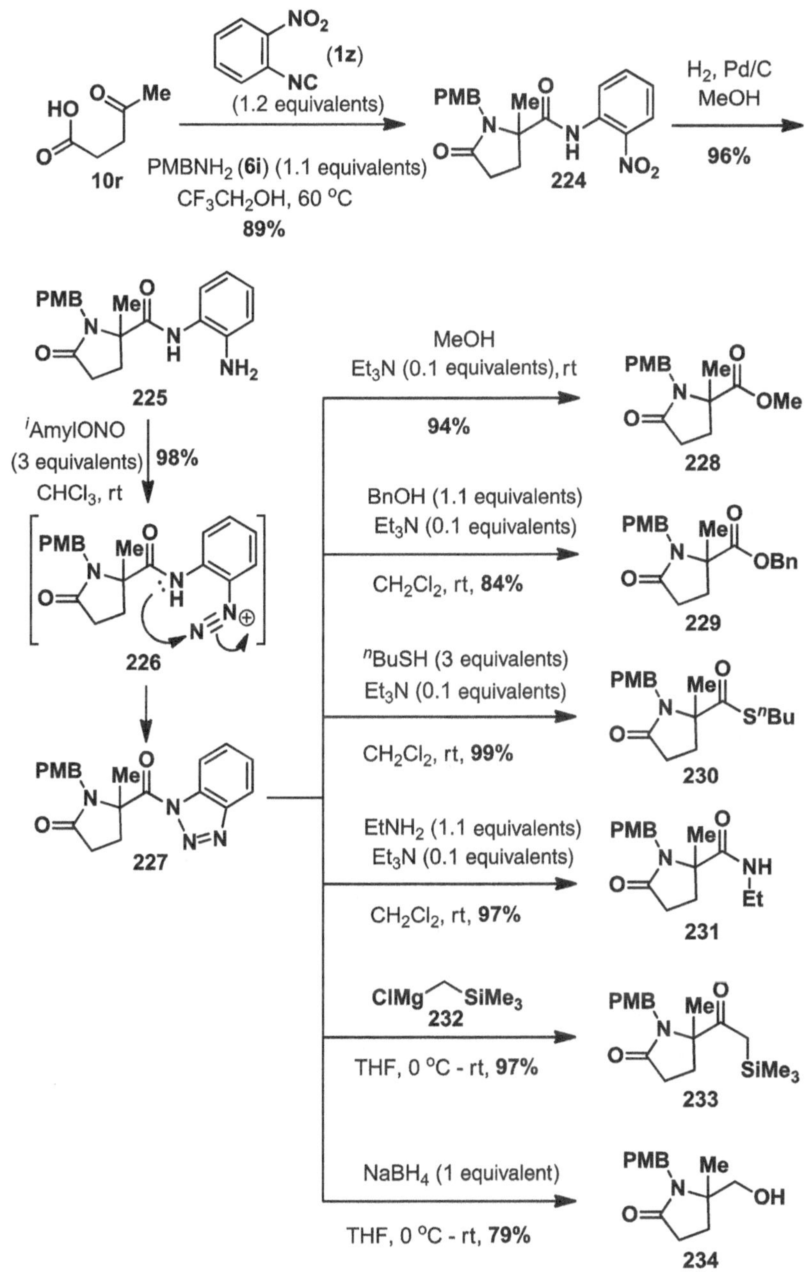

SCHEME 7.76 Utility of *N*-acylbenzotriazole **227** derived from U-4C-3CR using CIC **1z**.

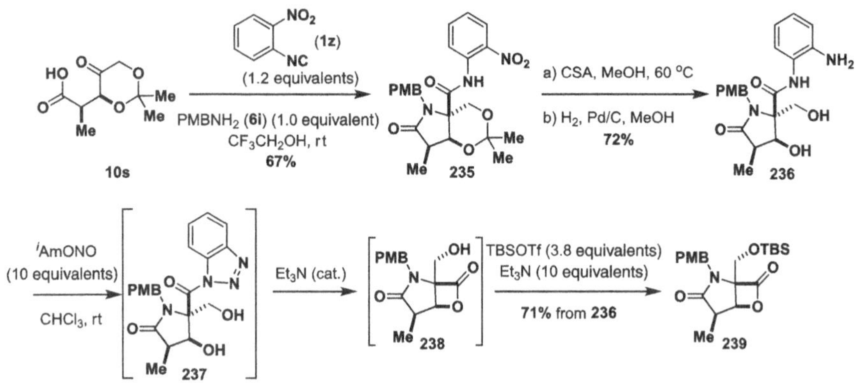

SCHEME 7.77 Application of CIC **1z** in the synthesis of fused γ-lactam β-lactone core **239** of omuralide (**207**).

possess potential analgesic properties. A closely related family member, acutumine (**241**), has been shown to inhibit the growth of T cells and enhance memory. In their synthetic strategy, they have synthesized the key anilide intermediate **247** from compound **246** using H_2O_2-mediated amide formation followed by the Goldberg reaction of 2-iodonitrobenzene and reduction of corresponding nitroanilide by $SnCl_2$ (Scheme 7.78). It is important to note that the base-mediated hydrolysis of **248** is quite difficult due to potential steric hindrance. However, tin-mediated chemoselective reduction followed by the treatment of isoamyl nitrite smoothly provided the *N*-acyltriazole intermediate **249**. Without isolation, compound **249** provided the methyl ester **250**, which after several steps afforded the propellene framework **251** of hasubanone (**241**).

Recently, Carreira and coworkers have utilized CIC **1z** in their U-4CR for the rapid synthesis of the core fragment of Orodin alkaloid Massadine (**252**, Figure 7.6), which was isolated from *Stylissa aff Massa* [91]. Reaction of norbornenone (**9v**), 2,4-dimethoxy benzylamine (DMB-NH_2, **6y**), *p*-TsOH, and **1za** provided the acyclic Ugi product **256** in a 79% yield. Removal of DMB group, Boc protection, and reduction, followed by treatment of *i*-amylnitrile afforded the *N*-acylbenzotriazole intermediate **257**. Borohydride reduction of **257** provided alcohol **258**. After several steps, the suitably substituted D-ring **259** of Massadine (**252**) was provided (Scheme 7.79) [91].

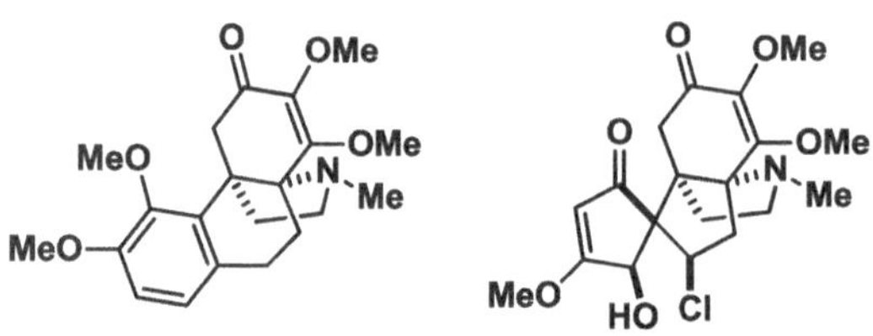

FIGURE 7.5 Hasubanan alkaloids.

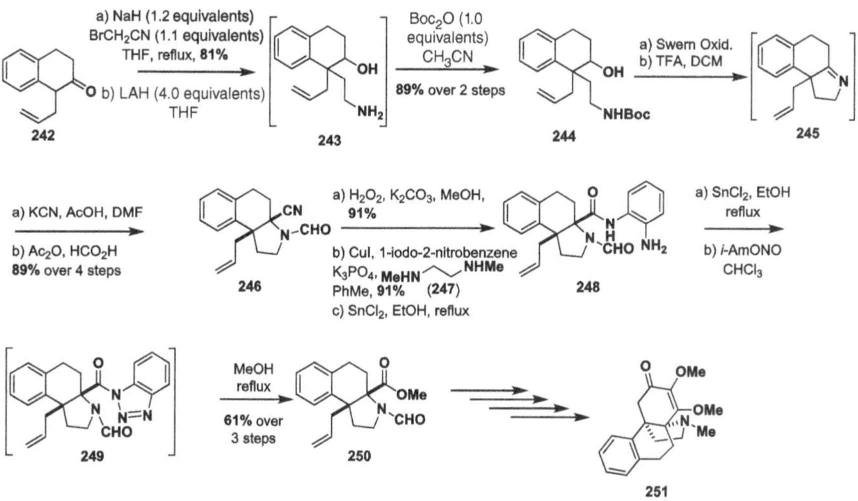

SCHEME 7.78 Synthesis of the propellane framework **251** of hasubanone **240**.

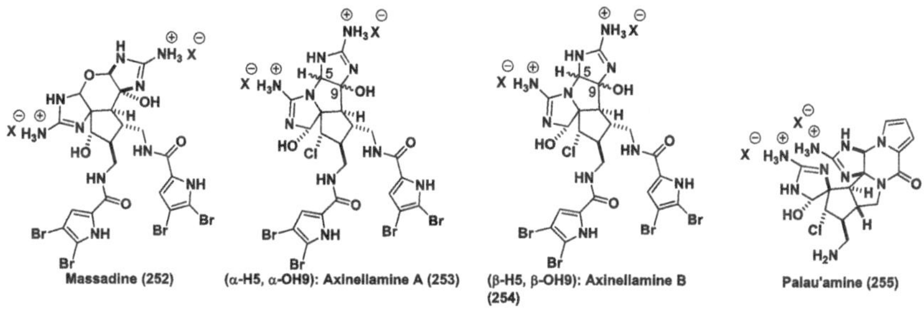

FIGURE 7.6 Orodin alkaloids (**252–255**).

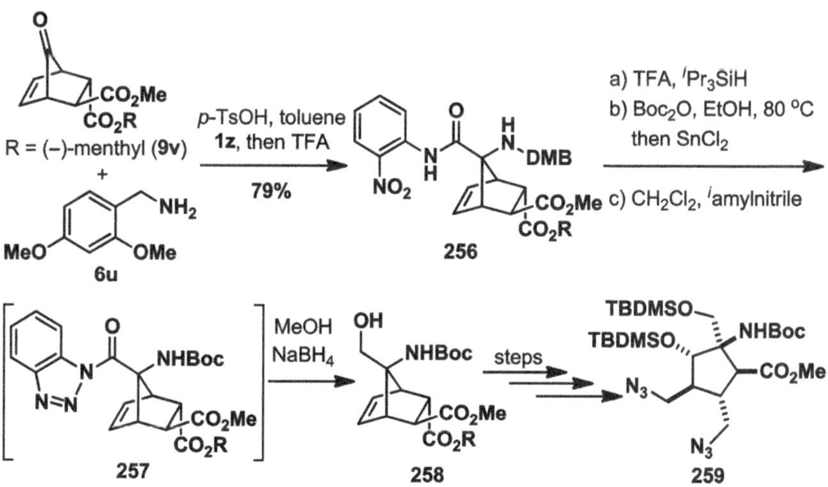

SCHEME 7.79 Synthesis of the core fragment **259** of massadine (**252**) using **1z**.

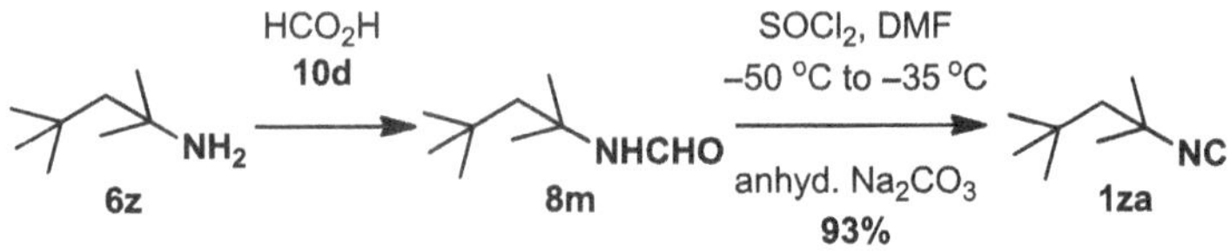

SCHEME 7.80 Synthesis of TMBIC (**1za**).

7.2.1.16 1,1,3,3-Tetramethylbutyl Isocyanide Walborsky and coworkers reported the first synthesis of 1,1,3,3-tetramethylbutyl isocyanide (TMBIC, **1za**) in 1969 [92]. It can be readily synthesized by the dehydration of corresponding formamide **8m** using a DMF (anhydrous) solution of thionyl chloride (which generates chlorodimethylforminium chloride *in situ*) and sodium carbonate (Scheme 7.80). TMBIC is commonly known as Walborsky's CIC.

Recently, the synthesis of 5-aminothiazoles **259** has been achieved by Thompson and Chen via the U-4CR using **1za** as a coupling partner [93]. The DMB group was removed by treating with a 20% TFA-CH_2Cl_2 solution to obtain **257**, which was treated with Lawesson's reagent and refluxed in toluene to afford **258**. Hydrolysis of the tetramethylbutyl (TMB) group can be achieved by a 1:1 mixture of TFA-CH_2Cl_2 within 20 min to obtain the final product **259** (Scheme 7.81). To expand the scope of their methodology, they used ammonia instead of DMB-NH_2 (as an ammonia replacement) in the U-4CR to obtain similar types of final products [94].

Dömling and coworkers have utilized TMBIC (**1za**) as a CIC to synthesize α-aminomethyl tetrazoles (**264**), which are isosteric to α-amino acids [45]. The TMB group can be removed by HCl/MeOH at room temperature (Scheme 7.82).

7.2.1.17 Tert-butyl Isocyanide It has long been known that TBIC (**1zb**) can be synthesized via the dehydration of the corresponding formamide. It is quite volatile and has an unpleasant odor. Traditionally, **1zb** has been used as a NCIC in MCRs. However, recently, Krasavin and coworkers have utilized TBIC (**1zb**) in the U-4CR to synthesize 5,6-dihydropyrazolo[1,5-*a*]pyrazine-4,7-diones **266** [95] and shown that

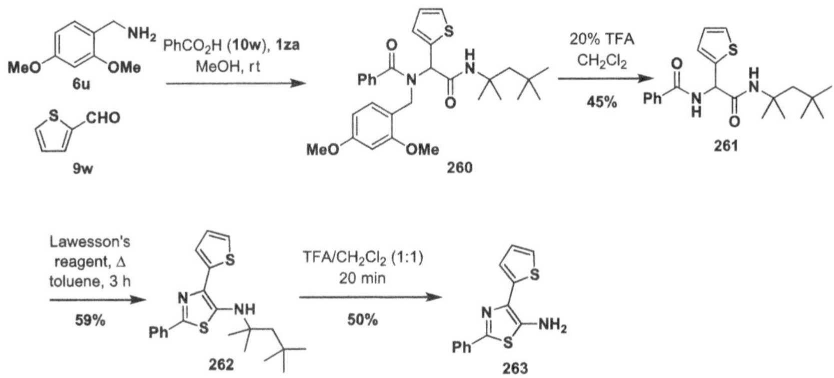

SCHEME 7.81 Synthesis of 5-aminothiazoles **263** using **1za**.

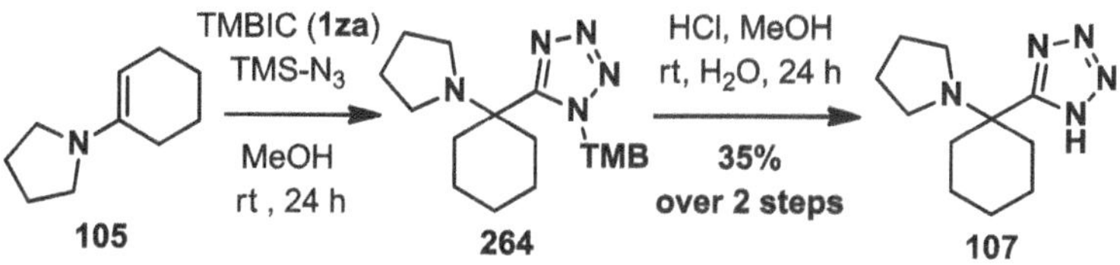

SCHEME 7.82 Synthesis of α-aminomethyl tetrazoles using TMBIC (**1za**).

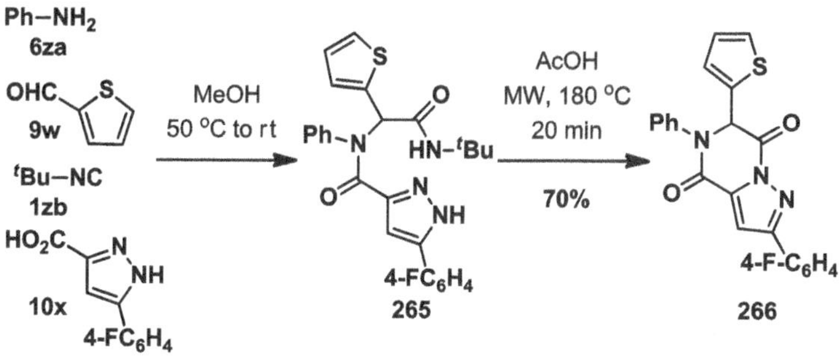

SCHEME 7.83 Synthesis of 5,6-dihydropyrazolo[1,5-*a*]pyrazine-4,7-diones **266**.

1zc can used as a CIC. The ^{t}Bu group can be cleaved by MWI using acetic acid as a solvent as well as a catalyst (Scheme 7.83). Lewis acids under MW irradiation can also cleave the *t*-butyl group [96].

More recently, Deprez and coworkers have also reported the use of **1zb** as a CIC in their synthesis of DKPs **268** (Scheme 7.84) [97].

7.2.1.18 Cyclohexyl Isocyanide Several methods are available for the synthesis of cyclohexyl isocyanide. However, it can be readily prepared from the corresponding formamide in the presence of $POCl_3$ or cyanuric chloride [98]. It is a colorless liquid having an offensive odor. Although cyclohexyl isocyanide (**1zc**) has been known for several years and commercially available, traditionally it has been used as a NCIC in MCRs. Recently, Yan and coworkers have reported that **1zc** can be used as a CIC in the presence of TFA in methanol under MWI. They employed this methodology for their fluorous synthesis of 1,4-BDZs **270** (Scheme 7.85), which can be further modified via the Suzuki coupling to obtain **272** [99]. Deprotection of cyclohexylamine has also been observed in the presence of Lewis acids [96].

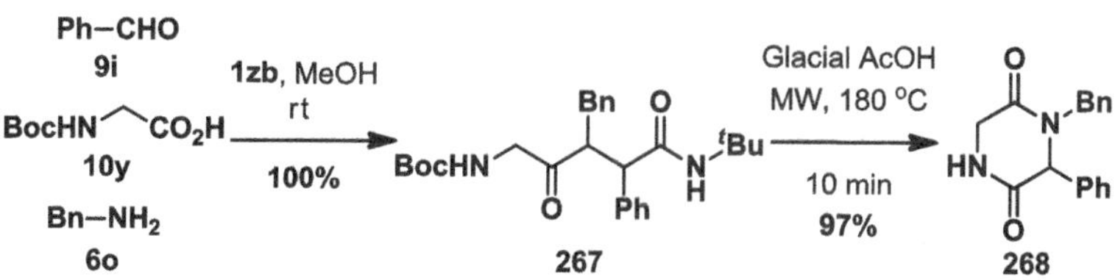

SCHEME 7.84 Synthesis of DKP **268** using **1zb**.

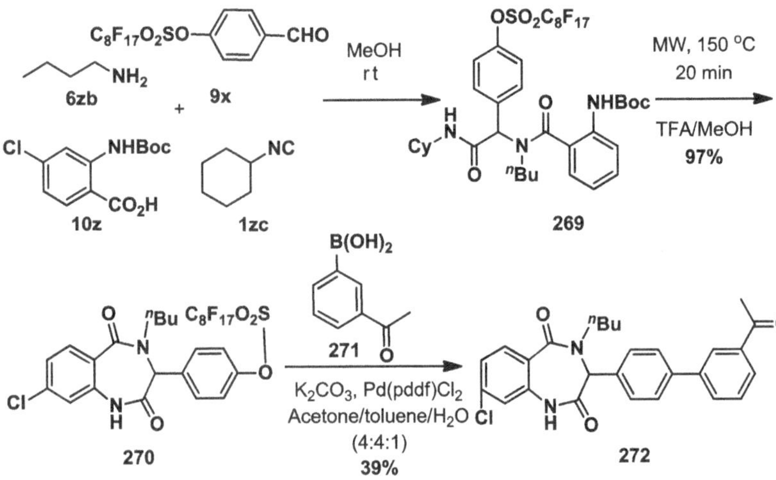

SCHEME 7.85 Synthesis of BDZs using CIC **1zc**.

7.2.1.19 Methyl 2-Isocyanoacetate Although methyl 2-isocyanoacetate (MICAc, **1zd**) is cheap and commercially available, it is relatively underexplored in combinatorial chemistry. However, Yan and coworkers have recently used it (**1zd**) as a CIC for their synthesis of 1,4-BDZs **274** via the U-4CR (Scheme 7.86) [99].

7.2.1.20 4-Isocyanopermethylbutane-1,1,3-triol Recently, Fukuyama and coworkers have developed the 2,4,4-trimethoxy-butylamine (**6zd**) group as a masked PG for amide bonds [100]. Inspired by the concept of the Fukuyama amide deprotection strategy, Wessjohann and coworkers have utilized 4-isocyanopermethylbutane-1,1,3-triol (IPBT, **1ze**) in their U-4CR. They have shown that **1ze** is highly reactive

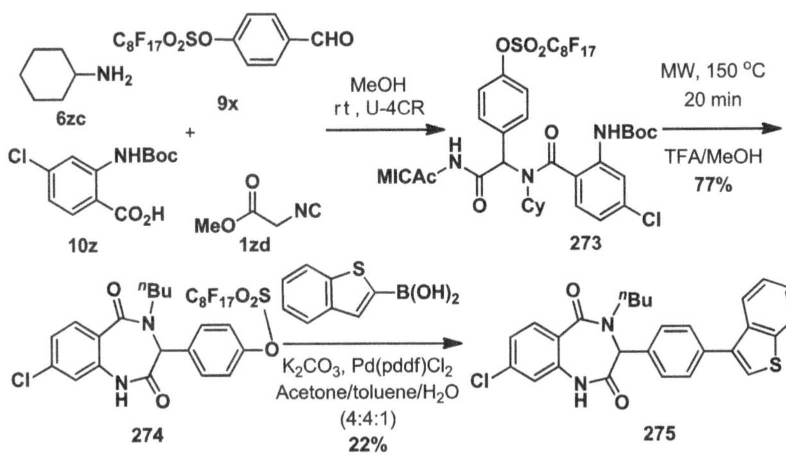

SCHEME 7.86 Synthesis of BDZs using CIC **1zd**.

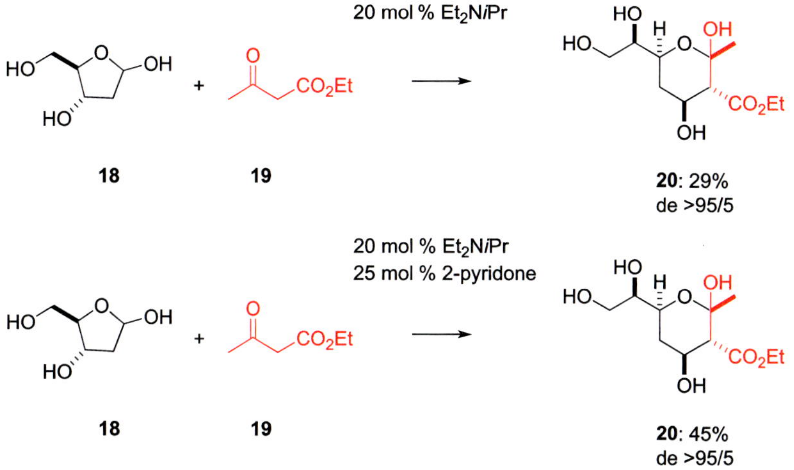

SCHEME 2.4 Base-catalyzed Knoevenagel reaction of ribose with ethyl acetoacetate.

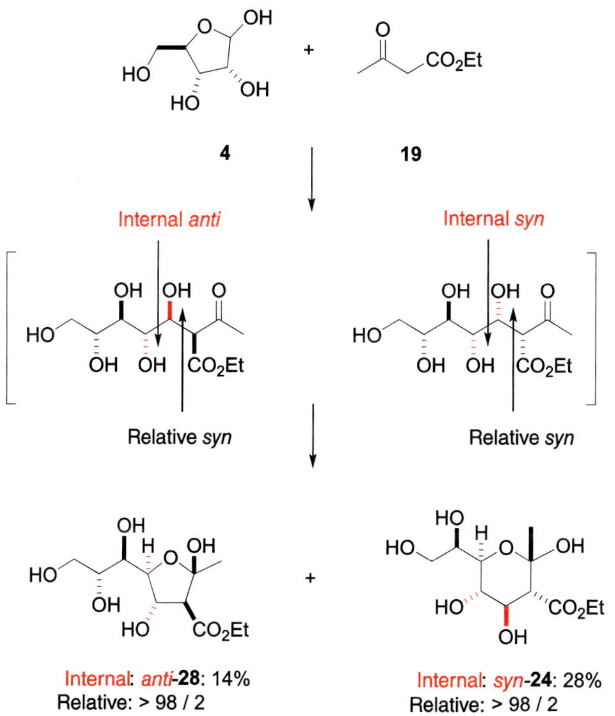

SCHEME 2.5 Configurative course of aldol reaction of D-ribose with ethyl acetoacetate.

Domino and Intramolecular Rearrangement Reactions as Advanced Synthetic Methods in Glycoscience, First Edition. Edited by Zbigniew J. Witczak and Roman Bielski.

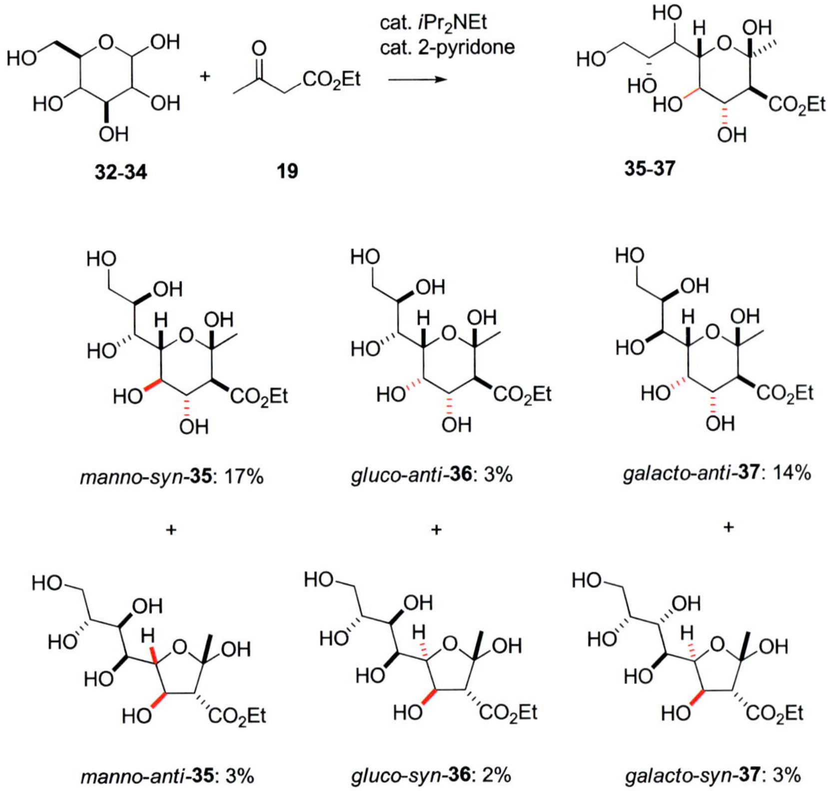

SCHEME 2.7 Chain elongation of D-hexoses with ethyl acetoacetate.

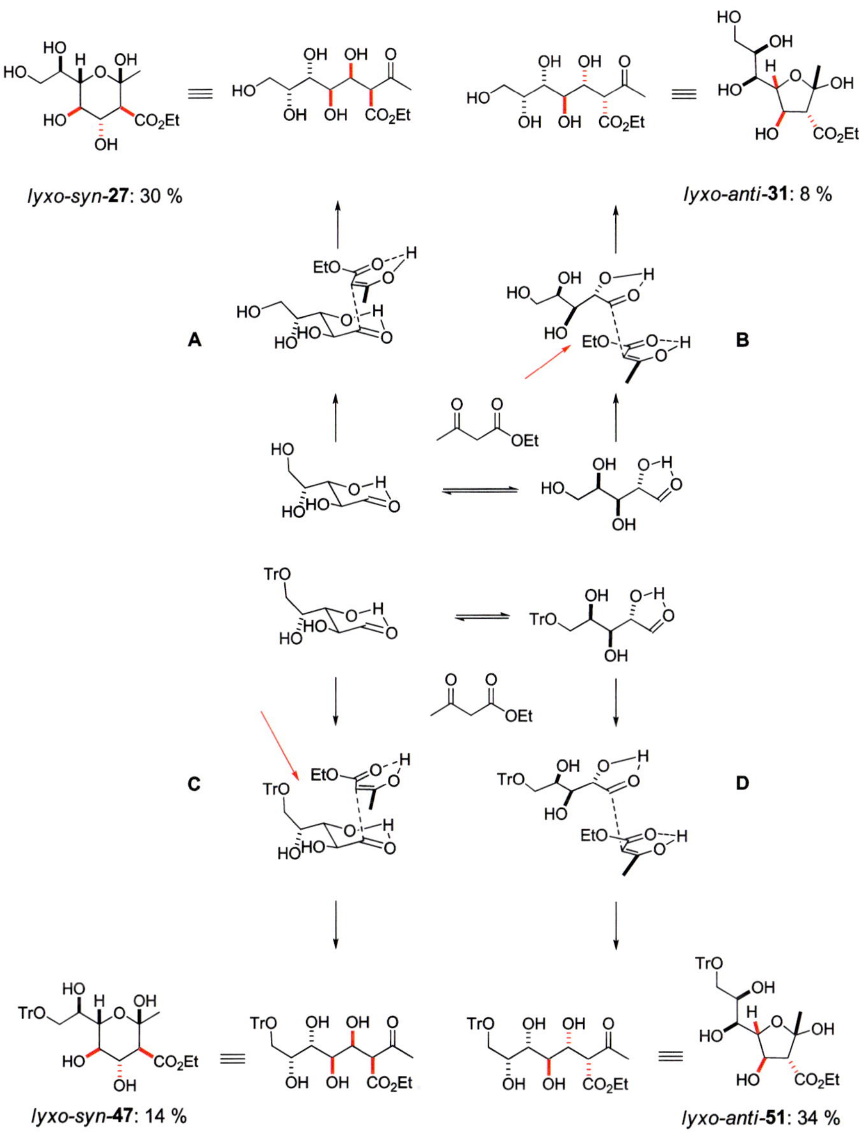

SCHEME 2.9 Configurative course of chain-elongation of lyxose **23** and 5-tritylated lyxose **42** with acetoacetic ester **19**.

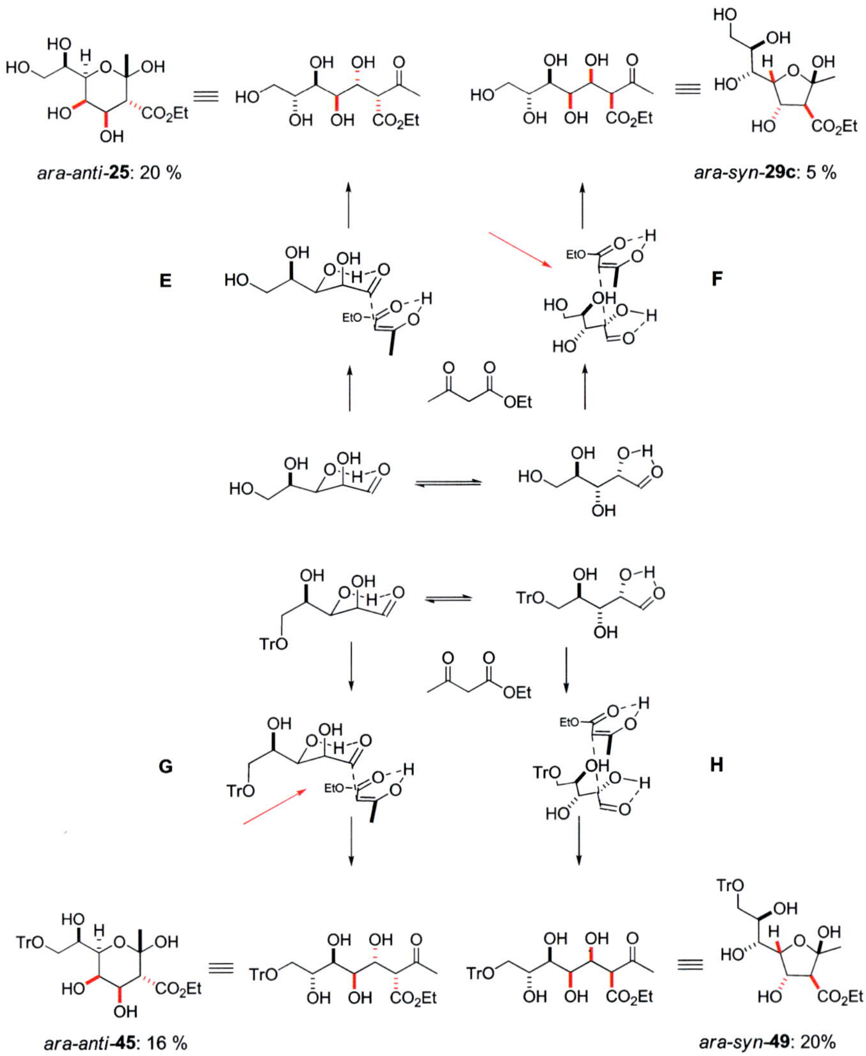

SCHEME 2.10 Configurative course of chain elongation of arabinose **21** and 5-tritylated arabinose **40** with ethyl acetoacetate **19**.

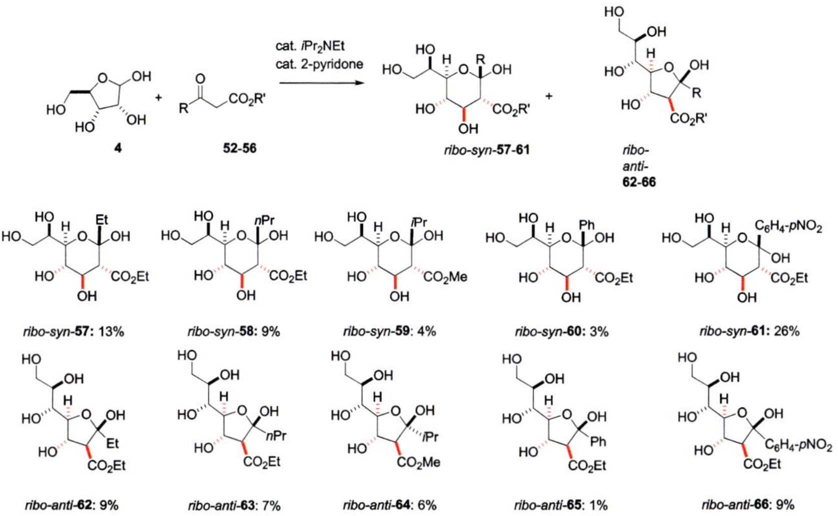

SCHEME 2.11 Chain elongation of D-ribose with different substituted acetoacetates.

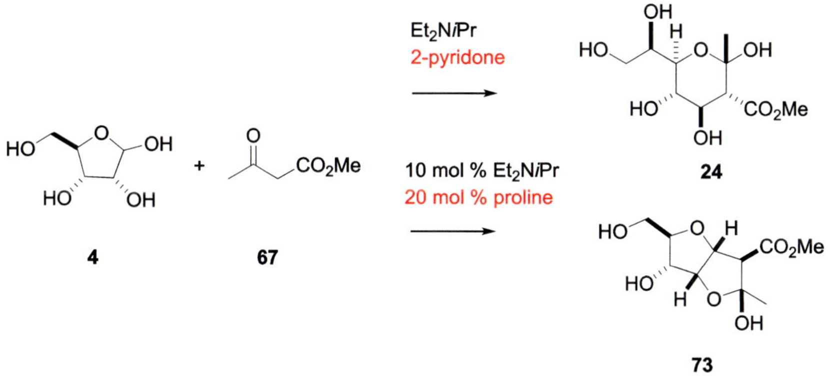

SCHEME 2.12 Different cascade reactions of D-ribose with acetoacetate.

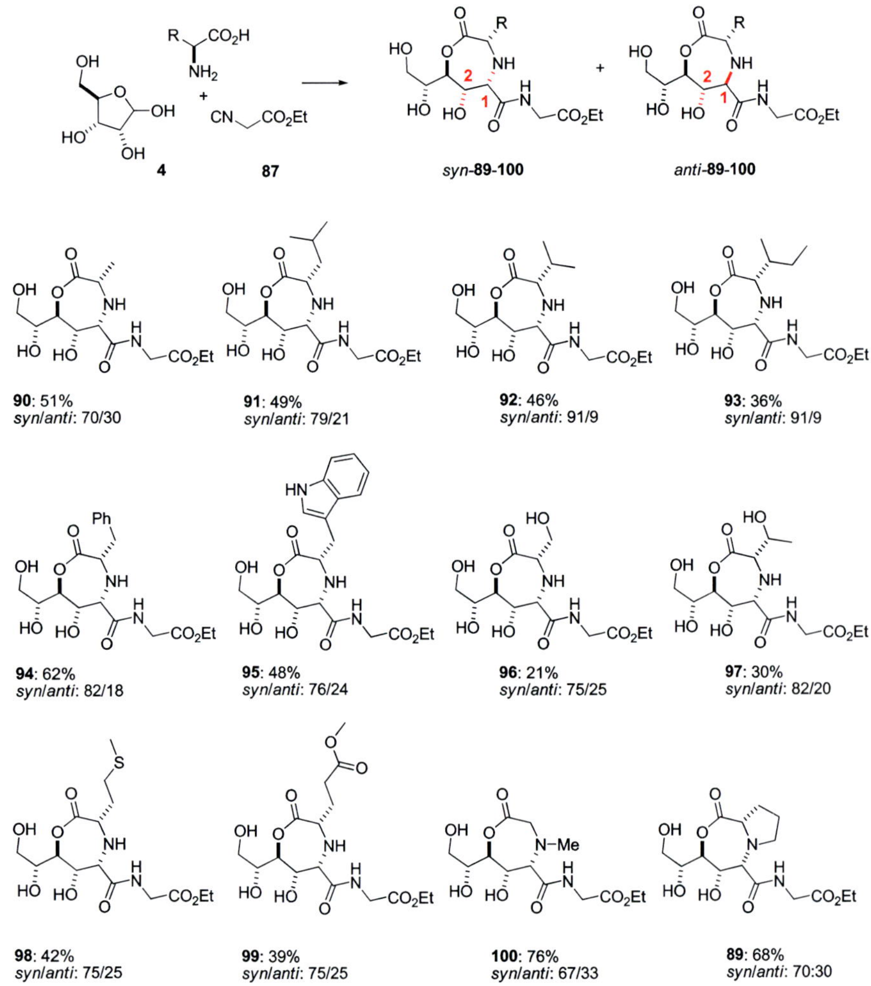

SCHEME 2.20 Multicomponent reactions of D-ribose and ethyl isocyanoacetate with L-amino acids.

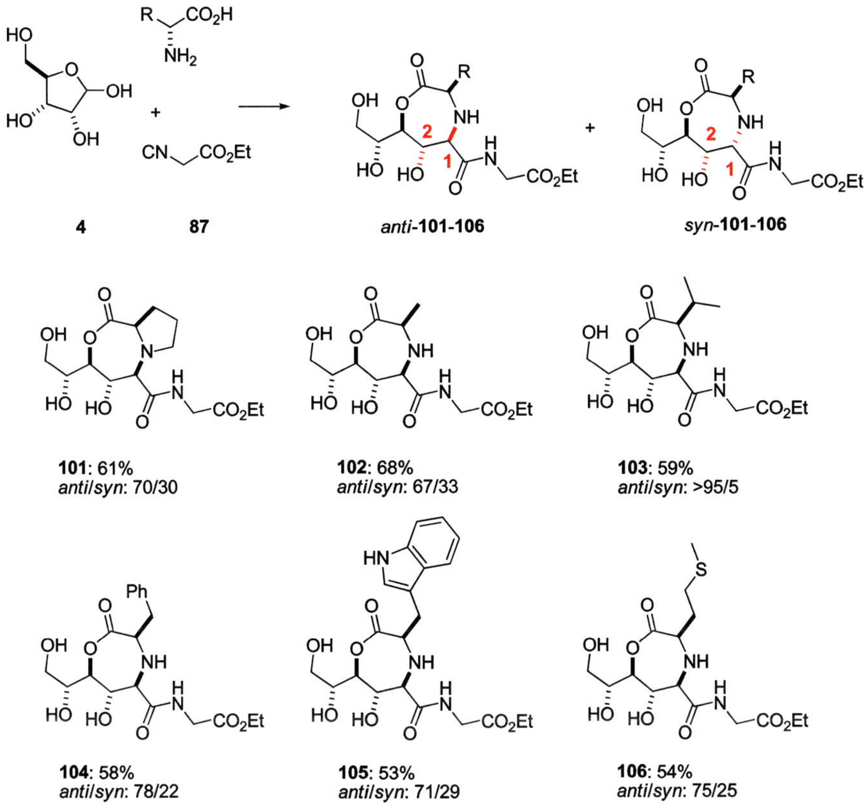

SCHEME 2.22 Multicomponent reactions of D-ribose and ethyl isocyanoacetate with D-configured amino acids.

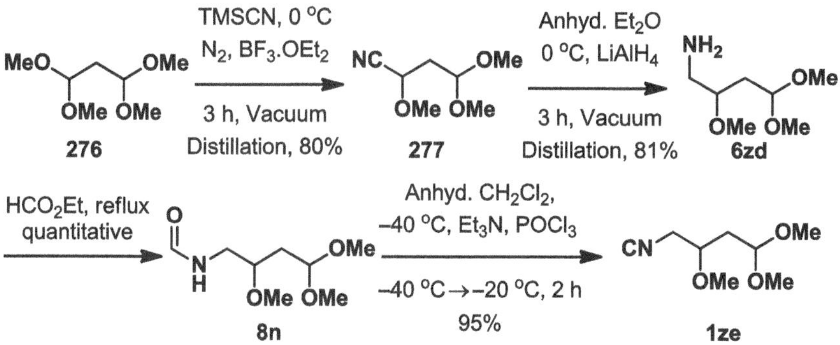

SCHEME 7.87 Synthesis of IPBT (**1ze**).

in the U-4CR and can be used as a CIC [101]. **1ze** can be synthesized as a light yellow oil from 1,1,3,3-tetramethoxypropane (**276**) in four steps (Scheme 7.87). IPBT is quite stable for several weeks when stored at −18°C.

The U-4CR of **10w**, **6zb**, **9d**, and **1ze** provided the Ugi product **278**, which under acidic conditions afforded the key intermediate *N*-acylpyrrole **279** (Scheme 7.88). In the presence of various nucleophiles, **279** afforded amides (**280** and **281**) and methyl ester **282** under refluxing conditions in good yields. In the presence of LiOH, **279** afforded the acid **283** at room temperature in an excellent yield. Employment of 1,8-diazabicyclo[5.4.0]undec-7-ene (DBU), $NaBH_4$, and Horner–Wadsworth–Emmons conditions provided the aldehyde **284**, alcohol **285**, and the α,β-unsaturated ester **286**, respectively. These synthetic transformations can be very useful; in particular, compound **286** is a valuable intermediate for the synthesis of 3-aza-bicyclo[3.1.0]hexane ring system, constrained amino acids, papain inhibitors, (±)-isocynodine, and (±)-isocynometrine [102].

7.2.1.21 1-Hydroxy-2-methyl-2-propyl Isocyanide Although 1-hydroxy-2-methyl-2-propyl isocyanide (**1zf**) has been known for a quite a while in its polymer-bound form [103], it was never explored for its potential as a CIC in a solution phase. Recently, Oikawa and coworkers have reported that **1zf** is odorless and can be used as a CIC in IMCRs (Scheme 7.89) [104].

The U-4CR of **9i**, **6i**, **10d**, and **1zg** afforded the acyclic Ugi product **287** in an appreciable yield. Treatment of **287** with catalytic amount of Lewis acid $Zn(OTf)_2$ and diethyl carbonate as an additive in the presence of a protic solvent such as ^{n}BuOH provided the ester **288** (Scheme 7.90).

The mechanism of solvolysis as proposed by Kita and coworkers works on a similar substrate [105]. Initial coordination of Zn^{2+} activates the amide carbonyl bond of **289**, which then undergoes an internal nucleophilic attack by the hydroxy group to form the *N*,*O*-acyl rearrangement product **291**. The activated ester carbonyl of **291** then undergoes a *trans*-esterification reaction with ^{n}BuOH to afford the desired ester **292**. Liberated 1-amino-2-methylpropan-2-ol (**293**) reacts with **294** to form the carbamte **295** (Scheme 7.91).

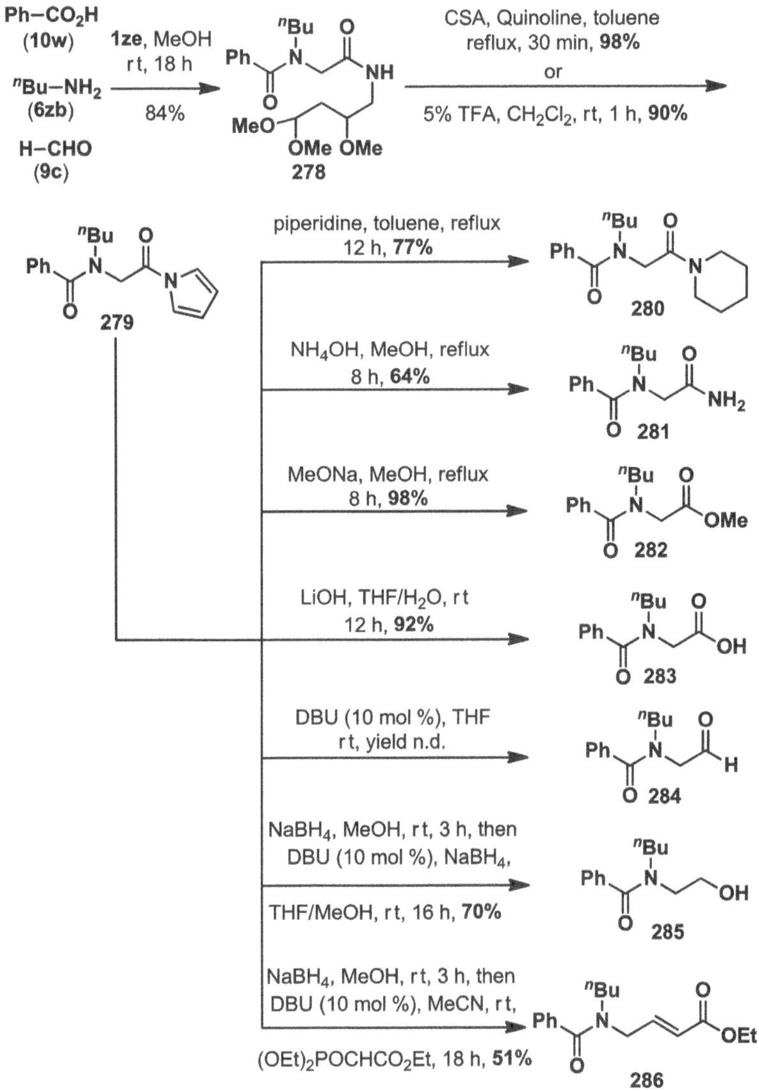

SCHEME 7.88 Synthetic application of *N*-acylpyrrole **279**.

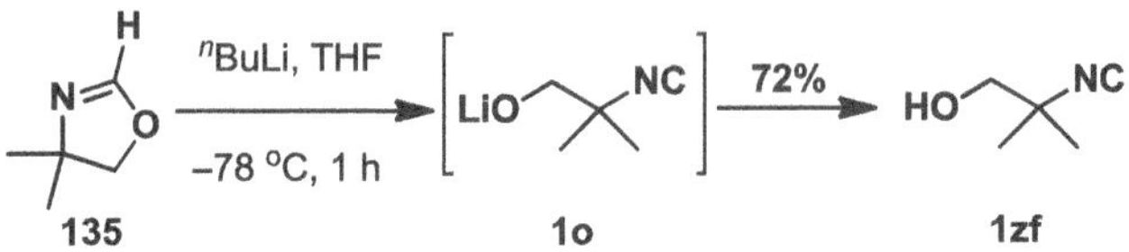

SCHEME 7.89 Synthesis of CIC **1zg**.

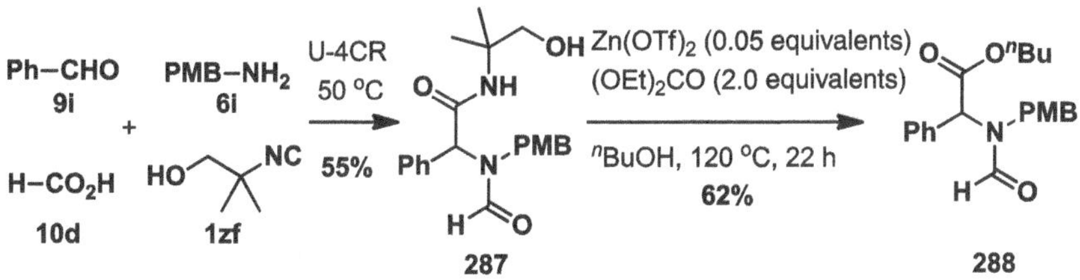

SCHEME 7.90 Application of CIC **1zf** in the U-4CR.

7.2.2 Resin-Bound CICs

Over the past number of years, solid-phase synthesis has become a synthetic mainstay as it avoids the tedious and costly parallel purification processes of the solution-phase products. Impurities can be removed just by washing with appropriate solvents and desired final products can be obtained by cleaving the resin link in high purity. Thus, MCRs as well as IMCRs involving CICs have found their application in solid-phase synthesis.

7.2.2.1 Safety-Catch Linker CIC Hulme and coworkers have developed a resin-bound CIC **1zg** by modification of the Wang resin and deployed in the U-4CR as a safety-catch-linker (SCL) [106]. Wang resin **296** was treated with 4-nitrophenylchloroformate in the presence of *N*-methyl morpholine to obtain the carbonate resin **297**. A nucleophilic displacement reaction of **297** with 4-(2-aminoethyl)aniline (**6ze**) in DMF forms the amino resin **6zf**. Formylation of **6zf** using formic acid and excess acetic anhydride followed by dehydration using triphenyl phosphine and triethylamine in CCl_4 provided an odorless resin-bound CIC **1zg** (Scheme 7.92). When stored at –20°C, **1zg** is stable for more than 6 months and, according to elemental analysis, it has a loading value of 0.8 mmol/g.

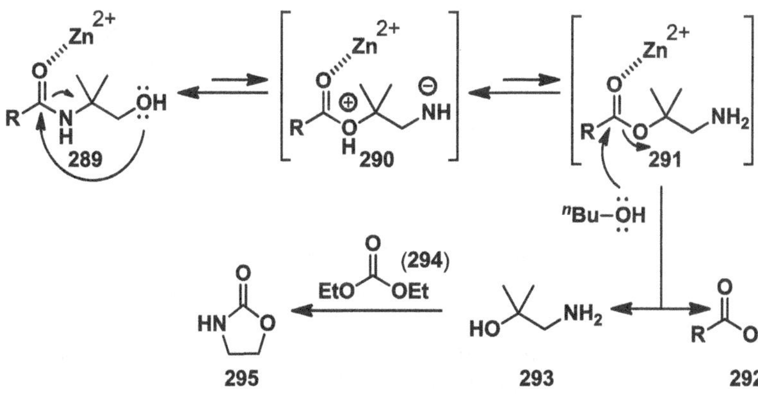

SCHEME 7.91 Mechanism of hydrolysis of the 1-hydroxy-2-methyl-2-propyl isocyanide group.

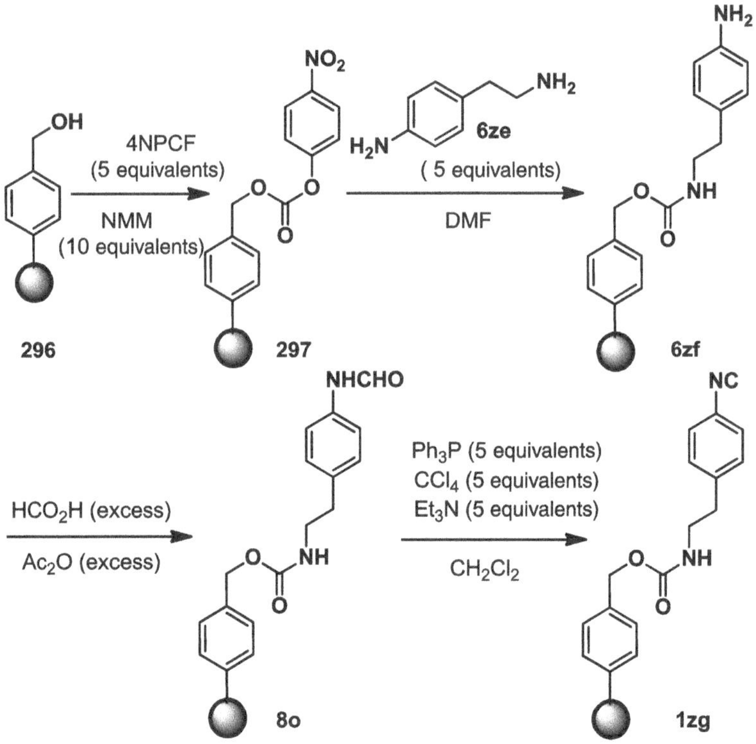

SCHEME 7.92 Synthesis of safety-catch linker CIC **1zg**.

The SCL CIC **1zg** has performed well in the U-4CR with a variety of coupling partners. The U-4CR products were obtained after 2–3 days. The acyclic U-4CR product **298** was Boc protected to obtain **299**. Treatment with sodium methoxide in a 1:1 mixture of MeOH/THF followed by either 10% AcCl/MeOH or 5% Et_2NH in dichloroethane or basic DOWEX generated biologically validated scaffolds BDZs **300**, DKPs **301**, ketopiperazines **302**, and dihydroquinoxalines **303** (Scheme 7.93). The methodology offers high-purity products and compatibility with a variety of side chain functionalities as well as "off-the-shelf" available aldehydes, primary amines, and carboxylic acids. The methoxide safety-catch clipping strategy and subsequent solution-phase cyclization offer similar advantages to tracelace linker strategies [107].

7.2.2.2 ***Resin Bound CyEIC*** Piscopio, at Amgen, reported a synthesis of odorless resin-bound CIC **1zh** based on Armstrong's CIC [108]. It can be prepared from 4-(4-hydroxyphenyl)-cyclohexanone using Merrifield as a base resin. Loading in the 0.7–0.8 mmol/g range was achieved and reactions can be monitored by on-bead Fourier transform infrared spectroscopy. Hulme and coworkers have employed the resin-bound CIC **1zh** in the so-called UDC (Ugi/De-Boc/Cyclization) strategy to synthesize bicyclic γ-lactam **304** [109]. The U-4CR of resin-bound CIC **1zi**, γ-keto acid **10za** and amine **6zg** affords resin-bound γ-lactam **304**. Acetyl chloride

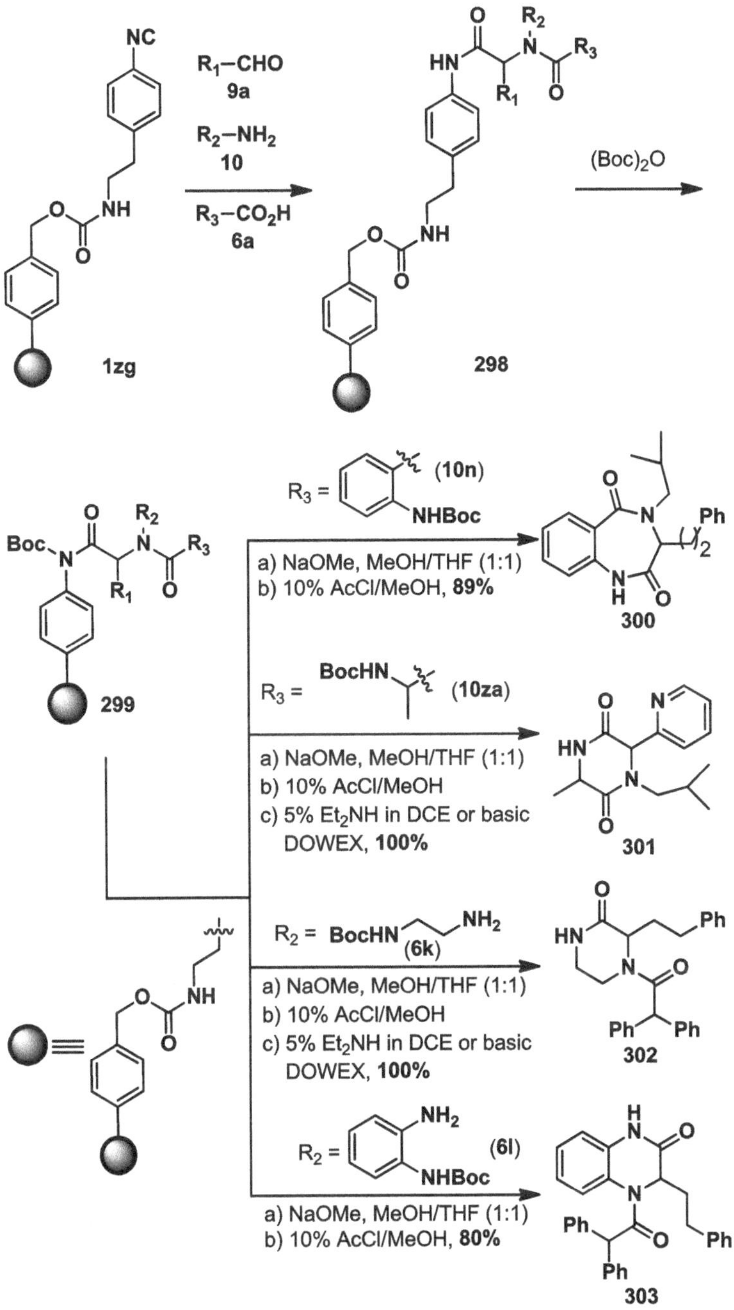

SCHEME 7.93 Application of the SCL CIC **1zg** in heterocycle synthesis.

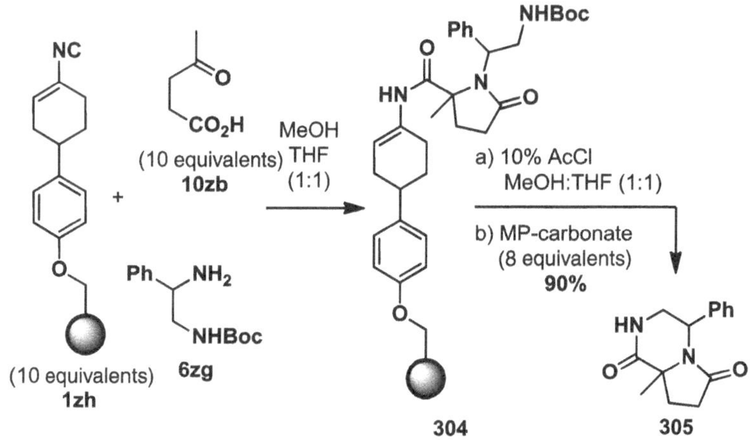

SCHEME 7.94 Synthesis of bicyclic γ-lactam **305** using resin bound CIC **1zh**.

promoted the cyclization of **304** and the presence of resin-bound proton scavengers (MP-carbonate or PS-*N*-methylmorpholine) provided the desired bicylic γ-lactam **305** in a 90% yield (Scheme 7.94).

7.2.2.3 Universal Rink Resin CIC Chen and coworkers utilized Hofheinz–Isenring's CIC concept and developed universal rink resin isocyanide **1zi** [110]. It can be easily synthesized from Fmoc-protected rink resin **306** by Fmoc deprotection using 20% piperidine followed by formylation using formic acid (5 equivalents), pyridine (1 equivalent), and diisopropylcarbodiimide (5 equivalents). The rink resin formamide **8p** can be converted to the desired CIC **1zi** by treating with $POCl_3$ and *N,N*-diisopropylethylamine (DIEPA) at 0°C (Scheme 7.95). The rink resin isocyanide **1zc** is odorless and stable under refrigeration over a period of 12 months. Furthermore, the process provides an excellent yield and quality of the product on a larger scale (50 g).

Chen and coworkers have utilized resin-bound CIC **1zi** in the U-4CR [108, 111] to synthesize a wide variety of biologically relevant heterocyclic scaffolds, for example, 3-acylamino imidazo[1,2-*a*]pyridines **308**, DKPs **310**, BDZs **312** and 5-substituted

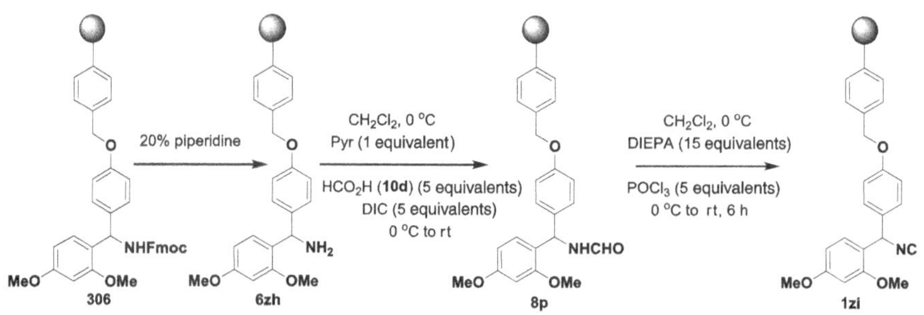

SCHEME 7.95 Synthesis of resin-bound CIC **1zi**.

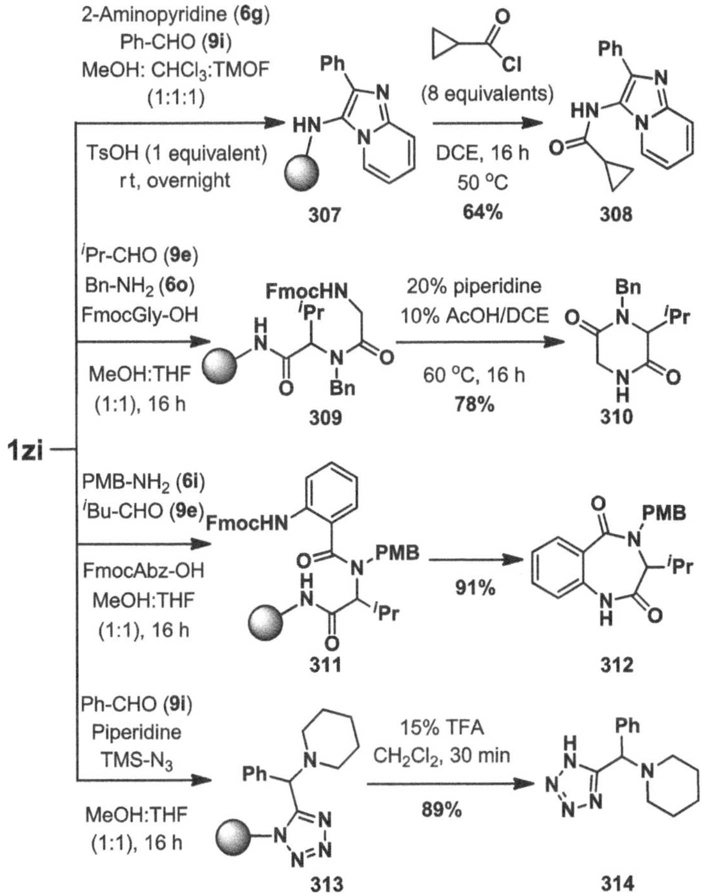

SCHEME 7.96 Application of resin-bound CIC **1zi**.

tetrazoles **314**, under a variety of deprotection and cyclization conditions in good to excellent yields (Scheme 7.96).

7.2.2.4 Universal Resin-Bound Carbonate CIC Kennedy and coworkers developed a novel, universal resin-bound carbonate CIC **1zj** on the basis of Lindhorst and Ugi's CIC **1p** [112]. Resin-bound chloroformate **315** can be generated by treating hydroxymethyl polystyrene twice with a 20% phosgene/toluene solution and washing extensively. The lithium alkoxide **1o** can be prepared by the treatment of *n*-butyl lithium with 4,4-dimethyl-2-oxazoline (**135**) in anhydrous THF at −78°C under a steady flow of argon and transferred via cannula to the solution of **315** in anhydrous THF at 0°C to afford **1zj** (Scheme 7.97). The resin is stable at room temperature and can be reproduced on a 200-g scale with a loading level of 0.87 mmol/g [112].

The resin-bound carbonate CIC **1zj** has been found to perform well with a variety of amines, aldehydes, and acids in the U-4CR to provide acyclic Ugi products **316** and

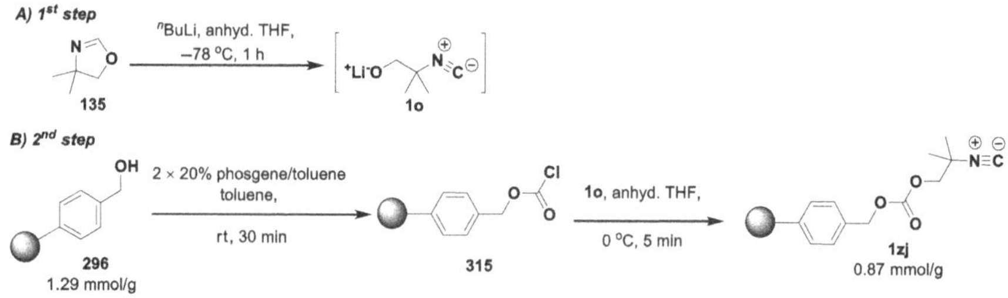

SCHEME 7.97 Synthesis of resin-bound CIC **1zj**.

317 in good to excellent yields. Treatment of KO^{t}Bu provides corresponding cyclic *N*-acyl urethanes **318** and **319**, respectively. The mechanism of formation of the *N*-acyl urethanes is similar to that of Lindhorst and Ugi's CIC (see Scheme 7.46). Removal of the *N*-acyl urethanes by NaOMe affords the corresponding methyl esters (**320** and **321**). TFA-catalyzed cyclization gives rise to DKPs (**322**) and BDZs (**323**), respectively (Scheme 7.98) [112]. Silica-based scavengers 1,3,5-benzenetricarboxylic acid-carbonate and isocyanate-3 were used to scavage unreacted amine and uncyclized product, respectively.

7.2.3 CIC Employed in the Ugi–Smile Reaction

Recently, Kaim and coworkers have developed a variation of the Ugi reaction, called the Ugi–Smile four-component reaction (US-4CR), by replacing the carboxylic acid component of the U-4CR with a phenol component [113]. Mechanistically, the US-4CR (Scheme 7.99) is quite similar to that of the U-4CR (Scheme 4). Condensation of the amine **6a** and aldehyde **9a** forms the imine **15**. Protonation of the imine **15** by the phenolic compound **326** activates the carbon atom of the C=N bond. The isocyanide **1c** then attacks the electrophilic carbon of the activated imine **15a**. The phenolate ion **327** attacks the electrophilic carbon of the isocyanide of the intermediate **16**. Smiles rearrangement [114] of the α-adduct **328** provides the Ugi–Smile product **329**.

Wessjohann and coworkers reported for the first time that IPBT (**1ze**) can be used as a CIC in the US-4CR. Reaction of *o*-nitrophenol (**326a**), **9e**, and **6zl** provided the Ugi–Smile intermediate **321**, which undergoes facile conversion to the corresponding acid **322** in the presence of CSA and quinoline or aqueous KOH under MWI (Scheme 7.100) [101].

7.2.4 CIC Employed in the Joullié–Ugi Reaction

Traditionally, preformed or *in situ*-formed acyclic imines (e.g., **15**; see Scheme 7.4) have been used in the U-4CR. However, in 1989, Joullié and coworkers developed a unique 3CR using pyrrole-based chiral cyclic imine **15g** (Figure 7.7). This version of the original U-4CR is commonly known as the Joullié–Ugi 3-CR (JU-3CR) [115]. The JU-3CR is one of the earliest diastereoselective versions of the original U-4CR and,

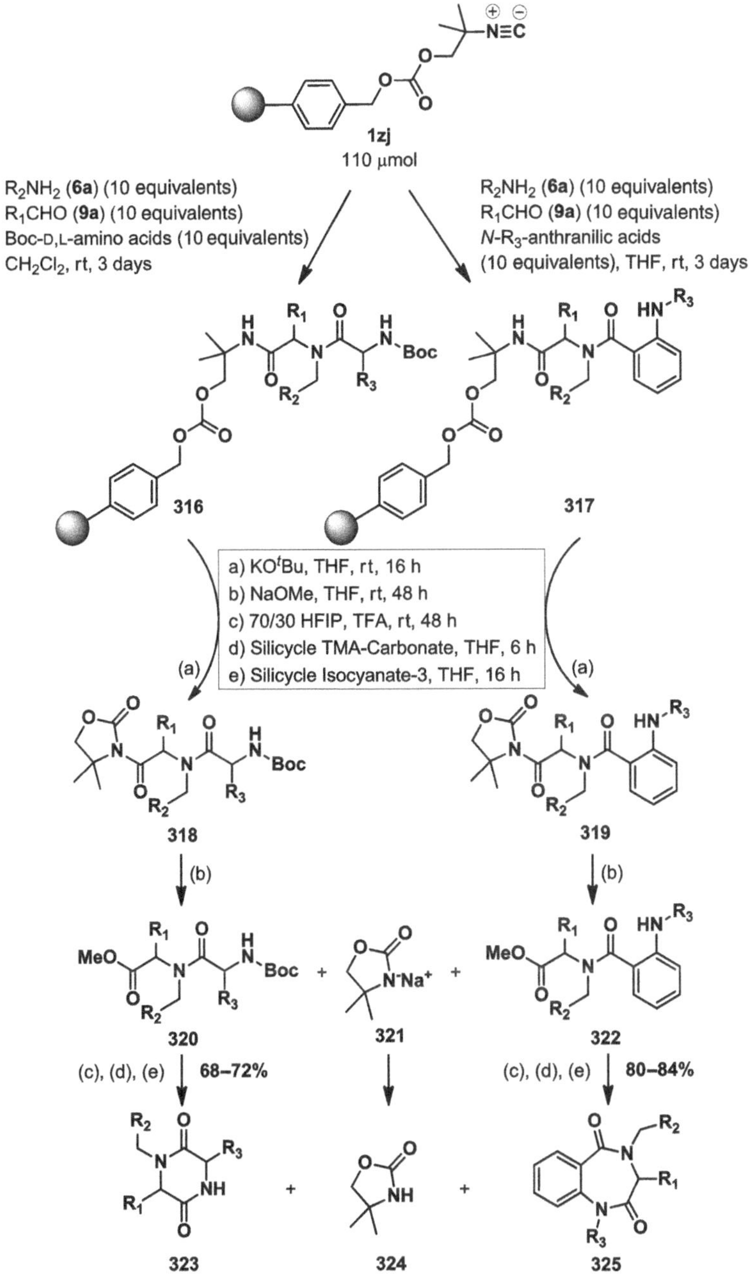

SCHEME 7.98 Application of resin-bound carbonate CIC **1zj**.

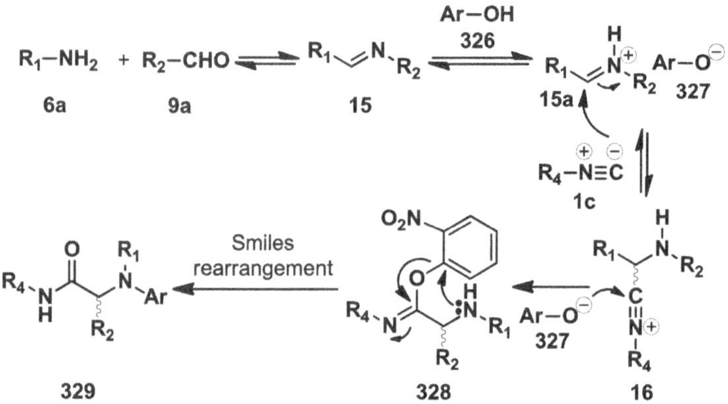

SCHEME 7.99 The Ugi–Smile reaction.

thereafter, has been utilized widely in the synthesis of glyco- and peptidomimetics [116].

7.2.4.1 p-Methoxyphenyl Isocyanide *p*-Methoxyphenyl isocyanide (**1zk**) has been known for quite a while as an almost odorless yellow solid. However, **1zk** has been found to be moisture sensitive and turns black upon repeated exposure. Banfi and coworkers have employed **1zk**, tert-butyldimethylsilyl (TBDMS)-protected chiral imine **15h**, and acid **10zc** in their JU-3CR to obtain the acyclic product **332** as a mixture of diastereomers. Boc protection followed by Cbz deprotection via hydrogenation and refluxing in ^{t}BuOH afforded the pyrrolodiazepinedione **334** in a good yield (Scheme 7.101) [117].

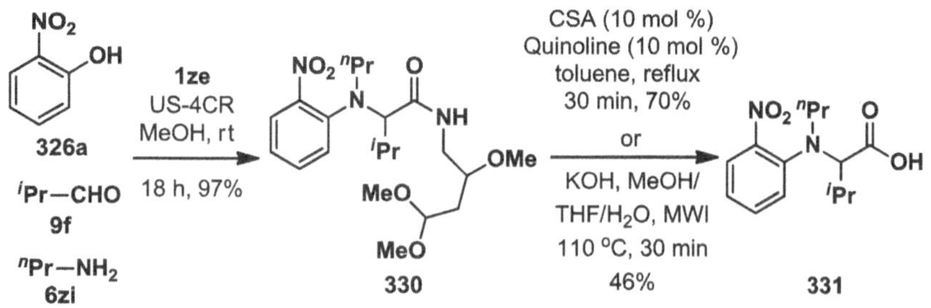

SCHEME 7.100 Application of **1ze** in the US-4CR.

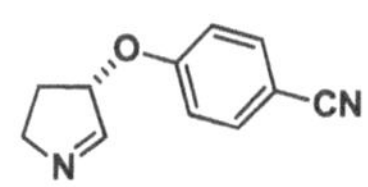

FIGURE 7.7 Chiral imine **15g** employed in the first JU-3CR.

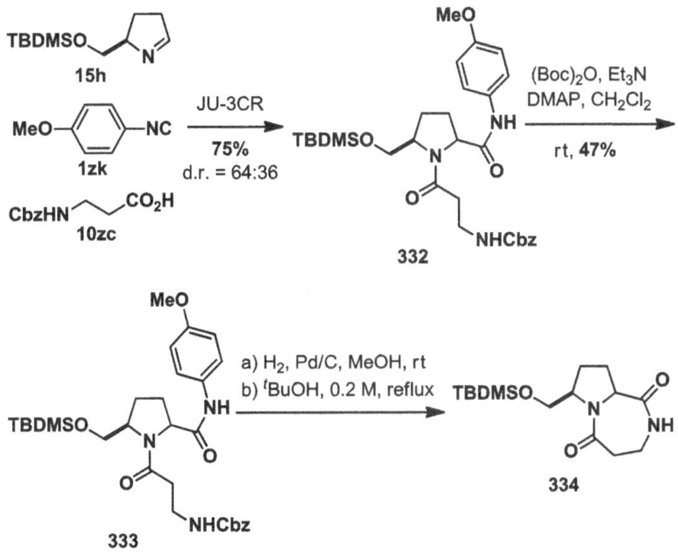

SCHEME 7.101 Application of CIC **1zk** in the synthesis of pyrrolodiazepinedione **334**.

7.2.5 CIC Employed in the Passerini Reaction

7.2.5.1 1-Isocyano-2-(2,2-dimethoxyethyl)benzene Rubinshtein and coworkers utilized CIC DMEBIC (**1x**; Scheme 7.67) in a P-3CR type reaction for the synthesis of side chain functionalized poly(α-hydroxy acid) (PHA) copolymers **341** [118]. This method enables the incorporation of functional groups in polylactic acid-based polymers (Scheme 7.102). This method exemplifies that IMCRs can be employed in polymer synthesis beyond their traditional application in a small molecule synthesis.

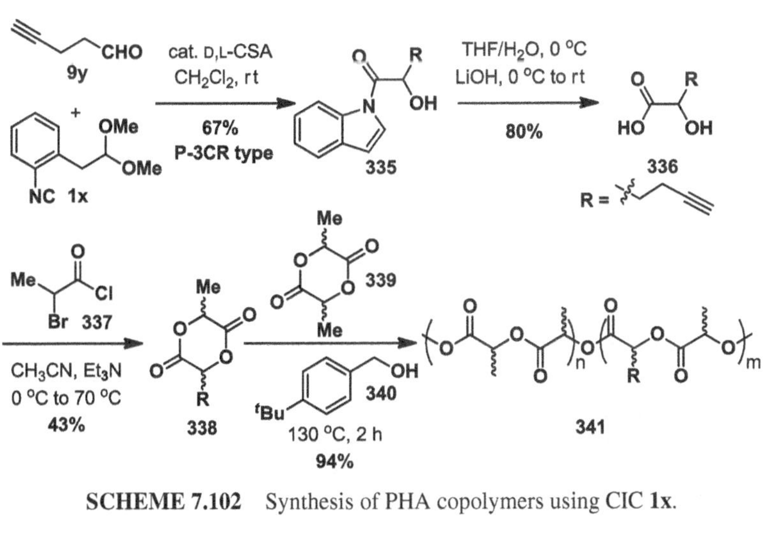

SCHEME 7.102 Synthesis of PHA copolymers using CIC **1x**.

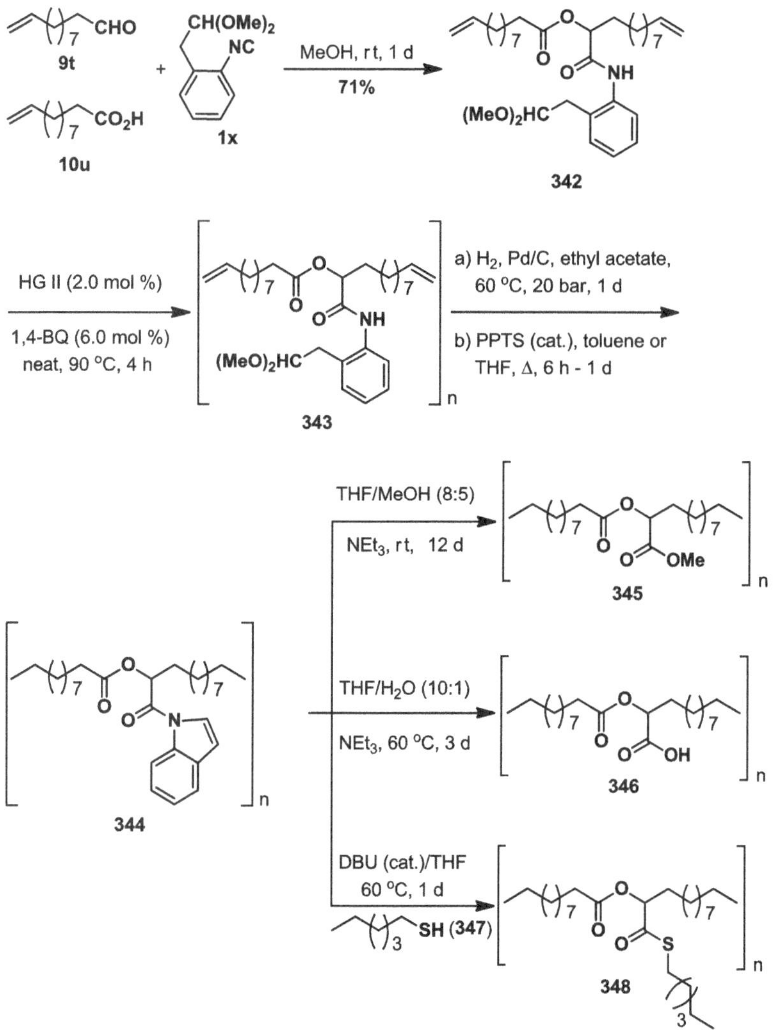

SCHEME 7.103 Application of CIC **1x** in polymer synthesis.

More recently, Kreye and coworkers have employed CIC **1x** in the original P-3CR for the synthesis of α,ω-diene monomer **343** (Scheme 7.103) [85]. Employing an acyclic diene metathesis polymerization strategy, the monomers were subsequently polymerized using the second-generation Hoveyda–Grubbs catalyst in the presence of 1,4-benzoquinone (1,4-BQ) as an additive to obtain compound **343**. Hydrogenation followed by the treatment with pyridinium *p*-toluenesulfonate provided the *N*-acylindole intermediate **344**. Under various conditions, **344** provided polymers containing amide and acid (**345**, **346**, and **348**) functionalities that can be useful for further modification of the side chain of the polymers. This method clearly exemplifies

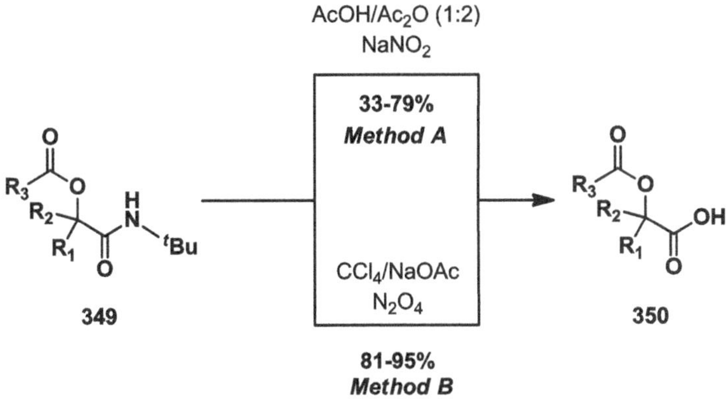

SCHEME 7.104 Application of CIC **1zb** in the P-3CR.

that judicious application of IMCRs beyond U-4CR can lead to a new era in polymer synthesis.

7.2.5.2 Tert-Butyl Isocyanide Recently, Ganem and coworkers have reported the first application of CIC in the original P-3CR. They have employed $NaNO_2$ (Method A) and N_2O_4 (Method B) to convert TBIC (**1zb**) derived secondary amide of the P-3CR product **349** into carboxylic acid derivatives **350** without affecting the existing sensitive ester group (Scheme 7.104) [119].

This method is quite similar to the method reported by Isenring and Hofheinz (Scheme 17) [36]. According to Le and coworkers [119], the sterically hindered Passerini product reacts sluggishly and gives lower yields when subjected to Method A. The authors proposed a mechanism (Scheme 7.105) that is quite similar to that proposed by Isenring and Hofheinz (Scheme 7.17).

7.2.5.3 4-Isocyanopermethylbutane-1,1,3-triol More recently, Filho and coworkers have utilized the CIC IPBT (**1ze**) in the P-3CR and employed milder

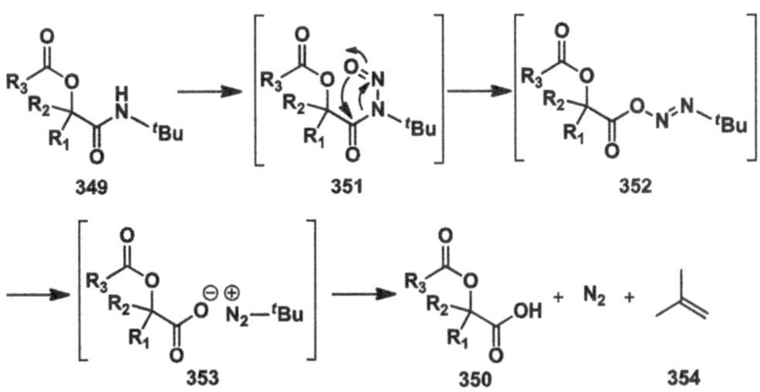

SCHEME 7.105 Mechanism of formation of acid-ester from the CIC **1zd**-derived Passerini product.

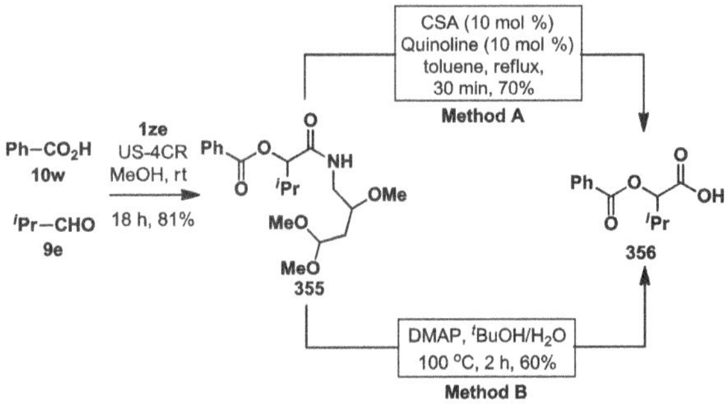

SCHEME 7.106 Application of **1ze** in the P-3CR.

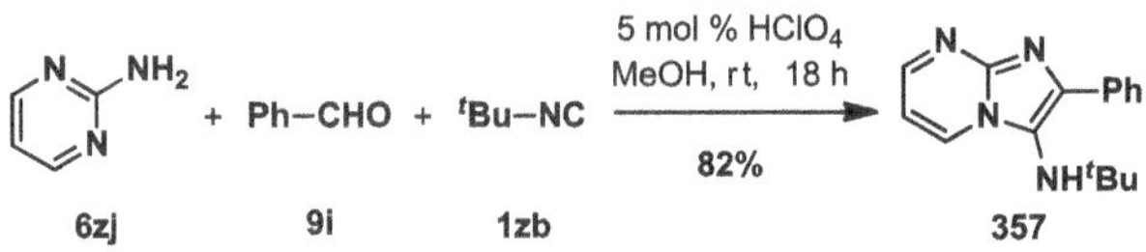

SCHEME 7.107 The Bienaymé reaction.

cleavage conditions [101]. Reaction of **10w**, **9e**, and **1ze** provided the Passerini product **355**. The isopropylbutyl (IPB) group was readily cleaved by using either CSA and quinoline or DMAP in a mixture of ^{t}BuOH and water to afford the desired benzyloxy ester **356** (Scheme 7.106).

7.2.6 CIC Employed in the Groebke–Blackburn–Bienaymé Reaction

In 1989, Groebke, Blackburn, and Bienaymé independently reported that isocyanides and aldehydes can react with amidines in protic solvents (e.g., MeOH) to provide heterocylic scaffolds containing a 3-amino-imidazole moiety (Scheme 7.107–7.109) [120]. Typically, an acid catalyst (e.g., $HClO_4$, $Sc(OTf)_3$, AcOH) is required to drive

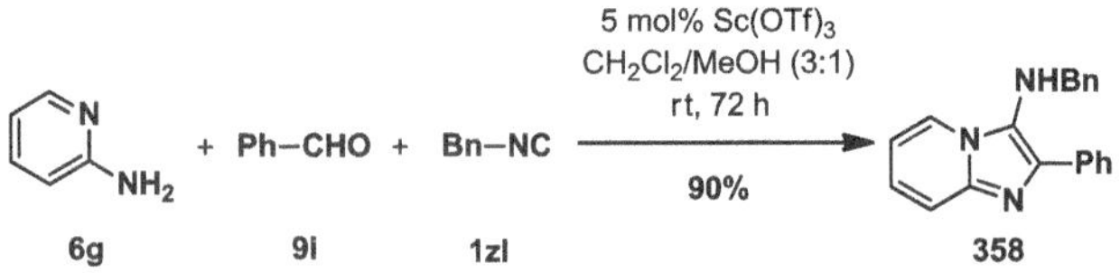

SCHEME 7.108 The Blackburn reaction.

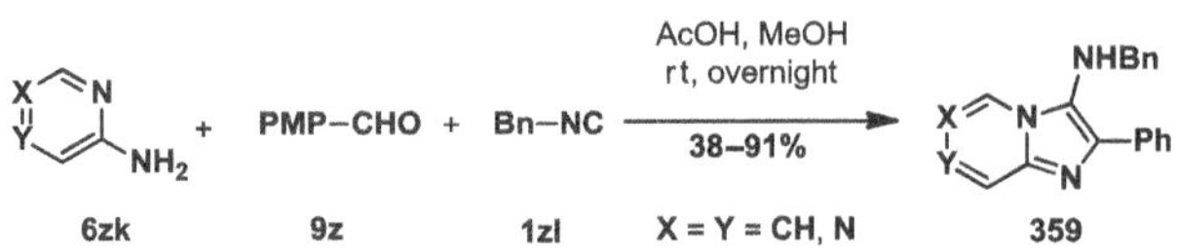

SCHEME 7.109 The Groebke reaction.

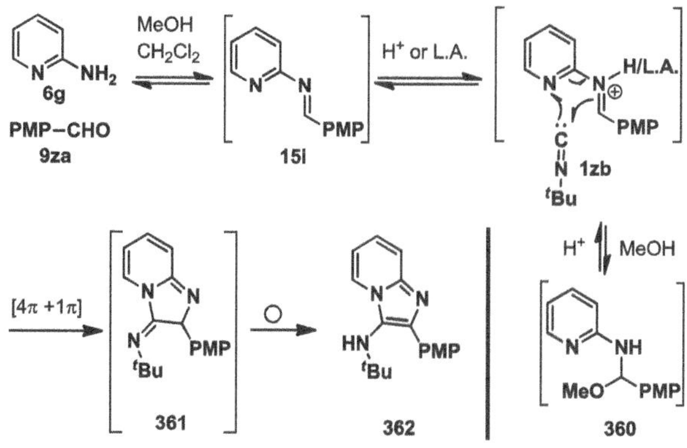

SCHEME 7.110 Commonly accepted mechanism of the GBB-4C-3CR.

the reaction in the forward direction. The amidines (e.g., 2-amino-pyridine, or pyrimidine or pyrazine) function as bifunctional substrates. Since this discovery, these types of reactions are commonly referred to as the Groebke–Blackburn–Bienaymé four-center three-component reaction (GBB-4C-3CR). The limitation of this reaction is that absence of nitrogen at the *ortho*-position of the pyridine ring does not provide any product. It is quite important to note that many clinically approved drugs and biologically active compounds contain the imidazo[1,2-*a*]pyridine/imidazo[1,2-*a*]pyrimidine/imidazo[1,2-*a*]pyrazine motifs. Despite the limitation, over the past several years, the GBB-4C-3CR has gained wide attention in the scientific community for new drug discovery initiatives [121].

The mechanisms proposed by these groups are quite similar. The most commonly accepted mechanism for the GBB-4C-3CR is depicted in Scheme 7.110. The amine **6g** and aldehyde **9z** condense to form imine **15i** which upon protonation (in the presence of either a Bronstead or Lewis acid) undergoes a nonconcerted [4π + 1π] cycloaddition with CIC **1zb** (which acts as a vinylidene carbenoid) to form **361**. Subsequent prototropic shift provides the final aromatic fused product **362**. The limitation of this reaction is that absence of nitrogen at the *ortho*-position of the pyridine ring does not provide any product.

7.2.6.1 1,1,3,3-Tetramethylbutyl Isocyanide Blackburn and coworkers have utilized Walborsky's CIC **1za** to synthesize 2-aminoimidazo[1,2-*a*]pyridines (**364**) [122]. The 3CR of 2-aminopyridine (**6g**), 4-methoxybenzaldehyde (**9z**), and CIC **1za** provided **363**. The 1,1,3,3-tetramethylbutyl group can be cleaved by using either a 1:1 solution of 12 N HCl/MeOH or a TFA/CH_2Cl_2 mixture to obtain the desired product **364** (Scheme 7.111). This method can also be used for solid-phase synthesis.

Recently, the synthesis of 5-aminothiazoles was achieved by Thompson and Chen via the GBB-4C-3CR using CIC **1za** (Scheme 7.112) [93]. The DMB group was removed by treating with a 20% TFA-CH_2Cl_2 solution to obtain **366**, which was treated with Lawesson's reagent and refluxed in toluene to afford compound **367**.

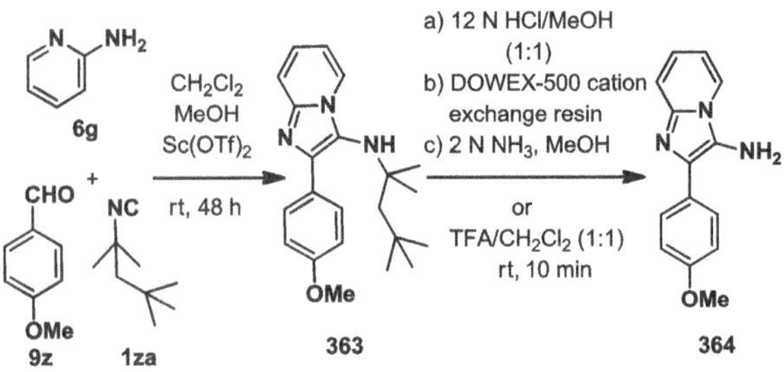

SCHEME 7.111 Synthesis of 2-aminoimidazo[1,2-*a*]pyridines (**364**).

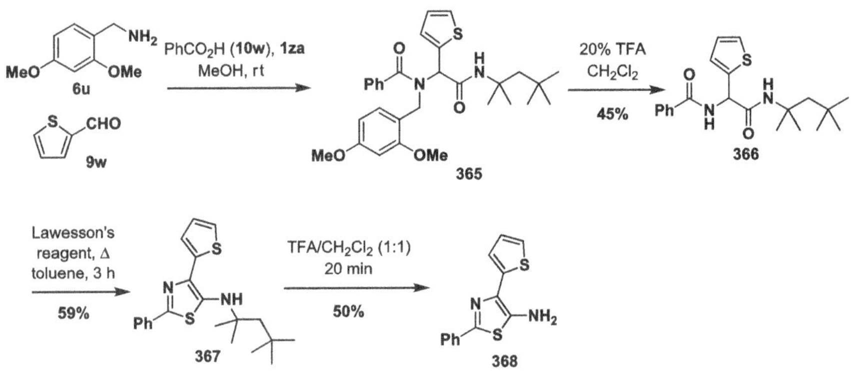

SCHEME 7.112 Synthesis of 5-aminothiazole **368** using **1za**.

Hydrolysis of the TMB group in the presence of a 1:1 mixture of TFA-CH_2Cl_2 for 20 min provided the desired 5-aminothiazole **368** in a moderate yield. To expand the scope of their methodology, they used ammonia instead of DMB-NH_2 (as an ammonia equivalent) in the GBB-4C-3CR to obtain similar scaffolds [94].

In yet another application, 2-phenylimidazo[1,2-*a*]pyrazin-3-amines (**370**) were synthesized via the GBB-4C-3CR using TMBIC [123]. Reaction of 2-aminopyrazine (**6zd**), benzaldehyde (**6i**), and **1zb** provided **368** in the presence of the promoter trimethylsilyl chloride. The TMB group can cleaved by 4 N HCl in dioxane and the 2-phenylimidazo[1,2-*a*]pyrazin-3-amines can be isolated as the corresponding HCl salts **369** (Scheme 7.113).

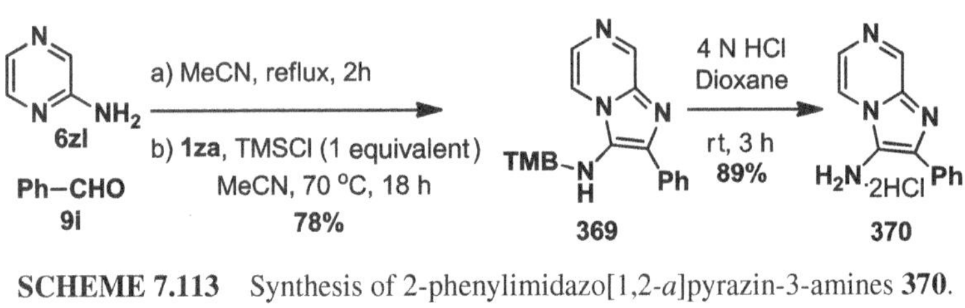

SCHEME 7.113 Synthesis of 2-phenylimidazo[1,2-*a*]pyrazin-3-amines **370**.

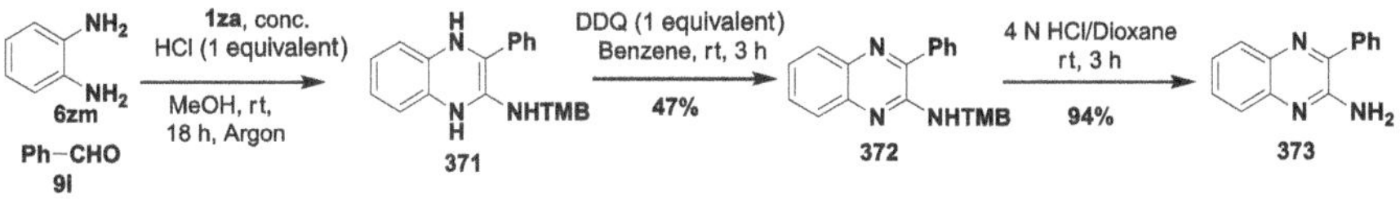

SCHEME 7.114 Synthesis of 2-aminoquinoxalines **373**.

Krasavin and coworkers have extended this methodology (Scheme 7.114) to synthesize 2-aminoquinoxalines [124]. In a variation of the GBB-4C-3CR, they have employed *o*-phenylenediamine (**6zm**), benzaldehyde (**9i**), and **1za** to obtain 1,4-dihydroquinoxaline **371**. Dichlorodicyano quinone-mediated oxidation in benzene provided the corresponding quinoxaline **372**, which was then hydrolyzed by 4 N HCl in dioxane to obtain the corresponding 2-aminoquinoxaline **373**. Other substrates also work well in this method.

A mechanism proposed by Krasavin and coworkers [124] is shown in Scheme 7.115. Nucleophilic attack of the CIC **1za** onto the activated carbon atom of the iminium group of **374** followed by an internal nucleophilic attack by the amine group of **6zm** onto the terminal carbon atom of the isocyanide provides **376**, which upon prototropic shift affords the desired product **372**.

7.2.6.2 Tert-Butyl Isocyanide Recently, Krasavin and coworkers have reported that TBIC (**1zb**) can be used as a CIC in the GBB-4C-3CR [123]. Reaction of 2-aminopyridine (**6g**), aldehyde **9za**, and **1zb** provided compound **377**, which upon hydrolysis by aqueous KOH afforded the acid **378**. The *tert*-butyl group from **378** can be removed by refluxing in TFA for 3 h to obtain the corresponding imidazo[1,2-a]pyridin-3-amine **379** (Scheme 7.116). The shortcomings of this methodology are the harsh conditions required to remove the *tert*-butyl group and the need to remove the TFA group for further library synthesis.

Recently, Guchhait and Madaan have reported the synthesis of *N*-fused imidazole-amines **380** and isoquinolinone-imidazoles **381** by employing 2-aminopyridine **6g**, **1zb**, and various aldehydes (e.g., **9zb**, **9zc**) in the GBB-4C-3CR (Scheme 7.117) [125].

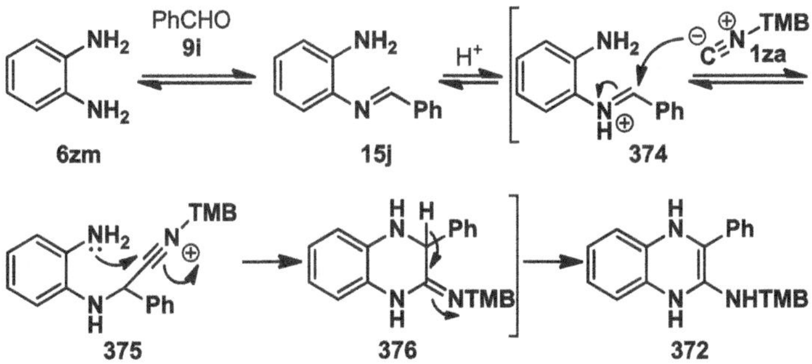

SCHEME 7.115 Mechanism of formation of the quinoxalines **372** from *o*-phenylenediamine **6zm**.

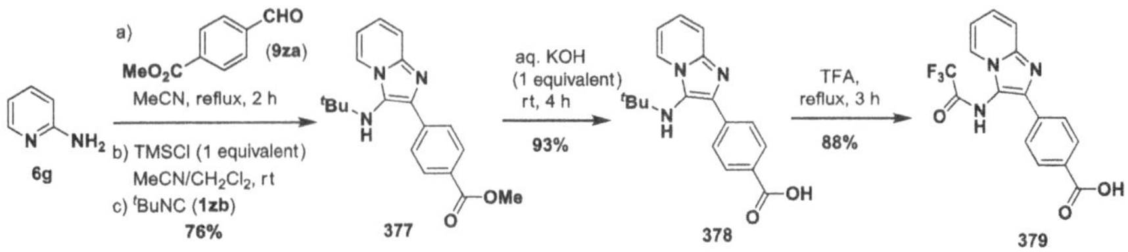

SCHEME 7.116 Synthesis imidazo[1,2-a]pyridin-3-amine **379**.

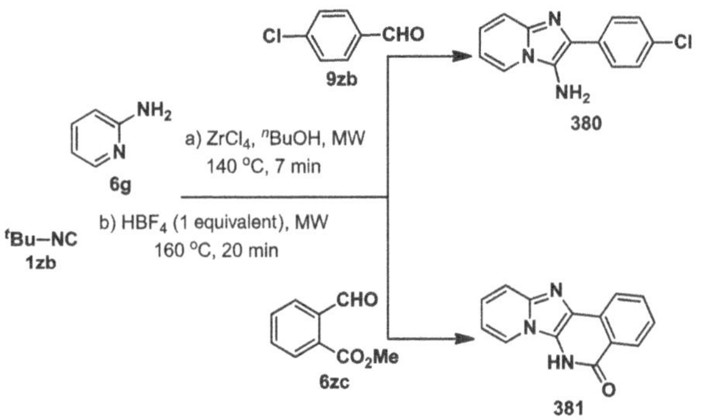

SCHEME 7.117 Synthesis of *N*-fused imidazole-amines (**380**) and isoquinolinone-imidazole (**381**) using **1zb**.

7.2.7 CIC Employed in the Diels–Alder Reaction

7.2.7.1 Tert-Butyl Isocyanide Recently, Neo and coworkers have employed **1zb** in their 3CR for the synthesis of polysubstituted anilines. In the initial step, **1zb** undergoes a [4π + 1π] hetero Diels–Alder (DA) reaction with the α,β-unsaturated carbonyl compound **382** to afford **384**. The polysubstituted furan **384**, which then undergoes an *in situ* [4π + 2π] DA reaction to form intermediate **385**, then undergoes dehydration followed by aromatization to provide the desired polysubstituted aniline **388** (Scheme 7.118). Compound **388** then undergoes de-*tert*-butylation in the presence of 5% $Y(OTf)_3$ in toluene at 100°C to provide compound **389** [126].

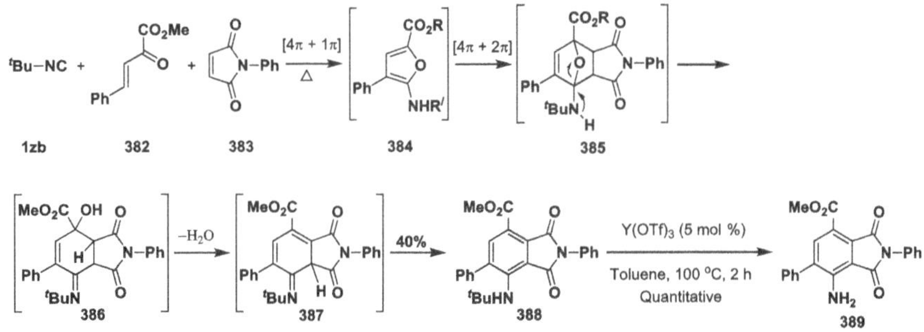

SCHEME 7.118 Use of **1zb** for the synthesis of polysubstituted aniline **389**.

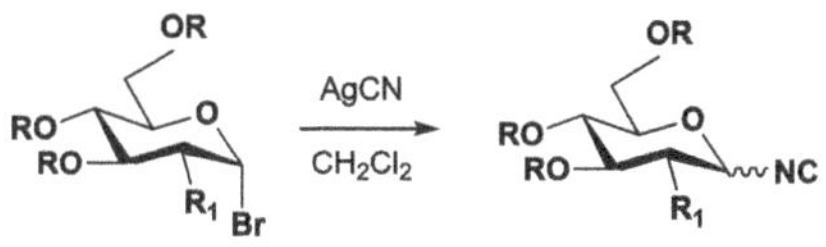

SCHEME 7.119 Preparation of monosaccharide isocyanides.

7.2.8 Monosaccharide Isocyanides Employed in the Ugi and Passerini Reaction

7.2.8.1 Glucosyl Isocyanides and Isocyanoglucoses for the Preparation of Glycopeptide Libraries Synthesis of monosaccharide isocyanides has become increasingly important in the construction of many heterocyclic carbohydrate systems. Gautier's method is commonly used to synthesize monosaccharide isocyanides (Scheme 7.119) [127].

The synthesis works best with benzylated halogeno sugars **390** and provides a mixture of anomeric (α + β) isocyanides **391**. The isocyanides can also be synthesized by desulfurization of glycosyl isothiocyanates [128].

Ziegler and coworkers explored the use of glycosyl isocyanides in Passerini and Ugi reactions to synthesize combinatorial libraries of glycopeptides (Scheme 7.120) [129]. The reactions were focused on the use of anomeric glycosyl isocyanides, namely 2,3,4,6-tetra-*O*-acetyl- and 2,3,4,6-tetra-*O*-benzyl-β-D-glucopyranosyl isocyanide. Passerini reactions of the isocyanides with *N*-Boc-glycinal, (*S*)-*N*-Boc-phenyl alaninal and propanal as the aldehyde component and acetic acid as the acid component indicated that the glycosyl isocyanides were sensitive to the steric effects of the aldehyde component, as witnessed by slow reaction time and lower yields. The reactions did give the corresponding product with little to no diastereoselectivity.

The Ugi reactions were carried out using *i*-butanal as the aldehyde component and *N*-Boc-glycine and *N*-propyl amine or (*S*)-serine in methanol. The reactions again proceeded with low diastereoselectivity and also a low yield.

Ziegler and coworkers then set about to use isocyanoglucoses in the Passerini and Ugi reactions (Scheme 7.121) [130]. They used 1,3,4,6-tetra-*O*-acetyl-2-deoxy-2-isocyano-β-D-glucopyranose and compared the results with the previously mentioned isocyanides. It was observed that the Passerini reactions proceeded in a better yield

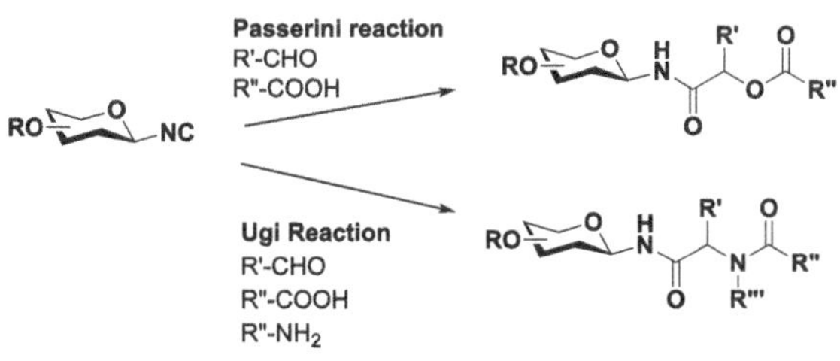

SCHEME 7.120 Products of Passerini and Ugi reactions of glycosyl isocyanides.

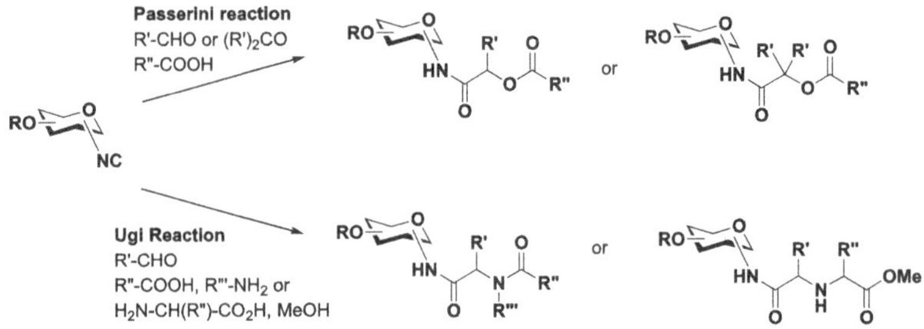

SCHEME 7.121 Products of Passerini and Ugi reactions of isocyanoglycoses.

and with faster reaction times using the isocyanoglucose. This indicated that the anomeric glucosyl isocyanides were more sensitive than the 2-isocyanoglucose.

The Ugi reactions, however, did not show any considerable differences between the anomeric glucosyl isocyanides and 2-isocyanoglucose. No higher reactivity was observed. Reactions times were still slow and yields continued to be low.

Although the reaction conditions did not yield optimal results, Ziegler did show that it was possible to achieve complex glycopeptide derivatives using anomeric glycosyl isocyanides or isocyanoglucoses.

7.2.8.2 Synthesis of Polypeptides and Glycopeptides via Type Ia, Type Ib, and Type IIb Isocyanide Coupling Reactions Recently, Danishefsky and coworkers set out to find improved methods for the synthesis of polypeptides and glycopeptides utilizing the reactivity of the isocyanide functional group [131]. Three coupling methodologies (type Ia, type IIb, and type Ib) for the formation of peptidyl and glycopeptidyl amide bonds were reported (Scheme 7.122).

Type Ia reactions involve isocyanide-based couplings using a carboxylic acid **10** and isocyanide **1b** to undergo a [1,3]-*O*→*N* acyl transfer to generate *N*-formyl amide products **392**. Glycosyl isocyanide **393** was coupled with aspartate **394** to yield glycosyl amino acid **395** in 85% (Scheme 7.123).

Type Ib reactions involve the coupling of isocyanides and thioacids under room temperature to generate dipeptide products. Thioacid **398** and isocyanide **399** were

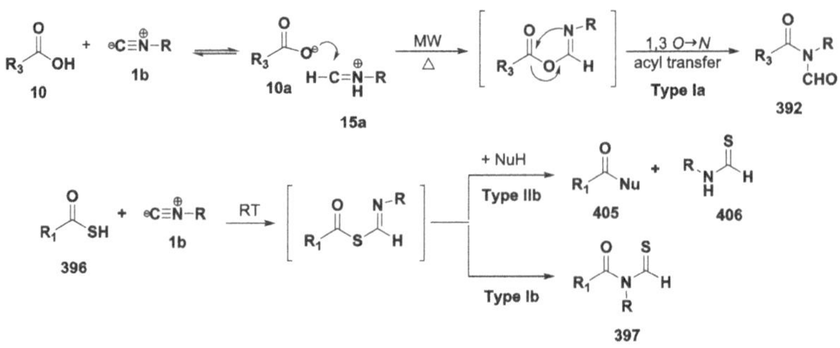

SCHEME 7.122 Type Ia, Ib, and IIb coupling methodologies.

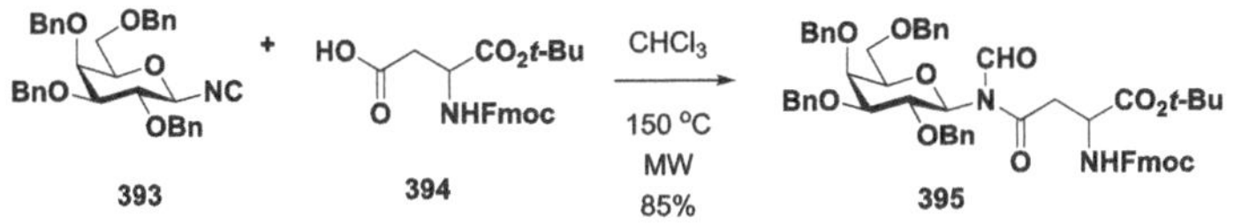

SCHEME 7.123 Synthesis of glycosyl amino acid **395** via a type Ia coupling.

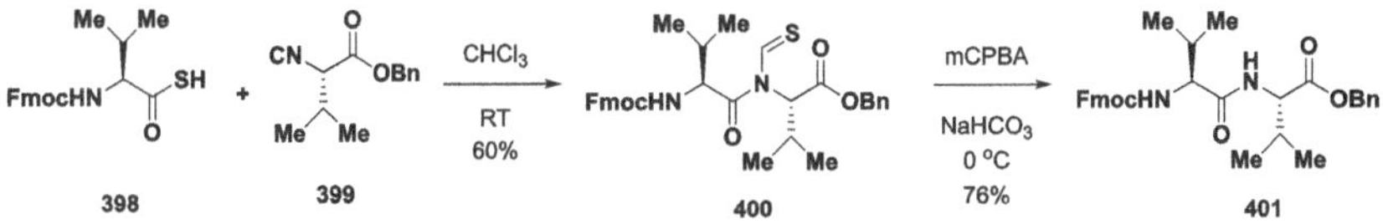

SCHEME 7.124 Synthesis of dipeptide **401** using thioacid **398** and isocyanide **399** via a type Ib coupling.

reacted at room temperature to afford dipeptide **400**, which underwent dethioformylation using *m*-CPBA and $NaHCO_3$ to yield dipeptide **401** in a 76% yield (Scheme 7.124). Glycan isocyanides were also proven to be useful type Ib coupling partners. Glycan isocyanide **402** was coupled with thioacid **403** in the presence of 2,6-dimethylthiophenol at room temperature to afford glycosyl amino acid **404** in an 88% yield (Scheme 7.125).

Type IIb reactions undergo a biomolecular acylation due to the presence of an external nucleophile and are useful with a variety of substrates, including bulky thioacids and amines. Along with the product **405**, a recyclable isocyanide-derived side product **406** is also formed. The coupling of thioacid **407** with bulky *tert*-butyl isocyanide **408** and aniline at room temperature afforded the amide adduct **409** in a 77% yield (Scheme 7.126).

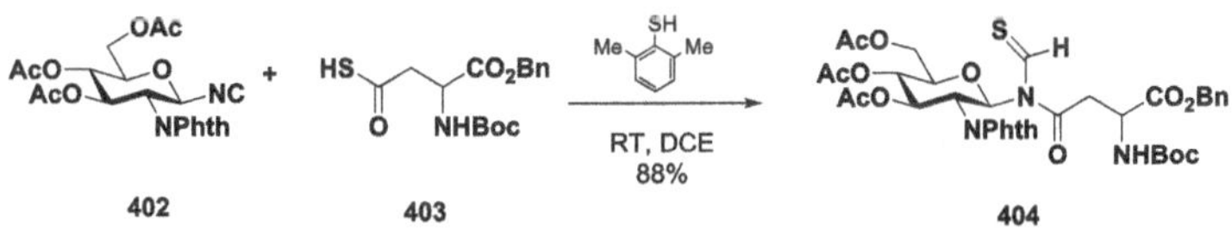

SCHEME 7.125 Synthesis of glycosyl amino acid **404** using thioacid **401** and glycan isocyanide **402** via a type Ib coupling.

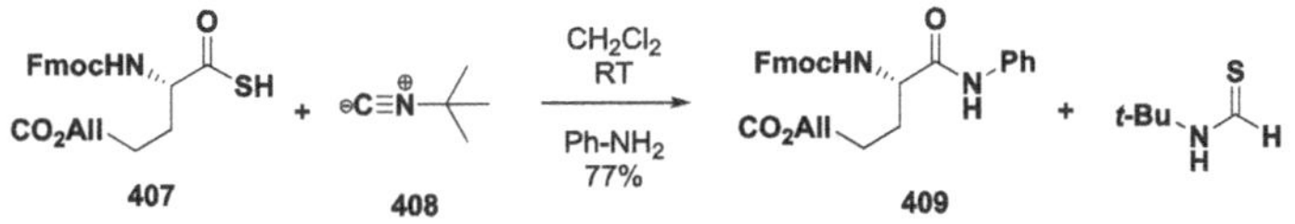

SCHEME 7.126 Type IIb coupling of thioacid **407** and isocyanide **408** for the synthesis of amino adduct **409**.

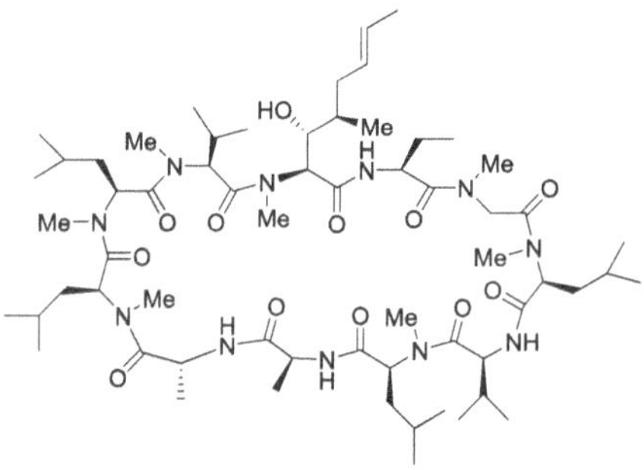

FIGURE 7.8 Cyclosporine A.

All three coupling methods were utilized by the Danishefsky group for the total synthesis of cyclosporine A (Figure 7.8), a cyclic peptide that contains seven sites of *N*-methylation.

The synthesis of constrained peptides and the formation of *N*-glycosyl asparagine glycopeptides were also reported.

7.2.9 Methyl isocyanide in the Preparation of the Hydroxy DKP Thaxtomin A

Andreana and coworkers have recently used **1d** in the synthesis of the natural herbicide thaxtomin A [132]. The required dipeptide was obtained under standard Ugi conditions on reaction with 4-nitroindolylacetaldehyde **410**, methyl amine **411**, 3-hydroxyphenylpyruvic acid **412**, and **1d**. Following the 4CR, the acyclic product was cyclized and epimerized with KOH at 70°C to afford thaxtomin A (±)-*syn*-**413** and its *anti*-diastereomer in a 4:1 ratio (Scheme 7.127). The cyclization presumably proceeds through deprotonation of the secondary amide and subsequent intramolecular attack onto the α-ketoamide functionality forming the hydroxy DKP core.

The Kunz diastereoselective strategy [68] has also been applied to this methodology by replacing **411** with peracetylated β-D-glycosylamine **414** to prepare an interesting sugar Ugi derivative. This approach provided control over the new stereocenter formed by the Ugi reaction, giving a 3:1 dr [133]. The acylic intermediate was then treated with NEt_3 to afford the anti-DKP **415** (Scheme 7.128).

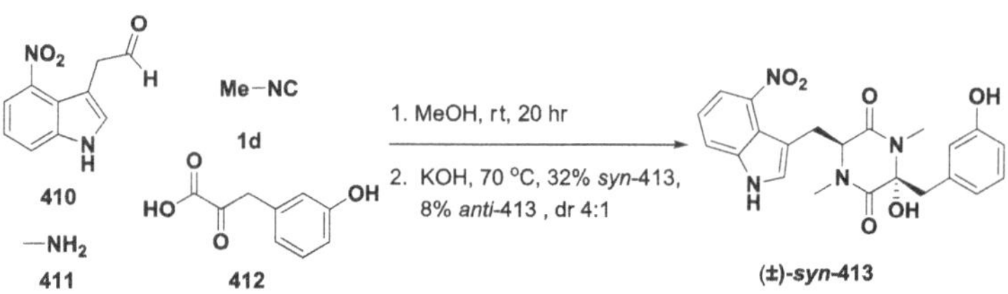

SCHEME 7.127 Use of 1d in the preparation of the hydroxy DKP thaxtomin A (±)-*syn*-**413**.

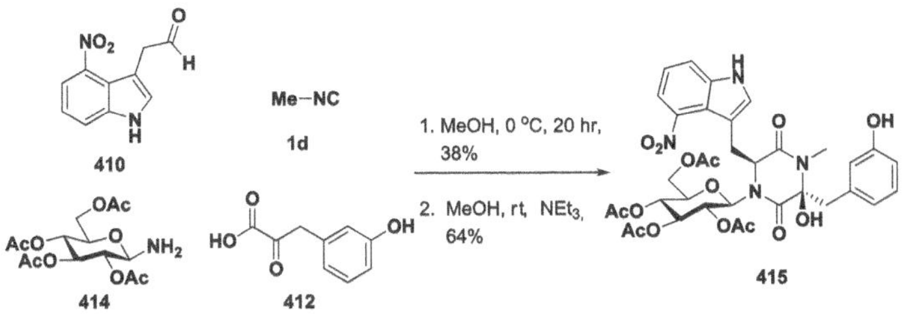

SCHEME 7.128 Kunz methodology for the preparation of sugar DKP derivative **415**.

7.3 CONCLUSIONS

The preceding survey of the recent developments in the synthesis of CICs and their application in small molecules and bioactive natural products reflects an active current interest in this highly versatile class of useful reagents. From a practical point of view, CICs have revolutionized the application of IMCRs, in particular that of U-4CR, due to their mild removal conditions that are compatible with many commonly used functional handles and their odorless, fragrant-like smell. Post-Ugi modification using CICs provides a method of transforming *N*-terminal secondary amides into carboxylic acids, esters, thioesters, ketones, aldehydes, and alcohols for further chemical diversification and thus helped remedy the lack of commercial isocyanide inputs.

Although the CICs were underexplored in the P-3CR, US-4CR, JU-3CR, GBB-4C-3CR, and DA reactions, in recent years there has been an impetus in the scientific community to apply CICs in other areas (e.g., material science) beyond their traditional application in a small molecule synthesis.

We anticipate that this review will stimulate further interest in the development of CICs in organic synthesis and hope that they will find application in the discovery of new therapeutic agents, as well as polymers, and accelerate the launch of new drugs.

REFERENCES

[1] (a) Lieke, W. *Justus Liebigs Ann. Chem.* **1859**, *112*, 316–321. (b) IUPAC Goldbook *isocyanides*. International Union of Pure and Applied Chemistry Compendium of Chemical Terminology Gold Book Version 2.3.3. 2014-02-24. http://goldbook.iupac.org/. (c) Moffat, J. B. *Chem. Phys. Lett.* **1978**, *55*, 125–130.

[2] (a) Sidgwick, N. V. *Chem. Rev.* **1931**, *9*, 77–88. (b) Brockway, L. O. *J. Am. Chem. Soc.* **1936**, *58*, 2516–2518. (c) Kessler, M. K.; Ring, H.; Trambarulo, R.; Gordy, W. *Phys. Rev.* **1950**, *79*, 54. (d) Costain, C. C. *J. Chem. Phys.* **1958**, *29*, 864–874. (e) Williams, R. L. *J. Chem. Phys.* **1956**, *25*, 656–659. (f) Pillai, M. G. K.; Cleveland, F. F. *J. Molec. Spect.* **1960**, *5*, 217.

[3] Tobisu, M.; Oshita, M.; Yoshioka, S.; Kitajima, A.; Chatani, N. *Pure Appl. Chem.* **2006**, *78*, 275–280.

[4] (a) Seebach, D. *Angew. Chem. Int. Ed. Engl.* **1979**, *18*, 239–258. (b) Hase, T. A. *Umpoled Synthons. A Survey of Sources and Uses in Synthesis*; John Wiley & Sons, Inc.: New York, 1987. (c) Dömling, A.; Ugi, I. *Angew. Chem. Int. Ed.* **2000**, *39*, 3168–3210.

[5] (a) Takei, F.; Tung, S.; Yanai, K.; Onitsuka, K.; Takahashi, S. *J. Organomet. Chem.* **1998**, *559*, 91–96. (b) van Beijnen, A. J. M. *Macromolecules* **1983**, *16*, 1679–1689. (c) Drenth, W.; Nolte, R. J. M. *Acc. Chem. Res.* **1979**, *12*, 30–35. (d) Millich, F. *Chem. Rev.* **1972**, *72*, 101–113.

[6] (a) Rüchardt, C.; Meier, M.; Haaf, K.; Pakusch, J.; Wolber, E. K. A.; Müller, B. *Angew. Chem. Int. Ed. Engl.* **1991**, *30*, 893–901. (b) Guillemard, H. *Ann. Chim.* **1908**, *14*, 363.

[7] (a) Kobyashi, Y.; Fukuyama, T. *J. Heterocycl. Chem.* **1998**, *35*, 1043–1056. (b) Josien, H.; Ko, S.-B.; Bom, D.; Curran, D. P. *Chem. Eur. J.* **1998**, *4*, 67–83. (c) Ryu, I.; Sonoda, N.; Curran, D. P. *Chem. Rev.* **1996**, *96*, 177–184.

[8] Rothe, R. *Pharmazie* **1950**, *5*, 190.

[9] (a) Gautier, A. *Justus Liebigs Ann. Chem.* **1867**, *142*, 289–294. (b) Hofmann, A. W. *Justus Liebigs Ann. Chem.* **1867**, *144*, 114–120.

[10] Hofmann, A. W. *Ber. Dtsch. Chem. Ges.* **1870**, *3*, 761–772.

[11] Kitano, Y.; Chiba, K.; Tada, M. *Tetrahedron Lett.* **1998**, *39*, 1911–1912.

[12] (a) Ugi, I.; Meyr, R. *Angew. Chem.* **1958**, *70*, 702–703. (b) Ugi, I.; Meyr, R. *Chem. Ber.* **1960**, *93*, 239–248.

[13] van Leusen, A. M.; Boerma, G. J. M.; Helmholdt, R. B.; Siderius, H.; Strating, J. *Tetrahedron Lett.* **1972**, 2367–2368.

[14] (a) Masutani, K.; Minowa, T.; Mukaiyama, T. *Chem. Lett.* **2005**, *34*, 1124–1125. (b) Grundmann, C. In *Methoden Org. Chem. (Houben-Weyl)*, 4th ed. 1952, Vol. E5, 1985; p. 1611. (c) O'Neil, I. A. In *Comprehensive Organic Functional Group Transformation*, Vol. 3; Katritzky, A. R., Meth-Cohn, D., Rees, C. W., Eds.; Pergamon: Oxford, 1995; p 963. (d) Lentz, D. *Angew. Chem.* **1994**, *106*, 1377–1393; *Angew. Chem. Int. Ed. Engl.* **1994**, *33*, 1315. (e) Witczak, Z. J. *Tetrahedron Lett.* **1986**, *27*, 155–158. (f) Hoffman, P.; Gokel, G.; Marquarding, D.; Ugi, I. In *Isonitrile Chemistry*; Ugi, I. Ed.; Academic Press: New York, 1971; p 9. (g) Ugi, I.; Fetzer, U.; Eholzer, U.; Knupeer, H.; Offermann, K. *Angew. Chem. Intl. Ed.* **1965**, *4*, 472–484.

[15] (a) Dömling, A.; Ugi, I. *Angew. Chem. Intl. Ed.* **2000**, *39*, 3168–3210. (b) Zhu, J.; Bienaymé, H. *Multicomponent Reactions*; Wiley-VCH: Weinheim, 2005. (c) Dömling, A. *Chem. Rev.* **2006**, *106*, 17–89.

[16] (a) Trost, B. M. *Angew. Chem.* **1995**, *107*, 285–307; *Angew. Chem. Intl. Ed.* **1995**, *34*, 259–281. (b) Wender, P. A.; Handy, S. T.; Wright, D. L. *Chem. Ind.* **1997**, *765*, 767–769.

[17] (a) Passerini, M. *Gazz. Chim. Ital.* **1921**, *51*, 126–129. (b) Passerini, M. *Gazz. Chim. Ital.* **1921**, *18*, 181–188.

[18] Maeda, S.; Komagawa, S.; Uchiyama, M.; Morokuma, K. *Angew. Chem. Int. Ed.* **2011**, *50*, 644–649.

[19] Ugi, I.; Meyr, R.; Fetzer, U.; Steinbrückner, C. *Angew. Chem.* **1959**, *71*, 386–388.

[20] (a) Akritopoulou-Zanze, I. *Curr. Opin. Chem. Biol.* **2008**, *12*, 324–331. (b) Santra, S.; Andreana, P. R. *Org. Lett.* **2007**, *9*, 5035–5038. (c) Schreiber, S. L. *Science* **2000**, *287*, 1964–1969. (d) Hulme, C.; Gore, V. *Curr. Med. Chem.* **2003**, *10*, 51–80. (e) Rikimaru, K.; Mori, K.; Kan, T.; Fukuyama, T. *Chem. Commun.* **2005**, 394–396. (f) Wang, X.; Wang, S.-Y.; Ji, S.-J. *Org. Lett.* **2013**, *15*, 1954–1957. (g) Du, Z.; Lu, Y.; Dai, X.; Zhang-Negrerie, D.; Gao, Q. *J. Chem. Res.* **2013**, *3*, 181–185. (h) Eckert, H. *Molecules* **2012**,

17, 1074–1102. (i) Dömling, A.; Wang, W.; Wang, K. *Chem. Rev.* **2012**, *112*, 3083–3135. (j) Ganem, B. *Acc. Chem. Res.* **2009**, *42*, 463–469. (k) Santra, S.; Andreana, P. R. *J. Org. Chem.* **2011**, *76*, 2261–2264. (l) Santra, S.; Andreana, P. R. *Angew. Chem. Int. Ed.* **2011**, *50*, 9418–9422.

[21] (a) Ugi, I. *J. Prakt. Chem.* **1997**, *339*, 499–516. (b) Rossen, K.; Pye, P. J.; DiMichele, L. M.; Volante, R. P.; Reider, P. J. *Tetrahedron Lett.* **1998**, *39*, 6823–6826. (c) Weber, L.; Waltbaum, S.; Broger, C.; Gubernator, K. *Angew. Chem.* **1995**, *107*, 2452–2454; *Angew. Chem. Int. Ed.* **1995**, *34*, 2280–2282. (d) Endo, A.; Yanagisawa, A.; Abe, M.; Tohma, S.; Kan, T.; Fukuyama, T. *J. Am. Chem. Soc.* **2002**, *124*, 6552–6554.

[22] (a) Agrestia, J. J.; Antipovc, E.; Abatea, A. R.; Ahna, K.; Rowata, A. C.; Barete, J. C.; Marquezf, M.; Klibanovc, A. M.; Griffiths, A. D.; Weitz, D. A. *Proc. Natl. Acad. Sci.* **2010**, *107*, 4004–4009. (b) Schonbrun, E.; Abate, A. R.; Steinvurzel, P. E.; Weitz, D. A.; Crozier, K. B. *Lab on a Chip* **2010**, *10*, 852–856.

[23] (a) Hulme, C.; Gore, V. *Curr. Med. Chem.* **2003**, *10*, 51–80. (b) Bienaymé, H.; Hulme, C.; Oddon, G.; Schmidtt, P. *Chem. Eur. J.* **2000**, *6*, 3321–3329.

[24] Ugi, I.; Fetzer, U.; Eholzer, U.; Knupfer, H.; Offermann, K. *Angew. Chem. Intl. Ed. Engl.* **1965**, *4*, 472–484.

[25] van Leusen, A. M.; Strating, J. *Q. Rep. Sulfur. Chem.* **1970**, *5*, 67.

[26] (a) van Leusen, A. M.; Hoogenboom, B. E.; Siderius, H. *Tetrahedron Lett.* **1972**, *13*, 2369–2372. (b) Possel, O.; van Leusen, A. M. *Heterocycles* **1977**, *7*, 77–80. (c) van Leusen, A. M.; Siderius, H.; Hoogenboom, B. E.; van Leusen, D. *Tetrahedron Lett.* **1972**, *13*, 5337–5340. (d) van Leusen, A. M.; Wildeman, J.; Oldenziel, O. H. *J. Org. Chem.* **1977**, *42*, 1153–1159. (e) van Leusen, D.; Flentge, E.; van Leusen, A. M. *Tetrahedron* **1991**, *47*, 4639–4644. (f) Barton, D. H. R.; Kervagoret, J.; Zard, S. Z. *Tetrahedron* **1990**, *46*, 7587–7598. (g) van Leusen, D.; van Echten, E.; van Leusen, A. M. *J. Org. Chem.* **1992**, *57*, 2245–2249. (h) Possel, O.; van Leusen, A. M.; *Tetrahedron Lett.* **1977**, *18*, 4229–4231. (i) Oldenziel, O. H.; van Leusen, A. M. *Tetrahedron Lett.* **1974**, *15*, 167–170. (j) van Leusen, D.; van Echten, E.; van Leusen, A. M. *Recl. Trav. Chim. Pays-Bas* **1992**, *111*, 469–477. (k) Moskal, J.; van Stralen, R.; Postma, D.; van Leusen, A. M. *Tetrahedron Lett.* **1986**, *27*, 2173–2176. (l) Moskal, J.; van Leusen, A. M. *J. Org. Chem.* **1986**, *51*, 4131–4139.

[27] van Leusen, D.; van Leusen, A. M. *Tetrahedron Lett.* **1984**, *25*, 2581–2584.

[28] (a) van Leusen, A. M.; Schaart, F. J.; van Leusen, D. *Recl. Trav. Chim. Pays-Bas.* **1979**, *98*, 258–262. (b) van Leusen, A. M.; Wildeman, J. *Recl. Trav. Chim. Pays-Bas* **1982**, *101*, 202–205.

[29] Yadav, J. S.; Reddy, P. S.; Joshi, B. V. *Tetrahedron* **1983**, *44*, 7243–4245.

[30] Mendiola, J.; Minguez, J. M.; Alvarez-Builla, J.; Varquero, J. J. *Org. Lett.* **2000**, *2*, 3253–3256

[31] (a) Perry, N. B.; Ettouati, L.; Litaudon, M.; Blunt, J. W.; Munro, M. H. G.; Parkin, S.; Hope, H. *Tetrahedron.* **1994**, *50*, 3987–3992. (b) Trimurtuhu, G.; Faulkner, D. J.; Perry, N. B.; Ettouati, L.; Litaudon, M.; Blunt, J. W.; Munro, M. H. G.; Jameson, G. B. *Tetrahedron* **1994**, *50*, 3993–4000.

[32] Di Santo, R.; Costi, R.; Forte, M.; Galeffi, C. *Arkivoc.* **2004**, *5*, 181–195.

[33] Tsuchida, K.; Ikeda, K.; Mizuno, Y. *Chem. Pharm. Bull.* **1980**, *28*, 2748–2752.

[34] (a) Mizuno, Y.; Endo, T.; Miyaoka, T.; Ikeda, K. *J. Org. Chem.* **1974**, *39*, 1250–1255. (b) Mizuno, Y.; Limn, W.; Tsuchida, K.; Ikeda, K. *J. Org. Chem.* **1972**, *37*, 39–42.

[35] Ugi, I.; Fetzer, U.; Eholzer, U.; Knupfer, H.; Offerman, K. *Angew. Chem.* **1965**, *77*, 492–504.

[36] Isenring, H. P.; Hofheinz, W. *Synthesis* **1981**, 385–387.

[37] (a) Isering, H. P.; Hofheinz, W. *Tetrahedron Lett.* **1983**, *24*, 4837–4838. (b) Isering, H. P.; Hofheinz, W. *Tetrahedron* **1983**, *39*, 2591–2597.

[38] (a) Aoki, H.; Sakai, H.; Kohsaka, M.; Konomi, T.; Hosoda, J.; Kubochi, Y.; Iguchi, E.; Imanaka, I. *J. Antibiot.* **1976**, *29*, 492–500. (b) Hosada, J.; Konomi, T.; Tani, N.; Aoki, H.; Imanaka, H. *Agric. Biol. Chem.* **1977**, *41*, 2013–2020.

[39] Kamiya, T.; Aoki, H.; Mine, Y. In *Chemistry and Biology of β-Lactam Antibiotics,* Vol. 2; Morin, R. B., Gorman, M., Eds.; Academic Press: New York, **1982**; pp 165–226.

[40] (a) Geller, J.; Ugi, I. *Chemica Scripta* **1983**, *22*, 85. (b) Obrecht, R.; Herrman, R.; Ugi, I. *Synthesis* **1985**, 400–402.

[41] Borthwick, A. D.; Davies, D. E.; Exall, A. M.; Hatley, R. J. D.; Hughes, J. A.; Irving, W. R.; Livermore, D. G.; Sollis, S. L.; Nerozzi, F.; Valko, K. L.; Allen, M. J.; Perren, M.; Shabbir, S. S.; Woolard, P. M.; Price, M. A. *J. Med. Chem.* **2006**, *49*, 4159–4170.

[42] Liddle, J.; Allen, M. J.; Borthwick, A. D.; Brooks, D. P.; Davies, D. E.; Edwards, R. M.; Exall, A. M.; Hamlett, C.; Irving, W. R.; Mason, A. M.; McCafferty, G. P.; Nerozzi, F.; Peace, S.; Philip, J.; Pollard, D.; Pullen, M. A.; Shabbit, S. S.; Sollis, S. L.; Westfall, T. D.; Woollard, P. M.; Wu, C.; Hickey, D. M. B. *Bioorg. Med. Chem. Lett.* **2008**, *18*, 90–94.

[43] Katritzky, A. R.; Chen, Y. X.; Yannakopolou, K.; Lue, P. *Tetrahedron Lett.* **1989**, *30*, 6657–6660.

[44] Katritzky, A. R.; Sutharchanadevi, M.; Urogdi, L. *J. Chem. Soc. Perkin Trans 1: Org. Bioorg. Chem.* **1990**, 1847–1851.

[45] Katritzky, A. R.; Cheng, D.; Musgrave, R. P. *Heterocycles* **1997**, *44*, 67–70.

[46] Dömling, A.; Beck, B.; Magnin-Lachaux, M. *Tetrahedron Lett.* **2006**, *47*, 4289–4291.

[47] (a) Herr, R. J. *Bioorg. Med. Chem.* **2002**, *10*, 3379–3393. (b) Patani, G. A.; LaVoie, E. J. *Chem. Rev.* **1996**, *96*, 3147–3176. (c) Venkatraman, S.; Lim, J.; Cramer, M.; Gardner, M. F.; James, J.; Alves, K.; Lingham, R. B.; Mumford, R. A.; Munoz, B. *Bioorg. Med. Chem. Lett.* **2005**, *15*, 4053–4056. (d) Johansson, A.; Poliakov, A.; Kerblom, A.; Wiklund, K.; Lindeberg, G.; Winiwarter, S.; Danielson, U. H.; Samuelsson, B.; Hallberg, A. *Bioorg. Med. Chem.* **2003**, *11*, 2551–2568. (e) Suzuki, T.; Matsuura, A.; Kouketsu, A.; Hisakawa, S.; Nakagawa, H.; Miyata, N. *Bioorg. Med. Chem.* **2005**, *13*, 4332–4342. (f) Raju, R.; Mortell, K.; Anandan, S; O'Dowd, H.; Gao, H.; Gomez, M.; Hackbarth, C.; Wu, C.; Wang, W.; Yuan, Z.; White, R.; Trias, L.; Patel, D. V. *Bioorg. Med. Chem. Lett.* **2003**, *13*, 2413–2418. (g) Jimonet, P.; Bohme, G. A.; Bouquerel, J.; Boireau, A.; Damour, D.; Debono, M. W.; Genevois-Borella, A.; Hardy, J. C.; Hubert, P.; Manfré, F.; Nemecek, P.; Pratt, J.; Randle, J. C. R.; Ribeill, Y.; Stutzmann, J.-M.; Vuilhorgne, M.; Mignani, S.; *Bioorg. Med. Chem. Lett.* **2001**, *11*, 127–132. (h) Campiani, G.; Morelli, E.; Gemma, S.; Nacci, V.; Butini, S.; Hamon, M.; Novellino, E.; Greco, G.; Cagnotto, A.; Goegan, M.; Cervo, L.; Valle, F. D.; Fracasso, C.; Caccia, S.; Mennini, T. *J. Med. Chem.* **1999**, *42*, 4362–4379. (i) Arvanitis, A. G.; Arvanitis, A. G.; Gilligan, P. J.; Chorvat, R. J.; Cheeseman, R. S.; Christos, T. E.; Bakthavatchalam, R.; Beck, J. P.; Cocuzza, A. J.; Hobbs, F. W.; Wilde, R. G.; Arnold, C.; Chidester, D.; Curry, M.; He, L.; Hollis, A.; Klaczkiewicz, J.; Krenitsky, P. J.; Rescinito, J. P.; Scholfield, E.; Culp, S.; De Souza, E. B.; Fitzgerald, L.; Grigoriadis, D.; Tam, S. W.; Wong, Y. N.; Huang, S.-M.; Shen,

H. L. *J. Med. Chem.* **1999**, *42*, 805–818. (j) Owens, A. P.; Harrison, T.; Moseley, J. D.; Swain, C. J.; Sadowski, S.; Cascieri, M. A. *Bioorg. Med. Chem. Lett.* **1998**, *8*, 51–56. (k) Ladduwahetty, T.; Ladduwahetty, T.; Baker, R.; Cascieri, M. A.; Chambers, M. S.; Haworth, K.; Keown, L. E.; MacIntyre, D. E.; Metzger, J. M.; Owen, S.; Rycroft, W.; Sadowski, S.; Seward, E. M.; Shepheard, S. L.; Swain, C. J.; Tattersall, F. D.; Watt, A. P.; Williamson, D. W.; Hargreaves, R. J. *J. Med. Chem.* **1996**, *39*, 2907–2914.

[48] Ugi, I.; Fetzer, U.; Unterstenhofer, G.; Behrenz, W.; Frohberger, P. E.; Scheinpflug, H. Patent Fr.1, 384, 209 Jan. 4, 1965; *Chem. Abstr.* **1965**, *63*, 6924e.

[49] Lacerda, R. B.; de Lima, C. K. F.; da Silva, L. L.; Romeiro, N. C.; Miranda, A. L. P.; Barreiro, E. J.; Fraga, C. A. M. *Bioorg. Med. Chem.* **2009**, *17*, 74–84.

[50] Mjalli, A. M. M.; Sarshar, S.; Baiga, T. J. *Tetrahedron Lett.* **1996**, *37*, 2943–2946.

[51] Weber, W. P.; Gokel, G. W.; Ugi, I. K. *Angew. Chem. Int. Ed. Engl.* **1972**, *11*, 530–531.

[52] Walborsky, H. M.; Ronman, P. *J. Org. Chem.* **1978**, *43*, 731–734.

[53] (a) Geller, J.; Ugi, I. *Chem. Scr.* **1983**, *22*, 85. (b) Ugi, I. *Angew. Chem. Int. Ed. Engl.* **1982**, *21*, 810–819. (c) Ugi, I.; Rosendahl, F. K. *Liebigs Ann. Chem.* **1963**, *666*, 65–67.

[54] Keating, T. A.; Armstrong, R. W. *J. Am. Chem. Soc.* **1995**, *117*, 7842–7843.

[55] Lown, J. W. In *1,3-Dipolar Cycloaddition Chemistry*, Vol. 1; Padwa, A., Ed.; Wiley-Interscience: New York; pp 653.

[56] Keating, T. A.; Armstrong, R. W. *J. Am. Chem. Soc.* **1996**, *118*, 2574–2583.

[57] Keating, T. A.; Armstrong, R. W. *J. Org. Chem.* **1996**, *61*, 8935–8939.

[58] Hulme, C.; Morrissette, M. M.; Volz, F. A.; Burns, C. J. *Tetrahedron Lett.* **1998**, *39*, 1113–1116.

[59] Strocker, A. M.; Keating, T. A.; Tempest, P. A.; Armstrong, R. W. *Tetrahedron Lett.* **1996**, *37*, 1149–1152.

[60] (a) Barton, D. H.; Bowles, T.; Husinec, S.; Forbes, J. E.; Llobera, A.; Porter, A. E. A.; Zard, S. Z. *Tetrahedron Lett.* **1988**, *29*, 3343–3346. (b) Baldwin, J. E.; Yamaguchi, Y. *Tetrahedron Lett.* **1989**, *30*, 3335–3338.

[61] (a) Maison, W.; Schlemminger, I.; Westerhoff, O.; Martens, J. *Bioorg. Med. Chem.* **2000**, *8*, 1343–1360. (b) Maison, W.; Schlemminger, I.; Westerhoff, O.; Martens, J. *Bioorg. Med. Chem.* **1999**, *9*, 581–584.

[62] (a) Egholm, M.; Berg, R. H.; Buchardt, O. *Science* **1991**, *254*, 1497–1500. (b) Nielsen, P. E.; Egholm, M. *Peptide Nucleic Acids*; Horizon Scientific: Wymondham, 1999.

[63] Xu, Z.; Dietrich, J.; Shaw, A. Y.; Hulme, C. *Tetrahedron Lett.* **2010**, *51*, 4566–4569.

[64] Lindhorst, T.; Bock, H.; Ugi, I. *Tetrahedron* **1999**, *55*, 7411–7420.

[65] Rikimaru, K.; Yanagisawa, A.; Kan, T.; Fukuyama, T. *Synlett* **2004**, 41–44.

[66] (a) Tokuyama, H.; Yokoshima, S.; Lin, S.-C.; Li, L.; Fukuyama, T. *Synthesis* **2002**, 1121–1123. (b) Tokuyama, H.; Yokoshima, S.; Yamashita, T.; Fukuyama, T. *Tetrahedron Lett.* **1998**, *39*, 3189–3192.

[67] Linderman, R. J.; Binet, S.; Petrich, S. R. *J. Org. Chem.* **1999**, *64*, 336–337.

[68] Kunz, H.; Pfrengle, W. *Tetrahedron* **1988**, *44*, 5487–5494.

[69] Cohen, E. *Ann. Rev. Entomol.* **1987**, *32*, 71–93.

[70] Plant, A.; Thompson, P.; Williams, D. M. *J. Org. Chem.* **2009**, *74*, 4870–4873.

[71] Hulme, C.; Chappeta, S.; Dietrich, J. *Tetrahedron Lett.* **2009**, *50*, 4054–4057.

[72] (a) Pirrung, M. C.; Ghorai, S. *J. Am. Chem. Soc.* **2006**, *128*, 11772–11773. (b) Pirrung, M. C.; Ghorai, S.; Ibarra-Rivera, T. R. *J. Org. Chem.* **2009**, *74*, 4110–4117.

[73] Van den Bogaert, A. M.; Neilssen, J.; Ovaere, M.; Van Meervelt, L.; Comperbolle, F.; De Borggraeve. *Eur. J. Org. Chem.* **2010**, 5397–5401.

[74] Ji, F.; Yi, W.-B.; Cai, C. *J. Heterocycl. Chem.* **2014**, *51*, 1287–1292.

[75] Kobayashi, K.; Yoneda, K.; Mizumoto, T.; Umakoshi, H.; Morikawa, O.; Konishi, H. *Tetrahedron Lett.* **2003**, *44*, 4733–4736.

[76] Batcho, A. D.; Leimbruger, W. *Org. Synth. Coll. Vol.* **1985**, *63*, 214–220.

[77] (a) Buller, M. J.; Gilley, C. B.; Nguyen, B.; Olshansky, L.; Fraga, B.; Kobayashi, Y. *Synlett* **2008**, 2244–2248. (b) Isaacson, J.; Gilley, C. B.; Kobayashi, Y. *J. Org. Chem.* **2007**, *72*, 3913–3916.

[78] Vamos, M.; Ozboya, K.; Kobayashi, Y. *Synlett* **2007**, 1595–1599.

[79] Gilley, C. B.; Buller, M. J.; Kobayashi, Y. *Org. Lett.* **2007**, *9*, 3631–3634.

[80] (a) Omura, S.; Fujimoto, T.; Otoguru, K.; Matsuzaki, K.; Moriguchi, R.; Tanaka, H.; Sasaki, Y. *J. Antibiot.* **1991**, *44*, 113–116. (b) Omura, S.; Matsuzaki, K.; Fujimoto, T.; Kosuge, K.; Furuya, T.; Fujita, S.; Nakagawa, A. *J. Antibiot.* **1991**, *44*, 117–118.

[81] (a) Fenteany, G. A.; Standaert, R. F.; Reichard, G. A.; Corey, E. J.; Schreiber, S. L. *Proc. Natl. Acad. Sci. U.S.A.* **1994**, *91*, 3358–3362. (b) Fenteany, G. A.; Standaert, R. F.; Lane, W. S.; Choi, S.; Corey, E. J.; Schreiber, S. L. *Science* **1995**, *268*, 726–731. (c) Richardson, P. G.; Anderson, K. C. *Clin. Adv. Hemato. Oncol.* **2003**, *1*, 596–600.

[82] Jochims, J. C.; Kobayashi, Y.; Skrzelewski, E. *Tetrahedron Lett.* **1974**, *7*, 571–574.

[83] (a) Isaacson, J.; Loo, M.; Kobayashi, Y. *Org. Lett.* **2008**, *10*, 1461–1463. (b) Isaacson, J.; Kobayashi, Y. *Angew. Chem. Int. Ed.* **2009**, *48*, 1845–1848.

[84] Sakai, R.; Oiwa, C.; Takaishi, K.; Kamiya, H.; Tagawa, M. *Tetrahedron Lett.* **1999**, *40*, 6941–6944.

[85] Kreye, O.; Trefzger, C.; Sehlinger, A.; Meier, M. A. R. *Macromol. Chem. Phys.* **2014**, *215*, 2207–2220.

[86] Baldoli, C.; Maiorana, S.; Licandro, E.; Zinzalla, G.; Perdicchia, D. *Org. Lett.* **2002**, *4*, 4341–4344.

[87] (a) Rose-Munch, F.; Alexakis, A.; Rose, E.; Mangeney, P.; Marek, I.; Serma, A.; Robert, F. *J. Am. Chem. Soc.* **1992**, *114*, 8288–8290. (b) Rose, E.; Rose-Munch, F. *Curr. Org. Chem.* **1999**, *3*, 445. (c) Davies, S. G.; McCarty, T. D. In *Comprehensive Organometallic Chemistry II*, Vol. 12; Oxford, 1995; pp 1039–1070. (d) Solladié-Cavallo, A. In *Advances in Metal-Organic Chemistry*, Vol. 1; Liebeskind, L. S., Ed.; JAI Press: Greenwich, 1989; p 99.

[88] Hahn, F. E.; Plumed, C. G.; Münder, M.; Lügger, T. *Chem Eur. J.* **2004**, *10*, 6285–6293.

[89] Gilley, C. B.; Kobayashi, Y. *J. Org. Chem.* **2008**, *73*, 4198–4204.

[90] Nishimura, S.; Matsunaga, S.; Shibazaki, M.; Suzuki, K.; Furihata, K.; van Soest, R. W. M.; Fusetani, N. *Org. Lett.* **2003**, *5*, 2255–2257.

[91] Chinigo, G. M.; Breder, A.; Carreira, E. M. *Org. Lett.* **2011**, *13*, 78–81.

[92] (a) Walborsky, H. M.; Nizik, G. E. *J. Org. Chem.* **1972**, *37*, 187–190. (b) Niznik, G. E.; Morrison, W. H., III; Walborsky, H. M. *Org. Synth.* **1971**, *51*, 31. (c) Walborsky, H. M.; Nizik, G. E. *J. Am. Chem. Soc.* **1969**, *91*, 7778–7778.

[93] Thompson, M. J.; Chen, B. *Tetrahedron Lett.* **2008**, *49*, 5324–5327.

[94] Thompson, M. J.; Chen, B. *J. Org. Chem.* **2009**, *74*, 7084–7093.

[95] Krasavin, M.; Tsirulnikov, S.; Nikulnikiv, M.; Sandulenko, Y.; Bukhryakov, K. *Tetrahedron Lett.* **2008**, *49*, 7318–7321.

[96] Maddirala, A. R.; Bourgault, J.-P.; Wainaina, M.; Santra, S.; Andreana, P. R. *Unpublished results* **2010**.

[97] Jida, M.; Soueidan, M.; Willand, N.; Agbossou-Niedercorn, F.; Pelinski, L.; Laconde, G.; Deprez-Poulain, R.; Deprez, B. *Tetrahedron Lett.* **2011**, *52*, 1705–1708.

[98] (a) Zuo, X.; Mi, N.; Fan, Z.-J.; Zheng, Q.-X.; Zhang, H.-K.; Wang, H.; Yang, Z.-K. *J. Agr. Food Chem.* **2010**, *58*, 2755–2762. (b) Crosignani, S.; Launay, D.; Linclau, B.; Bradley, M. *Mol. Divers.* **2003**, *7*, 203–210. (c) Isobe, T.; Ishikawa, T. *J. Org. Chem.* **1999**, *64*, 6984–6988. (d) Guirado, A.; Zapata, A.; Fenor, M. *Tetrahedron Lett.* **1992**, *33*, 4779–4782. (e) Wittman, R. *Angew. Chem.* **1961**, *73*, 219–220.

[99] Zhou, H.; Zhang, W.; Yan, B. *J. Comb. Chem.* **2010**, *12*, 206–214.

[100] Maehara, T.; Kanno, R.; Yokoshima, S.; Fukuyama, T. *Org. Lett.* **2012**, *14*, 1946–1948.

[101] Filho, R. A. W. N.; Stark, S.; Morejon, M. C.; Westermann, B.; Wessjohann, L. A. *Tetrahedron Lett.* **2012**, *53*, 5360–5363.

[102] (a) Chan, S.; Braish, T. F.; *Tetrahedron* **1994**, *50*, 9943–9950. (b) Hanessian, S.; Reinhold, U.; Nikoyic, S. *Tetrahedron Lett.* **1996**, *37*, 8967–8970. (c) Liu, S.; Hanzlik, R. P. *J. Med. Chem.* **1992**, *35*, 1067–1075. (d) Xie, X.; Lu, X. Y.; Xu, W. *J. Org. Chem.* **2001**, *66*, 6545–6550.

[103] Barluenga, J.; de Prado, A.; Santamaria, J.; Tomás, M. *Organometallics* **2005**, *24*, 3614–3617.

[104] Oikawa, M.; Sugamata, Y.; Chiba, M.; Fukushima, K.; Ishikawa, Y. *Synlett* **2013**, *24*, 2014–2018.

[105] Kita, Y.; Nishii, Y.; Higuchi, T.; Mashima, K. *Angew. Chem. Int. Ed.* **2012**, *51*, 5723–5726.

[106] Hulme, C.; Peng, J.; Morton, G.; Salvino, J. M.; Herpin, T.; Labaudiniere, R. *Tetrahedron Lett.* **1998**, *39*, 7227–7230.

[107] Plunkett, M. J.; Ellman, J. A. *J. Org. Chem.* **1995**, *60*, 6006–6007.

[108] Piscopio, A. Org. Poster 232, Las Vegas Am. Chem. Soc. Meeting, 1997.

[109] Hulme, C.; Ma, L.; Cherrier, M.-P.; Romano, J. J.; Morton, G.; Duquenne, C.; Salvino, J.; Labaudiniere, R. *Tetrahedron Lett.* **2000**, *41*, 1883–1887.

[110] Chen, J. J.; Golebioski, A.; McClenaghan, J.; Klopfenstein, S. R.; West, L. *Tetrahedron Lett.* **2001**, *42*, 2269–2271.

[111] Chen, J. J.; Golebioski, A.; McClenaghan, J.; Klopfenstein, S. R.; West, L. *Tetrahedron Lett.* **2002**, *43*, 4083–4085.

[112] Kennedy, A. L.; Fryer, A. M.; Josey, J. A. *Org. Lett.* **2002**, *4*, 1167–1170.

[113] (a) Kaim, L. E.; Griamud, L.; Olbe, J. *Angew. Chem. Int. Ed.* **2005**, *44*, 7961–7964. (b) Kaim, L. E.; Gizolme, M.; Grimaud, L. *Org. Lett.* **2006**, *8*, 5021–5023. (c) Kaim, L. E.; Gizolme, M.; Griamud, L.; Olbe, J. *Org. Lett.* **2006**, *8*, 4019–4021. (d) Kaim, L. E.; Gizolme, M.; Griamud, L.; Olbe, J. *J. Org. Chem.* **2007**, *72*, 4169–4180. (e) Olbe, J.; Kaim, L. E.; Gizzi, M.; Griamud, L. *Heterocycles* **2007**, *73*, 503–517.

[114] (a) Levy, A. A.; Rains, H. C.; Smiles, S. *J. Chem. Soc.* **1931**, 3264–3269. (b) Truce, W. E.; Ray, W. J.; Norman, O. L.; Eickemeyer, D. B. *J. Am. Chem. Soc.* **1958**, *80*, 3625–3629.

[115] Bowers, M. M.; Carroll, P.; Joullié, M. M. *J. Chem. Soc. Perkin Trans. I.* **1989**, 857–865.

[116] Chapman, T. M., Davies, I. G.; Gu, B.; Block, T. M.; Scopes, D. I. C. I; Hay, P. A.; Courtney, S. M.; McNeill, L. A.; Schofield, C. J.; Davis, B. G. *J. Am. Chem. Soc.* **2005**, *127*, 506–507.

[117] Banfi, L.; Basso, A.; Guanti, G.; Merlo, S.; Repetto, C.; Riva, R. *Tetrahedron* **2008**, *64*, 1114–1134.

[118] Rubinshtein, M.; James, C. R.; Young, J. L.; Ma, Y. J.; Kobayashi, Y.; Gianneschi, N. C.; Yang, J. *Org. Lett.* **2010**, *12*, 3560–3563.

[119] Le, H. V.; Fan, L.; Ganem, B. *Tetrahedron Lett.* **2011**, *52*, 2209–2011.

[120] (a) Blackburn, C.; Guan, B.; Fleming, P.; Shiosaki, K.; Tsai, S. *Tetrahedron Lett.* **1998**, *39*, 3635–3638. (b) Blackburn, C. *Tetrahedron Lett.* **1998**, *39*, 5469–5472. (c) Groebke, K.; Weber, L.; Fridolin, M. *Synlett* **1998**, 661–663. (d) Bienaymé, H.; Bouzid, K. *Angew. Chem. Int. Ed. Engl.* **1998**, *37*, 2234–2240.

[121] Devi, N.; Rawal, R. K.; Singh, V. *Tetrahedron* **2015**, *71*, 183–232.

[122] Blackburn, C.; Guan, B. *Tetrahedron Lett.* **2000**, *41*, 1495–1500.

[123] Sandulenko, Y.; Komarov, A.; Rufanov, K.; Krasavin, M. *Tetrahedron Lett.* **2008**, *49*, 5990–5993.

[124] (a) Krasavin, M.; Shkavrov, S.; Parchinsky, V.; Bukhryakov, K. *J. Org. Chem.* **2009**, *74*, 2627–2629. (b) Krasavin, M.; Parchinsky, V. *Synlett* **2008**, 645–648.

[125] Guchhait, S. K.; Madaan, C. *Org. Biomol. Chem.* **2010**, *8*, 3631–3634.

[126] Neo, A. G.; Bornadiego, A.; Díaz, J.; Marcaccini, S.; Marcos, C. F. *Org. Biomol. Chem.* **2013**, *11*, 6546–6555.

[127] Witczak, Z. *J. Carbohydr. Chem.* **1984**, *3*, 359–380.

[128] Witczak, Z. *Tetrahedron Lett.* **1986**, *27*, 155–158.

[129] Ziegler, T.; Schlomer, R.; Koch, C. *Tetrahedron Lett.* **1998**, *39*, 5957–5960.

[130] Ziegler, T.; Kaisers, H.; Schlomer, R.; Koch, C. *Tetrahedron* **1999**, *55*, 8397–8408.

[131] Wilson, R.; Stockdill, J.; Wu, X.; Vadola, P.; Park, P.; Danishefsky, S. *Angew. Chem. Int. Ed. Engl.* **2012**, *51*, 2834–2848.

[132] Bourgault, J. P.; Maddirala, A. R.; Andreana, P. R. *Org. Biomol. Chem.* **2014**, *12*, 8125–8127.

[133] Bourgault, J. P.; Maddirala, A. R.; Andreana, P. R. Manuscript in preparation, **2015**.

8

ADDING ADDITIONAL RINGS TO THE CARBOHYDRATE CORE: ACCESS VIA (SPIRO) ANNULATION DOMINO PROCESSES

DANIEL B. WERZ

Institut für Organische Chemie, Technische Universität Braunschweig, Hagenring 30, 38106 Braunschweig, Germany

8.1 INTRODUCTION

Domino reactions [1, 2] in carbohydrate chemistry commonly comprise a multitude of unusual transformations, leading often to products with a rearranged sugar backbone. In many cases, the original carbohydrate core that was employed as a substrate is not easily detectable any longer. Often several (protected) hydroxyl groups come into play as nucleophilic centers enabling a specific reaction, but hamper the structure elucidation difficult. It is not exaggerated to state that many of the domino processes involving the rearrangement of carbohydrates were found by serendipity since the multitude of reactive functionalities makes a prediction rather difficult. Besides these reactions, another class of domino processes uses carbohydrates as starting materials for the construction of small molecules without rearranging the complete core. In these latter cases, non-carbohydrate-specific domino reactions and (modified) carbohydrates as substrates to initiate the desired transformation were employed. In this short account, this is demonstrated by two major examples from our laboratory, how domino reactions were utilized (a) for the synthesis of spiroketals and (b) for the preparation of chromans or isochromans and derivatives thereof.

Domino and Intramolecular Rearrangement Reactions as Advanced Synthetic Methods in Glycoscience, First Edition. Edited by Zbigniew J. Witczak and Roman Bielski.

Spiroketals consist of a ring with an endocyclic oxygen that could be a furanose or pyranose, whereas chromans and isochromans need a pyranose as core structure to which a benzene is attached. Such an approach is particularly attractive since about half of the anticipated molecules are directly derived from a natural carbon source. In addition, the sugar commonly bears several residues such as hydroxyl groups that are available in different stereoisomers paving the way for a facile access to a plethora of structurally similar, but stereochemically diverse compounds. Such a strategy is ideally suited for the development and generation of compound libraries.

8.2 SPIROKETALS VIA A DOMINO OXIDATION/REARRANGEMENT SEQUENCE

Our adventures in the construction of [*n*,5]-spiroketals actually started with the desire to lock the 6-hydroxyl group of common pyranoses (e.g., in **1**) [3, 4]. An approach we envisioned was to use a Baeyer–Villiger reaction of the respective cyclopropyl aldehyde of type **2** to insert an oxygen atom between the carbonyl and the three-membered ring. Without thinking too much about the special properties of donor–acceptor (D–A) cyclopropanes [5, 6], we tried to generate cyclopropyl aldehydes such as **2**. However, the formation of an aldehyde created also a classical D–A system (in combination with the endocyclic oxygen) being prone for rearrangement due to its high ring strain and the weak bond between the donor and the acceptor [5,6] (Figure8.1).

Indeed, a domino process consisting of a combination of aldehyde formation and subsequent rearrangement enlarging the three-membered ring to a five-membered ring system took place and we exploited this method to generate spiroketals from monosaccharides [7]. The starting material, the cyclopropane derivative, is easily accessible via cyclopropanation of exocyclic enol ethers of type **4**. As precursors for the formation of the spiroannulated cyclopropyl derivatives, a series of five- and six-membered exocyclic enol ethers was prepared starting from the respective monosaccharides. Whereas unsubstituted six-membered exocyclic enol ethers tend to rearrange to the endocyclic isomers, the five-membered counterparts are more

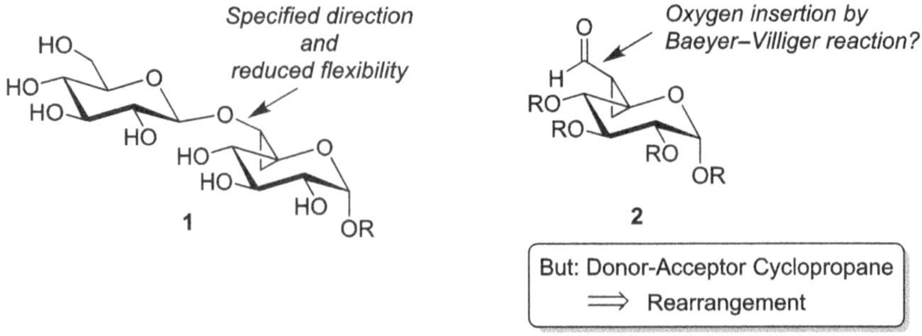

FIGURE 8.1 (left) Original idea to lock the 6-hydroxyl by a spiroannulated cyclopropane ring; and (right) failed idea to insert oxygen by Baeyer–Villiger reaction.

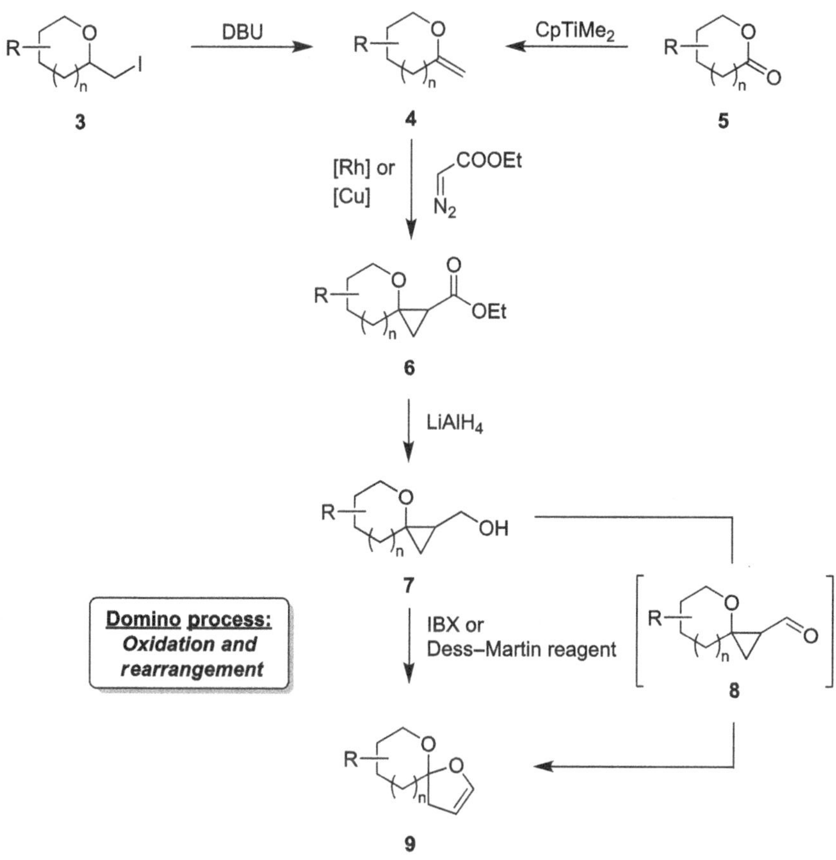

SCHEME 8.1 Synthesis of [*n*,5]-spiroketals **10** using exocyclic enol ethers **13** as starting materials; the last step involves the domino sequence.

stable in their exocyclic form. Thus, exocyclic glycals that are easily available by several routes were used for testing of the scope of this domino sequence. Either (hydrogen iodide) HI was eliminated from the exocyclic iodomethyl group next to the oxygen center in **3** by the action of 1,8-diazabicyclo[5.4.0]undec-7-ene or the corresponding fully protected lactones of type **5** were reacted with Tebbe or Petasis reagent [8] (Scheme 8.1). A high degree of substitution renders the six-membered cycles relatively stable in their exocyclic glycal form. These exocyclic glycals were treated with an excess of ethyl diazoacetate under rhodium or copper catalysis to afford the spiroannulated cyclopropane derivatives **6** that were commonly obtained as a mixture of several (up to four) different stereoisomers. The resulting cyclopropyl esters **6** were reduced with $LiAlH_4$ in tetrahydrofuran to yield the corresponding alcohols **7** in a good to quantitative yield [7].

The generation of D–A cyclopropane **8** (by oxidation of the hydroxy methyl group) was conducted in such a way that the oxidizing agent does not attack the newly generated enol ether **9**, which is also prone to oxidation. Intensive optimization studies proved that 2-iodoxybenzoic acid (IBX) (or Dess–Martin periodinane) was the reagent of choice. Nevertheless, in some cases, complete conversion could not be observed. Therefore, our attention focused toward the use of IBX in combination with different Lewis acids. It turned out that the choice of the Lewis acid is of utmost importance to increase the yield of the desired transformation. Whereas hard Lewis acids such as $Ti(Oi\text{-}Pr)_4$ and $SnCl_4$ lead only to traces of product and Lewis acids such as $BF_3{\cdot}OEt_2$ and $MgCl_2$ have almost no influence on the reaction outcome, soft Lewis acids such as $ZnCl_2$ and $Yb(OTf)_3$ increased the yield. A rationalization might be that the action of a Lewis acid is necessary to weaken the bond between the electron-donating substituent and the electron-withdrawing substituent in the three-membered ring. However, after the ring enlargement, the Lewis acid must not be able to attack the newly generated enol ether system. Thus, a successful outcome of the reaction results essentially as a compromise of two effects: an effective weakening of the D–A -substituted bond in the three-membered ring by coordinating the *in situ*-generated aldehyde moiety and the ability not to attack the enol ether system after its generation. Therefore, the kind of Lewis acid that has been added as a catalyst is crucial.

A variety of substrates were tested in the domino reaction and moderate to good yields of the corresponding spiroketals were obtained. In several cases, trace amounts of the highly unstable D–A cyclopropane derivatives bearing aldehyde moieties could be found as side products demonstrating that these entities are intermediates of the domino process. Figure 8.2 compiles a variety of spiroketals obtained by this method with the corresponding yields of the domino sequence. The method proved to be applicable for furanoses as well as pyranoses, as monosaccharides glucose, galactose, and mannose were employed. A wide choice of hydroxyl-protecting groups is tolerated as long as no oxidatively cleavable groups (such as PMB) are present.

However, with the use of Lewis acids, a loss of stereochemical information was encountered when diastereomerically pure cyclopropyl derivative was converted to the respective spiroketal. With Dess–Martin periodinane and (in most cases also with IBX) without adding any additional reagent, the stereochemical information at the spirocenter is almost conserved; that is, hydroxymethyl groups being below the plane of the six-membered ring are converted to the thermodynamically favored α-anomer, whereas a hydroxymethyl being above the plane of the ring led to the less thermodynamically favored β-anomer.

Such an outcome is plausible only if a concerted rather than a zwitterionic mechanism takes place when Dess–Martin periodinane is employed (Scheme 8.2). With IBX, such a clearcut scenario is not observable since as a highly polar reagent commonly DMSO as solvent was used, which stabilizes zwitterionic intermediates as well. Under the influence of Lewis acids, the mechanistic pathway most probably involves an open-chain intermediate and the product distribution mirrors the thermodynamic stability of the stereoisomeric products.

To afford fully deprotected sugar-derived spiroketals, hydrogenolysis using Pearlman's catalyst is able to cleave all benzyl groups and to reduce the double bond in a very good to quantitative yield. A proper analysis of the ^{13}C NMR spectra of protected and deprotected [6, 5]-spiroketals revealed an interesting tendency of the

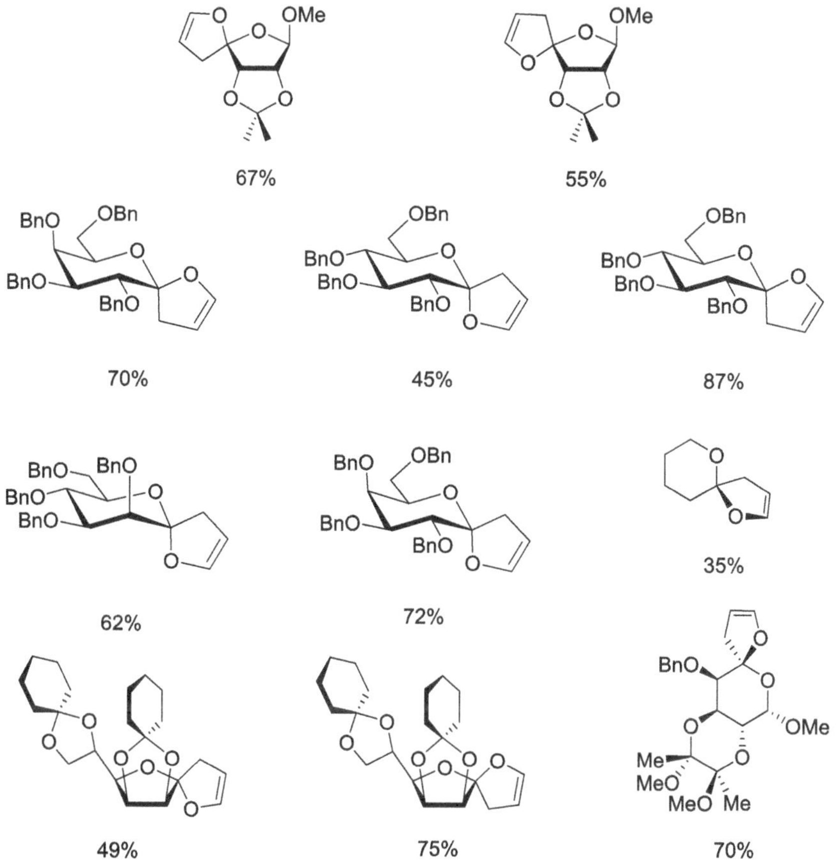

FIGURE 8.2 Compilation of different (carbohydrate-derived) spiroketals with respective yields of the domino process consisting of oxidation and ring enlargement.

chemical shift of the spiro carbon atom. All spiroketals that are anomerically (thermodynamically) favored showed a lower-frequency resonance than the corresponding stereoisomers that are not stabilized by the anomeric effect. In general, we found resonance frequencies of 109–111 ppm when anomeric effects are operating and resonance frequencies of 111–112 ppm without this effect in operation (in deuterated methanol). We assume that this slight difference in chemical shift may be attributed to the np(O)→σ*(C–O) interaction that increases the electron density around the spiro carbon atom whereas such an interaction cannot take place in the other anomer. As a result of the increased electron density, the spiro carbon atoms are more shielded. This observation may also be useful for the elucidation of the stereochemistry of unknown spiroketal structures and may give a further indication of their geometrical feature besides extensive spectroscopic investigations such as nuclear overhauser effect spectroscopy.

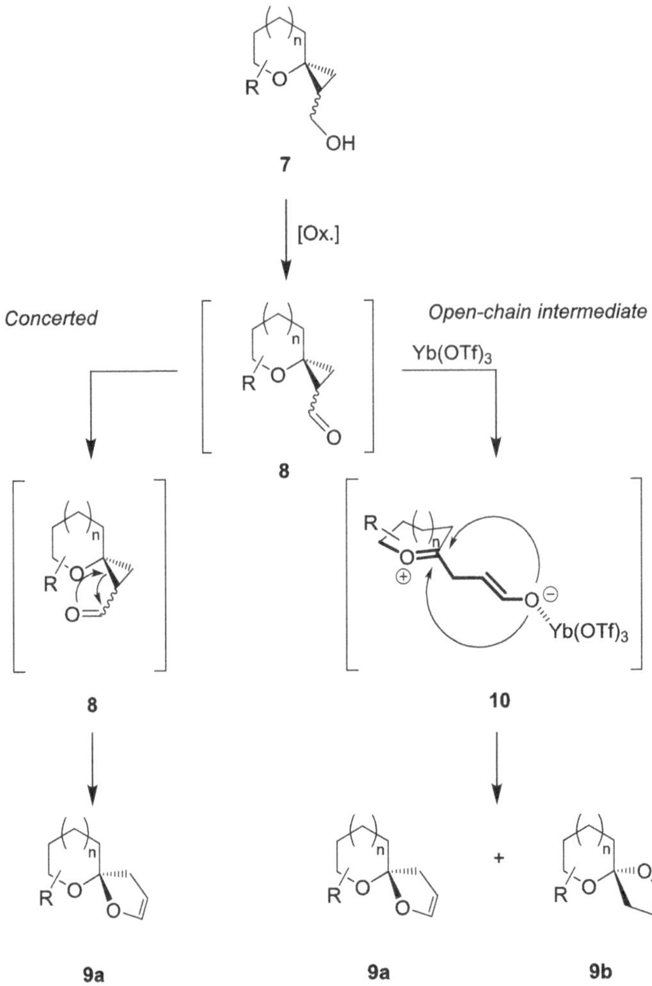

SCHEME 8.2 Potential mechanism of the spiroketal formation via a concerted pathway (left) or zwitterionic open-chain intermediate **10** (right).

8.3 CHROMANS AND ISOCHROMANS VIA DOMINO CARBOPALLADATION/ CARBOPALLADATION/CYCLIZATION SEQUENCE

In the second part of this account, our endeavors in the field of domino annulation chemistry with carbohydrates are in the focus. The annulation of a benzene moiety to the pyran system of carbohydrates leads to chromans and isochromans [9, 10]. Although biosynthetically not derived from carbohydrates, these structural motifs are widespread elements of different natural products with a variety of biological and pharmaceutical activities.

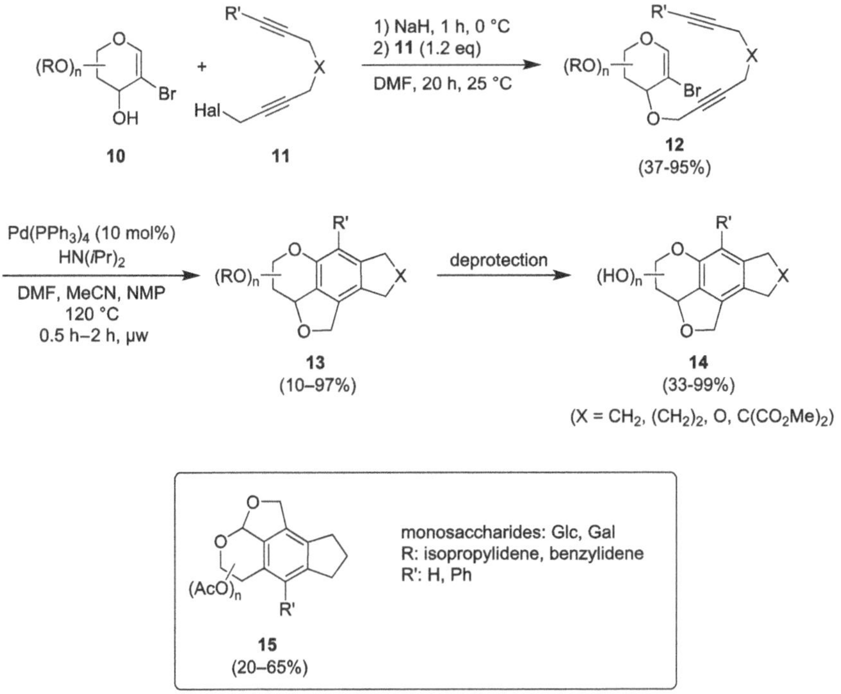

SCHEME 8.3 Chroman and isochroman synthesis via a domino carbopalladation approach.

The key step of this highly effective approach is a palladium-catalyzed domino carbopalladation sequence starting from suitably substituted 2-bromoglycals **10** [11,12]. These substrates are accessed from commercially available glycals via a bromination–elimination procedure. After the installation of suitable protecting groups at C-4 and 6-hydroxyls, attachment of various dialkyne chains **11** was carried out yielding a broad variety of different domino precursors **12** (Scheme 8.3). The synthesis of the corresponding propargylic halides **12** was achieved starting from symmetrical dialkynes, for example, from heptadiyne or octadiyne. The coupling was conducted either by nucleophilic substitution of the free 3-hydroxyl group with the propargylic halide or via Ferrier I reaction (in the case of isochroman precursors). Optimal conditions for the domino reaction comprise the use of a palladium(0) catalyst such as ($Pd(PPh_3)_4$) under microwave irradiation. In a similar approach, respective isochromans **15** were prepared. Attachment of a diynol to the pseudoanomeric center of the 2-bromoglycal set the stage for the Pd-catalyzed domino cyclization.

The domino cascade is initiated by an oxidative addition of the Pd-catalyst into the C–Br bond to form **18** and followed by two consecutive carbopalladation steps forming a 1,3,5-triene system **20** [13, 14]. The final step forms the benzene moiety and may be regarded as 6π-electrocyclization taking place in a disrotatory fashion, followed by β-hydride elimination. The latter reaction can easily take place since

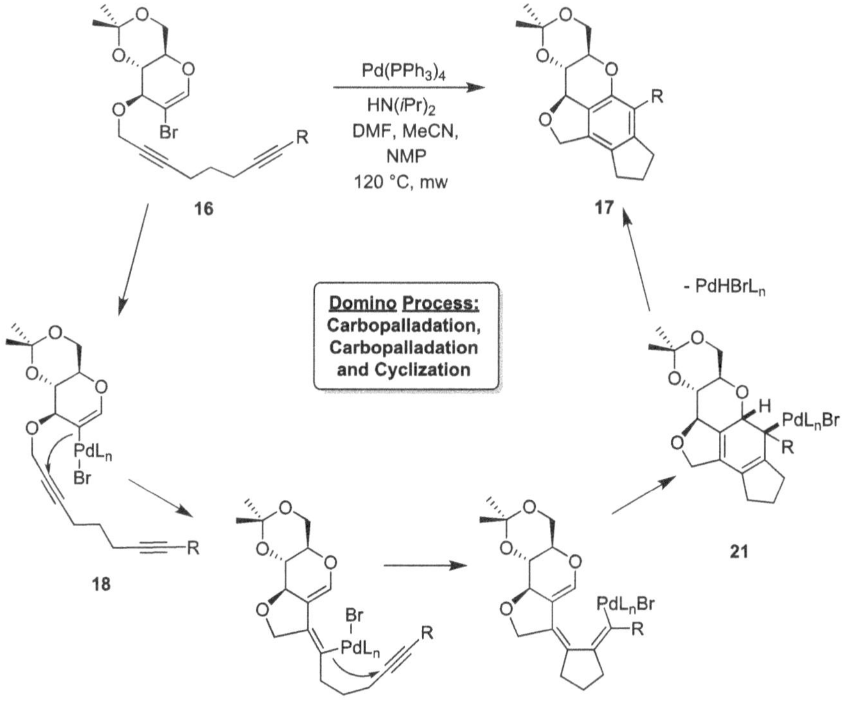

SCHEME 8.4 Mechanistic scenario of chroman formation by the domino sequence.

the electrocyclic ring-closure locates Pd and hydrogen on the same face of the six-membered ring. A mechanistic scenario is provided in Scheme 8.4.

In most cases, the domino reaction proceeds smoothly; different monosaccharide building blocks (galactose, glucose, rhamnose, and arabinose) as well as various substituted diyne chains were employed. Alkyl-, aryl-, and silyl-substituted triple bonds were successfully tested and even terminal acetylene units led to the desired products as Figure 8.3 demonstrates. The best yields were obtained when the diyne chain was attached via an ether linkage, and much poorer yields were obtained by an attachment via an ester linkage. A rationalization might be the reversal of polarity of the C=C triple bond disfavoring the first carbopalladation.

Our original studies showed limitations with respect to the substitution pattern of the chroman and isochroman framework. By omitting the tether between both triple bonds, it was possible to perform the second carbopalladation with an external alkyne **23** in an intermolecular fashion resulting in a more flexible approach [15]. The attachment of the first alkyne to the sugar moiety was performed as already described in Section 8.2. Best results for the domino reaction were obtained using a large excess (10- to 20-fold) of the external alkyne in a highly concentrated reaction mixture. The addition of the Fu's salt releasing the very electron-rich tris(*tert*-butyl)phosphine

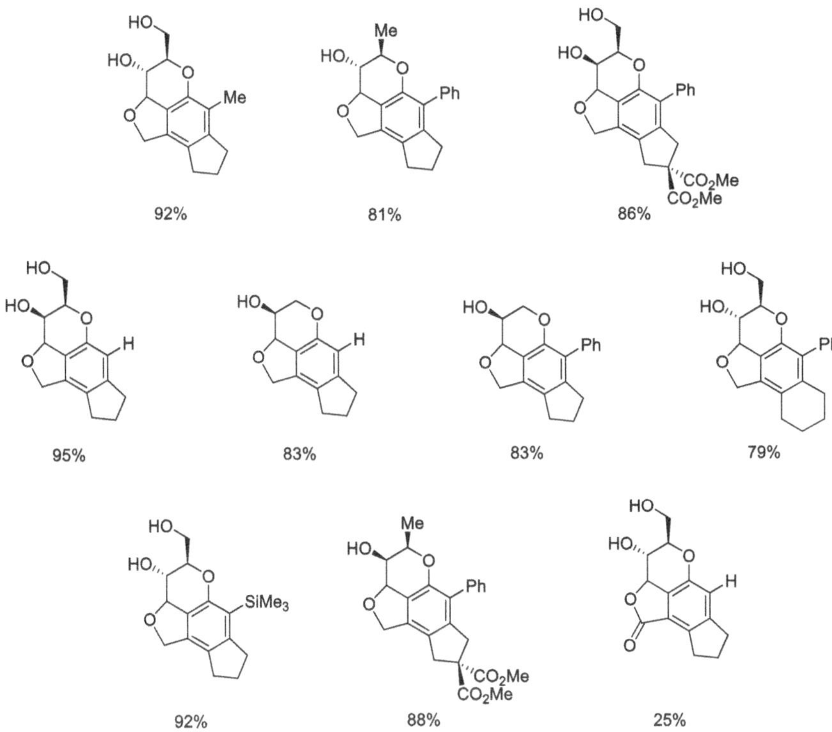

FIGURE 8.3 Compilation of chromans (in a deprotected form) generated by the domino carbopalladation/carbopalladation/cyclization sequence. Given are the yields of the domino sequence starting from the respective diynes.

proved to be essential; otherwise, only trace amount of the chroman was observed. The variation of the carbohydrate part and the differentiation of the attached alkyne as well as the external alkyne provided a broad range of different products. To demonstrate the generality and applicability of this approach besides chromans **24** also, isochromans **26** were accessed in such a way (Scheme 8.5). Glucose- and galactose-derived carbohydrates moieties as well as alkynes with silyl-, alkyl- and aryl-substitution were utilized. Good yields up to 60% were obtained for symmetrical external alkynes such as bis(trimethylsilyl)acetylene or 3-hexyne. The yield dramatically dropped down when sterically encumbered alkynes such as tolane were employed as reaction's diversified reagents.

The extension to a tandem-domino process opened the possibility to form biphenyl structures with a chiral axis [16]. Therefore, the same domino sequence as described before was applied to both sides of a linear C_2-symmetric precursor. The reacting centers are linked by a 1,3-butadiyne unit; thus, every separate process starts from its own vinyl bromide and dialkyne. The final cyclization step of the second domino reaction generates a chiral axis via the twofold benzene formation in the respective domino reaction (Scheme 8.6).

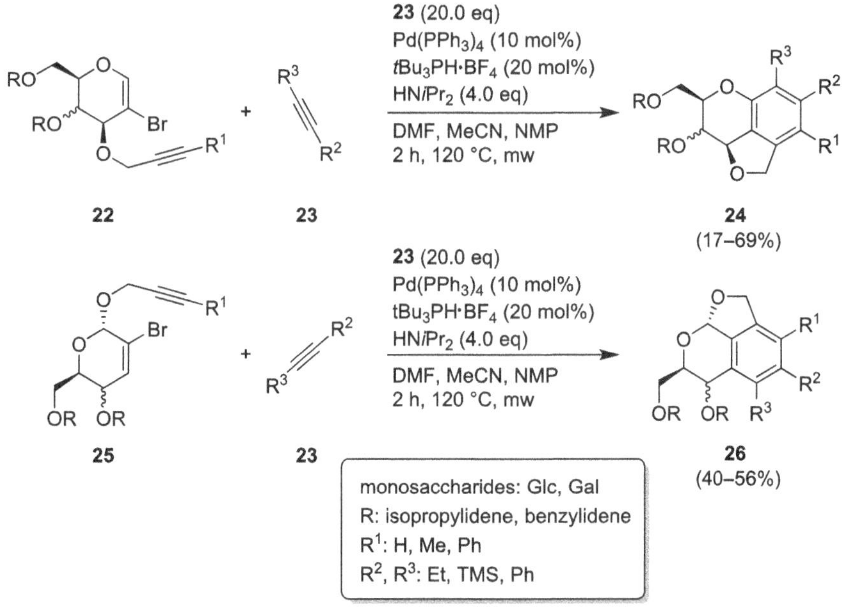

SCHEME 8.5 Intermolecular domino carbopalladation approach using external alkynes.

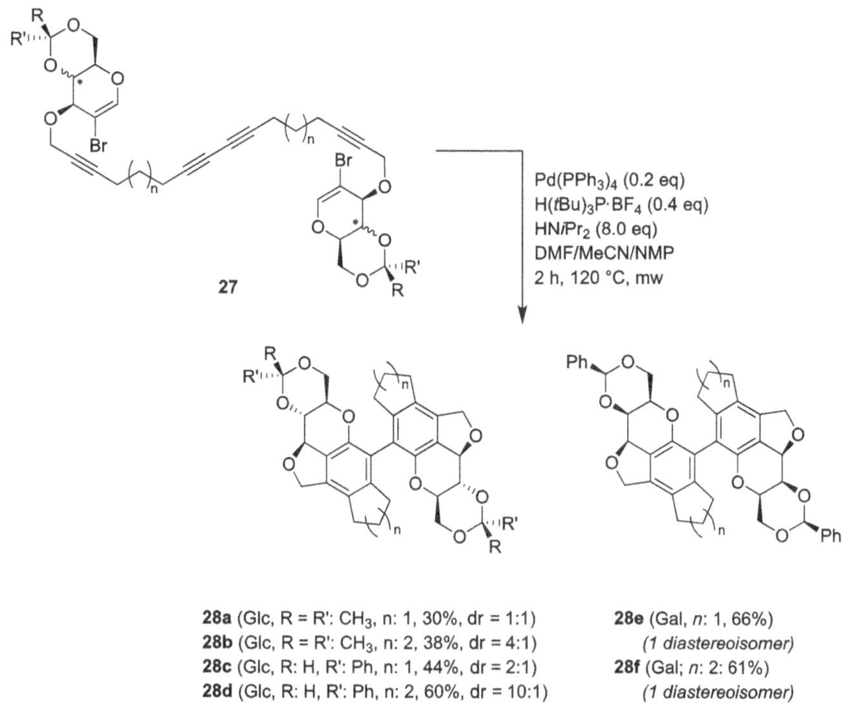

SCHEME 8.6 Tandem-domino process for the synthesis of carbohydrate-derived biphenyls.

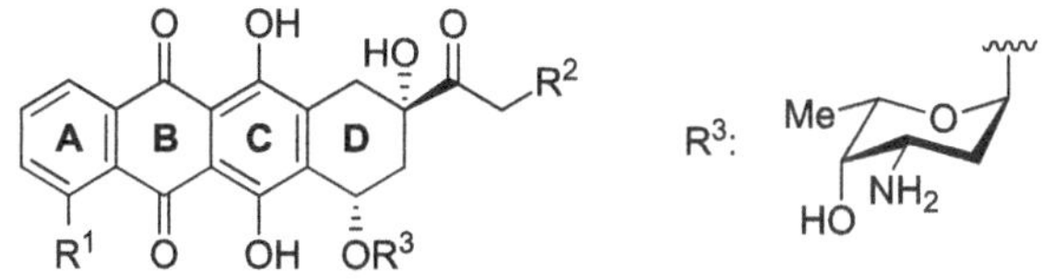

Daunomycin (**a**):	R^1 = OMe, R^2 = H
Doxorubicin (**b**):	R^1 = OMe, R^2 = OH
Idarubicin (**c**):	R^1 = H, R^2 = H

FIGURE 8.4 Several naturally occurring anthracycline antibiotics.

The tandem-domino approach to access carbohydrate-derived biphenyls **28** was investigated with glucose and galactose units as well as varying tether lengths. Best yields (61–66%) and diastereoselectivities (only one diastereoisomer) were obtained with galactose units bearing benzylidene protecting groups. In the case of glucose units protected with isopropylidene groups, only moderate yields and worse diastereoselectivities could be observed.

The success of these domino processes invoked us to investigate the synthesis of anthracycline mimics in a similar fashion. Anthracyclines belong to the natural product class of aromatic polyketides [17]. They were first isolated from the order of *Streptomycetales* by Brockmann in 1963, who described them as red to orange dyes [18]. Their structural features consist of a fourfold annulated ring system including two benzene moieties. The substitution of the D-ring bares several functional groups, that is, two alcohol groups and a glycosylated 2,6-dideoxy sugar (Figure 8.4). These carbohydrates functionalities are of highest importance for the biological and pharmaceutical activity of these natural products [19,20]. Their main application relies on the treatment of different types of cancer such as leukemia, lymphomas, and breast, uterine, ovarian, and lung cancers.

To prepare respective mimics from monosaccharides, the D-ring was exchanged by a 2-bromoglycal and instead of an aliphatic dialkyne chain an aromatic diyne **32** was employed. Furthermore, a silyl ether moiety should serve as tethering unit between both building blocks. The fourfold annulated ring system should be obtained in one single step, by applying the domino carbopalladation procedure as already used for the synthesis of chromans. This powerful transformation allows for the formation of the B- and C-ring in **30** and consequently the annulation of all four cycles as the retrosynthetic analysis depicted in Scheme 8.7 demonstrates [21,22].

Whereas the carbopalladation cascade as the key reaction did not cause any difficulties, several other transformations proved to be rather challenging. In a very first step, compounds of type **32** had to be synthesized (not shown). For this purpose, two different silylalkynyl groups were installed to appropriate benzylic iodides.

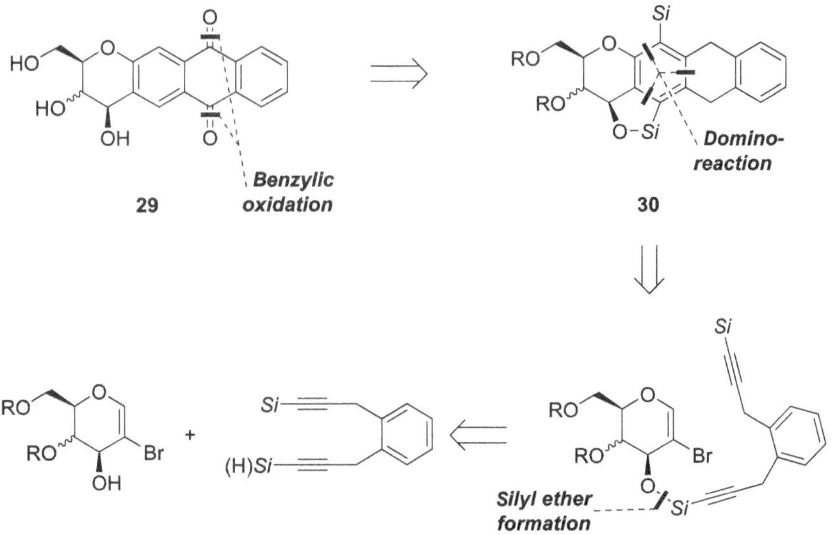

SCHEME 8.7 Retrosynthetic analysis of anthracycline aglycone mimics. Si: any silyl group.

In order to run an intramolecular and regiochemically defined carbopalladation process, the 2-bromoglycal and the dialkyne have to be linked to each other. This transformation proved to be a highly demanding task; a huge number of reaction conditions were explored, but only one led to the desired product in a sufficient yield. Therefore, the terminal silane **32** was converted into the respective sila bromide with a solution of elementary bromine in tetrachloromethane [23]. The so-formed reactive species was treated with the corresponding alcohol and triethylamine as base to furnish the coupling product **31**. Utilization of the already-known conditions for such a domino reaction provided the fourfold annulated ring system of type **33** in only one step and a very good yield. It is noteworthy that the best yields were obtained for the trimethylsilyl-substituted substrates.

The silyl groups at the newly formed benzene ring were cleaved by the action of acetyl chloride in anhydrous methanol and tetrabutylammonium fluoride to yield **34** and **35**, respectively. A Tamao-Fleming-like oxidation to furnish the respective phenol derivatives did not take place, probably due to the sp^2-hybridized carbon of the arene moiety. To install the anthraquinone moiety by oxidation of the benzylic positions, it was necessary to reprotect the alcohol functionalities. After *tert*-butyldimethylsilyl protection of the hydroxyl moieties to afford **36**, an iron(III)-catalyzed benzylic oxidation proceeded smoothly with yields up to 70%. Finally, hydrolysis with hydrochloric acid of the protecting groups generated the desired anthracycline derivatives **38** based on a carbohydrate skeleton [21, 22] (Scheme 8.8).

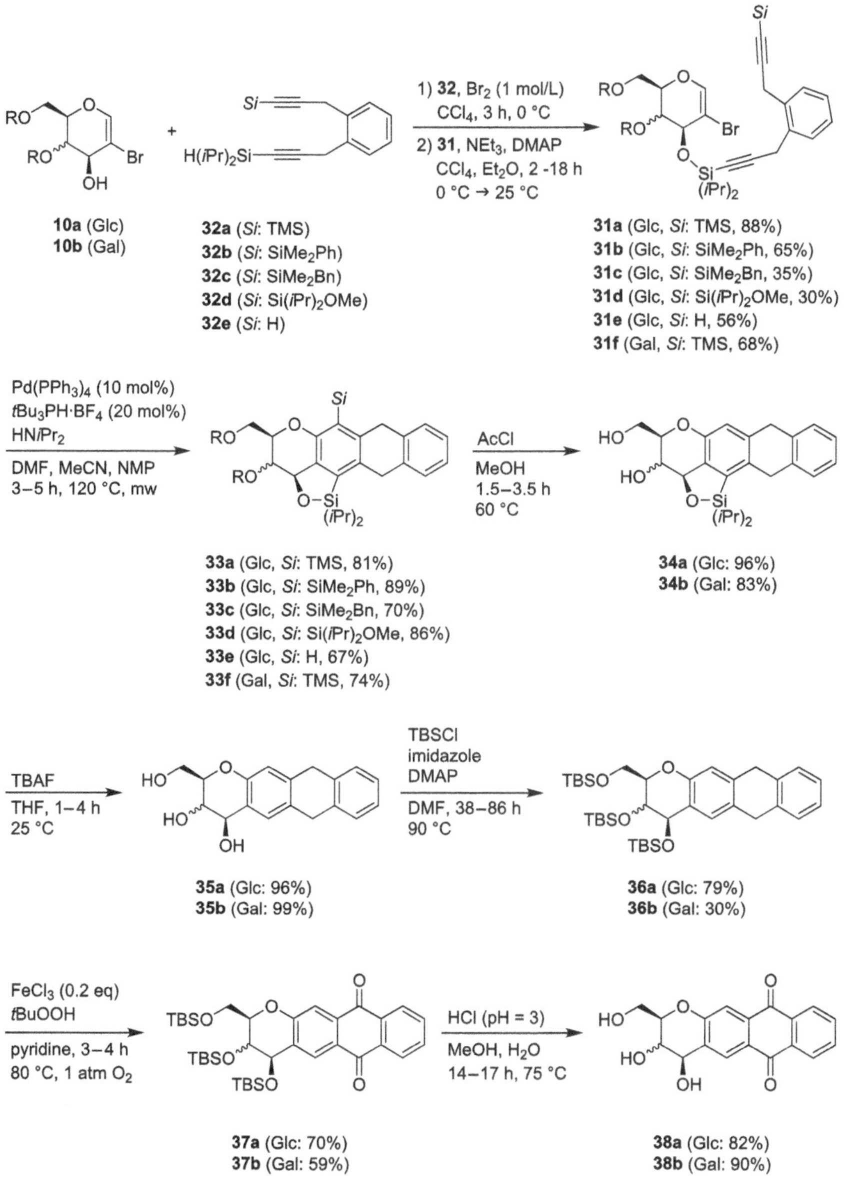

SCHEME 8.8 Silyl ether coupling, domino carbopalladation reaction, and derivatization to anthracycline derivatives.

REFERENCES

[1] Tietze, L. F.; Brasche, G.; Gericke, K. M. *Domino Reactions in Organic Synthesis*; Wiley-VCH: Weinheim, Germany, **2006**. (b) Tietze, L. F. *Chem. Rev.* **1996**, *96*, 115–136.

[2] *Domino Reactions, Concepts for Efficient Organic Synthesis*; Tietze, L. F., Ed.; Wiley-VCH: Weinheim, Germany, **2014**.

[3] In the meantime locking the 6-hydroxyl was achieved by a different approach: Brand, C.; Granitzka, M.; Stalke, D.; Werz, D. B. *Chem. Commun.* **2011**, *47*, 10782–10784.

[4] Brand, C.; Kettelhoit, K.; Werz, D. B. *Org. Lett.* **2012**, *14*, 5126–5129.

[5] Reissig, H.-U.; Zimmer, R. *Chem. Rev.* **2003**, *103*, 1151–1196.

[6] Schneider, T. F.; Kaschel, J.; Werz, D. B. *Angew. Chem. Int. Ed.* **2014**, *53*, 5504–5523.

[7] Brand, C.; Rauch, G.; Zanoni, M.; Dittrich, B.; Werz, D. B. *J. Org. Chem.* **2009**, *74*, 8779–8786.

[8] Petasis, N. A.; Bzowej, E. I. *J. Am. Chem. Soc.* **1990**, *112*, 6392–6394.

[9] Ellis, G. P.; Lockhart, I. M. *The Chemistry of Heterocyclic Compounds, Chromenes, Chromanones, and Chromones*; Wiley-VCH: New York, **2007**.

[10] Shen, H. C. *Tetrahedron* **2009**, *65*, 3931–3952.

[11] Leibeling, M.; Koester, D. C.; Pawliczek, M.; Schild, S. C.; Werz, D. B. *Nat. Chem. Biol.* **2010**, *6*, 199–201.

[12] Leibeling, M.; Koester, D. C.; Pawliczek, M.; Kratzert, D.; Dittrich, B.; Werz, D. B. *Bioorg. Med. Chem.* **2010**, *18*, 3656–3667.

[13] Meyer, F. E.; de Meijere, A. *Synlett* **1991**, 777–778.

[14] Blond, G.; Bour, C.; Salem, B.; Suffert, J. *Org. Lett.* **2008**, *10*, 1075–1078.

[15] Leibeling, M.; Milde, B.; Kratzert, D.; Stalke, D.; Werz, D. B. *Chem. Eur. J.* **2011**, *17*, 9888–9892.

[16] Leibeling, M.; Werz, D. B. *Chem. Eur. J.* **2012**, *18*, 6138–6141.

[17] Laatsch, H.; Fotso, S. *Top. Curr. Chem.* **2008**, *282*, 3–74.

[18] Brockmann, H. *Fortschr. Chem. Org. Naturst.* **1963**, *21*, 121–182.

[19] Leng, F.; Savkur, R.; Fokt, I.; Przewloka, T.; Priebe, W.; Chaires, J. B. *J. Am. Chem. Soc.* **1996**, *118*, 4731–4838.

[20] Menna, P.; Salvatorelli, E.; Gianni, L.; Minotti, G. *Top. Curr. Chem.* **2008**, *283*, 21–44.

[21] Leibeling, M.; Werz, D. B. *Beilstein. J. Org. Chem.* **2013**, *9*, 2194–2201.

[22] Leibeling, M.; Werz, D. B. *Top. Med. Chem.* **2014**; *12*, 1–21.

[23] Petit, M.; Chouraqui, G.; Aubert, C.; Malacria, M. *Org. Lett.* **2003**, *5*, 2037–2040.

9

INTRODUCTION TO REARRANGEMENT REACTIONS IN CARBOHYDRATE CHEMISTRY

Zbigniew J. Witczak and Roman Bielski
Department of Pharmaceutical Sciences, Wilkes University, Wilkes-Barre, PA, USA

9.1 INTRODUCTION

John McMurry [1] offers the following definition of a molecular rearrangement: "Rearrangement reactions occur when a single reactant undergoes a reorganization of bonds and atoms to yield an isomeric product." In most cases, the rearrangement causes the migration of a group from one position to another.

As the name suggests, a rearrangement is an intramolecular reaction whose product is an isomer of the starting material. There are several reasons why chemists take advantage of rearrangements whenever they can. First, while it is often unavoidable to introduce many atoms that will be later discarded, the rearrangement itself is a process offering the best possible atom economy; that is, all atoms present in the reactant are found in the product. Second, a rearrangement reaction does not require additional reagents employed in stoichiometric amounts. One can say that rearrangements are exceptionally economic (or cheap if you prefer). Many rearrangements are performed in the presence of a catalyst, but its amount is usually small and the catalyst can be often reused. Third, the rearrangement reactions often create new carbon–carbon bonds. It is always of interest since it enables the formation of a new carbon skeleton. Fourth, the result of many rearrangement reactions is a single product making the workup and product purification exceptionally easy. Fifth, most rearrangements, in which more than one stereoisomer is formed as possible product, are stereoselective;

Domino and Intramolecular Rearrangement Reactions as Advanced Synthetic Methods in Glycoscience, First Edition. Edited by Zbigniew J. Witczak and Roman Bielski.

that is, a single enantiomer or diastereoisomer is formed as the main or the only product.

Arguably, nature utilizes molecular rearrangements in the area of carbohydrates more often than in other areas. It is due to unique structures of carbohydrates containing multiple vicinal diol functionalities, five- or six-membered rings, and carbonyl groups. Moreover, polysaccharides differ substantially from other polymers of natural origin. Thus, carbohydrate chemists take advantage of rearrangement reactions probably more often than other organic chemists. Some of the referenced molecular rearrangement reactions were discovered when reacting monosaccharides. In addition, there are rearrangements that take place only on sugars.

Probably, the most often performed rearrangements in carbohydrate chemistry are:

- the Amadori rearrangement,
- the Ferrier rearrangement,
- the Claisen rearrangement, and
- the Overman rearrangement.

However, it must be emphasized that the Amadori rearrangement is a food chemistry process rarely taken advantage of in the chemical laboratory.

Here are a few examples of skillful use of rearrangements by carbohydrate chemists.

A typical rearrangement reaction proceeds with a sequential change of a functional group. All changes in atoms location/displacement are presumably and should be energy favored [2].

9.2 CLASSIFICATION

The rearrangement reactions can be classified or divided into several categories. For example, the mechanism of the process may be intramolecular or intermolecular. Other classifications look at the specific mechanism of the process. Also, the reactions may be classified according to the type of migration of specific groups. Some rearrangement reactions may take place at elevated temperatures, and others necessarily require the use of a catalyst or the use of a catalyst significantly improves the reaction yield. In some cases, the presence of a catalyst is the most dominant factor for the successful reaction outcome and the formation of energetically favored rearrangement products.

Among many examples of an intermolecular rearrangement, 3,3-sigmatropic rearrangement reaction is an important avenue leading to functionalized aminosugars. Ichikawa et al. [3,4] developed practicable synthetic approach to many aminofunctionalized sugars via the rearrangement of carbohydrate isocyanates.

An interesting example of [1, 3]-sigmatropic rearrangement has been reported by Oberdorfer et al. [5], who used an unsaturated carbohydrate synthon as the starting template. The overall yield, however, is highly dependent on the geometry of molecule and stereo-orientation of dominant functional groups at the strategic positions of the rearrangement template.

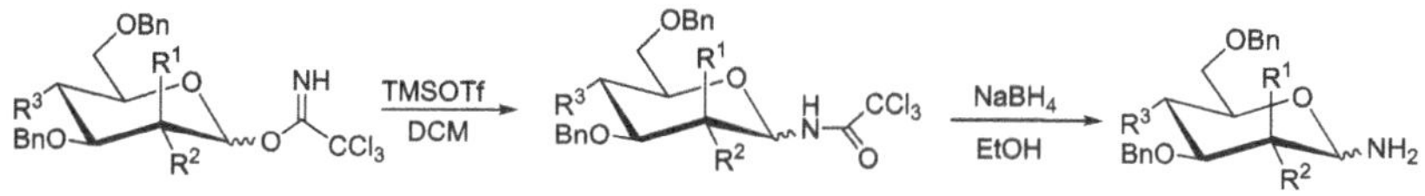

R¹= H, OBn, R² = OBn, H, R³ = OBn, (BnO)₄Glcα

SCHEME 9.1 Synthesis of N-protected glycosylamines employing the Chapman rearrangement.

One of the new rearrangement reactions is a dyotropic rearrangement [6,7]. It is an uncatalyzed process in which two D-bonds simultaneously migrate intramolecularly. This rearrangement is further classified as type 1 and type 2. Type 1 consists of reactions in which the two D-bonds interchange positions (e.g., rearrangement of nitrile oxide into isocyanate). In contrast, in type 2, the two D-bonds do not interchange positions.

9.3 CHAPMAN REARRANGEMENT

Motawia and coworkers [8] reported an interesting application of the Chapman rearrangement to the synthesis of *N*-protected glycosylamines. This approach starts with specifically functionalized glycosyl trichloroacetimidate protected with benzyl groups. The TMSOTF-induced rearrangement is a fast process and the reaction is completed within 1 h. The next step, the reductive removal of a trichloroacetyl group with sodium borohydride, is even faster and is completed within 15- to 30-min treatment. This synthetic strategy is illustrated in Scheme 9.1.

9.4 HOFMANN REARRANGEMENT

In a recent report by Santoyo-Gonzalez et al. [9], a very useful approach to the synthesis of important and rare 3-amino-3-cyano-3-deoxy sugars is an excellent Hofmann rearrangement. The Hoffman rearrangement usually involves the use of such reactants as aqueous NaOBr, Br_2/NaOMe in methanol, iodobenzene diacetate, and lead tetra-acetate. In the septanose series of C-3 functionalized carbamates, the reaction was carried out with an excess of lead tetra-acetate, using tert-butyl alcohol to trap the intermediate isocyanates. Treating of the formed rearrangement product with sodium methoxide in methanol gave cyclic carbamate, which was formed by a selective intramolecular nucleophilic displacement of the tert-butoxycarbonyl group by the alkoxide anion formed at C-2 during the reaction. The approach starts with specifically protected series of sugars and is illustrated in Scheme 9.2.

9.5 COPE REARRANGEMENT

The Cope rearrangement constitutes a universal tool for the synthesis of new carbon skeletons.

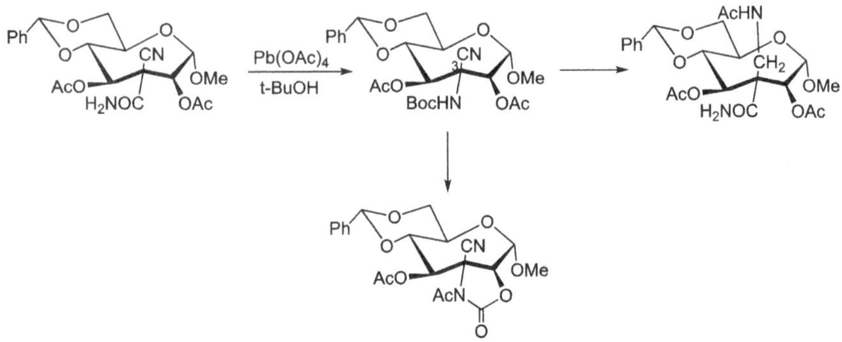

SCHEME 9.2 Synthesis of 3-amino-3-cyano-3-deoxy sugars via the Hofmann rearrangement.

Interestingly, the aza-Cope rearrangement is a specific modification, where 1-amino-substituted sugars could be easily prepared via the 3,3-sigmatriopic rearrangement of Schiff base (aza-Cope rearrangement). This modification of the aza-Cope rearrangement was employed to the synthesis of chain-extended aminosugar derivatives as reported by Deloisy and Kunz [10] and depicted in Scheme 9.3.

9.6 FERRIER REARRANGEMENT

Recent review by Gomez et al. [11] covers very extensively all important aspects of the Ferrier rearrangement. The stereoselectivity of the reactions and the mechanism of catalyst action effects on the yields of the rearrangement products were discussed. The practical aspects of selected rearrangement reactions with their detailed synthetic utilization and application were described as well.

Among many reports of the classical Ferrier rearrangement, Ghosh and coworkers [12] reported the synthesis of 2-*C*-acetoxymethyl glycals as convenient starting templates for the rearrangement. The rearrangement reaction with aliphatic

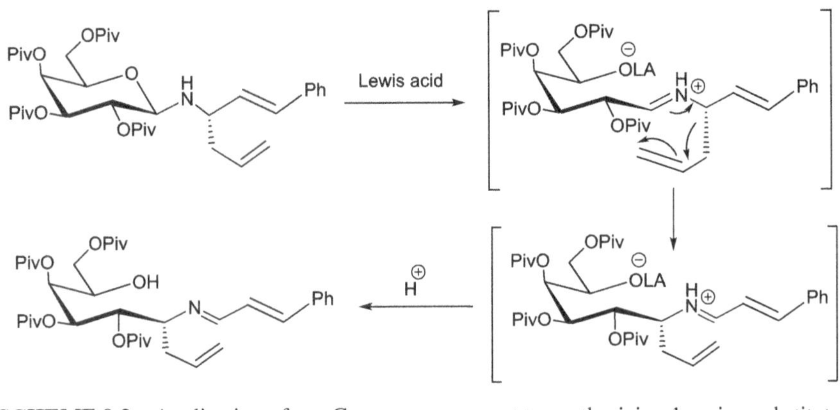

SCHEME 9.3 Application of aza-Cope rearrangement to synthesizing 1-amino-substituted sugars.

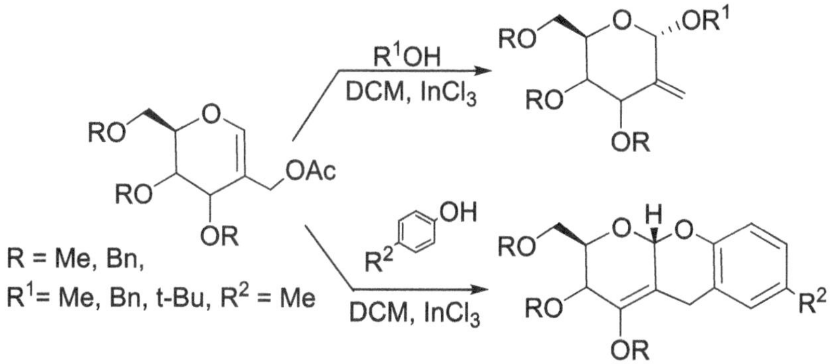

SCHEME 9.4 Synthesis of 2-*C*-methylene-α-D-glucosides and aromatic pyranopyrans *via* the Ferrier rearrangement.

nucleophiles, induced by indium trichloride ($InCl_3$), affords exclusively corresponding 2-*C*-methylene-α-D-glucosides in an 83% yield. When substituted phenols were used as nucleophiles, only aromatic pyranopyrans were formed in an excellent yield (83–96%). The cyclization products were presumably formed via the initial exocyclic Ferrier rearrangement followed by a concomitant $InCl_3$-catalyzed intramolecular cyclization as illustrated in Scheme 9.4.

Other important, practical approaches to *C*-glycosides and *S*-thioglycosyl derivatives were reported by Descotes et al. [13] and Osborn et al. [14].

9.7 CLAISEN REARRANGEMENT

The Claisen rearrangement [15] is one of the most useful approaches to many *C*-linked functionalized carbohydrate derivatives. Werschkun and Thiem [16] reviewed the applications of the Claisen rearrangement in carbohydrate chemistry, its synthetic accessibility as well as generality of the approach and its synthetic utilization.

The Claisen rearrangement offers a very useful approach to synthesizing rare sugars functionalized at strategically important positions, which could be used as intermediate templates to more complex natural products [15].

Fairbanks and coworkers [17] reported an elegant approach to α-(1-6)-*C*-disaccharides by tandem the Tebbe methylenation and Claisen rearrangement. The synthesis starts from specifically functionalized uronic acids, which were esterified with an *allo* glycal and then underwent the Tebbe methylenation. The thermal Claisen rearrangement of intermediate vinyl ethers produced α-(1-6)-*C*-disaccharides in good yields and with a complete control of anomeric stereochemistry, as depicted in Scheme 9.5. Interestingly, the corresponding *galacto* stereoisomer is totally unreactive under the same reaction condition. This fact additionally supports the claim that stereochemistry plays a significant role in the reaction mechanism.

Other important example of the Claisen rearrangement reaction is the synthesis of a functionalized template during a long asymmetric synthetic route to natural product, tetrodotoxin, accomplished by Isobe [18]. The rearrangement reaction is the

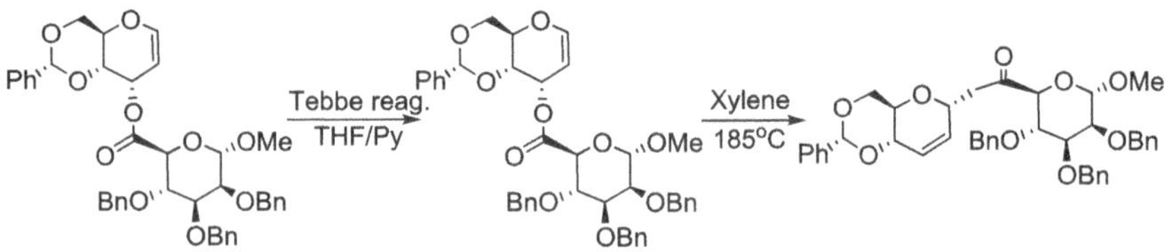

SCHEME 9.5 Fairbanks's approach to α-(1-6)-*C*-disaccharides by tandem the Tebbe methylenation and Claisen rearrangement.

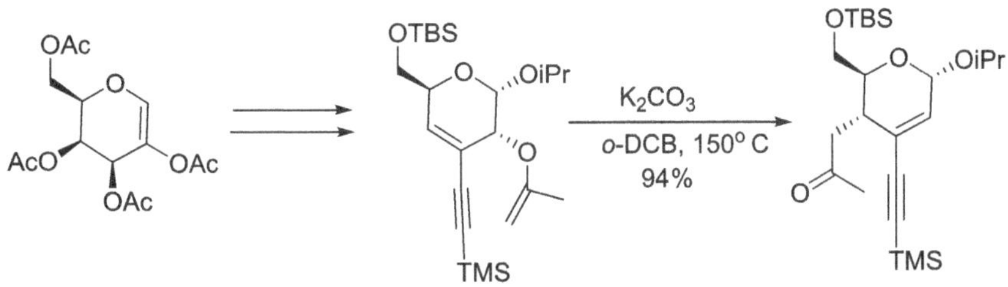

SCHEME 9.6 Isobe's use of the Claisen rearrangement during the total synthesis of tetrodotoxin.

key step and was carried out under thermal condition in the presence of potassium carbonate (K_2CO_3) in *o*-DCB (*o*-dichlorobenzene) to afford the rearrangement product in an amazingly high yield (94%). The sequential synthetic route is depicted in Scheme 9.6.

9.8 OVERMAN REARRANGEMENT

This rearrangement [19] offers a very easy way to synthesize allylic amines. An example of a strategy employing the Overman rearrangement is depicted in Scheme 9.7. The first step, which controls the stereoselectivity of the reaction, is highly dependent on the overall geometry of the core molecule to be functionalized. Other aspects affecting diastereoselectivities during the rearrangement of unsaturated carbohydrate scaffolds for peptide–carbohydrate hybrids are reported [20–22]. Herradon

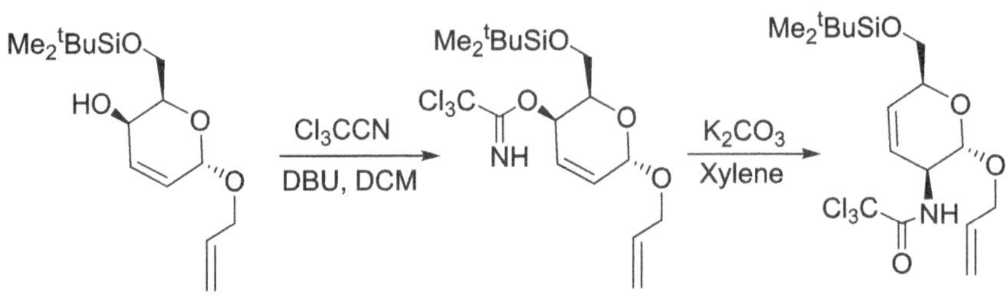

SCHEME 9.7 Synthesis of sugar allylic amines employing the Overman rearrangement.

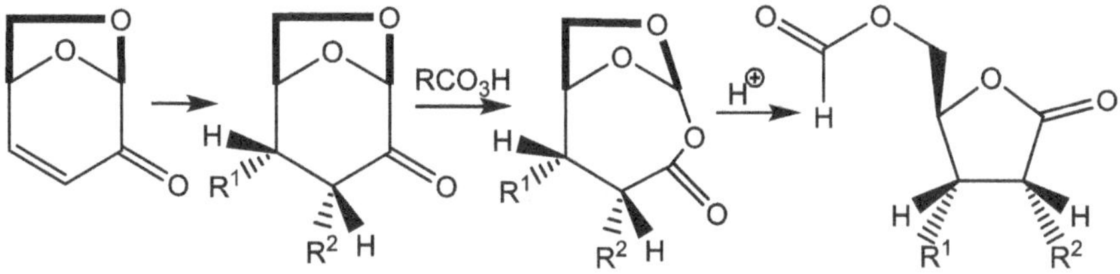

SCHEME 9.8 Ebata's rare example of the ring contraction as a result of the Baeyer-Villiger rearrangement.

and coworkers [21] explored this methodology by using a stereocontrolled synthesis of an unsaturated carbohydrate having two amino groups, one of them masked as an azide. The obtained rearrangement scaffold was used further to synthesize two peptides functionalized with aromatic groups as potential calpain inhibitors.

9.9 BAEYER–VILLIGER REARRANGEMENT

The specific reaction described here is an interesting example of a ring contraction when using peracids. Levoglucosenone reacts with peracids with the formation of addition products followed by the ring expansion with unusual transformation to butanolides as reported by Ebata and coworkers [23–25]. Evidently, this is a peculiar case of a Baeyer–Villiger rearrangement with a subsequent acid-catalyzed recyclization as illustrated in Scheme 9.8.

Another interesting example of the rearrangement has recently been reported by Samet and coworkers [26], who describe a reaction of levoglucosenone with sulfur ylides to form unusual cyclopropane derivatives. These annulated cyclopropanes were smoothly converted by 30% aqueous hydrogen peroxide in acetic acid to cyclopropanecarboxylic acids. The final acid hydrolysis (MeOH/HCl) converted them into synthetically valuable cyclopropanated lactones. This case of the Baeyer–Villiger rearrangement reaction is depicted in Scheme 9.9.

9.10 RING CONTRACTION

Ring contraction reactions of functionalized carbohydrates have become one of very useful tools for methodical syntheses of diversified molecules as reported by Redlich [27].

Diethylaminosulfur trifluoride-assisted ring contraction [28] has been the fluorination reaction of anhydrosugars. The 1,6:3,4-dianhydropyranoses yielded products of skeletal rearrangement resulting from migration of the tetrahydropyran oxygen (educts of D-*altro* and D-*talo* configuration) or of the 1,6-anhydro bridge oxygen (D-*allo*, D-*galacto*). The major products 1,6:2,3-dianhydropyranoses were compounds arising from nucleophilic substitution, with configuration at C4 either retained (D-*talo*, D-*gulo*) or inverted (D-*manno*), or from C6 migration (D-*allo*). The minor products in the 1,6:2,3-series resulted from migration of the tetrahydropyran oxygen (D-*gulo*)

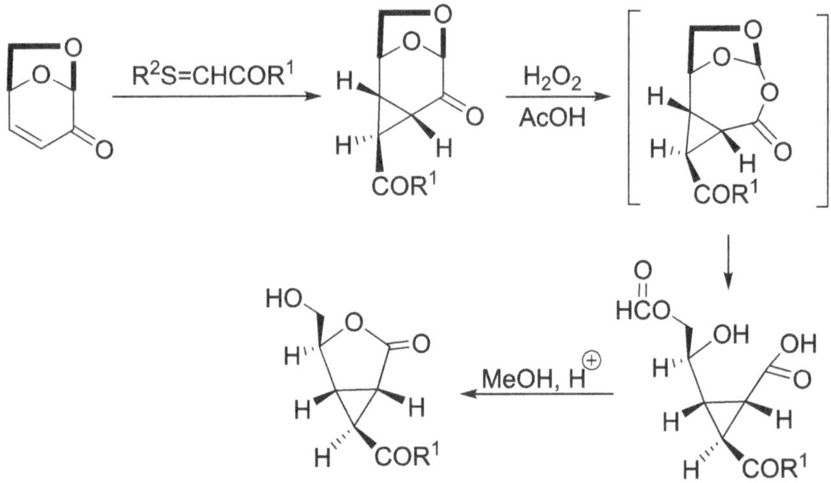

SCHEME 9.9 Samet's synthesis of cyclopropanated lactones via the Baeyer-Villiger rearrangement.

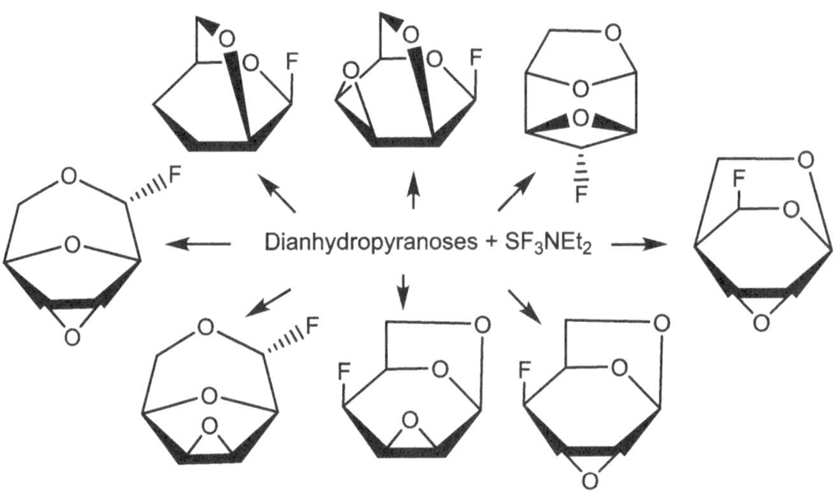

SCHEME 9.10 Various rearranged products formed as a result of treating dianhydropyranosess with diethylaminosulfur trifluoride.

or the oxirane oxygen (D-*manno*), or from nucleophilic substitution with retention of configuration (D-*manno*) as illustrated in Scheme 9.10.

9.11 CONCLUSIONS

Recent applications of the rearrangement reactions in the synthesis of many functionalized carbohydrates are in accordance with many great developments in organic synthesis of the last decade. Some rearrangement reactions in carbohydrate chemistry presented in this review have been applied to the synthesis of important target molecules exhibiting biological activities and diverse pharmacological responses.

Generally, rearrangement approaches provide mild reaction conditions for the formation of new bonds such as C–C, C–N, C–S, and this type of new bond formation has been exploited in many rearrangements in synthetic carbohydrate chemistry.

We hope and strongly believe that the topic of rearrangement reactions in synthetic carbohydrate chemistry will grow steadily, delivering new avenues for many synthetic methodologies leading to diverse classes of functionalized carbohydrates. In addition, the rearrangement reactions that use synthetic scaffolds from carbohydrate frameworks as building blocks will continue to be an important part of synthetic methodologies when synthesizing various molecules of biological interest. Rearrangements in carbohydrate chemistry deliver important contributions to the development of glycosciences.

REFERENCES

[1] McMurry, J. *Organic Chemistry*; Thomson-Brooks/Cole: Belmont, CA, **2008**; pp. 138.

[2] For review on rearrangement in carbohydrate chemistry, see (a). S. Koto, Rearrangement Reactions, in *Glycoscience, Chemistry and Chemical Biology*, Vol. 1, Springer: Heidelberg, 2001; pp 785–874. (b). *Topics in Current Chemistry*, Vol. 215; Springer: Heidelberg, 2001.

[3] Ichikawa, Y.; Kobayashi, C.; Isobe, M. *J. Chem. Soc. Perkin Trans. I* **1996**, 377.

[4] Ichikawa, Y.; Osada, M.; Ohtani, II.; Isobe, M. *J. Chem. Soc. Perkin Trans. I* **1997**, 1449.

[5] Oberdorfer, F.; Haeckel, R.; Lauer, G. *Synlett* **1998**, 449.

[6] Reetz, M. T. *Angew. Chem Int. Ed.* **1972**, *11*, 129–130.

[7] Tylor, X. *J. Chem. Soc. Perkin 1* **1985**, 1181.

[8] Larsen, K.; Olsen, C. E.; Motawia, M. S.; *Carbohydr. Res.* **2008**, *343*, 383–387.

[9] Santoyo-Gonzalez, F.; Berenguel, A. V.; Mateo, F. H.; Mendoza, P. G. *Carbohydr. Res.* **1991**, *209*, 131–143.

[10] Deloisy, S.; Kunz, H. *Tetrahedron Lett.* **1998**, *39*, 791–794.

[11] Gomez, A. M.; Lobo, F.; Uriel, C.; Lopez, J. C. *Eur. J. Org. Chem.* **2013**, 7221–7262.

[12] Ghosh, R.; Chakraborty, A.; Maiti, D. K.; Puranik, V. G. *Tetrahedron Lett.* **2005**, *46*, 8047–8051.

[13] De Freitas Filho, J. R.; Srivastava, R. M.; Soro, Y.; Cottier, L.; Descotes, G. *J. Carbohydr. Chem.* **2001**, *20*, 561–568.

[14] Ellis, D.; Norman, S. E.; Osborn, H. M. I. *Tetrahedron* **2008**, *64*, 2832–2854.

[15] (a) Majumdar, K. C.; Nandi, R. K. *Tetrahedron* **2013**, *69*, 6921–6957. (b) Martín Castro, A. M. *Chem. Rev.* **2004**, *104* (6), 2939–3002.

[16] Werschkun, B.; Thiem, J. *Top Curr. Chem.* **2001**, *215*, 293–325.

[17] Chambers, D. J.; Evans, G. R.; Fairbanks, A. J. *Tetrahedron* **2005**, *61*, 7184–7192.

[18] Ohyabu, N.; Nishikawa, T.; Isobe, M. *J. Am. Chem. Soc.* **2003**, *125*, 8798–8895.

[19] Overman, L.E., Carpenter, N. E. *Org. React.* **2005**, *66*, 2–123.

[20] Jaunzeme, I.; Jirgensons, A. *Tetrahedron* **2008**, *64*, 5794–5799.

[21] Montero, A.; Mann, E.; Herradon, B. *Tetrahedron Lett.* **2005**, *46*, 401–405.

[22] Montero, A.; Benito, E.; Herradon, B. *Tetrahedron Lett.* **2010**, *51*, 277–280.

[23] Koseki, K.; Ebata, T.; Kawakami, H.; Matsushita, H.; Koseki, K.; Naoi, Y.; Itoh, K. *Heterocycles* **1990**, *31*, 423–426.

[24] Ebata, T.; Matsumoto, K.; Yoshikoshi, H.; Koseki, K.; Kawakami, H.; Matsushita, H. *Heterocycles* **1990**, *31*, 1585–1588.

[25] Matsumoto, K.; Ebata, T.; Koseki, K.; Okano, K.; Kawakami, H.; Matsushita, H. *Bull. Chem. Soc. Jpn.* **1995**, *68*, 670–672.

[26] Samet, A. V.; Lutov, D. N.; Konyuskin, L. D.; Strelenko, Y. A.; Semenov, V. V. *Tetrahedron Asymmetry* **2008**, *19*, 691–694.

[27] Redlich, H. *Angew. Chem. Int. Ed.* **1994**, *33*, 1345–1347.

[28] Karban, J.; Císařová, I.; Strašák, T.; Červenková Šťastná, L.; Sýkora, J. *Org. Biomol. Chem.* **2012**, *10*, 394–403.

10

REARRANGEMENT OF A CARBOHYDRATE BACKBONE DISCOVERED *"EN ROUTE"* TO HIGHER-CARBON SUGARS

SŁAWOMIR JAROSZ, ANNA OSUCH-KWIATKOWSKA, AGNIESZKA GAJEWSKA, AND MACIEJ CIEPLAK

Institute of Organic Chemistry, Polish Academy of Sciences, Kasprzaka, Warsaw, Poland

10.1 INTRODUCTION

Rearrangement processes in sugar chemistry are very common and known since a long time. They can be divided into two main topics: (1) those that change the arrangement of the substituents in the sugar unit but leave the skeleton untouched and (2) those that modify this skeleton. This chapter discusses predominantly such transformations that change the sugar skeleton.

The first topic is reviewed very shortly, showing only the most important reactions having the great impact on the chemistry of sugars, as well as on the stereoselective organic synthesis. Such transformations are described in many excellent reviews.

There are many useful transformations dealing with the rearrangement of the orientation of substituents, transposition of the double bond (more precisely the "unsaturation"), etc., which change the properties and the character of the molecule but leave the skeleton of the sugar untouched.

Domino and Intramolecular Rearrangement Reactions as Advanced Synthetic Methods in Glycoscience, First Edition. Edited by Zbigniew J. Witczak and Roman Bielski.

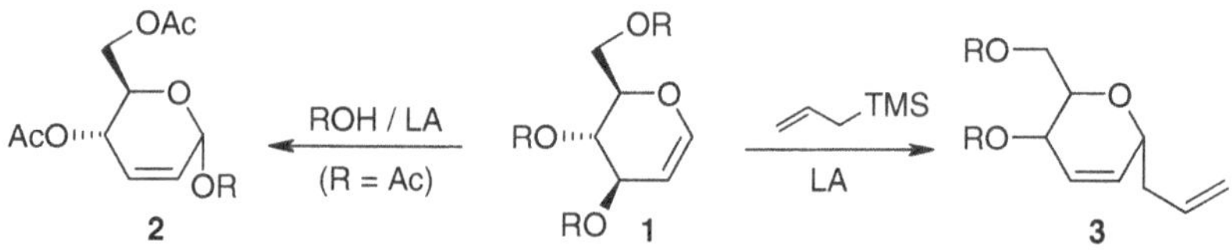

FIGURE 10.1 The Ferrier-I rearrangement of glycals.

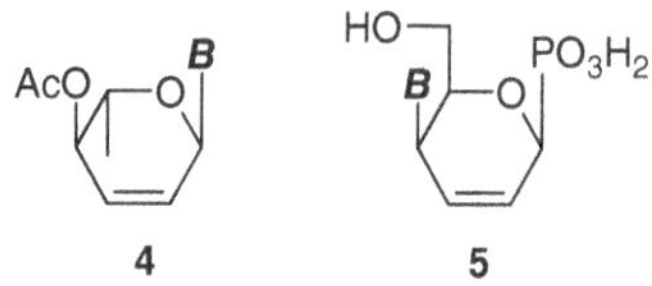

FIGURE 10.2 Biologically important derivatives via the Ferrier-I rearrangement.

10.2 REARRANGEMENTS WITHOUT CHANGING THE SUGAR SKELETON

The most important of such processes is the reaction of glycals with alcohols catalyzed with Lewis acids known as Ferrier-I rearrangement [1, 2]; an example is shown in Figure 10.1. Reaction of glycals with allylsilanes catalyzed by Lewis acids is also a convenient method for the synthesis of *C*-allyl glycosides (e.g., **3**) [2, 3].

1-Aza- and 1-thia-Ferrier-I rearrangements allow to obtain the corresponding nitrogen and sulfur derivatives. This process was used for the preparation of many biologically important molecules, such as 2,3-unsaturated pyranosyl nucleosides **4** or phosphorylated pyranosyl nucleoside **5** (Figure 10.2) [4].

The transposition of a double bond, an example of which is presented in Figure 10.3, is commonly applied in organic synthesis. The [3,3] sigmatropic rearrangement of ester **6** leads to amide **7** [4,5], while the branched saccharide **9** is readily obtained from **8** [4].

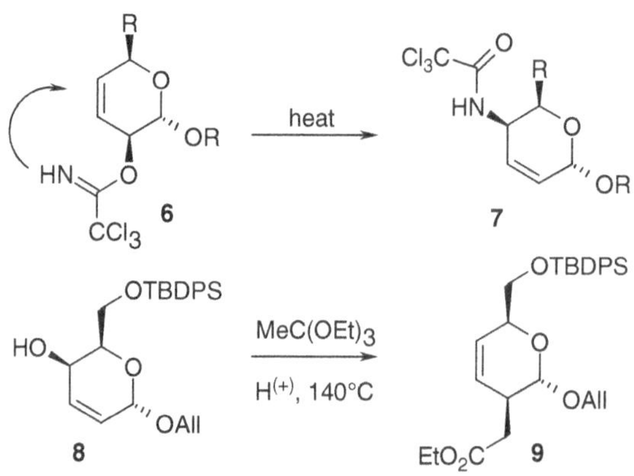

FIGURE 10.3 From allylic alcohols to allyl amines and branched sugars.

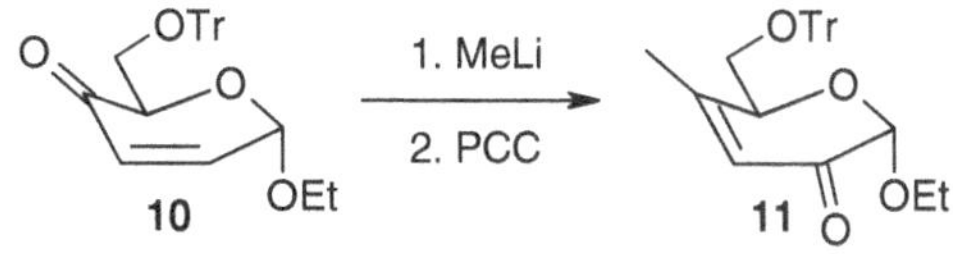

10 11

FIGURE 10.4 Transposition of the double bond in unsaturated hexose.

Rearrangement of enone **10**, shown in Figure 10.4 [6], provided another enone **11**, convenient starting material in the stereoselective synthesis of Prelog–Djerassi lactone [7].

Very well-known interconversion between epoxides [8] (Figure 10.5; **12** ↔ **13**) may be regarded also as a useful type of the rearrangement without changing the skeleton of the sugar. Another synthetically useful transformation is represented by a rearrangement of sugar oxiranes to allylic alcohols induced by a strong base; an example (**14** → **15**) is shown in Figure 10.5 [4, 9].

Next "classical" example of the rearrangement with the creation/transposition of the double bound, applied in the synthesis of biologically important molecules, is shown in Figure 10.6 [4].

10.3 REARRANGEMENTS CONNECTED WITH THE CHANGE OF SUGAR UNIT(S)

One of the most common methods of the formation of the carbocyclic ring from the "normal" sugar ring is undoubtedly the Ferrier-II rearrangement discovered by Ferrier in 1979 [10]. It is based on an $HgCl_2$-catalyzed rearrangement of unsaturated sugars with the C5–C6 (for pyranosides) or C4–C5 (for furanosides) *exo*-cyclic double bond. This method is discussed in many excellent reviews [11–13]. Such unsaturated

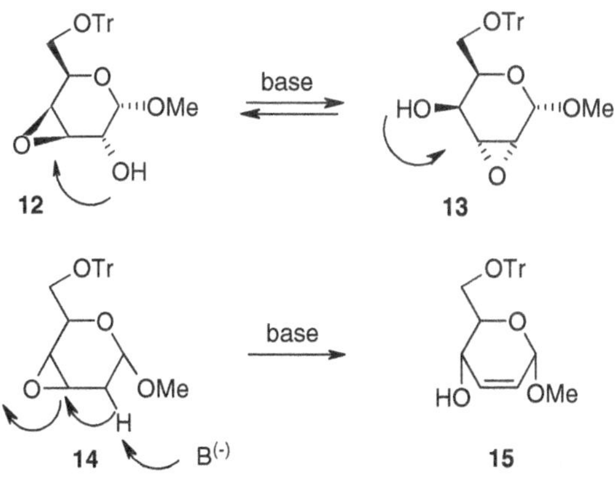

12 13 14 15

FIGURE 10.5 Interconversion between epoxides and base-catalyzed rearrangement of epoxides.

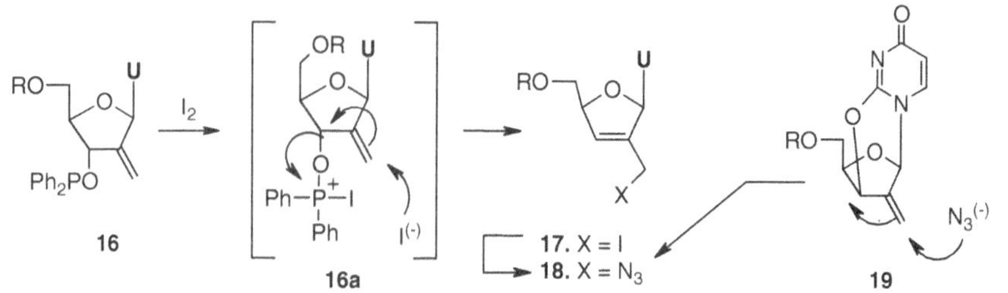

FIGURE 10.6 Preparation of biologically important derivatives by transposition of a double bond.

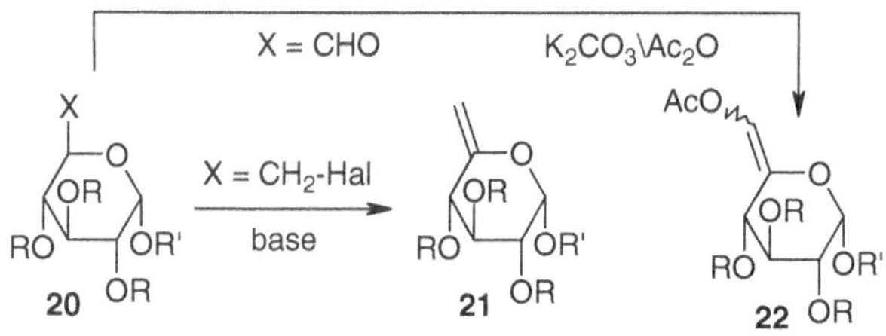

FIGURE 10.7 Synthesis of unsaturated sugars with the *egzo* double bond.

derivatives may be prepared by action of AgF on 6-halogeno-pyranosides (**20**; X = CH_2Hal) [14]; much more convenient and cheaper is, however, elimination of the HX by DBU [15] or sodium hydride [16]. The acyloxy derivatives such as **22** are prepared by treatment of terminal aldehydosugars (**20**; X = CHO) with K_2CO_3/Ac_2O (Figure 10.7) [4].

Treatment of **21** or **22** with Hg(II) salts induces the rearrangement of the sugar skeleton. An example of such transformation providing carbasugars **24** is shown in Figure 10.8. This process is not selective and may give a mixture of polyhydroxylated cyclohexanones [11]. Modification of such transformation was proposed by Sinaÿ and coworkers, who applied triisobutylaluminum (TIBAL) [17, 18] as the reagent inducing the rearrangement for the stereoselective preparation of polyhydroxylated

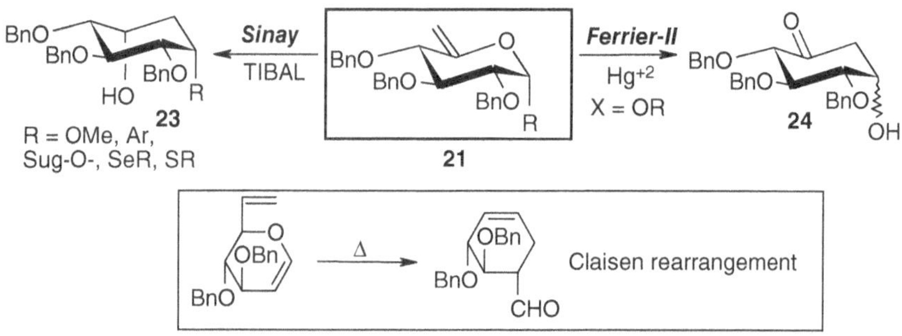

FIGURE 10.8 Synthetic strategies to carbasugars.

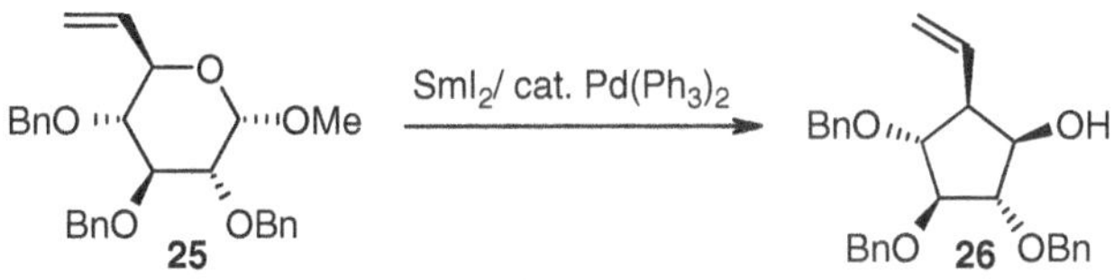

FIGURE 10.9 Application of SmI_2 in the preparation of carbocycles.

carbocycles. For example, only one stereoisomeric derivative **23** was obtained from **21** upon treatment with TIBAL (Figure 10.8).

The carbocyclic skeleton is also available by a modified Claisen rearrangement of unsaturated sugars as shown in Figure 10.8 [19]. Recently, the general methods for the preparation of carbasugars by a variety of methods were reviewed by Spanish scientists [20].

Next paragraphs briefly discuss the main common reactions proceeding with the modification of the sugar skeleton. The most typical examples are shown to illustrate these processes. Various types of rearrangements in sugar chemistry up to 2001 are reported in an excellent review by Koto [21].

First topic deals with the processes of the contraction of the sugar ring(s). Samarium iodide [22] is especially useful to induce such rearrangement [23]. For example, treatment of **25** with SmI_2 and catalytic amounts of $Pd(PPh_3)_2$ afforded polyhydroxylated cyclopentane **26** (Figure 10.9) [24].

Zirconium-mediated ring contraction of vinyl sugars is also a useful method for the preparation of carbocycles [25].

Second topic deals with the processes of the expansion of the sugar ring(s). Rearrangement of 1,2-cyclopropanosugars, such as **27** or **29**, is a method for the preparation of polyhydroxylated derivatives with seven-membered ring(s) (Figure 10.10) [4, 26].

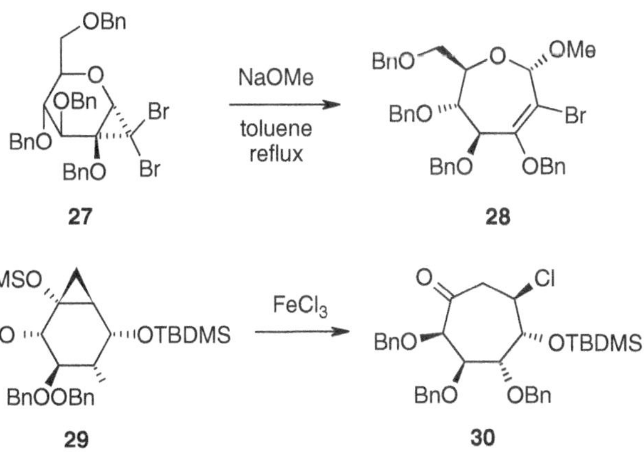

FIGURE 10.10 Seven-membered sugars and carbasugars by rearrangement of cyclopropane derivatives.

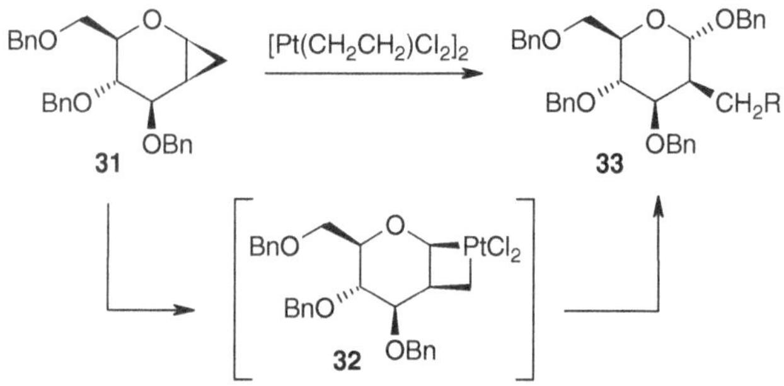

FIGURE 10.11 Synthesis of branched sugars from cyclopropanated derivatives.

The ring opening of 1,2-cyclopropanated sugars not only is the way to "expanded" sugars, but can also provide the branched derivatives as proposed by Madsen [27]; an example is shown in Figure 10.11.

10.4 REARRANGEMENTS CHANGING THE STRUCTURE OF A SUGAR SKELETON

This chapter shows selected examples of the rearrangement of the sugar skeleton leading to the open-chain derivatives.

In 1989, Polish scientists described an interesting rearrangement of the adducts of unsaturated sugar lactones with derivatives of hydroxylamine [28]. The addition of *N*-benzylhydroxylamine to unsaturated sugar lactone **34** provided **35**, which spontaneously rearranged to the open-chain derivative **36** (Figure 10.12), used further in the synthesis of negamycin lactone and iminosugars [29].

The most unexpected rearrangement of a sugar core was discovered three decades ago by Mootoo and Fraser-Reid. Working on a stereocontrolled synthesis of ansamycin antibiotics, by so-called pyronopyran methodology [30], they observed that bromination of complex derivative **37** did not provide the expected bromohydrin **39** but resulted in the rearrangement of the carbohydrate skeleton! (Figure 10.13) [31,32].

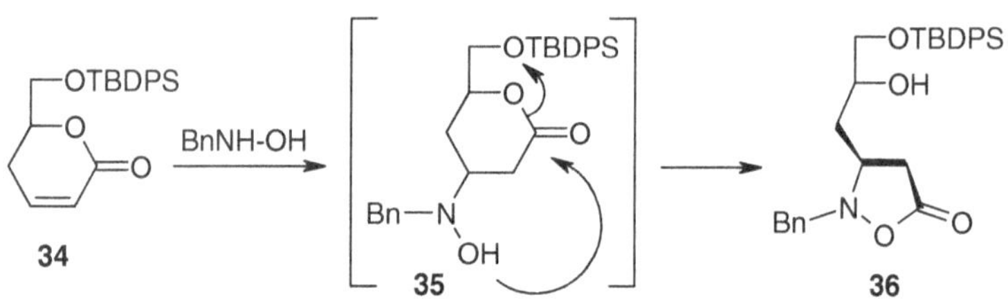

FIGURE 10.12 Synthesis of the precursor of negamycin lactone **36**.

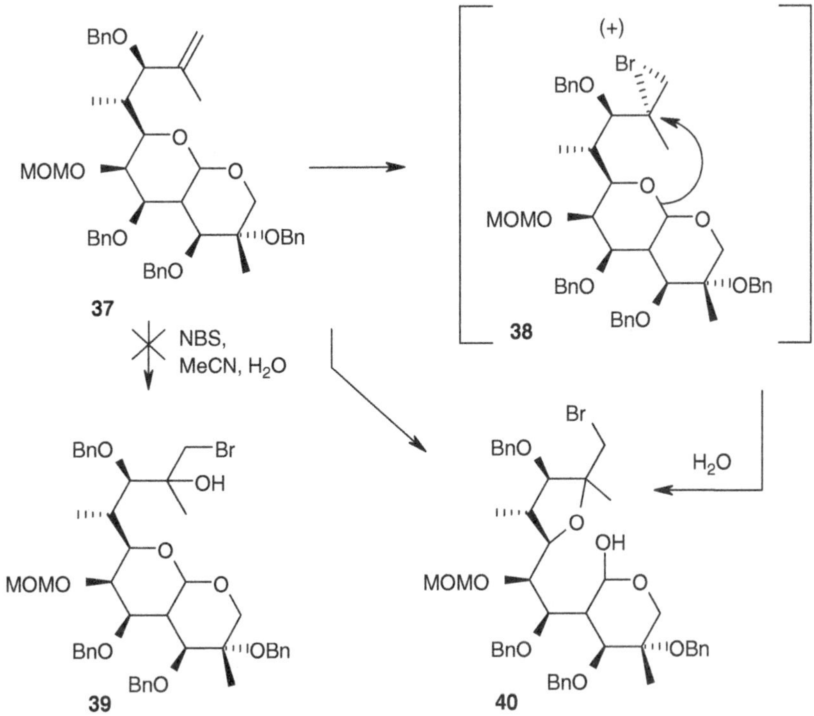

FIGURE 10.13 Unexpected rearrangement of pyranopyrans that open a route to “pentenyl” glycoside methodology.

This observation opened a route to so-called “pentenyl glycoside methodology” [32, 33], which is now one of the most important methods of the creation of the glycosidic bond.

One of the most synthetically useful procedures of the preparation of highly advanced synthons is undoubtedly the Vasella reaction [34]. It is based on the reductive fragmentation of 6-deoxy-6-halogenopyranosides with zinc in ethanol and provides the open-chain unsaturated sugars in a 60–80% yield (Figure 10.14); *n*-butyllithium may induce the same process.

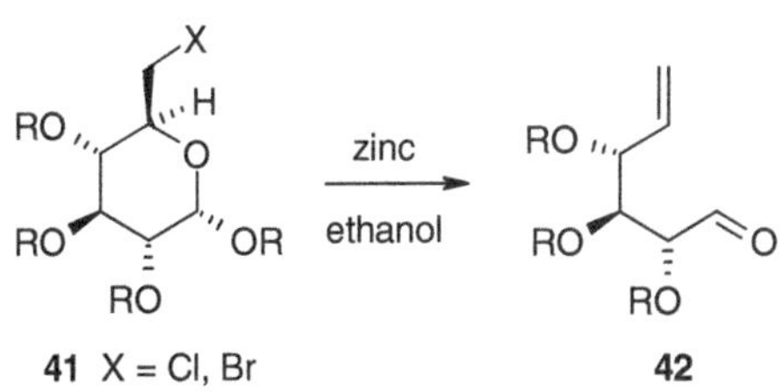

FIGURE 10.14 Preparation of unsaturated sugars by Vasella procedure.

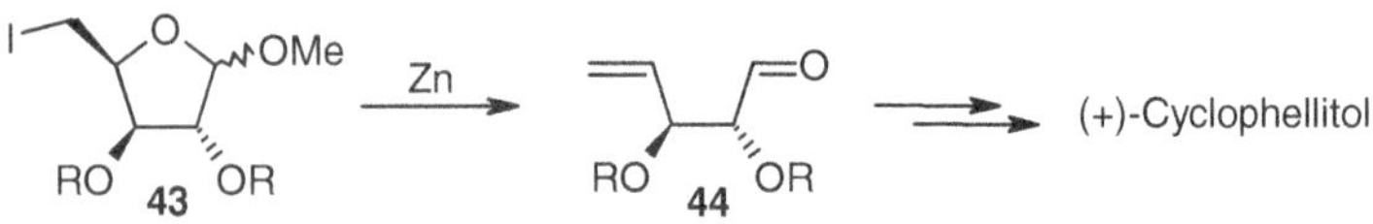

FIGURE 10.15 Synthesis of complex polyhydroxylated products utilizing the Vasella reaction.

Modification of the original procedure, proposed by Fürstner [35], used zinc on graphite.

The Vasella reaction is commonly applied to prepare the key intermediates used in the synthesis of biologically important molecules such as (+)-cyclophellitol prepared by Madsen [36] from the "Vasella" product **44** (Figure 10.15). The same strategy was employed in the preparation of seven-membered azasugars [37].

The eight-membered aza-sugars, as well as the bicyclic products of this type, were also prepared from the Vasella product [38].

10.5 REARRANGEMENT OF THE SUGAR SKELETON DISCOVERED *EN ROUTE* TO HIGHER-CARBON SUGARS

10.5.1 Synthesis of Higher-Carbon Sugars by the Wittig-Type Methodology

We have proposed the convenient methodology for the preparation of higher-carbon sugars (HCSs), containing more than 10-carbon atoms in the chain, already in the 1980s. That time, only limited examples of the preparation or isolation of such complex products were reported [39].

Our method proposed the coupling of two properly activated sugar subunits either by a Wittig-type methodology or by acetylene methodology [40,41]. First paper, on the coupling by Wittig method (*route a* in Figure 10.16), appeared in 1986 [42]. The

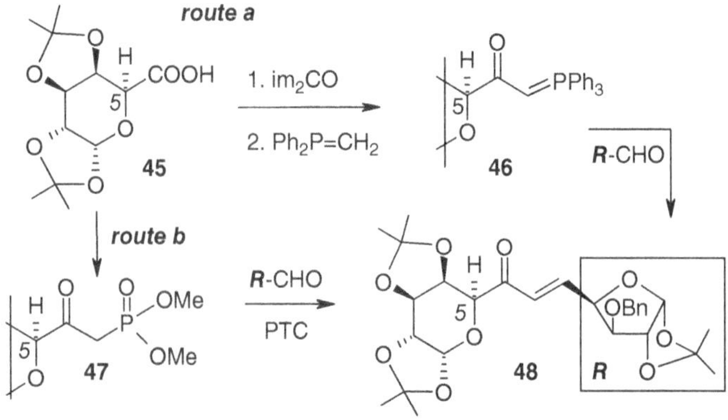

FIGURE 10.16 Synthesis of higher sugar enones by reaction of sugar aldehydes with stabilized sugar ylids or sugar phosphonates.

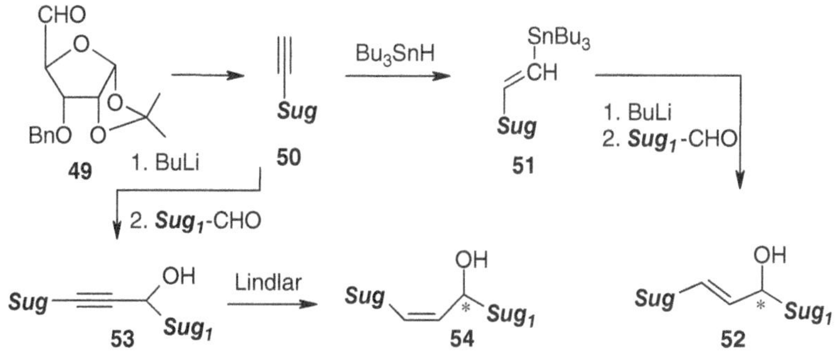

FIGURE 10.17 Synthesis of *E*- and *Z*- higher-sugar allylic alcohols from the common precursor.

improvement, replacing the stabilized Wittig reagent (e.g., **46**) for the more reactive phosphonate (e.g., **47**; *route b*), was reported by us few years later [41, 43].

10.5.2 The Acetylene/Vinyltin Methodology in the Synthesis of HCS

We introduced also the vinyltin methodology to the HCS field, which allowed to couple two sugar subunits with high efficiency. For example, sugar aldehyde **49** was converted into acetylene **50** and further into vinyltin derivative **51** *acc.* to our earlier described procedure [43]. Replacement of the tin moiety for lithium and subsequent reaction with sugar aldehydes (***Sug₁***-CHO) afforded the *E*-higher sugar allylic alcohols **52** in a good yield. Alternatively, **50** was reacted with ***Sug₁***-CHO to give propargylic alcohols **53**, easily converted into Z-allylic alcohols **54** (Figure 10.17) [41, 44].

However, the application of the vinyltin/acetylene methodology (such as **50** → **51**) may be not straightforward. For example, during reaction of (similar to **50**) acetylene **55** with Bu_3SnH, we observed rather unusual rearrangement. Partial reduction of the triple bond afforded the vinyltin radical **57a**, which underwent a rearrangement to another radical **57b**. Cyclization of this intermediate was faster than reduction and provided product **58** instead of the expected vinyltin (Figure 10.18) [45].

10.5.3 The Allyltin Methodology in the Synthesis of HCS

Application of allyltin derivatives in organic synthesis is well documented [46–48]. Although the reports on the use of "normal" allyltins appeared already in the 1970s, the sugar analogs were introduced to the chemical literature almost two decades later. In 1988, Mortlock and Thomas reported the first preparation of such compound derived from glyceraldehyde [49]. Soon after that first paper, we proposed the synthesis of more complex organometallics from **D**-hexoses [41, 50].

These compounds were designed as the key intermediates in the synthesis of HCSs containing more than 10-carbon atom in the chain. Reaction of organometallic **59**

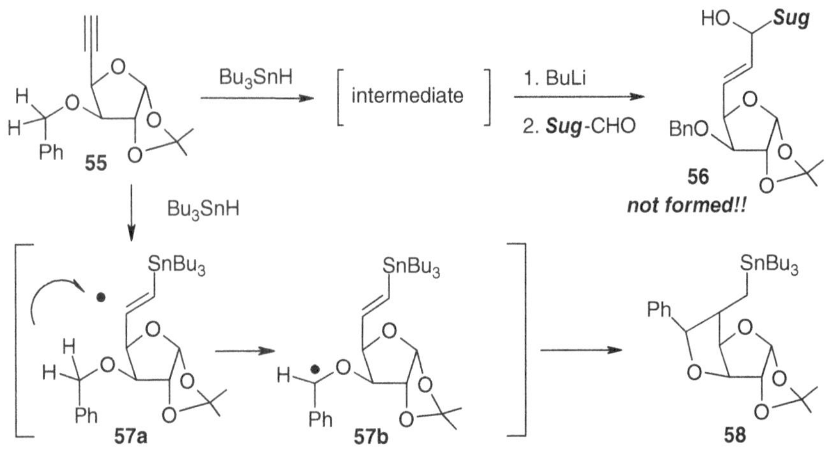

FIGURE 10.18 Unexpected rearrangement of **D**-xylo configurated sugar acetylene **55**.

with sugar aldehyde **60** catalyzed with a Lewis acid should provide higher-sugar skeleton **61**. However, under the standard conditions (slow addition of LA) only the *rearranged* product **62** was formed [41, 50, 51]! Later, we succeeded to optimize the reaction conditions allowing for the efficient preparation of the coupling product(s) in a good yield (Figure 10.19) [41, 52].

10.5.3.1 Application of Sugar Allyltins in Stereocontrolled Organic Synthesis

The dienoaldehyde formed by decomposition (better to say: controlled fragmentation) of the primary sugar allyltin (such as **59**) had always the *E*-geometry across the internal double bond, *regardless* of the configuration (*E* or *Z*) of the starting sugar

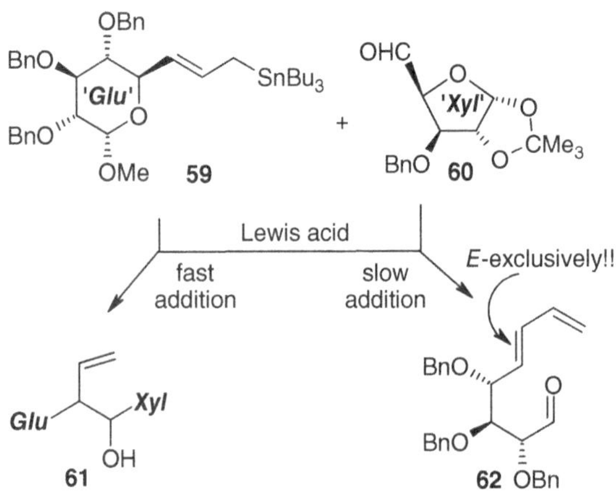

FIGURE 10.19 Application of sugar allyltins in the synthesis of HCS. Unexpected skeletal rearrangement of these organometallics.

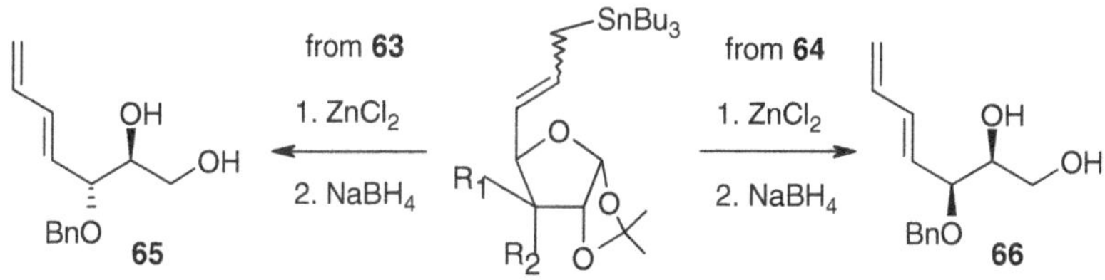

63. R_1 = OBn; R_2 = H
64. R_1 = H; R_2 = OBn

FIGURE 10.20 Synthesis of optically pure dienes from sugar allyltins.

organometallic [41, 53]. This observation allowed us to propose the convenient synthesis of various, configurationally different conjugated dienes, exclusively with the *E*-geometry across the internal double bond, from the corresponding sugar allyltins (examples are shown in Figure 10.20) [41].

Dienoaldehyde **62** was also successfully applied in the stereocontrolled synthesis of complex polyhydroxylated bicyclic derivatives as shown in Figure 10.21 [41].

Our methodology of the application of sugar allyltins in the stereocontrolled synthesis of polyhydroxylated hydrindanes (such as **67**) or decalins (e.g., **69**), although very convenient, had one main disadvantage. Fragmentation of the primary sugar allyltins provided only the *E*-dienoaldehyde, *regardless* of the configuration across the double bond in sugar organometallic. These dienes could be converted further *only* into *trans*-hydrindanes or *cis*-decalins, since the intramolecular Diels–Alder reaction of the corresponding intermediate trienes (**67a**, **69a**) proceeded exclusively via the *endo*-transition state that forced these structures [41].

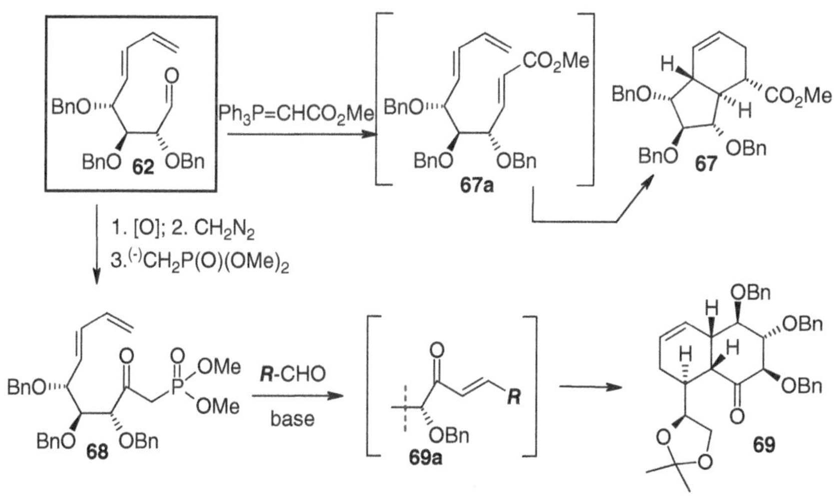

FIGURE 10.21 Synthesis of polyhydroxylated precursors of decalins and hydrindanes from the rearranged product **62**.

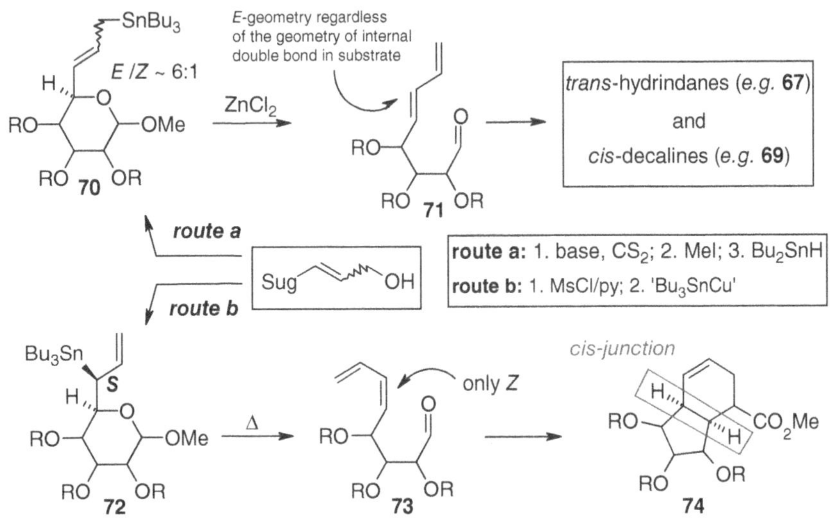

FIGURE 10.22 Conversion of sugar allyltins into pure geometrical isomers of the dienoaldehydes.

To prepare efficiently the "opposite" isomers, that is, *cis*-hydrindanes and *trans*-decalins, the access to the *Z*-dienoaldehyde was required. We have found that such *Z*-dienoaldehyde is available by thermal decomposition of the secondary sugar allyltin (**72** in Figure 10.22). Its reaction with $Ph_3P{=}CHCO_2Me$ followed by cyclization afforded the *cis*-hydrindane **74** [41, 54].

How does this reaction, that is, formation of the *Z*-dienoaldehyde, occur? Treatment of the **D**-*manno*-derived allylic mesylate **75** with the Lipshutz reagent ("Bu_3SnCu") afforded the secondary allyltin **76** as the predominant product [53]; small amounts (<1%) of the epimer at the C-6 was also formed (Figure 10.23). We proved that thermal rearrangement of the secondary sugar allyltins proceeded via the concerted mechanism as shown in Figure 10.23 [41, 54].

By the allyltin methodology, a number of bicyclic derivatives (examples are shown in Figure 10.24) were prepared [41, 55–58].

10.5.4 Rearrangement of the Structure of HCS

The Wittig-type coupling of two sugar subunits allowed us to prepare a number of different HCS enones. They can be stereoselectively reduced with zinc borohydride to the HCS allylic alcohols with the *R*-configuration at the newly created stereogenic center [41]. The question, however, arose: how to prepare the opposite *S*-isomer? It seemed reasonable that this might be done by an inversion of the configuration at the stereogenic carbinol center. The results, however, were not satisfactory.

Compound **79(*R*)** (for its preparation, see Figure 10.25) was used as a model for the S_N2 process that should provide the opposite stereoisomer **79(*S*)**.

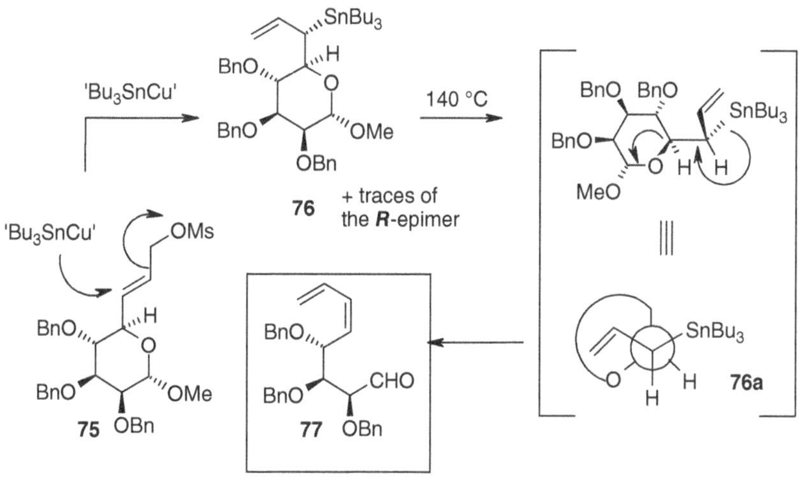

FIGURE 10.23 Synthesis and rearrangement of secondary sugar allyltins.

However, reaction of this derivative with triflic anhydride in basic media provided exclusively unexpected cyclic product **80** as the only isomer. The epimeric derivative **79(*S*)** gave, under the same conditions, also cyclic product that differed in the configuration at the five-membered ring. This process was, therefore, highly stereoselective (Figure 10.26) [59].

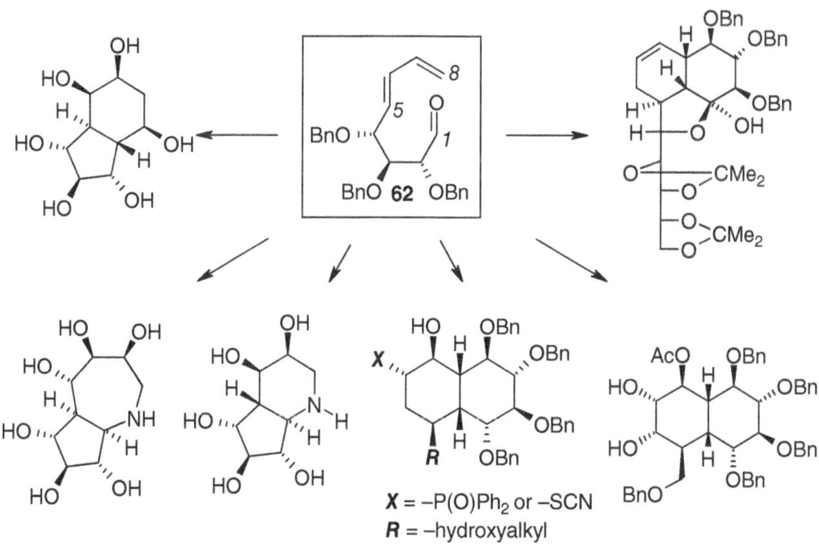

FIGURE 10.24 Examples of highly oxygenated bicyclic mimetics obtained by allyltin methodology.

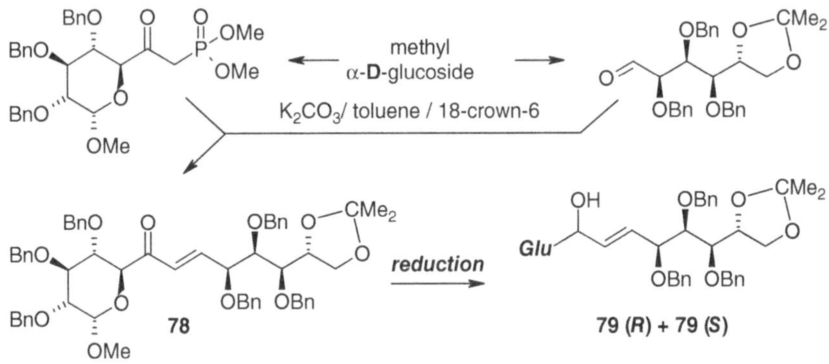

FIGURE 10.25 Preparation of the C-13 higher-sugar allylic alcohols.

Although debenzylative cyclizations in sugar chemistry are known, they usually proceed in acidic media. Very often, strong acids (e.g., 2,4-dinitrobenzenesulfonic acid) should be used to achieve this transformation [60].

The high selectivity observed in the rearrangement of **79** is rather an exception than a rule. For example, treatment of alcohol **82**, a close analog of **79**, with triflic anhydride in pyridine provided a mixture of both stereoisomeric tetrahydrofurans [61].

Under these conditions, alcohol **82(*R*)** gave two epimeric products **83** and **84** in a 3.4:1 ratio. Treatment of the second isomer **82(*S*)** with Tf_2O/Py afforded the same products in a similar proportion; again, tetrahydrofuran **83** predominated.

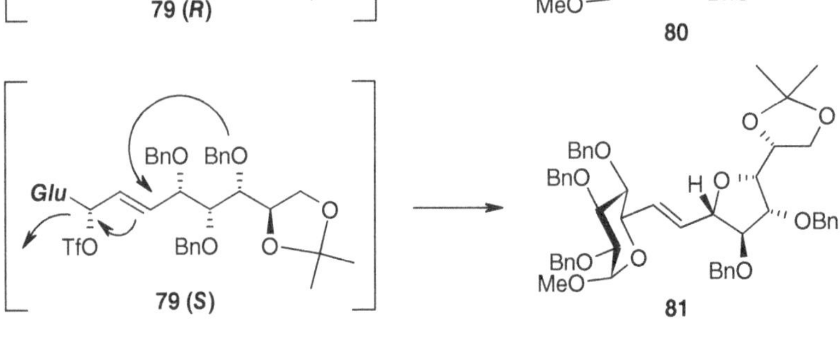

FIGURE 10.26 Rearrangement of the C-13 higher-sugar allylic alcohols.

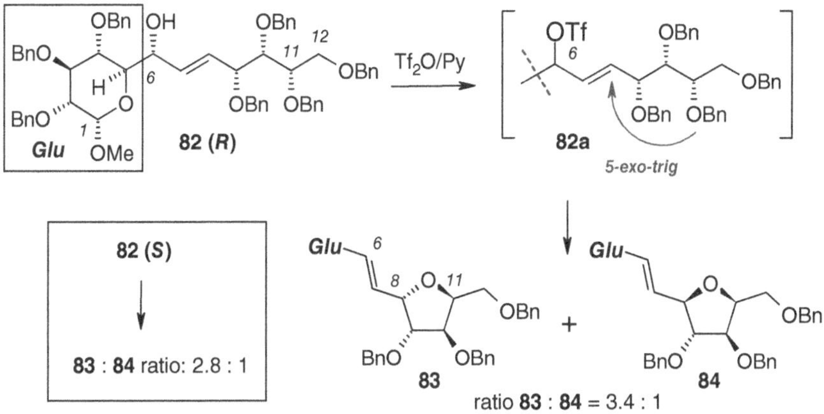

FIGURE 10.27 Nonselective rearrangement of HCS allylic alcohols.

We have already published an example of the model preparation of the C-21 monosaccharide [43]; however, it was not possible to deprotect the important function(s). The efficient preparation of such large analogs required a concise study, which include also the protecting groups and which might be *selectively* removed. Such selective deprotection should allow for further manipulation at the skeleton of a higher sugar. Enone **87** was selected as the model for the study of the selective removal of the isopropylidene blocks from various positions of the HCS skeleton (Figure 10.28).

Reaction of aldehyde **85** with phosphonate **86** should provide enone **87**. After proper functionalization of the C_{11}–C_{13} bridge followed by removal of the isopropylidene blocks (and subsequent reduction), the C18 linear higher-sugar alditols will be available. We did obtain the desired enone **87**, however in a moderate yield. Besides the expected product, we isolated also two other compounds that were identified as alcohols: **88** and **89** (Figure 10.28). The removal of the benzyl block from either position C7 or C8 occurred under the basic conditions [62]!

Removal of benzyl block(s) in acidic media (especially by strong acids) is known. However, such process in basic media is very strange.

How can we explain this phenomenon? The possible route to these strange products is shown in Figure 10.29. The primary adduct **90** undergoes the "normal" reaction with the elimination of the phosphate moiety providing the expected enone **87**.

However, we have to consider also another possibility. The open-chain adduct **90** is in an equilibrium with the structure **90a**. The attack of the oxygen from the C7 (6-exo-tet) or C8 (5-exo-tet) provides two hemiacetals **91** and **92**. Their reaction with phosphonate **86** afforded the corresponding products **89** and **88** with one of the positions: C7 or C8 free (Figure 10.29).

Although this explanation may not be very convincing, we do not have any alternative. Presence of the intermediate **90** is necessary for this transformation. Treatment of aldehyde **85** with the same mild phase transfer catalysis reagent, but without the presence of **86**, was completely ineffective; the aldehyde remained unchanged.

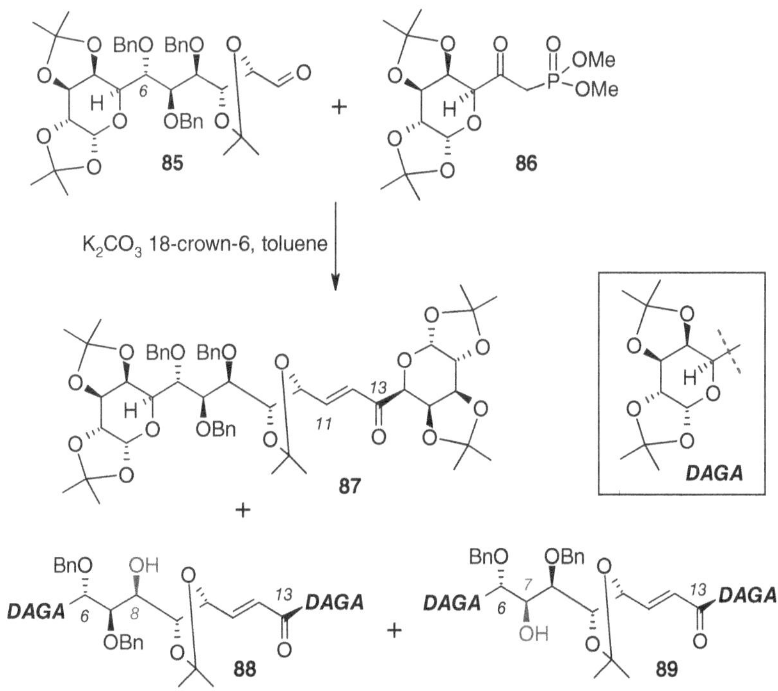

FIGURE 10.28 Synthesis of an 18-carbon sugar. Unexpected removal of the benzyl block under basic conditions.

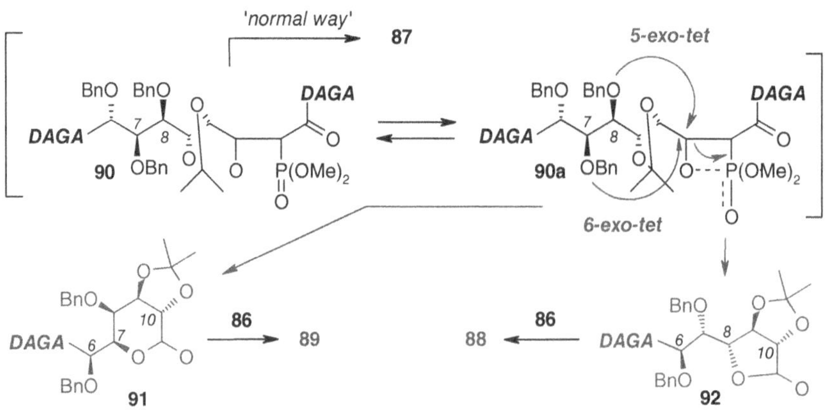

FIGURE 10.29 Proposed mechanism for the removal of the benzyl group under the phase transfer catalysis conditions.

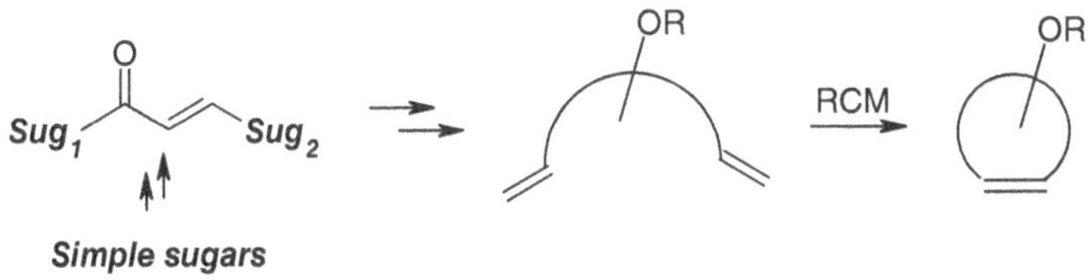

FIGURE 10.30 Approach to polyhydroxylated derivatives with large rings.

10.5.5 Synthesis of Polyhydroxylated Carbocyclic Derivatives with Large Rings

Synthesis of polyhydroxylated carbocyclic derivatives with large rings (>10 C-atoms) is a real challenge in organic synthesis. One of the possible routes was the RCM cyclization of the properly activated, at both terminal positions, HCSs. This assumption seemed to be realistic, based on the previous results reported by Fürstner and colleagues, who used the RCM for the synthesis of complex large-ring systems [63]. The idea of our synthesis is shown in Figure 10.30.

Preparation of an HCS by a Wittig-type methodology and subsequent standard transformations into a linear derivative should give the di-olefin ready for the RCM cyclization.

We initiated our synthesis from known C12 alcohol **93**, readily obtained by a Wittig-type methodology as shown in Figure 10.31 [41]. Our strategy required hydrolysis of the methyl glycoside followed by the reduction of the resulting hemiacetal to triol **94a**; further standard transformations should provide diol **94b**.

This compound can be oxidized to di-aldehyde **95** and further converted into di-olefin **96**. Such activated derivatives should (potentially) cyclize under either McMurry (compound **95**) or RCM (for **96**) conditions.

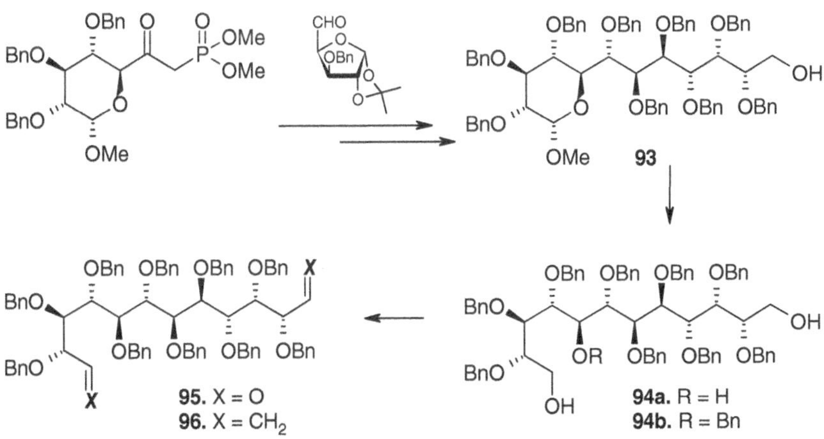

FIGURE 10.31 Synthetic strategies to polyhydroxylated di-olefin and di-aldehyde.

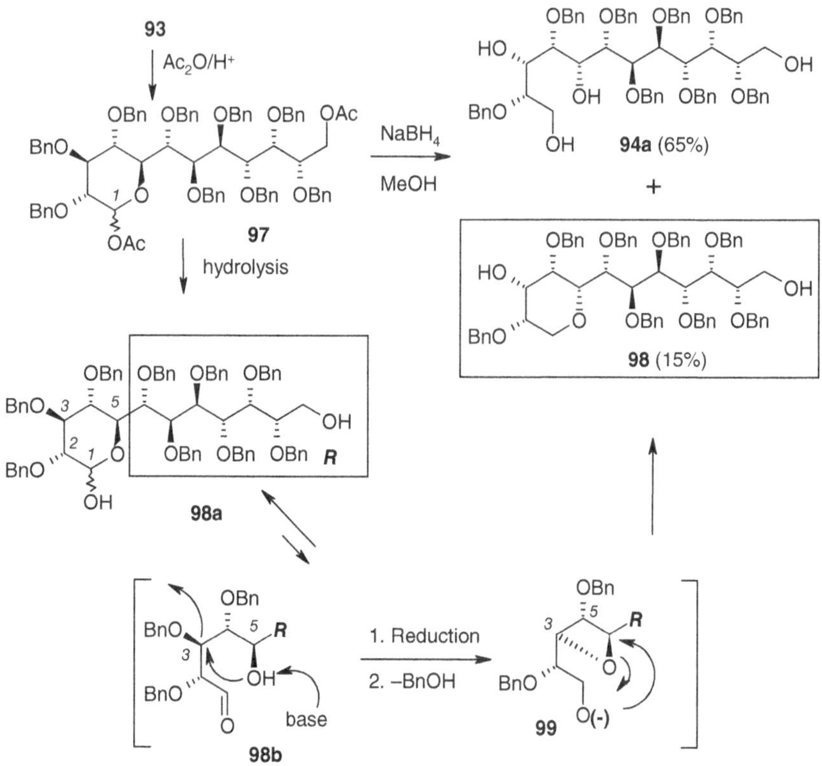

FIGURE 10.32 Unusual rearrangement of HCS diacetate **97** under the basic conditions.

Removal of the protecting group from the C1 position in compound **93** was performed by acetolysis, a method commonly used in sugar chemistry. Treatment of **93** with acetic anhydride and H_2SO_4(cat.) afforded the expected mixture of di-acetates **97**. We decided to remove the acetates under the basic conditions with simultaneous reduction of the resulting hemiacetal at the C1 position. Thus, compound **97** upon treatment with $NaBH_4$ in THF/MeOH afforded the desired diol **94a**, but a substantial amount of the rearranged product **98** was also formed (Figure 10.32) [64].

The rationalization of the formation of this unexpected product is shown in Figure 10.32. Removal of the acetate group from the C1 position under the basic conditions provided hemiacetal **98a**, which is in equilibrium with aldehyde alcohol **98b**. Reduction of **98a** yields the expected triol **94a**. However, there is a possibility of the attack of the oxy-anion from the C-5 position at the C3 center with the elimination of benzyl alcohol. The resulting oxetane **99** is then opened by the primary alkoxylate (formed after reduction of the carbonyl group), providing finally the anhydrosugar **98**. Although we do not have the hard proof for this mechanism, it seems reasonable; we have isolated the single stereoisomer of **98**, which is possible only for concerted process. This transformation (elimination of benzyl alcohol) proceeds *before* reduction of the carbonyl group, since the desired triol **94a** is stable under these reaction conditions; treatment of **94a** with $NaBH_4$ in methanol did not provide the rearranged product **98**.

10.6 CONCLUSIONS

This chapter describes the rearrangement(s) of the higher-sugar skeleton observed in the processes designed for the efficient preparation of complex polyhydroxylated derivatives, having more than 10 carbon atoms in the chain. One of these unexpected transformations, controlled fragmentation of sugar allyltins, was found to be synthetically very useful and opened a route to the stereocontrolled synthesis of polyhydroxylated carbo- and heterobicyclic derivatives, which may be regarded as sugar mimetics.

Other transformations are rather curiosity and have no synthetic potential (at least now). However, several interesting observations were noted, which are not reported for more simple derivatives (i.e., in "normal" sugar chemistry). The elimination of benzyl alcohol under the basic conditions represents undoubtedly such "crazy" transformation.

ACKNOWLEDGMENTS

The authors thank for the support from Grant of National Science Centre UMO-2012/05/B/ST5/00377.

REFERENCES

[1] Ferrier, R. J. *Adv. Carbohydr. Chem.* **1969**, *24*, 199–266.

[2] *Recent review*: Gomez, A. M.; Lobo, F.; Uriel, C.; Lopez, J. C. *Eur. J. Org. Chem.* **2013**, 7221–7262.

[3] Danishefsky, S. J.; De Ninno, S.; Lartey, P. *J. Am. Chem. Soc.* **1987**, *109*, 2082–2089.

[4] These types of rearrangements were discussed. In *Glycoscience—Chemistry and Chemical Biology*, Vols. I–III; Fraser-Reid, B., Tatsuta, K., Thiem, J., Eds.; Springer-Verlag, 2008; S. Jarosz, M. Nowogródzki, chapter 2.4. pp 271–303, and chapter 2.6, pp 343–374; *see also references therein.*

[5] Břjstrup, M.; Fanefjord, M.; Lundt, I. *Org. Biomol. Chem.* **2007**, *5*, 3164–3171.

[6] Fitzsimmons, B. J.; Plauman, D. E.; Fraser-Reid, B. *Tetrahedron Lett.* **1979**, 3925–3928.

[7] Jarosz, S.; Fraser-Reid, B. *Tetrahedron Lett.* **1981**, *22*, 2533–2534.

[8] Buchanan, J. G. *Methods Carbohydr. Chem.* **1972**, *6*, 135–141.

[9] Wang, Y.; Li, Q.; Cheng, S.; Wu, Y.; Guo, D.; Fan, Q.-H.; Wang, X.; Zhang, L.-H.; Ye, X.-S. *Org. Lett.* **2005**, *7*, 5577–5579.

[10] Ferrier, R. J. *J. Chem. Soc. Perkin Trans 1* **1979**, 1455–1458.

[11] Ferrier, R.; Middleton, S. *Chem. Rev.* **1993**, *93*, 2779–2831.

[12] Ferrier, R. J.; Hoberg, J. O. *Adv. Carbohydr. Chem. Biochem.* **2003**, *58*, 55–119.

[13] Meek, S. J.; Harrity, J. P. A. *Tetrahedron* **2007**, *63*, 3081–3092.

[14] Blair, M. G. *Methods Carbohydr. Chem.* **1963**, *2*, 415–414; *see also ref. 4.*

[15] Mirza, S.; Molleyres, L.-P.; Vasella, A. *Helv. Chim. Acta* **1985**, *68*, 988–996.

[16] Barton, D. H. R.; Auge-Dorey, S.; Camara, J.; Dalko, P.; Delaumeny, J. M.; Gero, S. D.; Quiclet-Sire, B.; Stuetz, P. *Tetrahedron* **1990**, *46*, 215–230.

[17] Das, S. K.; Mallet, J.-M.; Sinaÿ, P. *Angew. Chem. Int. Ed. Engl.* **1997**, *36*, 493–496.

[18] Du Roizel, B.; Sollogoub, M. S.; Pearce, A. J.; Sinaÿ, P. *Chem. Comm.* **2000**, 1507–1508.

[19] Suami, T.; Ogawa, S. *Adv. Carbohydr. Chem. Biochem.* **1990**, *48*, 21–90.

[20] Arjona, O.; Gomez, A. M.; Lopez, J. C.; Plumet, J. *Chem. Rev.* **2007**, *107*, 1919–2036.

[21] Koto, S. *Rearrangement Reactions. "Glycoscience—Chemistry and Chemical Biology,"* Vols. *I–III*; Fraser-Reid, B., Tatsuta, K., Thiem J., Eds.; Springer-Verlag, **2001**; chapter 2.9, pp 417–464.

[22] Mollander, G. H.; Harris, C. R. *Chem. Rev.* **1996**, *96*, 307–338.

[23] Kan, T.; Nara, S.; Ozawa, T.; Shirahama, H.; Matsuda, F. *Angew. Chem.* **2000**, *112,* 363–365.

[24] Aurreocoechea, J. M.; Lopez, B. *Tetrahedron Lett.* **1998**, *39*, 2857–2860.

[25] Hanzawa, Y.; Ito, H.; Taguchi, T. *Synlett.* **1995**, 299–305.

[26] Ganesh, N. V.; Jayaraman, N. *J. Org. Chem.* **2007**, *72*, 5500–5504.

[27] Beyer, J.; Skaanderup, P. R.; Madsen, R. *J. Am. Chem. Soc.* **2000**, *122*, 9575–9583.

[28] Panfil, I.; Maciejewski, S.; Bełżecki, C.; Chmielewski, M. *Tetrahedron Lett.* **1989**, *30*, 1527–1528.

[29] Rabiczko, J.; Chmielewski, M. *J. Org. Chem.* **1999**, *64*, 1347–1351, *and references therein.*

[30] *Recent review on pyranopyran methodology*: Giuliano, R. *Current Org. Chem.* **2014**, *18*, 1686–1700.

[31] Mootoo, D. R.; Fraser-Reid, B. *J. Chem. Soc., Chem. Commun.* **1986**, 1570–1571.

[32] *Account*: Fraser-Reid, B.; Udodong, U. E.; Wu, Z.; Ottoson, H.; Merrit, J. R.; Rao, C. S.; Roberts, C.; Madsen, R. *Synlett* **1992**, 927–942.

[33] *review*: Fraser-Reid, B.; Lopez, J. C.; Gomez, A. M.; Uriel, C. *Eur. J. Org. Chem.* **2004**, 1387–1395.

[34] Bernet, B.; Vasella, A. *Helv. Chim. Acta* **1979**, *62*, 1990–2016.

[35] Fürstner, A. *Angew. Chem. Int. Ed. Engl.* **1993**, *32*, 164–189, *and the literature cited therein.*

[36] Hansen, F. G.; Bundgaard, E.; Madsen, R. *J. Org. Chem.* **2005**, *70*, 10139–10142.

[37] *See, for example*: Padar, P.; Bokros, A.; Paragi, G.; Forgo, P.; Kele, Z.; Howarth, N. M.; Kovacs, L. *J. Org. Chem.* **2006**, *71*, 8669–8672, *and references cited therein.*

[38] Sletten, E. M.; Bertozzi, C. R. *Org. Lett.* **2008**, *10*, 3097–3099.

[39] *First paper on this subject:* Paulsen, H.; Roden, K.; Sinwell, V.; Koebernick, W. *Angew. Chem., Int. Ed. Eng.* **1976**, *15*, 439–440; *see also ref. 4.*

[40] *Review*: Jarosz, S. *J. Carbohydr. Chem.* **2001**, *20*, 93–107.

[41] *Review*: Jarosz, S. *Curr. Org. Chem.* **2008**, *12*, 985–994.

[42] Jarosz, S.; Mootoo, D. R.; Fraser-Reid, B. *Carbohydr. Res.* **1986**, *147*, 59–68.

[43] Jarosz, S. *Tetrahedron Lett.* **1994**, *35*, 7655–7658.

[44] Jarosz, S. *J. Carbohydr. Chem.* **1993**, *12*, 1149–1160.

[45] Jarosz, S.; Kozłowska, E. *Polish J. Chem.* **1994**, *68*, 539–545.

[46] Yamamoto, Y.; Asao, N. *Chem. Rev.* **1993**, *93*, 2207–2293.

[47] Jarosz, S.; Gaweł, A. *Eur. J. Org. Chem.* **2005**, 3415–3432.

[48] Motoyama, Y.; Sakakura, T.; Takemoto, T.; Shimozono, K.; Aoki, K.; Nishiyama, H. *Molecules* **2011**, *16*, 5387–5401.

[49] Mortlock, S. V.; Thomas, E. J. *Tetrahedron Lett.* **1988**, *29*, 2479–2483.

[50] Jarosz, S.; Fraser-Reid, B. *J. Org. Chem.* **1989**, *54*, 4011–4013.

[51] Kozłowska, E.; Jarosz, S. *J. Carbohydr. Chem.* **1994**, *13*, 889–898.

[52] Jarosz, S.; Skóra, S.; Szewczyk, K.; Ciunik, Z. *Tetrahedron Asymmetry* **2001**, *12*, 1895–1905.

[53] Jarosz, S. *Tetrahedron* **1997**, *53*, 10765–10774.

[54] Jarosz, S.; Szewczyk, K.; Zawisza, A. *Tetrahedron Asymmetry* **2003**, *14*, 1709–1713 and 1715–1723.

[55] Nowogródzki, M.; Malik, M.; Jarosz, S. *Tetrahedron Asymmetry* **2012**, *23*, 1501–1511.

[56] Magdycz, M.; Jarosz, S. *Tetrahedron Asymmetry* **2013**, *24*, 1402–1411.

[57] Magdycz, M.; Jarosz, S. *Tetrahedron Asymmetry* **2013**, *24*, 1412–1416.

[58] *Recent review*: Jarosz, S.; Nowogrodzki, M.; Magdycz, M.; Potopnyk, M.A. *RSC Spe Periodic Rep Carbohydr. Chem.* **2012**, *37* (chapter 11), 303–325.

[59] Jarosz, S.; Gajewska, A.; Luboradzki, R. *Tetrahedron Asymmetry* **2008**, *18*, 1385–1391.

[60] Cribiu, R.; Borbas, K. E.; Cumpstey, I. *Tetrahedron* **2009**, *65*, 2022–2031.

[61] Potopnyk, M. A.; Cmoch, P.; Cieplak, M.; Gajewska, A.; Jarosz, S. *Tetrahedron Asymmetry* **2011**, *22*, 780–786.

[62] Cieplak, M.; Jarosz, S. *Tetrahedron Asymmetry* **2011**, *22*, 1757–1762.

[63] Fürstner, A.; Thiel, O. R.; Kindler, N.; Bartkowska, B. *J. Org. Chem.* **2000**, *65*, 7990–7995, *and references cited therein.*

[64] Osuch-Kwiatkowska, A.; Jarosz, S. *Tetrahedron Asymmetry* **2013**, *24*, 468–473.

11

NOVEL LEVOGLUCOSENONE DERIVATIVES

Roman Bielski and Zbigniew J. Witczak

Department of Pharmaceutical Sciences, Wilkes University, Wilkes-Barre, PA, USA

11.1 INTRODUCTION

Levoglucosenone is one of the most attractive chiral synthons deriving from carbohydrates [1]. While it is a highly functionalized compound, taking advantage of a single functionality at a time is relatively easy. The presence of the 1,6-anhydro bridge freezes the conformation containing the (unusual) axial substituents on carbon atoms 1 and 5 and enables the formation of a single diastereoisomer as a result of most reactions. Levoglucosenone's applications as a chiral synthon were recently reviewed [2]. The authors divided the possible applications according to the structure of final products. Since levoglucosenone is relatively easily manufactured from cellulose and similar sources, its production is sustainable and rather nonexpensive. The only chiral centers (in the carbohydrate nomenclature carbon atoms 1 and 5) are involved in the relatively stable anhydro ring. Of course, at the desired moment during the synthesis, the 1,6-anhydro system can be hydrolyzed to reconstruct the hydroxyl functionalities in positions 1 and 6.

The functionalization schemes of levoglucosenone can be grouped into a few categories. The following list offers the most useful categories with selected examples.

Domino and Intramolecular Rearrangement Reactions as Advanced Synthetic Methods in Glycoscience, First Edition. Edited by Zbigniew J. Witczak and Roman Bielski.

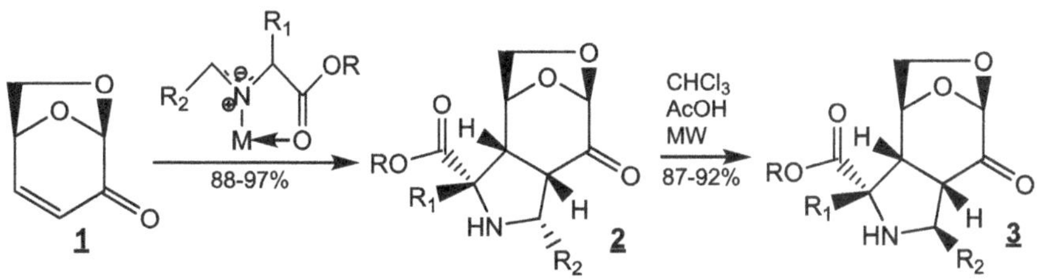

SCHEME 11.1 Example of cycloaddition of azomethine ylide to levoglucosenone.

11.2 ADDITIONS TO THE DOUBLE BOND OF THE ENONE SYSTEM LEADING TO THE FORMATION OF NEW RINGS

They include:

(a) A 1,3-dipolar cycloaddition of azomethine ylides [3a] and nitrones [3b]. The products of the former cycloaddition are promising organocatalysts for iminium ion-based asymmetric Diels–Alder reactions.
(b) Diastereoselective Diels–Alder (diene) additions [4].

Scheme 11.1 shows a spectacular example of the cycloaddition of azomethine ylides.

11.3 REDUCTIONS OF THE CARBONYL GROUP FOLLOWED BY VARIOUS REACTIONS OF THE FORMED ALCOHOL

Due to the steric hindrance of the anhydro moiety, the reduction produces the threo product as the dominant or only product. This alcohol has been transformed to various products that were subsequently rearranged.

For example, one of us [5] developed a method enabling the effective transformation of levoglucosenone (**1**) into isolevoglucosenone (**7**) via the 2,3-sigmatropic rearrangement of allylic selenide. The approach is shown on Scheme 11.2.

Matsumo *et al.* [6] reduced the carbonyl group of levoglucosenone and reacted the formed alcohol with thionyl chloride in pyridine. The resulting reaction mixture consisted of three allylic chlorides. The 2-chloro compound with inversed configuration at C-2 (**8**) was a major product. When this compound was reacted with benzyl alcohol (in the presence of NaH), the rearranged benzyl ether (**11**) was isolated in an excellent yield. Scheme 11.3 shows the process.

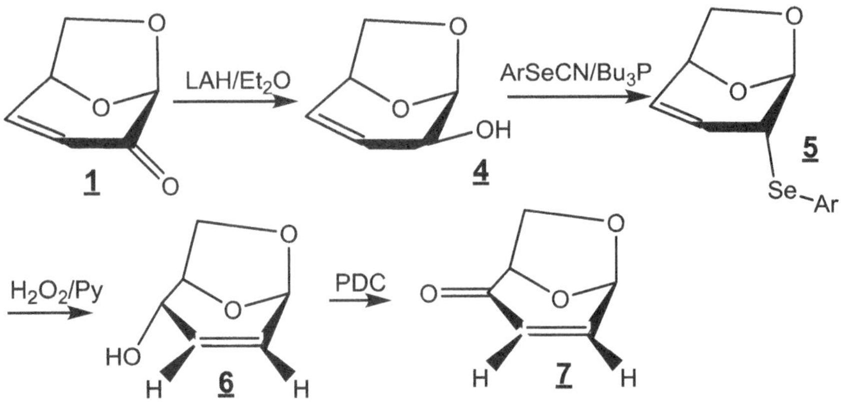

SCHEME 11.2 Effective transformation of levoglucosenone into isolevoglucosenone.

11.4 FUNCTIONALIZATION OF THE CARBONYL GROUP BY FORMING CARBON-NITROGEN DOUBLE BONDS (OXIMES, ENAMINES, HYDRAZINES)

Valeev and coworkers [7] tried to synthesize amino derivatives of levoglucosenone. The most direct approach seemed to comprise the formation of oximes or hydrazones that, upon reduction, should give the desired products. However, the reduction turned out to be problematic. The researchers synthesized the amino compounds by transforming the reduction product of levoglucosenone into the mesylate (or tosylate) and performing the ammonolysis that gave the desired SN2 substitution product in a good yield.

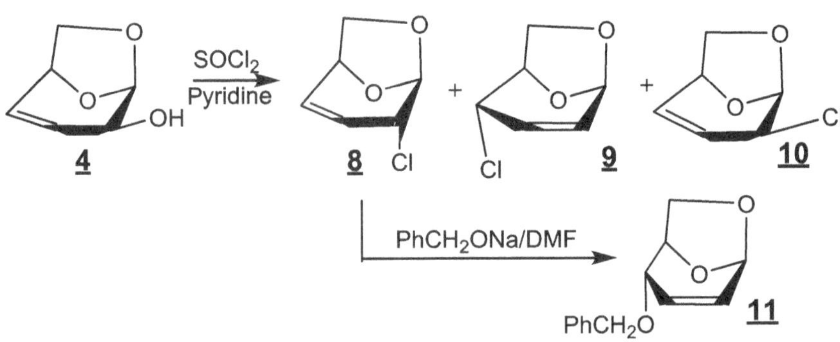

SCHEME 11.3 Interesting allylic rearrangement of levoglucosenone epimer carrying chloro substituent.

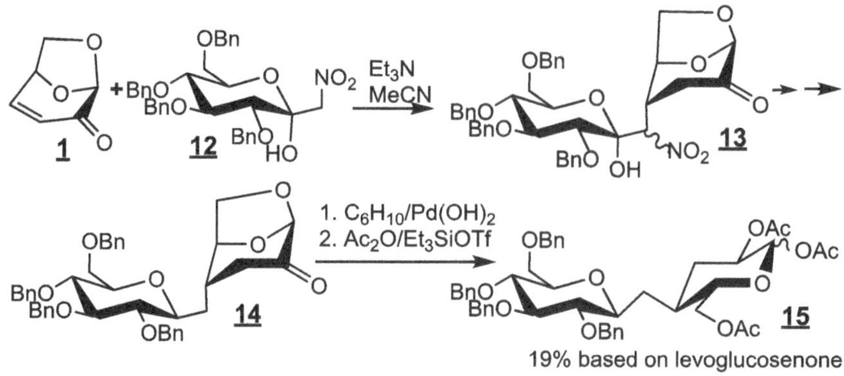

SCHEME 11.4 Michael addition to levoglucosenone producing carbon-carbon bonds.

11.5 ADDITIONS (BUT NOT CYCLOADDITIONS) (PARTICULARLY MICHAEL ADDITIONS) TO THE DOUBLE BOND OF THE ENONE

The literature on Michael additions to levoglucosenone is abundant. However, arguably there are two types of such additions that offer exceptionally interesting products. One category involves the formation of a carbon–carbon bond [8]. A wonderful example of applying the discussed approach is shown on Scheme 11.4. The other category (addition of thiols) leads to the formation of a sulfur–carbon bond [9] (Scheme 11.5). Both processes have been taken advantage of to produce very interesting oligosaccharides.

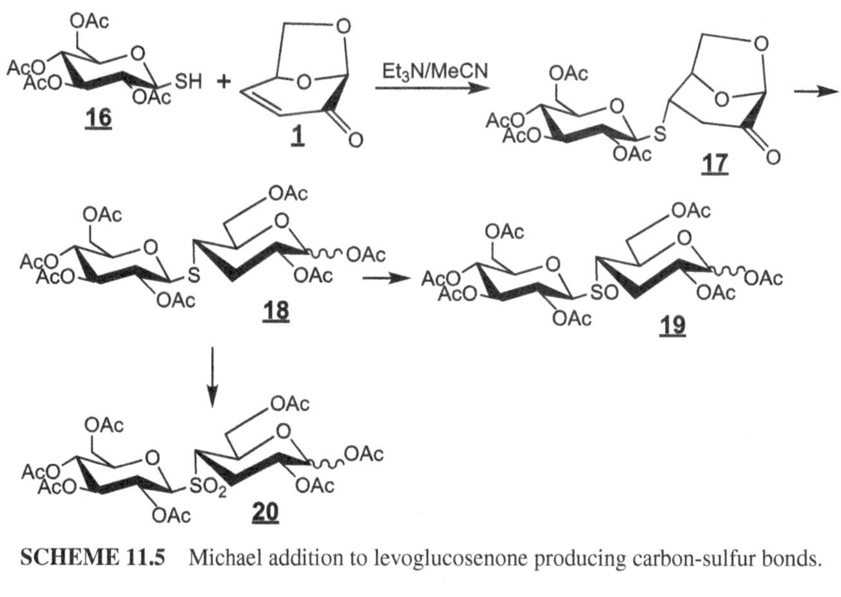

SCHEME 11.5 Michael addition to levoglucosenone producing carbon-sulfur bonds.

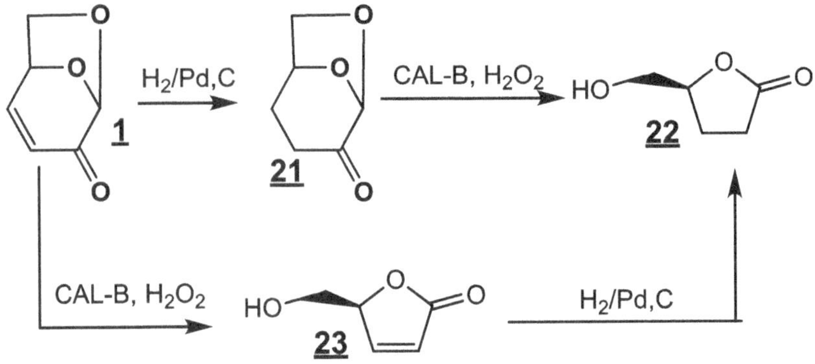

SCHEME 11.6 Example of chemoenzymatic syntheses starting from levoglucosenone.

11.6 ENZYMATIC REACTIONS OF LEVOGLUCOSENONE

Levoglucosenone has been employed as a substrate for successful enzymatic syntheses. For example, recently Allais and coworkers described [10] a very effective chemoenzymatic synthesis of optically pure (*S*)-γ-hydroxymethyl-α,β-butenolide (**23**), using a two-step sequence involving a lipase-mediated Baeyer–Villiger oxidation and an acid hydrolysis from levoglucosenone. Another valuable product these authors transformed levoglucosenone into is (*S*)-γ-hydroxymethyl-γ-butyrolactone (**22**). Both products in an enantiomerically pure form have been used as chiral synthons in a variety of total syntheses. The discussed compounds are shown on Scheme 11.6.

11.7 HIGH-TONNAGE PRODUCTS FROM LEVOGLUCOSENONE

Some time ago an Australian company, Circa (http://circagroup.com.au/), entered the market of converting waste biomass into advanced biochemical materials. One of the products of their interest is levoglucosenone. Since then, they have developed and promoted development of various products deriving from this green compound. Recently, their CEO, Tony Duncan, presented recent accomplishments of the company and their collaborators in finding novel applications of levoglucosenone [11]. They include such diverse compounds as polar aprotic solvent, cyrene, 5-hydroxymethyl-2-furfural, and 1,6-hexanediol. Interested readers should check Tony Duncan's presentation.

In addition, it is worth mentioning that levoglucosenone and isolevoglucosenone have been used as a chiral auxiliary. For example, it has been demonstrated that levoglucosenone and isolevoglucosenone undergo isonitrone cycloaddition with a high level of double asymmetric induction. Also, it has been taken advantage of in the successful parallel kinetic resolution experiments of racemic nitrones [12].

11.7.1 Overman and Allylic Xanthate Rearrangement

Recently, we became interested in Claisen rearrangements in which one or more of the main core carbon atoms were replaced with heteroatoms. We call these processes Hetero-Oxa-Claisen rearrangements to emphasize that the central atom of the process is the oxygen atom like in the classical Claisen rearrangement. There are several processes belonging to this category. They include the Overman rearrangement and the allylic rearrangement of xanthates, thiocarbamates, and other compounds containing C=S. The driving force of the rearrangements is the formation of the thermodynamically stable carbonyl group.

We thought that it would be interesting to examine the application of these rearrangements to transforming levoglucosenone to modified isolevoglucosenone. Since the discussed reactions belong to the Claisen rearrangements, they are [3,3]-sigmatropic rearrangements and require a chair-like transition state. Thus, it was clear that a known [13] xanthate (**24**) will not be able to enter the allylic rearrangement. However, its diastereoisomer with the reversed configuration at position 2 should.

By the way, Spanavello and coworkers [13] describe a very elegant synthesis of acetylated *D*-allal starting from xanthate derived from levoglucosenone (Scheme 11.7). The reaction of levoglucosenone reduction product (**4**) with a base, carbon disulfide, and methyl iodide gave the expected xanthate (**24**), which was oxidized (*cis*-hydroxylated) with osmium tetroxide and *N*-morpholine oxide to form a fully substituted product. The hydroxyl groups of xanthate were protected to give (**25**) and the reaction with trimethylsilyl triflate gave peracetylated 1,3-oxathiolane-2-thione (**26**) in an excellent yield. This cyclic dithiocarbonate upon treatment with trimethyl phosphite (Winter–Corey reaction) gave the desired allal (**27**) in a very good yield.

To synthesize alcohol with the reversed configuration at C-2, we started with the reduction of levoglucosenone to form the well-known alcohol (**4**) (Scheme 11.8). It was reacted with mesyl chloride to produce mesylate (**28**). The nucleophilic substitution was performed in refluxing toluene (or xylene) using sodium benzoate as a nucleophile and a nonnucleophilic phase transfer catalyst (tetrabutylammonium hydrogen sulfate) to give a very good yield of the expected benzoate (**29**). The hydrolysis of the

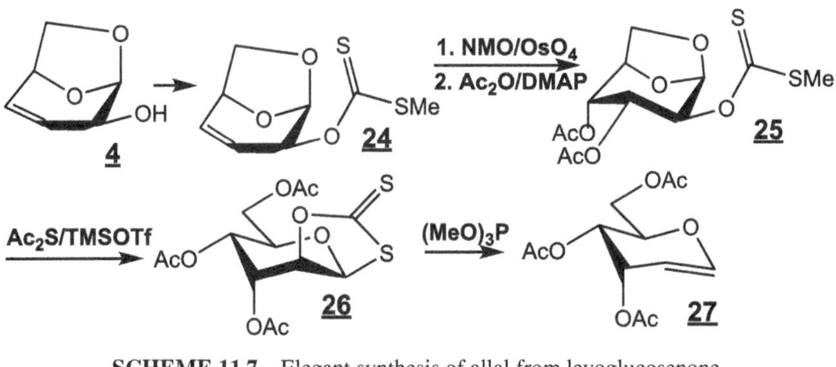

SCHEME 11.7 Elegant synthesis of allal from levoglucosenone.

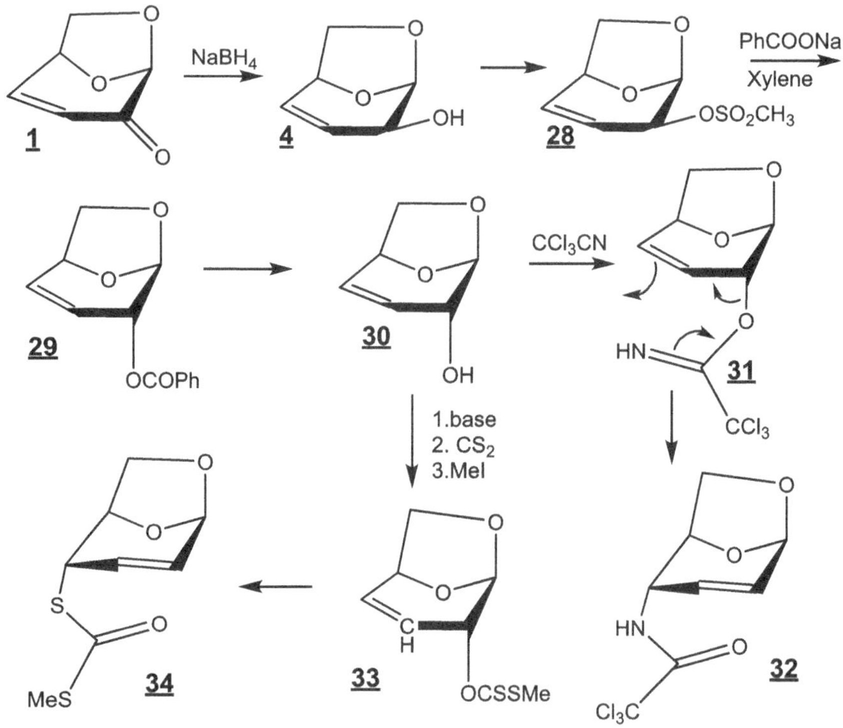

SCHEME 11.8 Synthesis of thio and amino substituted isolevoglucosenone derivatives from levoglucosenone.

benzoate group forms the alcohol (**30**) with the inversed configuration at C-2. The reaction of this alcohol with trichloroacetonitrile produces the expected trichloroacetimidate (**31**), which rearranges to the substituted 3-aminoproduct (**32**) when heated. Similarly, the analogous xanthate (**33**) produces the 3-thio-product (**34**) when heated. The more detailed description of these transformations will be published elsewhere.

11.8 CONCLUSIONS

Levoglucosenone is an exceptionally useful, carbohydrate-derived chiral synthon that can enter a very large number of diverse reactions. The resulting products are high-value modified carbohydrates, nonsugar derivatives, other synthons, etc. In addition, several total syntheses take advantage of this synthon. Furthermore, there is a growing interest in compounds such as the solvent and cyrene that can be produced from levoglucosenone but are much less costly. The synthetic route leading from alcohol (**4**) derived from levoglucosenone, via alcohol (**30**), to 1,6-anhydroderivatives bearing amino (**32**) or thio (34) substituents at the position 2 seems attractive. The process takes advantage of the Overman or allylic xanthate rearrangement.

REFERENCES

[1] *Levoglucosenone and Levoglucosans: Chemistry and Applications*; Z. J. Witczak, Ed.; ATL Press, Inc. Science Publishers: Mount Prospect, IL, **1994**.

[2] Sarotti, A. M.; Zanardi, M. M.; Spanevello, R. A.; Suárez, A. G. *Curr. Org. Synth.* **2012**, *9*, 439–459.

[3] (a) Sarotti, A. M.; Spanevello, R. A.; Suárez, A. G.; Echeverría, G. A.; Piro, O. E. *Org. Lett.* **2012**, *14*, 2556–2559. (b) Müller, Ch.; Gomez-Zurita Frau, M. A.; Ballinari, D.; Colombo, S.; Bitto, A.; Martegani, E.; Airoldi, C.; van Neuren, A. S.; Stein, M.; Weiser, J.; Battistini, C.; Peri, F. *Chem. Med. Chem* **2009**, *4*, 524–528.

[4] (a) Sarotti, A. M.; Spanevello, R. A.; Suárez, A. G. *Org. Lett.* **2006**, *8*, 1487–1490. (b) Sarotti, A. M.; Spanevello, R. A.; Suárez, A. G. *Arkivoc* **2011**, 31–37. (c) Sarotti, A. M.; Spanevello, R. A.; Suárez, A. G. *Tet. Lett.* **2004**, *45*, 8203–8206. (d) Sarotti, A. M.; Spanevello, R. A.; Suárez, A. G. *Tet. Lett.* **2005**, *46*, 6987–6990. (e) Valeev, F. A.; Krasnoslobodtseva, O. Y.; Salikhov, S. M.; Tolstikov, G. A. *Chem. Sustainable Dev.* **2008**, *16*, 699–703.

[5] Witczak, Z. J.; Kaplon, P.; Kolodziej, M. *J. Carbohydr. Chem.* **2002**, *21*, 1–6.

[6] Matsumoto, M.; Ishikawa, H.; Soya, Y.; Ozawa, T. *Heterocycles* **1994**, *55*, 2377–2381.

[7] Valeev, F. A.; Kalimullina, L. K.; Salikhov, S. M.; Shitikova, O. V.; Tsypysheva, I. O.; Safarov, M. G. *Chem. Nat. Compounds* **2004**, *40*, 521–525.

[8] Awad, L.; Demange, R.; Zhu, Y.-H.; Vogel, P. *Carb. Res.* **2006**, *341*, 1235–1252.

[9] Witczak, Z. J.; Kaplon, P.; Dey, P. M. *Carb. Res.* **2003**, *338*, 11–18.

[10] Flourat, A. L.; Peru, A. A. M.; Teixeira, A. R. S.; Brunissen, F.; Allais, F. *Green Chem.* **2015**, *17*, 404–412.

[11] http://www.biovale.org/documents/York_June_2014_copy_21.pdf

[12] Cardona, F.; Lalli, D.; Faffi, C.; Goti, A.; Brandi, A. *J. Org. Chem.* **2008**, *73*, 1999–2002.

[13] Giordano, E. D. V.; Frinchaboy, A.; Suárez, A. G.; Spanevello, R. A. *Org. Lett.* **2012**, *14*, 4602–4605.

12

THE PREPARATION AND REACTIONS OF 3,6-ANHYDRO-D-GLYCALS

VIKRAM BASAVA, EMI HANAWA, AND CECILIA H. MARZABADI
Department of Chemistry and Biochemistry, Seton Hall University, South Orange, NJ, USA

12.1 INTRODUCTION

Anhydrosugars comprise a large class of carbohydrate derivatives characterized by both the positions of ring carbons from which they are formed and the size of the resulting ring [1–5]. The most common anhydrosugars, or intramolecular anhydrides, are formed between the anomeric carbon and a hydroxyl group at another ring carbon in the sugar. These derivatives are frequently referred to as the glycosans [1]. Ring sizes for anhydrosugars can vary; the smaller anhydro rings are generally considered to be less stable. The less stable derivatives include the oxirane and oxetane anhydrosugars, while the tetrahydrofuran (THF) and tetrahydropyran anhydrosugars show greater stability [2].

A common procedure for the synthesis of anhydrosugars is by activation of one of the hydroxyl groups as a leaving group (e.g., tosylate, triflate, halogen), while a second hydroxyl group acts as a nucleophile in basic medium [3]. For secondary hydroxyl groups, this results in inversion of configuration. Alternatively, acid-catalyzed dehydration can also lead to the formation of anhydrosugars [6].

The properties of anhydrosugars make them useful as protecting groups that, under the appropriate conditions, can enable them to serve as reactive intermediates for the formation of new C–C and C–X bonds. These characteristics have made them useful precursors for the formation of a variety of carbohydrate derivatives and medicinally important compounds. For example, anhydrosugars have been utilized in

Domino and Intramolecular Rearrangement Reactions as Advanced Synthetic Methods in Glycoscience, First Edition. Edited by Zbigniew J. Witczak and Roman Bielski.

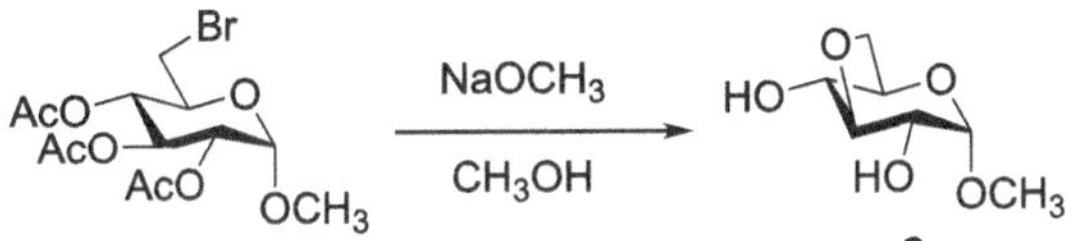

1 2

SCHEME 12.1 Formation of a 3,6-anhydrosugar in the presence of added base.

the construction of nucleoside derivatives [7,8], *C*-glycosides [9,10], peptidomimetics [11], and in the synthesis of *N*-glycosides [12].

Among the six-membered ring anhydrosugars (tetrahydropyrans), the 3,6-anhydro pyranoses are the most well-known analogs. These compounds have been isolated as constituents from red algae [13] and 3,6-anhydro-L-galactose is a constituent of the biopolymer agar [14]. In the laboratory, these compounds are typically formed from the 6-halo or 6-*O*-tosyl sugars via treatment with base. For example, methyl 3,6-anhydro-β-D-glucopyranoside **2** was formed from methyl 2,3,4-tri-*O*-acetyl-6-bromo-6-deoxy-β-D-glucopyranose **1** via treatment with barium hydroxide or sodium methoxide (Scheme 12.1). The corresponding galactose and mannose derivatives were also prepared in a similar manner [15]. Attempted displacement of methyl 6-*O*-*p*-toluenesulfonate-D-galactopyranose with sodium azide also afforded the corresponding 3,6-anhydrosugar [16].

There have been fewer reports describing the synthesis of the 3,6-anhydroglycals. The 3,6-anhydroglucal was prepared by Tucker and colleagues from 3,4-di-*O*-acetyl-6-*O*-*p*-toluenesulfonyl-D-glucal using either sodium methoxide of De-Acidite FF IP (−OH) resin [17]. Fraser-Reid and colleagues also reported the formation of this compound as a by-product from the lithium aluminum hydride reduction of 6-*O*-tosyl-D-glucal [18]. More recently, Crich and Richie isolated anhydroglycal **4** as a product when attempting to benzylate the 3-*O*-silylated glucal **3** (Scheme 12.2). Migration of a silyl group from the 3- to 4-position of the glucal was also observed to occur [19].

Several reducing agents have been utilized to remove tosylate esters of sugars and other polyols. In particular, lithium aluminum hydride and lithium triethylborohydride have been most extensively used [20–23]. When secondary alcohol tosylates are used, epoxide intermediates are frequently involved. Reduction is observed to occur through either C–O or O–S bond cleavage [24]. In this chapter, we discuss our results from the reduction of 6-*O*-tosylates of D-glucal and D-galactal with lithium aluminum hydride in THF.

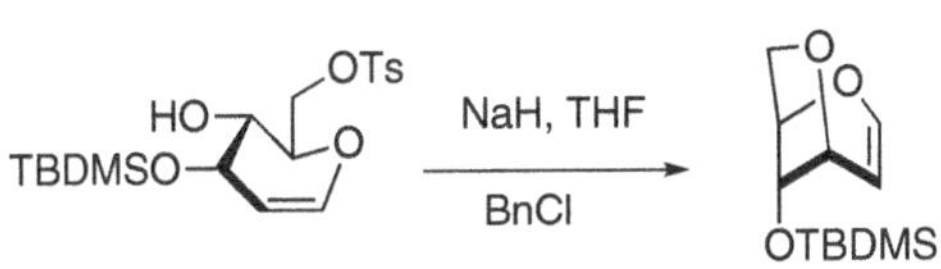

3 4

SCHEME 12.2 Formation of a 3,6-anhydro-D-glucal with benzylation reaction conditions.

To our knowledge, a study of the selectivity of double-bond addition reactions to 3,6-anhydroglycals has not been reported to date. We have explored a series of these reactions and also describe the outcome of these reactions in terms of yield and selectivity for products.

12.2 PREPARATION OF 3,6-ANHYDRO-D-GLUCAL UNDER REDUCTIVE CONDITIONS

We were interested in the formation of the 6-deoxyglycal (D-rhamnal) as an intermediate in our synthesis of a novel trisaccharide building block. We reasoned that the reduction of a 6-*O*-*p*-toluenesulfonate-D-glucal would provide us access to the 6-deoxyglycal. Previously, Fraser-Reid and colleagues [18] reported that treatment of an equimolar mixture of 6-*O*-tosyl-D-glucal **5** and lithium aluminum hydride in THF at reflux (8 h) gave a 1:1 mixture of D-rhamnal **6** to 3,6-anhydroglycal **7** (Scheme 12.3). These two products were readily isolable by column chromatography (SiO_2). We reasoned that using these conditions would allow us to obtain modest amounts of D-rhamnal **6** that could be used in subsequent steps in our reaction sequence.

6-*O*-Tosyl-D-glucal **5** was prepared from D-glucal by reaction with 1.5 equivalents tosyl chloride in a mixture of pyridine in dichloromethane (56%). Reduction of the tosyl glucal **5** was carried out using the previously reported conditions [18]. Our initial attempts at forming the 6-deoxyglycal **6** under these conditions failed to give substantial amounts of product. However, an increase in the concentration of lithium aluminum hydride by approximately sixfold while refluxing in THF gave interesting results. None of the desired D-rhamnal **6** was obtained; only the 3,6-anhydrosugar **7** was isolated from the reaction mixture in an 82% yield [25].

Previously, only one-dimensional nuclear magnetic resonance (NMR) spectroscopic data had been reported for the 3,6-anhydroglucal **7** in the literature [17]. We carried out additional two-dimensional NMR experiments (^{1}H-^{1}H COSY, NOESY, and HMQC) to fully characterize the compound. The proposed anhydro structure was supported by the observation of long-range coupling between the vinylic proton (C2) and the bridgehead proton (C4; 1.5 Hz). We were also able to obtain crystals for 3,6-anhydroglucal **7** and an x-ray structure was obtained [26].

The crystal structure showed interesting H-bonding networks both within and between unit cells. Each bridgehead hydroxyl group formed a hydrogen bond exclusively with the ring oxygen contained within the six-membered ring of the

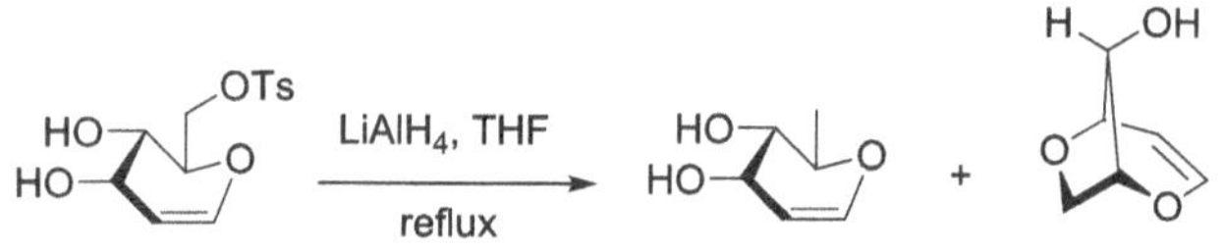

5 **6** **7**

SCHEME 12.3 Lithium aluminum hydride reduction of 6-*O*-tosyl-D-glucal 5.

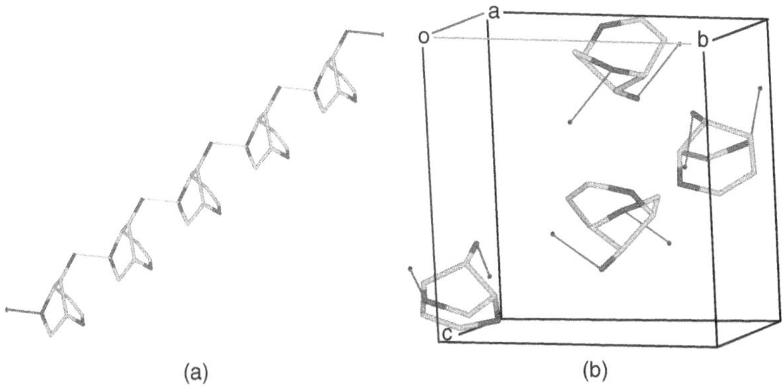

FIGURE 12.1 X-ray diffraction data for compound X. (a) Showing H-bonding between molecules in the lattice. (b) Unit cell showing H-bonding with other cells.

adjacent bicycle (Figure 12.1). No H-bonding with the five-membered ring portion was observed [27].

12.3 ADDITION REACTIONS OF 3,6-ANHYDRO-D-GLUCAL

Although the proposed structure of 3,6-anhydro-D-glucal **7** had been previously reported, very little had been reported about further transformations of this glycal [28]. So we investigated a series of glycal addition reactions for this compound [25]. The iodoglycosylation reaction [29, 30] was explored as a polar addition reaction for the glycal double bond. A free-radical promoted addition of thiophenol using ceric ammonium nitrate (CAN) was also investigated [31, 32]. We were interested in exploring the effects of the anhydrosugar structure on the stereoselectivity of these reactions.

Iodoglycosylation reactions have been used successfully for the preparation of 2-deoxy-2-iodoglycosides from glycals. The 2-iodo group can then be cleaved under reductive conditions using reagents such as tin hydride/azobisisobutyronitrile to give the 2-deoxyglycoside [33]. For tri-*O*-acetyl-D-glucal, the major products favored are the *trans* products from the diaxial addition of first iodine and then the alcohol to the vinyl ether double bond (α-manno). This selectivity is well explained because of the large steric bulk of the iodine atom and because of the large polarizability of the halogen, which enables it to accommodate a bridging, participatory structure between C1 and C2 of the glycal. This blocks *syn* attack by the incoming alcohol nucleophile [29].

For our bicyclic anhydro structure **7**, treatment with methanol and with 1.5 equivalents of *N*-iodosuccinimide (NIS) in acetonitrile gave an inseparable mixture of diastereomeric α-manno, β-manno, and β-gluco-2-deoxy-2-iodomethylglycosides **8a–c** (1:1.5:5.2 by proton NMR integration) in a 93% yield (Scheme 12.4). Contrary

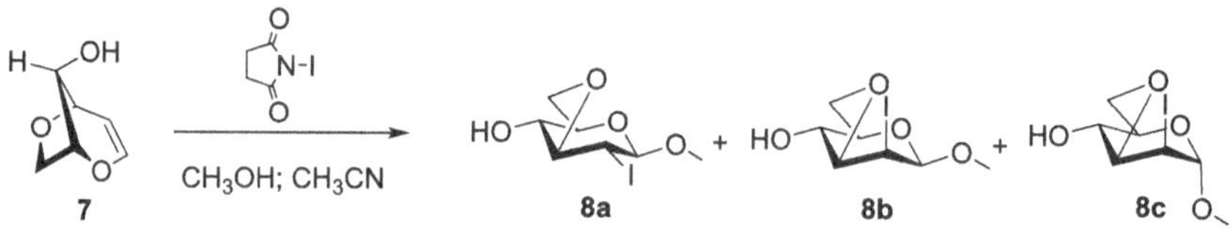

SCHEME 12.4 Iodomethoxylation of anhydroglucal **7**.

to what is observed for tri-*O*-acetyl-D-glucal [29], the diequatorial *trans* product predominated (β-gluco, **8c**) from the reaction mixture. Presumably, axial approach by iodine and methanol are blocked due to the bridging anhydro structure.

A larger alcohol nucleophile was then used in this reaction: 1,2:3,4-Di-isopropylidene-D-galactopyranose was used instead of methanol. With this bulkier nucleophile, none of the expected disaccharide was observed. Instead, the dianhydrosugar **9** was isolated as the major product (48%). Presumably, this product was formed by an intramolecular reaction of the bridgehead alcohol with sugar iodonium ion intermediate. None of the external nucleophile was able to react because of the sterically crowded environment.

In fact, when the anhydroglucal **7** was placed in anhydrous acetonitrile with NIS and only activated 4 Å molecular sieves, an improved yield of 76% of dianhydrosugar **9** was obtained. Previously, Brimacombe and colleagues observed the formation of similar dianhydrosugar when the 3,6-anhydro-D-glucal **7** was left in anhydrous deuterochloroform [28].

The CAN-catalyzed glycosylation of glycals is believed to occur through a single electron transfer mechanism to afford both glycosides and Ferrier glycosidation products with a high degree of alpha selectivity [31]. Treatment of anhydroglucal **7** with catalytic CAN in acetonitrile and an excess of thiophenol afforded a mixture of thiophenylglycosides **10a,b** (α:β 1.85:1) in a 65% yield (Figure 12.2). In addition, Ferrier-rearranged products **11a,b** (2.5:1 α:β; 30%) resulting from breakdown of the anhydrosugar were also isolated.

Sodium iodide is believed to further enhance the alpha selective addition of nucleophiles under these conditions [32]. Addition of sodium iodide to the CAN/acetonitrile/thiophenol reaction mixture with the anhydroglycal **7** gave a diminished amount of glycoside **10a,b** (35%, α:β, 2:1) and increased amounts of Ferrier product (65%). Only the α-Ferrier product **11a** was isolated.

12.4 PREPARATION OF 6-*O*-TOSYL-D-GALACTAL AND REDUCTION WITH LITHIUM ALUMINUM HYDRIDE

The selective 6-*O*-tosylation of D-galactal was also conducted using tosyl chloride and pyridine. In this case, better yields of the desired 6-*O*-tosyl sugar were obtained when only 1.1 equivalents of tosyl chloride were used in the reaction mixture (72%). With excess tosyl chloride, 4,6-di-*O*-tosyl-D-galactal was formed as a by-product. Reduction with excess lithium aluminum hydride under refluxing conditions was

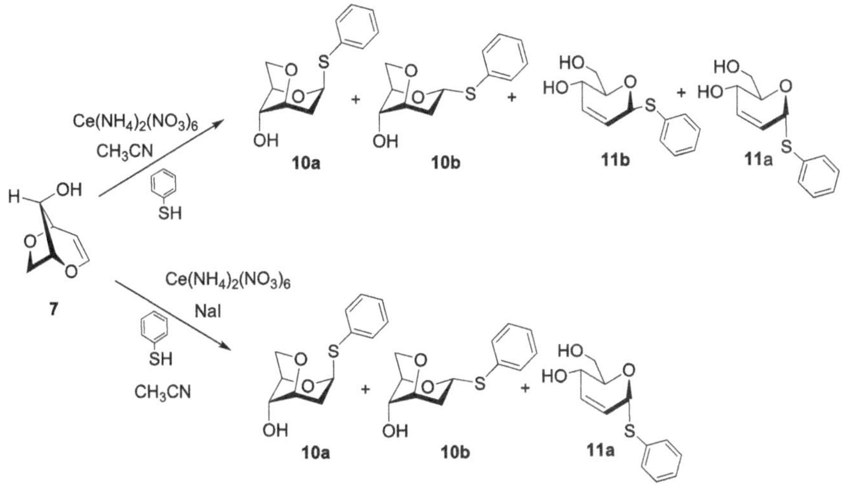

FIGURE 12.2 Reaction of 3,6-anhydro-D-glucal 7 with CAN and thiophenol with and without added sodium iodide.

also conducted. A variety of reaction conditions were tried in hopes of optimizing the reaction, including those previously described for the D-glucal 6-*O*-tosylate. Complex mixtures of products were formed with higher concentrations of lithium aluminum hydride and with longer reaction times at reflux. When the reduction was carried out with 3 equivalents of lithium aluminum hydride under reflux for 3 h, a 2:1 mixture of 6-deoxy-D-galactal (D-fucal) **13** and the 3,6-anhydro-D-galactal **14** was obtained in a 72% yield (Scheme 12.5). In this case, the anhydrosugar **14** was isolated as the minor product. With longer reflux times, more products were observed including higher-molecular-weight products presumably from dimerization of the glycal.

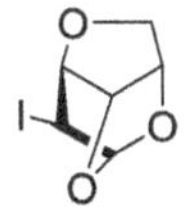

9

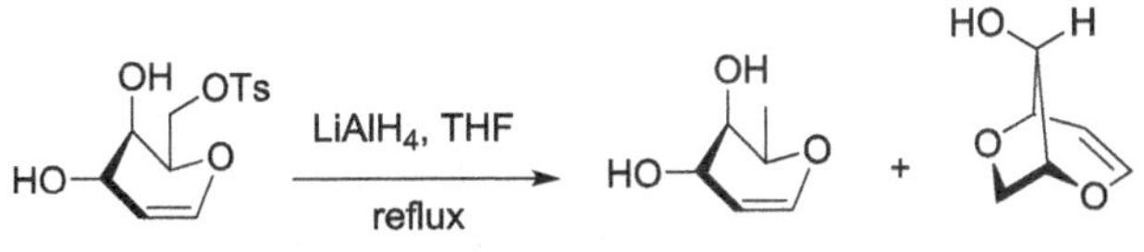

12 **13** **14**

SCHEME 12.5 Lithium aluminum hydride reduction of 6-*O*-tosyl-D-galactal 12.

12.5 CONCLUSIONS

The 6-*O*-tosylate derivatives of D-glucal and D-galactal have been prepared and subjected to lithium aluminum hydride reduction in THF. The conditions required for the reduction to go to completion and the distribution of products obtained for the two different glycals vary significantly. With the glucal tosylate, more forcing conditions (reflux, longer reaction time, more reducing agent) were required to cause conversion of starting material to product. The major product from this reaction was the anhydroglycal **7**. None of the 6-deoxyglycal was observed in this reaction sequence.

On the contrary, with 6-*O*-tosyl-D-galactal, the $LiAlH_4$ was more facile. After 3 h at reflux, none of the tosyl glycal **12** was detected by thin layer chromatography (SiO_2; 10% methanol in chloroform, added triethylamine). Of the two major products isolated, the 6-deoxyglycal **13** was the major product (48%). Anhydrogalactal **14** was isolated in only 24% yield.

Spectroscopic and crystallographic data were obtained for anhydroglucal **7.** The x-ray structure showed a highly organized arrangement of sugars with an extensive H-bonding network occurring exclusively between the bridgehead hydroxyl and the ring oxygen on the adjacent sugar through the six-membered ring only (Figure 12.1).

Glycal addition reactions of 3,6-anhydro-D-glucal were conducted. Iodomethoxylation with NIS afforded three major diastereomeric methyl glycosides. The major addition product was the β-gluco diastereomer. Less of the expected diaxial isomer was observed, presumably because of steric constraints in the axial plane. An unusual dianhydrosugar was formed when a primary sugar nucleophile was employed or when the reaction was carried out in dry solvent with the exclusion of a nucleophile. This product can be explained by an intramolecular addition of the bridgehead hydroxyl group to the anhydroglycal double bond.

The CAN-catalyzed addition of thiophenol to anhydroglucal **7**, in both the absence and presence of sodium iodide, was studied. In both cases, the α-glycosides and Ferrier-rearranged α-glycosides were the major products. The Ferrier products were the major products when sodium iodide was added.

Interesting 3,6-anhydroglycals have been prepared for glucose and galactose via the 6-*O*-tosyl glycals and their lithium aluminum hydride reductions. We are currently investigating the corresponding reactions of the C-3 epimeric sugar, D-allal for comparison. In the future, we also plan to look at these reductions in L-sugar glycals. A small sampling of glycal addition reactions has been carried out on 3,6-anhydro-D-glucal. Other polar and ionic addition reactions will be studied in due course.

REFERENCES

[1] Cerny, M. *Adv. Carbohydr. Chem. Biochem.* **2003**, *58*, 121–198.
[2] Ball, D. H.; Parish, F. W. *Adv. Carbohydr. Chem. Biochem.* **1969**, *24*, 139–147.
[3] Williams, N. R. *Adv. Carbohydr. Chem. Biochem.* **1970**, *25*, 109–179.
[4] Cerny, M.; Stanek J., Jr. *Adv. Carbohydr. Chem. Biochem.* **1977**, *34*, 23–177.

[5] Ferrier, R. J.; Middleton, S. *Chem. Rev.* **1993**, *93*, 2779–2831.

[6] Haworth, W. N.; Owen, L. N.; Smith, F. J. *J. Chem. Soc.* **1941**, 88–102.

[7] Reist, E. J.; Spencer, R. R.; Baker, B. R. *J. Org. Chem.* **1958**, *23*, 1757–1760.

[8] Serrano, J. A.; Jimenez, M.; Roman, E. *J. Carbohydr. Chem.* **1997**, *16*(7), 1051–1059.

[9] Wipf, P.; Pierce, J. B.; Zhuang, N. *Org. Lett.* **2005**, *7*(3), 483–485.

[10] Singh, I.; Seitz, O. *Org. Lett.* **2006**, *8*(19), 4319–4322.

[11] Chakaraboty, T. K.; Ghosh, S.; Ramana Rao, M. H. V.; Kunwar, A. C.; Cho, H.; Ghosh, A. K. *Tetrahedron Lett.* **2000**, *41*(51), 10121–10125.

[12] Gallant, M.; Link, J. T.; Danishefsky, S. J. *J. Org. Chem.* **1993**, *58*(2), 343–349.

[13] For example: Ermak, I. M.; Reunov, A. V.; Lapshina, L. A.; Byankina, A. O.; Bratskya, Y. S.; Sokolova, E. V. *Chem. Nat. Compounds* **2013**, *49*, 593–595.

[14] Sousa, A. M.; Borges, J.; Silva, A. F.; Goncalves, M. P. *Carbohydr. Polym.* **2013**, *96*(1), 163–171.

[15] Misra, P. A.; Mathad, T. V.; Raj, K.; Baduri, P.A.; Tiwari, R.; Srivastova, A.; Mehrotra, P. K. *Bioorg. Med. Chem.* **2001**, *9*, 2763–2772.

[16] Lewis, B.; Smith, F.; Stephen, A. M. 2,5- and 3,6-Anhydrosugars and their derivatives. In *Methods in Carbohydrate Chemistry*, Vol. II; Academic Press: New York, **1963**; p 174.

[17] Brimacombe, J. S.; Da'Aboul, I.; Tucker, L. C. N. *Carbohydr. Res.* **1971**, *19*, 276–280.

[18] Fraser-Reid, B.; Kelly, D. R; Tulshian, D. B.; Ravi, P. S. *J. Carbohydr. Chem.* **1983**, *2*, 105–114.

[19] Crich, D.; Ritchie, T. J. *Carbohydr. Res.* **1990**, *197*, 324–326.

[20] Tipson, R. *Adv. Carbohydr. Chem.* **1953**, *8*, 107–215.

[21] Ball, D. H.; Parrish, F. W. *Adv. Carbohydr. Chem.* **1968**, *23*, 233–280.

[22] Wolfrom, M. L.; Szarek, W. A. In *The Carbohydrates*, 2nd edition; Pigman, W., Horton, D., Eds.; Academic Press: New York, **1972**; p 232.

[23] Baer, H. H.; Hanna, R. H. *Carbohydr. Res.* **1982**, *110*, 19–41.

[24] Baer, H. H.; Mekarska-Falicki, M. *Can. J. Chem.* **1985**, *63*, 3043–3053.

[25] Basava, V.; Gorun, S. M.; Marzabadi, C. H. *Carbohydr. Res.* **2014**, *391*, 106–111.

[26] The crystal structure coordinates for compound **7** were deposited at the Cambridge Crystallographic Database (CCDC988697) and can be accessed at www.ccdc.cam.ac.uk.

[27] The authors thank Prof. Sergiu Gorun (Seton Hall University) for obtaining the crystal structure of compound **7**.

[28] Brimacombe, J. S.; Da'Aboul, I.; Tucker, L. C. N.; Calvert, N.; Ferrier, R. J. *Carbohydr. Res.* **1973**, 254–256.

[29] Lemieux, R. U.; Fraser-Reid, B. *Can. J. Chem.* **1965**, *43*, 1460–1476.

[30] Lemieux, R. U.; Hendriks, K. B.; Stick, R. V.; James, K. *J. Am. Chem. Soc.* **1975**, *97*, 4056–4062.

[31] Paul, S.; Jayaraman, N. *Carbohydr. Res.* **2004**, *339*, 2197–2204.

[32] Roush, W. R.; Narayan, S.; Bennett, C. E.; Briner, K. *Org. Lett.* **1999**, *1*(6), 895–897.

[33] Horton, D.; Rodemayer, G.; Saiki, H. *Carbohydr. Res.* **1977**, *59*, 607–611.

13

RING EXPANSION METHODOLOGIES OF PYRANOSIDES TO SEPTANOSIDES AND STRUCTURES OF SEPTANOSIDES

Supriya Dey, N. Vijaya Ganesh, and N. Jayaraman

Department of Organic Chemistry, Indian Institute of Science, Bangalore, Karnataka, India

13.1 INTRODUCTION

Ring expansions of a cycloalkyl moiety to the higher homologue provide an attractive and easy access to medium ring cyclic structures. Homologations of one or more carbon atoms are exercised whether targeting synthesis of monocyclic, bicyclic, and tricyclic systems of importance in natural and unnatural products. Ring expansion has become an important synthetic methodology in C—C and C-heteroatom bond-forming reactions in general. A number of review articles appear periodically that update progress in ring expansion-based synthetic methodologies, concerning both C—C and C-heteroatom bond formations [1–3]. In focusing this chapter, methodology development in ring expansions through one-carbon homologation is chosen, with a particular emphasis to discuss such methodologies relevant to sugar chemistry. Ring expansion on cyclic sugars to the corresponding higher homologues is an upcoming area of interest, although methodologies for the preparation of medium ring sugars are rather limited currently.

Ring expansions of cyclic structures to the corresponding higher homologues are performed generally by (i) σ-bond rearrangement as a result of the presence of a carbonium ion center exocyclic to the ring structure (Figure 13.1a) and (ii)

Domino and Intramolecular Rearrangement Reactions as Advanced Synthetic Methods in Glycoscience, First Edition. Edited by Zbigniew J. Witczak and Roman Bielski.

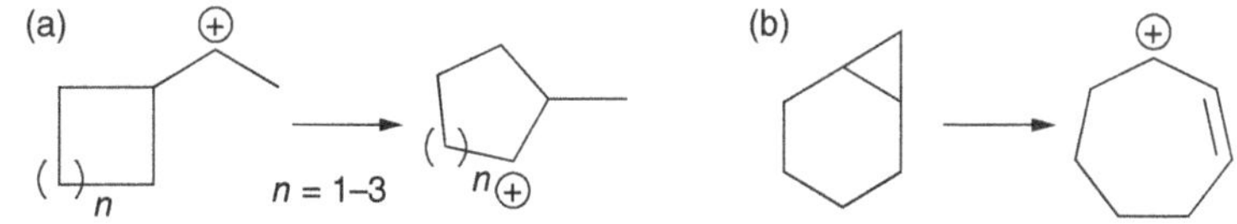

FIGURE 13.1 Ring expansions through (a) σ-bond rearrangement to an exocyclic carbonium ion center and (b) endocyclic ring opening.

bond reorganization of endocyclic σ-bond of a bicyclic system (Figure 13.1b). σ-Bond rearrangement due to the presence of proximal carbonium ion center might be attributed to the tendency of less-substituted carbonium ion migration to become more substituted carbonium ion center, namely, the 1,2-shift. The latter case of a bond reorganization results predominantly due to relief of ring strain or angular strain. Relief of an angular strain could amount to ~28 kcal mol^{-1}, in the case of the strain energy associated with cyclopropanes [4].

Among one carbon homologations to cyclic structures, involvement of the cyclopropyl moiety is the most common, as a result of facile formation of cyclopropyl rings, coupled with their reactivities associated with ring strain and bond reorganizations. The most practiced reactions involving cyclopropanes are (i) rearrangements as in vinyl cyclopropanes, (ii) protic and Lewis acid-induced ring opening, (iii) transition metal-catalyzed ring opening, and (iv) thermal rearrangements. Whereas nonfunctionalized cyclopropanes undergo ring-opening reactions, further fine-tuning the reactivity, stereo- and regiochemical outcomes involve installation of the cyclopropyl ring with desired functionalities. Such attributes become even more pertinent in bicyclic systems, where the presence of the cyclopropyl moiety and subsequent ring opening leads to one carbon insertion and formation of one carbon higher homologue of the cyclic structure. The reactivities of the cyclopropyl moiety are analogous to those of the double bonds, for example, activations, rearrangements, and electrophilic and nucleophilic reactivities.

Ring expansions utilizing cyclopropyl moiety are resourceful in the case of structural modifications of carbohydrates. Existing predominantly as glycosides in natural systems, the ring structures provide immense opportunities to introduce a multitude of modifications on the basic building blocks, namely, the monosaccharides. Available in varying constitutions and configurations from natural sources, five-membered furanoses and six-membered pyranoses act as versatile synthons for modifications of monosaccharides. One carbon homologations are one of the early examples of synthetic manipulations on furanoses and pyranoses, for example, the cases of cyanohydrin formation by Kiliani method, Grignard, Wittig, and aldol reactions of aldoses, all allowing to ascend by at least one carbon of an aldose. Although homologations are performed through aforementioned methods, subsequent cyclization of the newly formed aldose to the corresponding hemiacetals is not immediately obvious. Thus, an aldohexose to aldoheptose might still retain conformationally a more stable pyranosidic cyclic form of aldoheptose than a ring-expanded seven-membered cyclic form. Conversely, in a first report, Stevens and coworkers detected a small percentage

(<2%) of seven-membered cyclic sugar formation, which was in equilibrium with pyranose and acyclic form in an aldohexose (*vide infra*). Interest in seven-membered cyclic sugars, coined as septanoses and septanosides, gained importance since then. Development and application of newer synthetic methodologies to such unnatural sugars were reported periodically, as a result of such structures representing a general class of oxepanes additionally [5].

Larger cyclic rings are more flexible; thus, a study of the conformations of the seven-membered cyclic sugars creates an interest, through solid-state, solution-phase, and computational techniques. With substituents at all or most carbons, the interest is to identify the effect of such substituents on the conformational features adopted by cyclic structures. The number of possible low-energy conformers is large in the solution phase as compared with that observed in the solid state, due to very low free energy differences (<1 kcal mol^{-1}) among closely related conformers. Computational techniques, implemented with appropriate force fields and molecular dynamics simulation protocols, help derive the low-energy conformers, in conjunction with results from solid-state and solution-phase structural determinations [6].

The purpose of this chapter is to focus on ring expansion methodologies to the preparation of these structures, in addition to analyzing their conformational features as revealed by recent reports. It is also important from an overall point of view of progress in this field that septanose formation through cyclization of linear precursors is also presented before discussion on ring expansion methodologies and subsequent structural analysis.

13.2 SYNTHESIS OF SEPTANOSIDES

The majority of known methods of septanosides belong to (i) hemiacetal and acetal formation of a precursor containing aldehyde and appropriately positioned hydroxyl group; (ii) epoxidations; (iii) Knoevenagel condensation of sugar aldehyde with active methylene synthons; (iv) Baeyer–Villiger oxidation of inositol derivatives; (v) electrophile-induced cyclizations; (vi) acid-catalyzed cyclizations; (vii) metal-catalyzed cyclizations; (viii) Nicolas–Ferrier rearrangement of substituted glycals; and (ix) ring opening of 1,2-cyclopropanated sugars. Many of these methods provide either partially functionalized oxepane derivatives or fully functionalized septanoses.

13.2.1 Synthesis of Septanosides via Hemiacetal Formation

The first report of the formation of a septanose could be traced back to the work of Micheel and Suckfüll in 1933 [7, 8]. Pyranose derivative **1**, upon treatment with acetic anhydride (Ac_2O), in the presence of pyridine, led to the formation of α/β-mixture of 6-deoxy *per-O*-acetyl-D-galactoseptanose **2** (Scheme 13.1a). Viability of the cyclization of acyclic precursors was reiterated further by Anet, and Stevens and coworkers; for example, treatment of D-glucose with acetone in the presence of mineral acid afforded septanoside **3**, although in yields less than 5% (Scheme 13.1b) [9, 10].

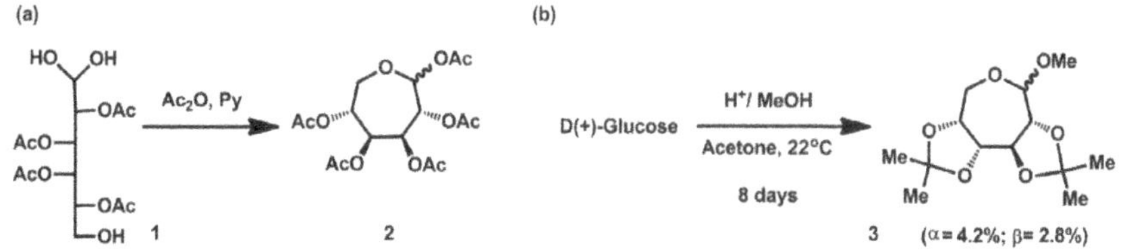

SCHEME 13.1 Formation of (a) septanose [7, 8] and (b) septanoside by cyclization of acyclic precursors [9, 10].

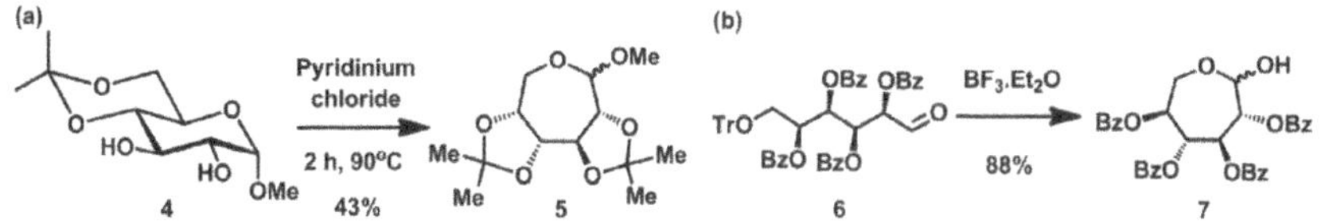

SCHEME 13.2 Formation of (a) septanoside from an isopropylidene protected pyranoside **4** [11] and (b) septanose from trityl-functionalized aldohexose **6** [12].

The yields of the septanoside increased moderately when *in situ*-generated acyclic form underwent cyclization to the required derivative, as demonstrated by Gelas and coworkers [11]. Acid-sensitive isopropylidene and acetal functionalities in pyranoside derivative **4** facilitated pyranoside ring opening in the presence of pyridinium hydrochloride at 90°C and afforded the septanoside derivative **5**, in a 43% yield (Scheme 13.2a). On the contrary, yield of septanose formation increased even further upon activation of the primary hydroxyl group as in aldohexose **6** with trityl-functionality, followed by treatment with Lewis acid to afford septanose **7** (Scheme 13.2b) [12].

Such cyclization methods found general utility as in the case of the preparation of nucleoside analogue **10** from an aldoheptose **8** and septanose **9**, reported by Gosselin and coworkers (Scheme 13.3a) [13]. In this instance, 5-deoxy septanoside was obtained. On the contrary, a 2-deoxy septanoside derivative was obtained by oxidation of alcohol **11**, followed by oxidation to aldehyde and acid-catalyzed cyclization to afford **12** (Scheme 13.3b) [14].

An interesting observation of septanoside formation was reported by Murphy and coworkers, when studying epoxidations of 6-deoxy hex-5-enopyranosides [15]. Thus, epoxidation of **13** for a prolonged period in the presence of oxone and 1,1,1,-trifluoroacetone, followed by hydrolysis, afforded hexos-5-ulose **14**, which, upon

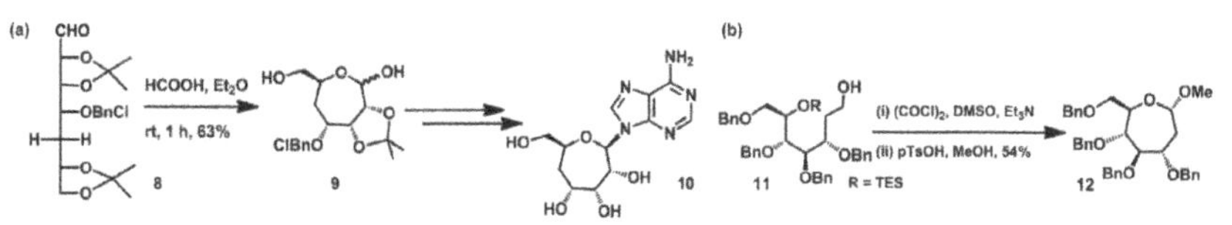

SCHEME 13.3 Cyclization of linear precursors to (a) septanose **9** [13] and (b) septanoside **12** [14].

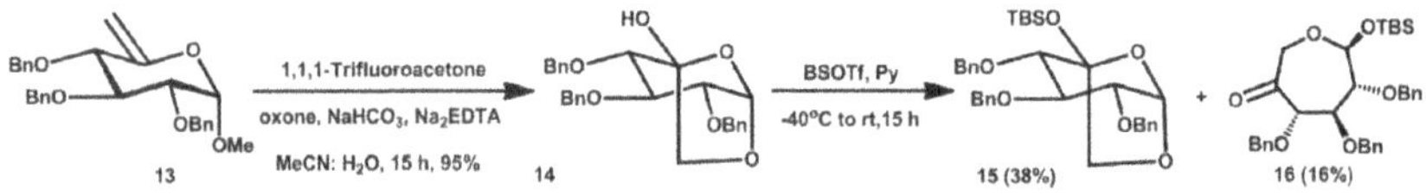

SCHEME 13.4 Reactions of 6-deoxy-5-enopyranoside to prepare septanoside **16** [15].

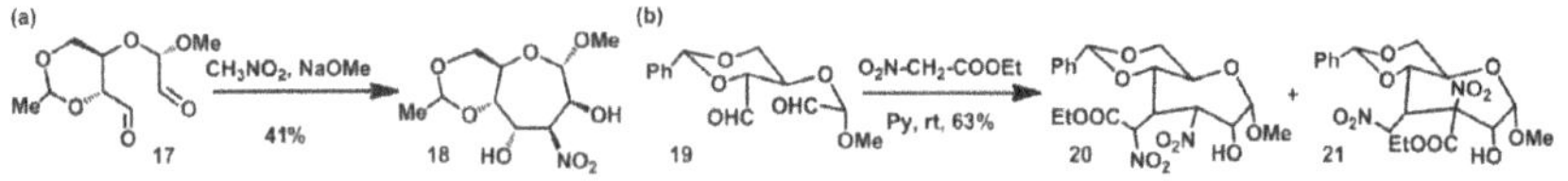

SCHEME 13.5 Knoevenagel condensation reactions to prepare septanosides [16, 17].

treatment with *tert*-butyldimethylsilyl trifluoromethanesulfonate/pyridine, afforded septanoside derivative **16**, as one of the products along with **15** (Scheme 13.4) [15].

13.2.2 Knoevenagel Condensation

Whereas cyclization of aldohexose and aldoheptose proceeded septanoside formation, implementing aldol-type condensation was also devised to derive septanoside systems. An early report by Baschang illustrated that a dialdehyde, such as **17**, obtained by periodate oxidation of a pyranoside, upon Knoevenagel condensation afforded 3-deoxy-3-nitroseptanoside **18** (Scheme 13.5a) [16]. On the contrary, such a reaction on dialdehyde **19** with ethyl nitroacetate afforded *bis*-nitro-septanoside derivatives **20** and **21** (Scheme 13.5b) [17].

13.2.3 Baeyer–Villiger Oxidation of Cyclohexanone Derivatives

The Baeyer–Villiger oxidation of a cyclohexanone to the septanoside derivative may be considered as a first instance of ring expansion of a six-membered ring system to the seven-membered ring system. Scheme 13.6 illustrates such a reaction to prepare septanosides. The oxidation of **22** led to septanoside **23** as a result of regioselective insertion of oxygen. On the contrary, such a regioselectivity was less pronounced in the case of oxygen insertion with *bis*-isopropylidene derivative **24**, which afforded septanouloside derivatives **25** and **26** [18].

13.2.4 Electrophile-Induced Cyclization

Examples of electrophile-induced cyclization are known to prepare septanosides in recent years. Köver and coworkers reported the first example of preparation of

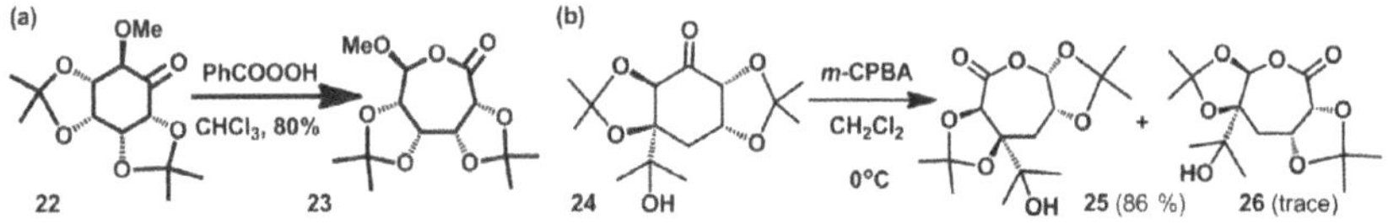

SCHEME 13.6 Baeyer-Villiger oxidation to prepare septano-uloside derivatives [18].

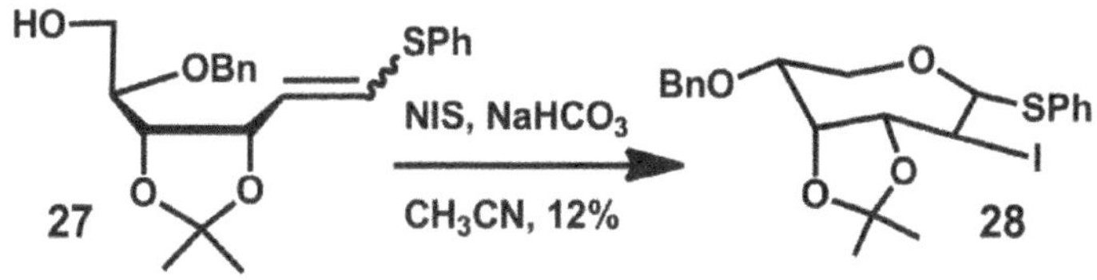

SCHEME 13.7 Vinyl sulfide synthon to prepare thioseptanoside [19].

septanosides via electrophile-induced cyclization [19]. Vinyl sulfide **27** upon activation underwent *in situ* cyclization to afford thioseptanoside derivative **28** (Scheme 13.7).

13.2.5 Metal-Catalyzed Cyclization

Metal-catalyzed reactions were employed for the synthesis of various septanoside derivatives, broadly categorized as (i) ring-closing metathesis and (ii) photo-cycloisomerization. van Boom and coworkers reported the novel method of ring-closing metathesis reaction of diene **29** in the presence of Grubbs' first-generation catalyst to furnish oxepine **30** (Scheme 13.8a) [20]. A similar strategy was applied to the synthesis of *tetra-O*-benzyl oxepine **32** (Scheme 13.8b) from a diene precursor **31**, using Schrock catalyst [21].

The ring-closing metathesis approach was used beneficially to prepare isomeric seven-membered oxepine derivatives by Schmidt and Biernat. Thus, with the use of Grubbs' first-generation catalyst, the metathesis of **33** afforded oxepine **34**, whereas a tandem ring-closing metathesis–isomerization of the same substrate afforded glycal **35** (Scheme 13.9) [22].

A photo-cycloisomerization reaction was developed as a novel method to prepare oxepine derivatives, wherein alkynol substrates were found to undergo *endo*-selective cyclization leading to septanose glycals [23]. Thus, photo-cycloisomerization of acetonide protected alkyne diol **36**, in the presence of $W(CO)_6$ and Et_3N, under irradiation of light (350 nm) at 65°C, afforded glycal **37** (Scheme 13.10a). Such a methodology was also investigated on furanose derivative **38** to secure oxepinone **39** (Scheme 13.10b) [24].

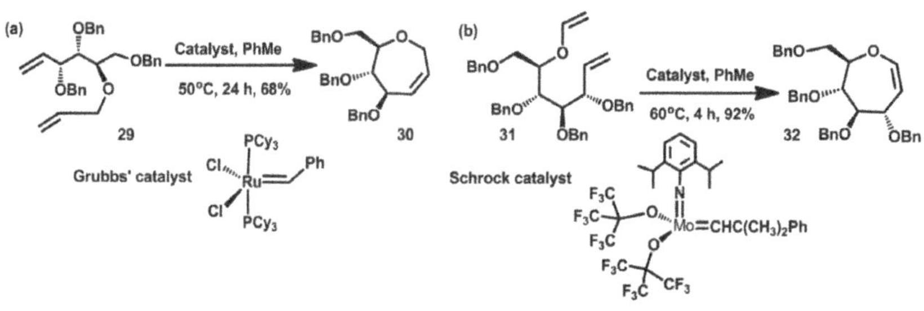

SCHEME 13.8 Metal-catalyzed ring-closing metathesis reactions to prepare oxepines **30** and **32** [20,21].

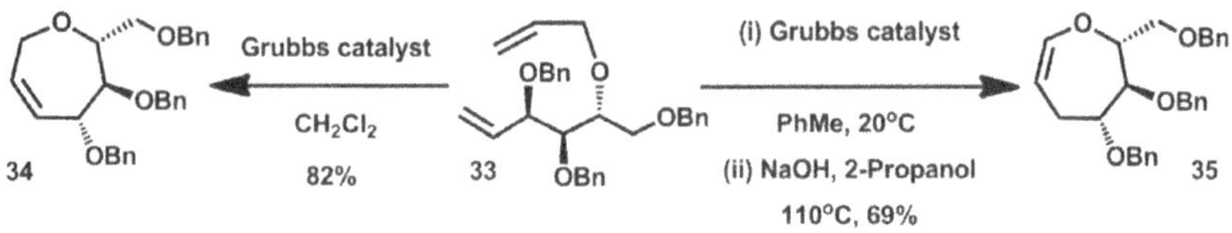

SCHEME 13.9 Ring-closing metathesis reactions leading to oxepine **34** and **35** [22].

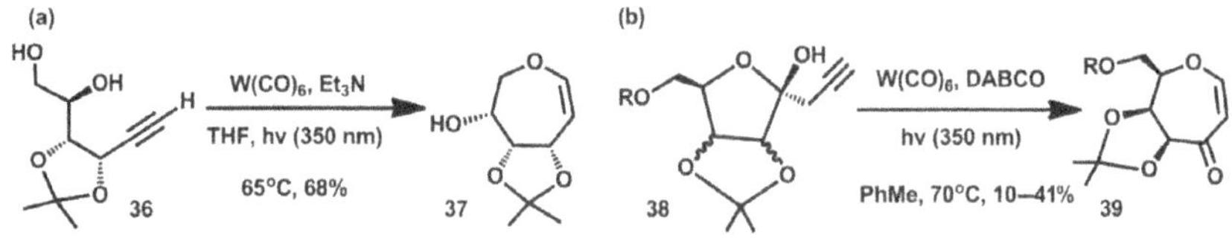

SCHEME 13.10 Photo-cycloisomerizations to prepare (a) septanose glycal **37** [23] and (b) oxepinone **39** [24].

13.2.6 Nicolas–Ferrier Rearrangements

López and coworkers expanded the possibility of synthesizing sugar oxepanes through devising a new methodology of Nicholas–Ferrier reaction [25]. This reaction pertains to the combination of Nicholas reaction, which relies on the addition of a nucleophile to propargyl cation dicobalt hexacarbonyl complex and oxocarbenium ion of a sugar moiety. The resulting "Nicholas pyranosidic oxocarbenium ion" derives additional stability through the dicobalt hexacarbonyl propargyl moiety (Scheme 13.11). Thus, reaction of $Co_2(CO)_8$ with glycal-derived ene-yne **40** led to an adduct **41**, which generated an active Nicolas–Ferrier cation **I** or **II**. The reactivity of this adduct in the presence of $BF_3{\cdot}OEt_2$ led to oxepanes **42**, in addition to C3-branched pyranosides. The formation of oxepane derivatives was due to (i) 1,6-hydride shift, (ii) electrophilic addition, and further (iii) pyranoside ring opening.

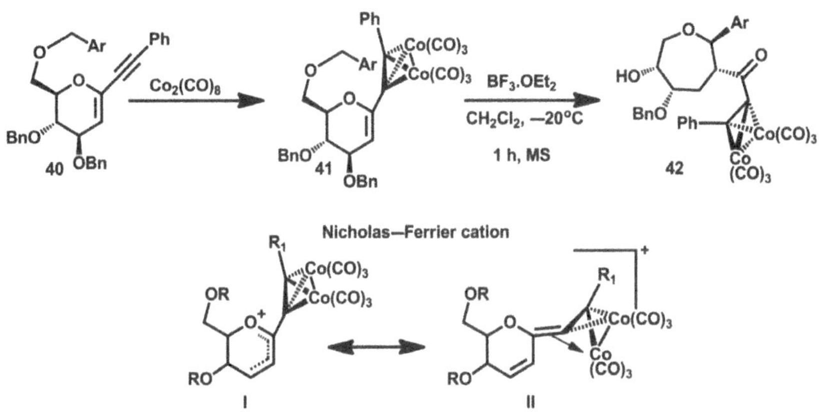

SCHEME 13.11 Nicolas-Ferrier rearrangement to prepare oxepane **42** [25].

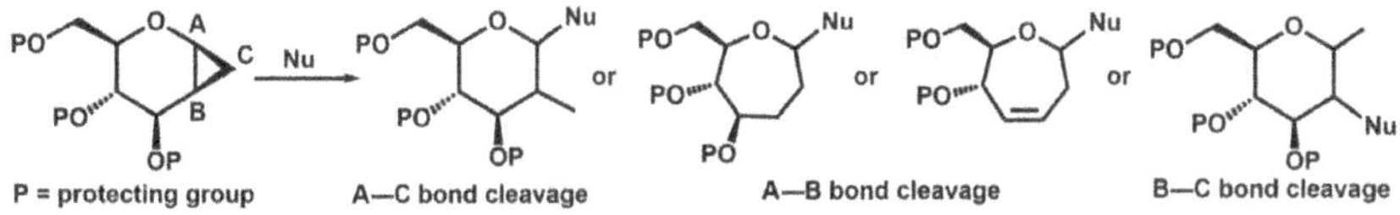

SCHEME 13.12 Pathways by which a bicyclic framework to undergo ring-opening reactions.

13.2.7 Ring Opening of Carbohydrate-Derived Cyclopropanes

Having delineated features of cyclization methodologies, the following discussion shall focus on ring expansion methodologies to prepare septanosides. Formation of a [4.1.0] system comprising a six-membered and three-membered ring-fused bicyclic framework provides an excellent opportunity for ring expansion of a six-membered system to the next higher homologue. Possible pathways by which a bicyclic framework may undergo ring-opening reactions are (i) A—C bond cleavage, leading to substitution at C-1; (ii) A—B bond cleavage, leading to substitution at C-1; and (iii) B—C bond cleavage, leading to substitution at C-3 (Scheme 13.12).

13.2.8 Ring Opening of Glycal-Derived 1,2-Cyclopropane

In a first instance, Hoberg and coworkers demonstrated the ring-opening reaction of Simmon–Smith cyclopropanes, under acidic conditions. The ring-opening reaction led to oxepine formation through a Ferrier rearrangement. The ring-opening reaction of D-galactal-derived cyclopropane **43** with an acetate group at C-3 was performed in the presence of trimethylsilyl trifluoromethanesulfonate (TMSOTf) (20 mol %) and a nucleophile in MeCN led to afford oxepane **44**, as an anomeric mixture of 2:1 (α:β) (Scheme 13.13) [26]. A change of protecting group led to a different product under similar reaction conditions. The treatment of the cyclopropane derivative **45** in the presence of TMSOTf (10 mol %) and trimethylsilyl cyanide yielded a bicyclic derivative **46** in a good yield (Scheme 13.13b).

A similar ring-opening reaction was reported by Sridhar and coworkers to form acyclic thio-derivative **48** from **47**, which was used beneficially to react with acceptor alcohols, in the presence of glycosylation promoters, to afford septanoside containing disaccharide such as **49** (Scheme 13.14) [27].

The ring-opening reaction of functionalized 1,2-cyclopropyl carbohydrates is also known, apart from that of unfunctionalized derivatives. Yu and Pagenkopf demonstrated the formation of oxepine **51** through Ferrier rearrangement-based ring opening

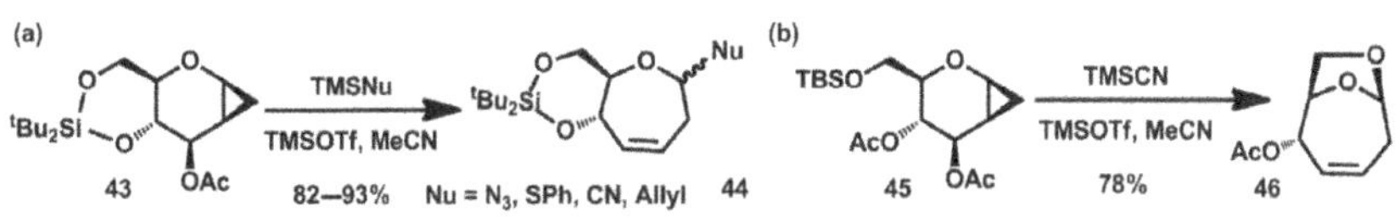

SCHEME 13.13 Ring-opening reaction of glycal derived 1,2-cyclopropane to prepare oxepine **44** and **46** [26].

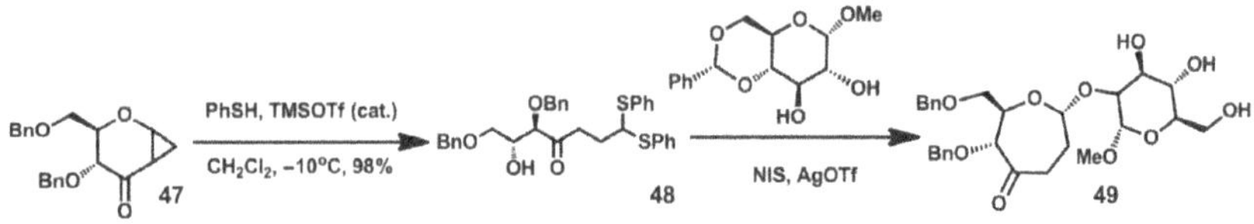

SCHEME 13.14 Cyclization of an acyclic thio-derivative **48** to septanoside containing disaccharide **49** [27].

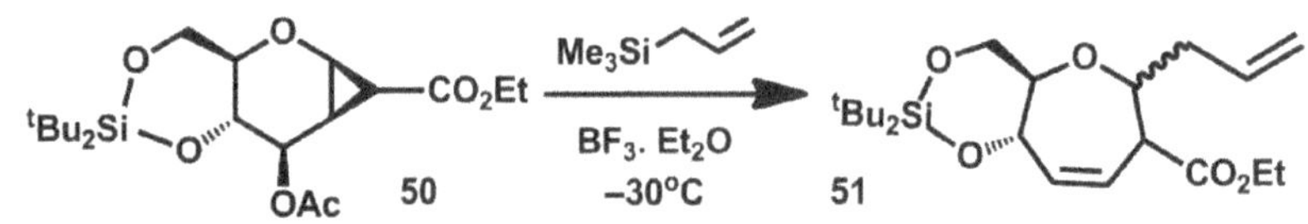

SCHEME 13.15 Ring-opening of an ester functionalized 1,2-cyclopropane derivative **50** to oxepine **51** [28].

of ester-containing 1,2-cyclopropane derivative **50** with allyl trimethylsilane, in the presence of $BF_3{\cdot}Et_2O$ (Scheme 13.15) [28].

An interesting example of ring expansion reaction of 1,2-cyclopropyl carbohydrates having *C*-3 *exo*-methylene through σ-Ferrier rearrangement was disclosed by Chandrasekaran and coworkers [29]. An elegant feature of this reaction is that it not only leads to ring-expanded oxepane, but also permits functionalization at *C*-4 of the newly formed oxepane. D-Glucal/galactal-derived vinylcyclopropane **53** was used for the ring expansion reaction using chloramine-T hydride **54** and phenyltrimethylammonium tribromide in MeOH. The ring expansion led to oxepane **55**, as an anomeric mixture (3:2 = α:β) (Scheme 13.16).

Among functionalization of the cyclopropyl moiety, presence of *gem*-dihalosubstituent was found to provide facile reactivities. The ring-opening reaction of *gem*-dihalogenated 1,2-cyclopropane derivatives under acidic and basic media was conducted. It was observed that under basic medium, the ring opening of glycal-derived *gem*-dibromo **56** cyclopropane derivative provided *C*-2 branched pyranosides **57**, as and 1:1 anomeric mixture, rather than anticipated bromo-oxepine, reflecting it to be a case of A—C bond cleavage (Scheme 13.17) [30a]. The reaction proceeded nucleophilic addition and 1,2-hydrogen shift, leading to the pyranoside **57**, with exocyclic bromomethylene substituent at *C*-2.

On the contrary, Harvey and coworkers also demonstrated that the reactivity of such *gem*-dihalocyclopropanated glycals could be switched to an A—B bond cleavage when the reaction was conducted under Lewis acid conditions [30b]. Thus, the ring opening of **58** using AgOAc or $AgNO_3$ furnished the halo-oxepine product. The intermediate

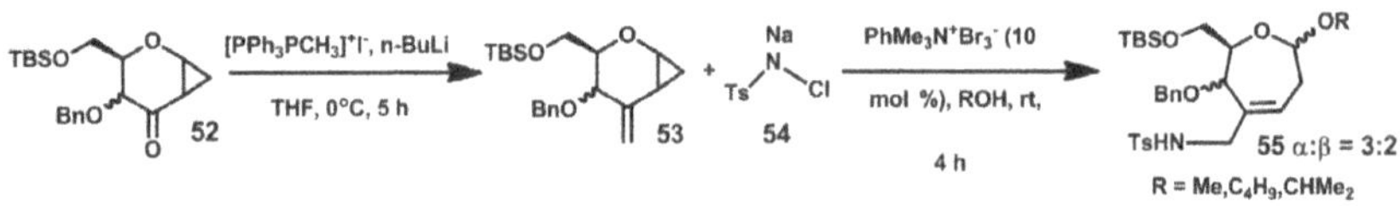

SCHEME 13.16 Formation of *C*-4 functionalized oxepine **55** [29].

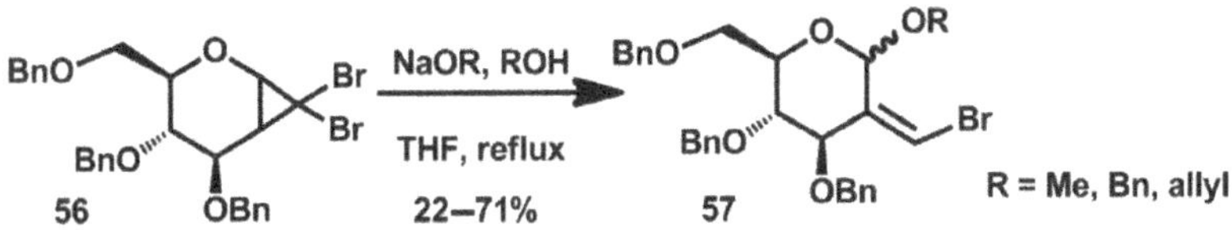

SCHEME 13.17 Ring-opening of glycal derived *gem*-dibromocyclopropane **56** under basic medium [30].

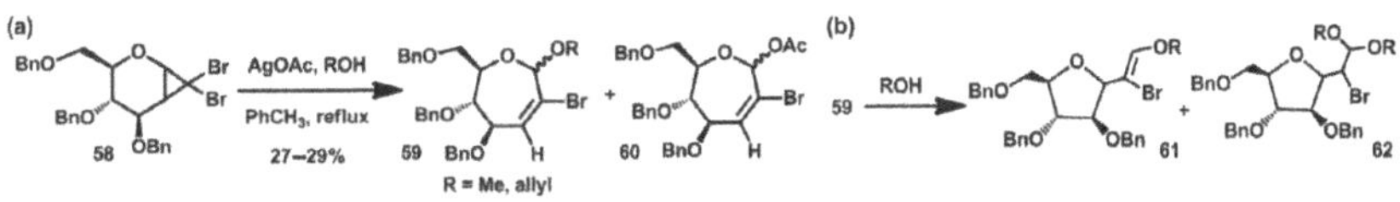

SCHEME 13.18 (a) Acid-catalyzed ring-opening of glycal derived *gem*-dibromocyclopropane derivative **58** and (b) rearrangement of halo-oxepines to *C*-furanosides **61** and **62** [30].

carbocation formation appeared to facilitate A—B bond cleavage in the presence of the halophilic reagent. It was observed that D-glucal derived *gem*-dihalo-1,2-cyclopropane furnished an anomeric mixtures of bromo-oxepine **59** (Scheme 13.18a) whereas D-galactal-derived cyclopropane led to only α-configured bromo-oxepine. Acetate also acted as a competitive nucleophile to afford anomeric acetates **60**. It was demonstrated further that the newly formed halo-oxepines were not stable under the reaction conditions and would tend to undergo further rearrangements, leading eventually to the formation of *C*-furanosides **61** and **62** (Scheme 13.18b) [30b]. These observations provided the first instance of realizing the reactivities of halo-oxepines, wherein a concomitant role of endocyclic oxygen lone pairs of electrons to initiate the rearrangement of the vinyl halide moiety was observed.

13.2.9 Ring Opening of Oxyglycal Derived 1,2-Cyclopropane

Introduction of an oxygen functionality at *C*-2 leads to the 1,2-unsaturated sugar being a 2-hydroxy glycal or oxyglycal. The ring opening of such oxyglycal-derived *gem*-dihalo-1,2-cyclopropanes is found very effective in ring expansion reactions [31a]. The ring expansion reaction could be performed in both basic and Lewis acidic media. The sequence of reactions initiating from oxyglycal **63** is shown in Scheme 13.19. D-Oxyglycal **63** was functionalized using *in situ*-generated *gem*-dibromo carbene.

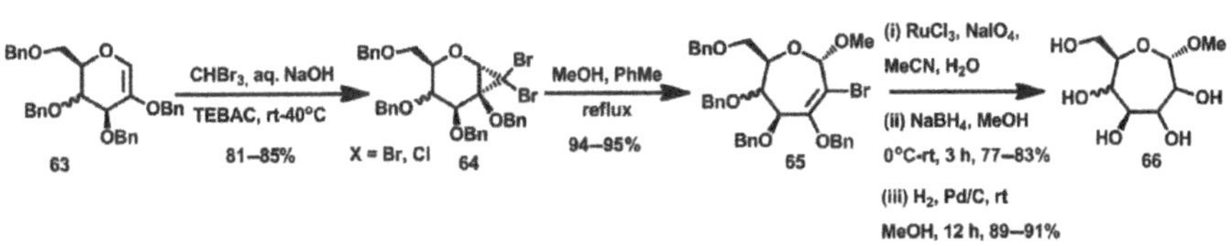

SCHEME 13.19 Ring-opening of oxyglycal derived *gem*-dibromocyclopropane **64** to septanoside **66** [31a].

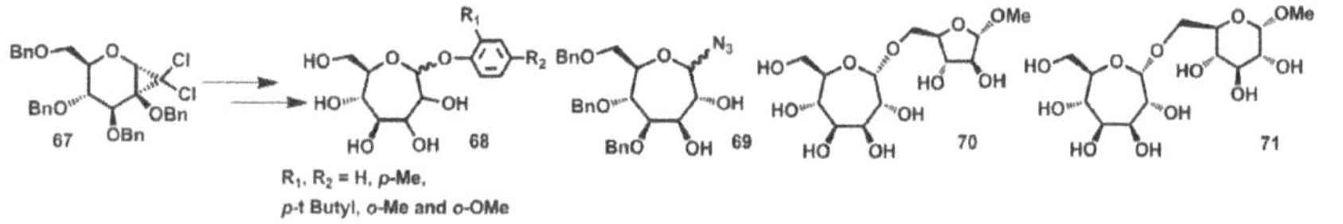

SCHEME 13.20 Ring-opening of oxyglycal derived *gem*-dichlorocyclopropane **67** to phenyl (**68**), azido- (**69**) septanosides, and septanoside-containing disaccharides **70** and **71** [31b].

The carbene insertion was found to be stereoselective and afforded cyclopropanated derivative **64**, which upon treatment with NaOMe afforded bromo-oxepine **65**, as an α-anomer exclusively. Oxidation of the vinyl halide **65** through *in situ*-generated RuO_4 furnished a diketone intermediate, which was reduced subsequently to a diol and deprotection of benzyl-protecting groups led to the isolation of septanoside **66**. The ring-opening reaction of the D-oxyglycal-derived *gem*-dichloro-1,2-cyclopropane **67** was also performed with phenol, azide, and sugar alcohols (Scheme 13.20) using above sequence of reactions, to afford the corresponding septanosides **68–71** [31b].

Appropriately protected furanoside and pyranoside were chosen *as* nucleophiles in the ring opening of **67** to generate septanoside-containing disaccharides **70** and **71**, respectively [31b].

The use of oxyglycal as a resourceful synthon to prepare trisaccharides was also demonstrated. Thus, lactose-derived cyclopropnated oxyglycal **72** was subjected to the sequence of reactions leading to formations oxepine **73**, diol **74**, and the trisaccharides **75** and **76** (Scheme 13.21) [31c].

Following the synthesis of septanoside containing di- and trisaccharides, disaccharide constituted with septanoside moiety at both reducing and nonreducing ends was also synthesized, through the above ring expansion methodology [32]. For synthesis of diseptanoside, the ring-opening reaction of oxyglycal-derived *gem*-dichloro-1,2-cyclopropane **67** was performed with 6-hydroxy oxyglycal **77**, in the presence of K_2CO_3 and 18-C-6 (cation) in refluxing PhMe, to afford 1,7-linked chloro-oxepine

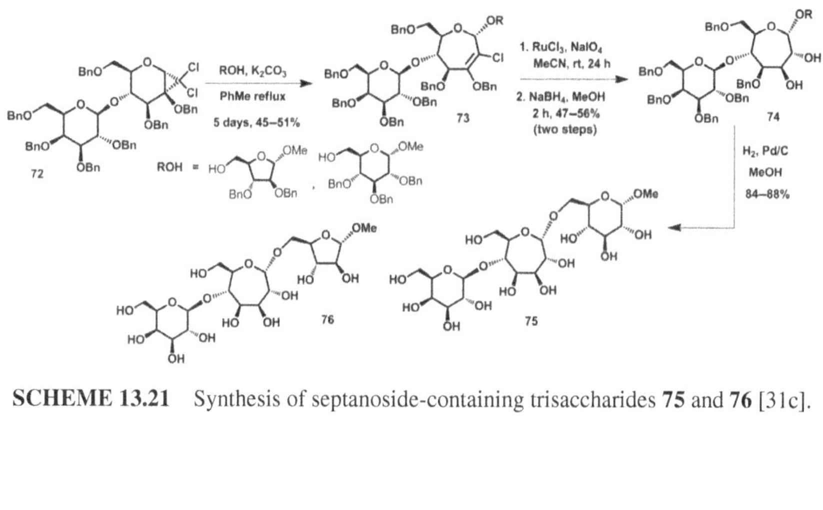

SCHEME 13.21 Synthesis of septanoside-containing trisaccharides **75** and **76** [31c].

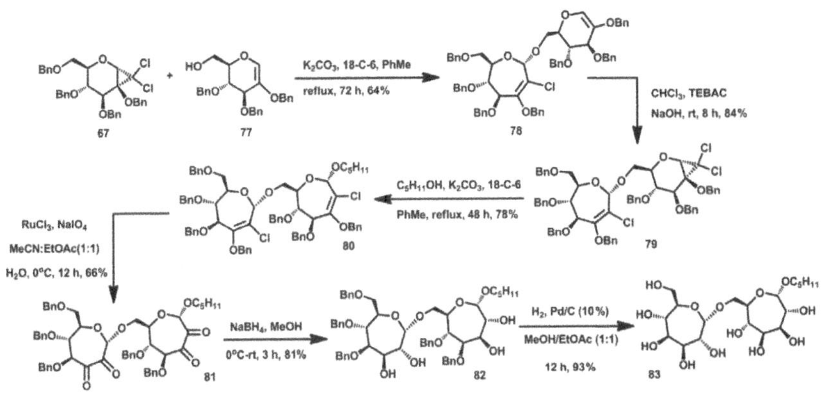

SCHEME 13.22 Synthesis of a diseptanoside **83** [32].

containing disaccharide **78** (Scheme 13.22). The ring-opening reaction was found to be stereoselective and only α-anomer was observed. Disaccharide **78** contains a chloro-oxepine moiety at the nonreducing end and an oxyglycal in the reducing end, connected through α-1,7-glycosidic bond. The reducing end double bond was functionalized through *in situ*-generated dichlorocarbene using $CHCl_3$, NaOH, and triethyl benzyl ammonium chloride (cation) to afford **79**, without affecting the chlorovinyl ether moiety present at the nonreducing end. Ring opening of cyclopropyl moiety using *n*-pentanol, in the presence of K_2CO_3 and refluxing PhMe, afforded 1,7-linked α-dichloro-oxepine **80**, as α-anomer exclusively. $RuCl_3$-$NaIO_4$-mediated oxidation of **80** to tetra-ketone **81** and followed by $NaBH_4$-reduction afforded **82**. The reduction was diastereoselective and only one isomer was obtained. The removal of *O*-benzyl protecting groups subsequently using Pd/C, H_2 furnished free hydroxyl group containing α-1,7-linked D-*glycero*-D-*galacto* septanoside **83**.

The above experiments establish that the reactivities of cyclopropanated oxyglycals differ from those of cyclopropanated glycals [30]. In the latter case, only an A—C cleavage (cf. Scheme 13.17) occurs eventually to afford the pyranoside with a newly generated exocyclic substituent, whereas in the former case, an A—B cleavage occurs to afford the septanoside, under base-promoted conditions during the ring-opening reaction [31, 32]. It was thus of further interest to study the pattern of ring-opening reactions of cyclopropanated oxyglycals under acidic conditions, as that exercised in the case of cyclopropanated glycals.

The ring-opening reactions of oxyglycal-derived *gem*-dihalo-1,2-cyclopropane in Lewis acid medium were studied [32], for example, the reaction of **67** with nucleophiles in the presence of Lewis acids, namely, BF_3-Et_2O, AgOAc, TMSOTf, and Tf_2O. Among these acids, silver salt was found to promote the reaction. Thus, the use of AgOAc (1 molar equivalent), in the presence of an alcohol, in refluxing PhMe for 48–72 h, led the reaction to afford halo-oxepines **84** and **85** (Scheme 13.23). A range of nucleophiles, including primary, secondary, and benzyl alcohols, were used for the ring-opening reaction. In these reactions, anomeric acetate was also obtained to an

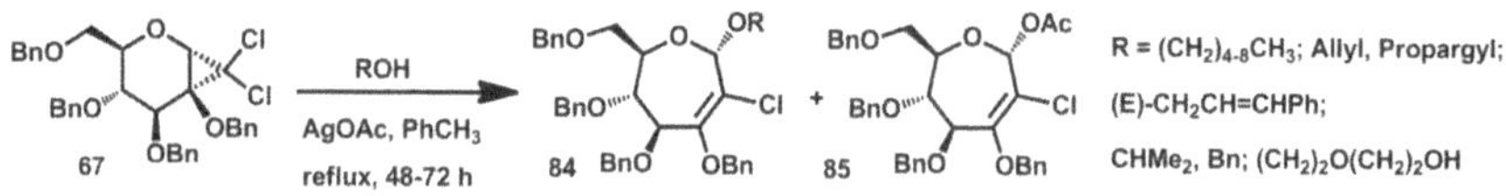

SCHEME 13.23 Acid-catalyzed ring-opening of glycal-derived *gem*-dichlorocyclopropane derivative **67** to chloro-oxepines **84** and **85** [32].

extent. Furthermore, when the reaction was performed with AgOAc and NaOAc, in the absence of an alcohol, the *O*-acetyl chloro-oxepine **85** was obtained exclusively.

Comparison of the above reactions with that of glycal-derived cyclopropanes illustrates that acid-mediated ring opening of *gem*-dihalo-cyclopropanated oxyglycals affords halo-oxepines exclusively and as stable products. Such a reaction in the case of *gem*-dihalo-cyclopropanated glycals tends to undergo further rearrangements, leading to the formation of branched furanosides, as shown in Scheme 13.18. The rearrangement to *C*-furanoside occurred most likely due to isomerization of less-substituted *endo*-cyclic double bond at *C*2=*C*3 of oxepine to *C*1=*C*2 unsaturated vinyl ether, whereas oxyglycal-derived chloro-oxepine **84** would not undergo such an isomerization. The observation that oxyglycal-derived 1,2-cyclopropanes afforded only chloro-oxepine was further confirmed by the following sequence of reactions: (i) $RuCl_3$-$NaIO_4$-mediated oxidation, (ii) $NaBH_4$ reduction, and (iii) Pd/C-mediated hydrogenolysis to afford *n*-pentyl α-D-*glycero*-D-*galacto*-septanoside [32].

13.2.10 Functionalization of Oxepines

The ring expansion of oxyglycal-derived *gem*-dihalo 1,2-cyclopropanate oxyglycals provides stable seven-membered halo-oxepines, that are thus suitable precursors for further functionalizations. In a study, vinyl bromide of the bromo-oxepine was used for the metal-catalyzed cross-coupling reactions, namely, Heck, Suzuki, and Sonogashira reactions, so as to generate 2-deoxy-2-*C*-alkyl/ary/alkynyl septanosides [33].

Heck and Suzuki coupling reactions with bromo-oxepine **65** afforded 2-deoxy-2-alkyl/aryl septanosides. For example, Heck reaction of **65** with acrylates and styrene led to a high yield of coupled products groups as **86**, whereas the use of α-methyl styrene furnished **87**, in which the *exo*-cyclic double bond isomerized to a terminal double bond. Suzuki coupling reaction with substituted phenylboronic acid led to 2-deoxy-2-*C*-aryl septanoside derivatives **88** in good yields. On the contrary, Sonogashira coupling reaction of **65** with acetylenes furnished 2-deoxy-2-*C*-alkenyl septanoside **89** (Scheme 13.24).

One of the Heck-coupled products was subjected to a hydrogenolysis (Pd/C, H_2) to afford D-*manno*-sept-3-uloside **90**. Subsequent $NaBH_4$-mediated reduction led to hydroxyl group containing 2-deoxy-2-C-alkyl septanoside **91**, in an excellent yield as a single epimer (Scheme 13.25) [33].

Apart from metal-catalyzed cross-coupling reactions, hydrogenolysis of halo-oxepine was used as a key step for the synthesis of 2-chloro-2-deoxy septanoside

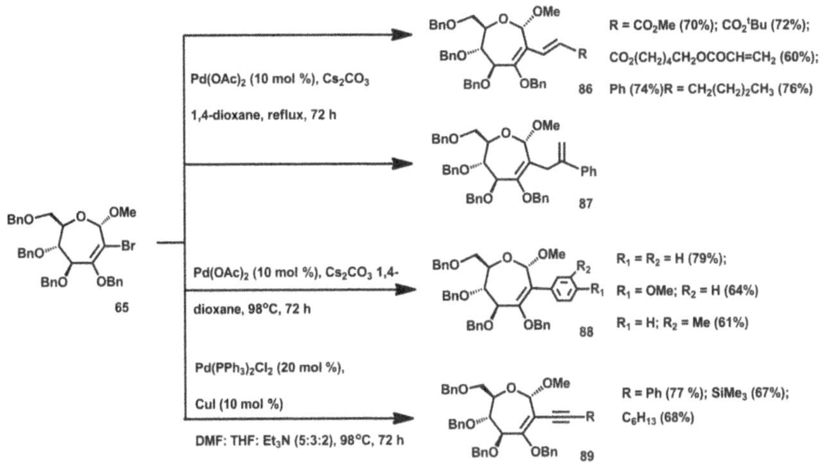

SCHEME 13.24 Metal-catalyzed cross-coupling reactions on bromo-oxepine **65** [33].

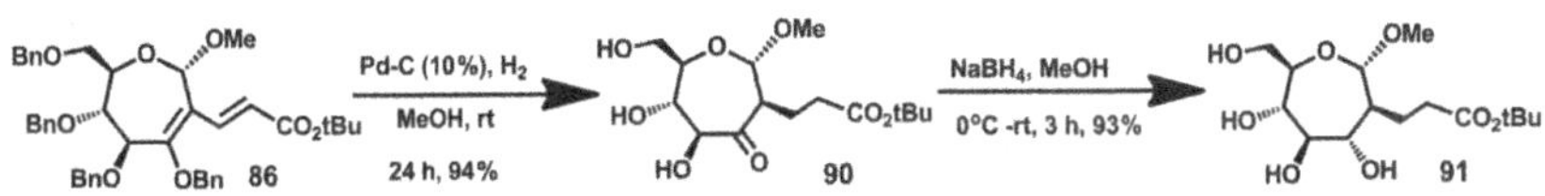

SCHEME 13.25 Synthesis of 2-deoxy-2-*C*-alkyl septanoside **91** [33].

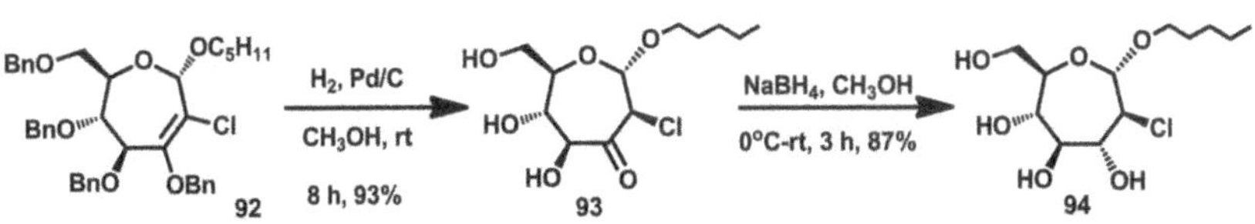

SCHEME 13.26 Synthesis of 2-chloro-2-deoxy septanoside derivative **94** [32].

derivative, as shown in Scheme 13.26 [32]. *n*-Pentyl chloro-oxepine **92** was subjected to hydrogenolysis (H_2, Pd/C), which furnished *n*-pentyl-2-chloro-2-deoxy-α-D-*manno*-sept-3-uloside **93**, which was reduced further, to secure free hydroxyl group containing *n*-pentyl-2-chloro-2-deoxy-α-D-*glycero*-D-*ido*-septanoside **94**.

13.3 STRUCTURE AND CONFORMATION OF SEPTANOSIDES

Studies on conformations of septanosides in the solid-state, solution-phase, and computational methods are still limited, unlike in the case of naturally occurring furanosides and pyranosides for which finer details of the conformations are well elucidated [34]. Being a larger ring system, septanosides are more flexible, leading to attenuation of more complex pseudorotational equilibria, with energy barriers among

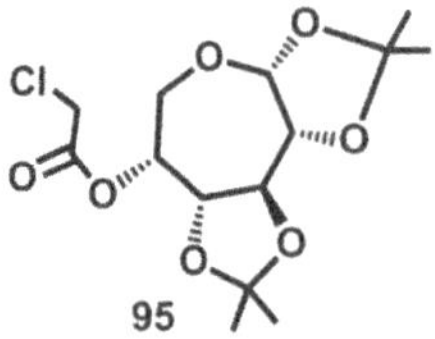

95

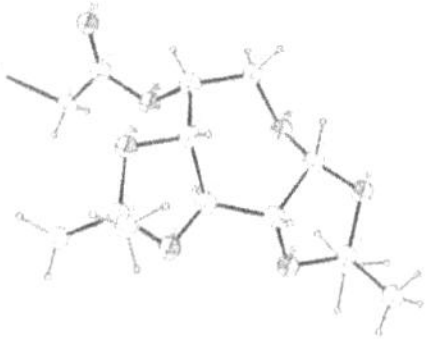

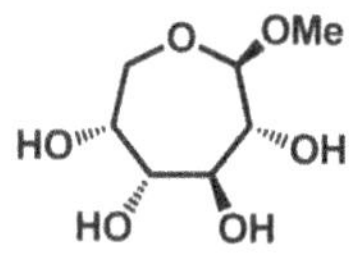

97

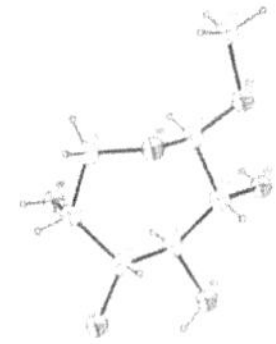

FIGURE 13.2 Molecular and solid-state structures of septanosides **95** and **97**.

conformations being minimal [35, 36]. Four conformations of septanosides can be classified having distinct energy barriers, namely, chair, twist-chair, boat, and twist-boat. These conformations generally follow principles that were laid out in the case of cycloheptanes and seven-membered oxepanes [6].

13.3.1 Solid-State Structures and Conformations

Single-crystal X-structural determinations reveal that septanosides adopt predominantly twist-chair conformation in the majority of the cases. Variations of the conformation result depending on the nature of the protecting groups and ring flexibility. Septanoside derivatives having benzylidene- and isopropylidene-protecting groups lead the molecule being bicyclic and preferred conformation in these instances are the chair conformation. 1,2:3,4-Di-*O*-isopropylidene-5-*O*-chloro-acetyl-α-D-glucoseptanoside **95** (Figure 13.2), one of the early crystal structures reported by Sundaralingam and coworkers, shows that the molecule adopts a chair conformation [37]. The chair conformation was due to a degree of rigidity imposed by *O*-isopropylidene rings. 3-*O*-Acetyl-1,2:4,5-di-*O*-isopropylidene-α-D-glucoseptanose (**96**) also was observed to adopt a chair conformation, as reported by Stevens and coworkers [38].

The distorted twist orientation of *O*-isopropylidene rings flattened the septanoside ring by positioning C5 toward the plane consisting of C6, O6, and C3, thereby leading to a chair conformation. Interestingly, septanosides that are not bound by cyclic protecting groups also exhibit a chair conformation; for example, methyl β-D-glucoseptanoside **97** adopted a chair conformation, with conformational descriptor being $^{2}C_{5,6}$ [39].

Twist-chair conformations were calculated to be the most favorable conformations for cycloheptane systems. Systematic theoretical study by Hendrickson on seven-membered cycloheptanes in mid-1960s [40,41] and Strauss in 1977 established twist-chair conformation as the ground state lower energy conformer whereas twist-boat as the highest energy conformer [42]. There are several instances when septanoside ring adopted a twist-chair conformation. The twist-chair conformation of the septanoside derivative **98** adopting $^{4,5}TC_{6,0}$ is shown in Figure 13.3.

A combined x-ray and neutron diffraction studies showed that 5-*O*-acetyl-1,2:3,4-di-*O*-isopropylidene-α-D-*gluco*-septanose (**99**) adopted a twist-chair conformation $^{4,5}TC_{6,0}$ [43]. Torsional angles are evaluated to identify the ring conformation. Thus, torsion angles of −31.1° for O6—C1—C2—C3 and −46.0° for C1—C2—C3—C4 of

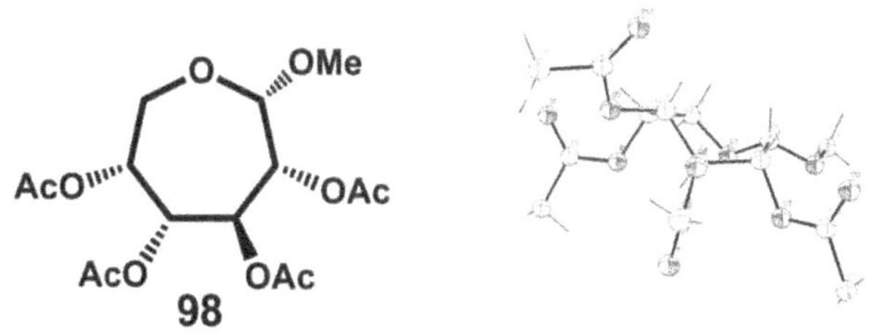

FIGURE 13.3 Molecular and solid-state structures of septanoside **98**.

led to conclude that methyl 2,3,4,5-*tetra*-*O*-acetyl-α-D-glucoseptanoside **98** adopted a twist-chair conformation $^{4,5}TC_{6,0}$ [44]. Another common twist-chair conformation adopted by septanoside ring is $^{5,6}TC_{3,4}$ in the solid state (Figure 13.4). The solid-state structure of *n*-pentyl 2-chloro-2-deoxy-α-D-*manno*-sept-3-uloside **100** was determined, from which the torsion angle analysis was evaluated [45]. Torsion angles of 48.7° for C6–O6–C1–C2 and 39.6° for O6–C1–C2–C3 indicated least-square plane passing through O6, C1, and C2. The twist-chair $^{5,6}TC_{3,4}$ was thus assigned as C5, and C6 atoms are above the mean plane whereas C3 and C4 are below the plane. Crystal structure analysis showed that an array of intermolecular O–H···O and C–H···O hydrogen-bonding interactions stabilized the molecular packing in **100**. The hydroxyl groups O4–H4, O5–H5, and O7–H7 acted both as hydrogen bond donors and as acceptors. Furthermore, O3, O5, and H4 (O4) participated in a three-centered hydrogen bond, in which O4–H4 was a bifurcated donor for an intramolecular hydrogen bond in O4–H4···O3 and an intermolecular hydrogen bond in O4–H4···O5 (*x*, *y* − 1, *z*) (Figure 13.4). The presence of potential C–H···O interactions [46,47], along with O–H···O interactions, was also observed. The molecular packing indicated that a bilayer arrangement of the molecules with sugar head group is separated from the lipophilic alkyl chains. The twist-chair conformation $^{5,6}TC_{3,4}$ was also observed in the case of β-D-glucoseptanose pentaacetate **101** in which axis of symmetry passes

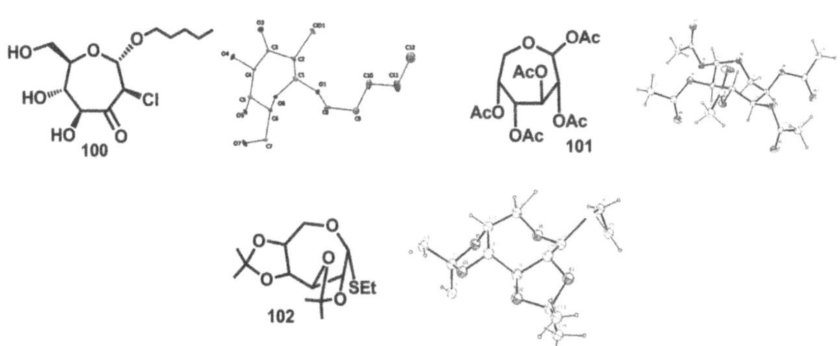

FIGURE 13.4 Twist-chair ($^{5,6}TC_{3,4}$) conformation of septanosides **100–102** in the solid state.

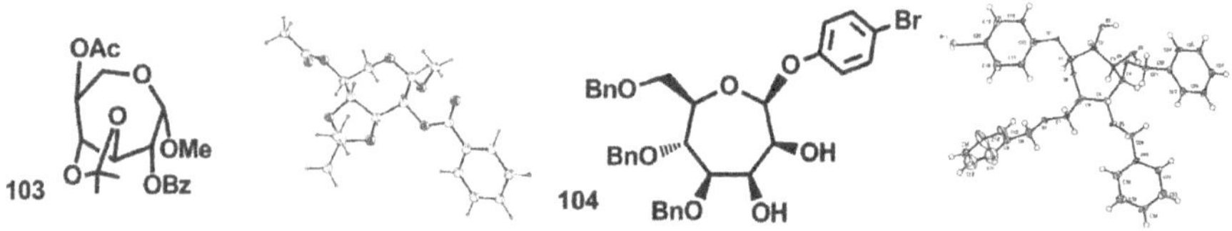

FIGURE 13.5 Molecular and solid-state structures of septanosides **103** and **104**.

through C1 [48]. The torsion angles of 40.5° and 45.0° for C6–O6–C1–C2 and O6–C1–C2–C3, respectively, indicated that least-square plane passed through O6, C1, and C2.

The twist-chair conformation was also seen intact when thiol substitution was introduced at anomeric center with 2,3,4,5-di-*O*-isopropylidene protecting group, namely, ethyl 2,3:4,5-di-*O*-isopropylidene-1-thio-β-D-glucoseptanoside **102** [49]. Positive signs and closer torsion angles of 53.5° for C6-O6-C1-C2 and 35.7° for O6-C1-C2-C3 led to assign the conformational descriptor $^{5,6}TC_{3,4}$.

There are few instances wherein septanoside adopted a $^{0,1}TC_{2,3}$ conformation. Figure 13.5 provides examples of septanoside derivatives with this conformation [50,51]. An analysis of the torsion angles C3—C4—C5—C6 of −49° and C4—C5—C6—O6 of −32° of needle-type orthorhombic crystal of **103** showed that mean plane passed through C4, C5, and C6 atoms. Thus, the conformation was assigned as $^{0,1}TC_{2,3}$, describing that O6 and C1 are above the plane whereas C2 and C3 below the plane [50].

p-Bromophenyl 4,5,7-*tri-O*-benzyl-β-D-*glycero*-D-*talo*-septanoside **104** also exhibited $^{0,1}TC_{2,3}$, even when there is no similarity between **103** and **104**, other than that both are septanosides. Torsion angles for C3—C4—C5—C6 and C4—C5—C6—O6 of **104** were −25.7° and −49.0°, respectively. Molecular packing of the **104** in the crystal was dominated by O−H···O, C−H···π, and π···π interactions. Two independent intermolecular O−H···O hydrogen-bonding interactions between the translated neighbor along crystallographic “a” axis stabilized the packing. A three-centered hydrogen bond formed between O2, O4, and H4 (O3). An intramolecular hydrogen bonding between O2−H3A···O3 was also observed, in which, O2−H3A acted as a donor and O3 acted as an acceptor. Apart from hydrogen bonding, a large pool of aromatic interactions was observed between the aromatic rings [51].

Solid-state structures of septanosides showed that it adopted a twist-chair and chair conformations in the majority of the examples. Yet there are reports on septanosides adopting boat conformation, even when the boat conformation is known for its higher energy, as compared with chair and twist-chair conformers. Single-crystal x-ray structural determination of the needle-shaped 5-*O*-acetyl-1,2:3,4-di-*O*-isopropylidene-α-D-galactoseptanose **105** indicated that it adopted a boat conformation, even when the corresponding α-D-glucoseptanose **99** adopted a twist-chair conformation, $^{4,5}TC_{6,0}$ [52].

The boat conformation of **105** was assigned on the basis of the matching of the observed torsion angles within the septanoside with that of oxepane

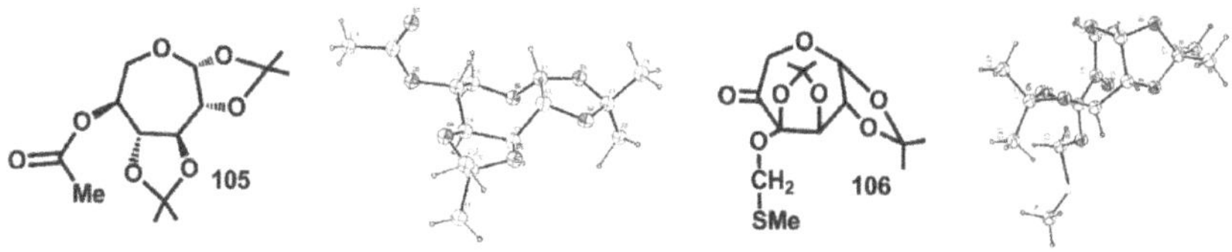

FIGURE 13.6 Molecular and solid-state structures of **105** and **106**.

boat (C1—C2—C3—C4, −74.0°; C2—C3—C4—C5, 34°; C3—C4—C5—C6, 56.0°; C4—C5—C6—O6, −53.0°; C5—C6—O6—C1, −47.0°; C6—O6—C1—C2, 86.0°; O6—C1—C2—C3, −2.0°) [52]. On the contrary, 1,2:3,4-di-*O*-isopropylidene-4-*C*-methylthio methoxy-β-L-*arabino*-hexoseptanos-5-ulose **106** also adopted a boat conformation in the solid state. An analysis of torsion angles (C2—C3—C4—C5, −1.0°; C3—C4—C5—C6, −60.2°; and C4—C5—O6—C1 23.1°) involving the seven-membered ring led to assign $^{3,4,O}B(L)$ conformation of **106** with two dioxolane rings in envelop forms (Figure 13.6) [53].

13.3.2 Solution-Phase Conformations

Conformation of septanosides in the solution phase is more diverse and complex, as compared with the solid-state structural determinations. The NMR technique of proton–proton coupling constants is used predominantly in the structural determinations in solution phase.

Figure 13.7 shows examples of septanosides, where the chair conformation was reported for septanoside derivatives, on the basis of ^3J and ^4J coupling constants. Specifically, the chair conformations were determined to be $^5C_{1,2}$ through observation of coupling constants $J_{3,4}$ = 10.16 Hz, $J_{4,5}$ = 9.44 Hz, $J_{5,6a}$ = 10.86 Hz, and $J_{5,6b}$ = 4.42 Hz. In the case of $^2C_{5,6}$ of **108** and **109**, a common coupling constants $J_{6a,6b}$

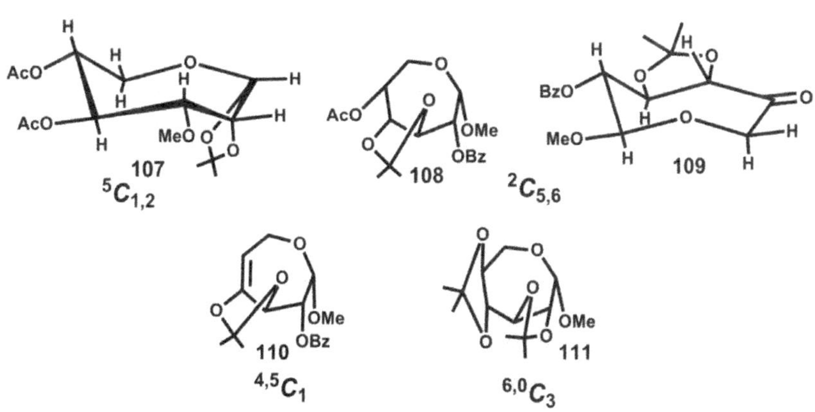

FIGURE 13.7 Possible chair conformations of **107–111** in the solution phase.

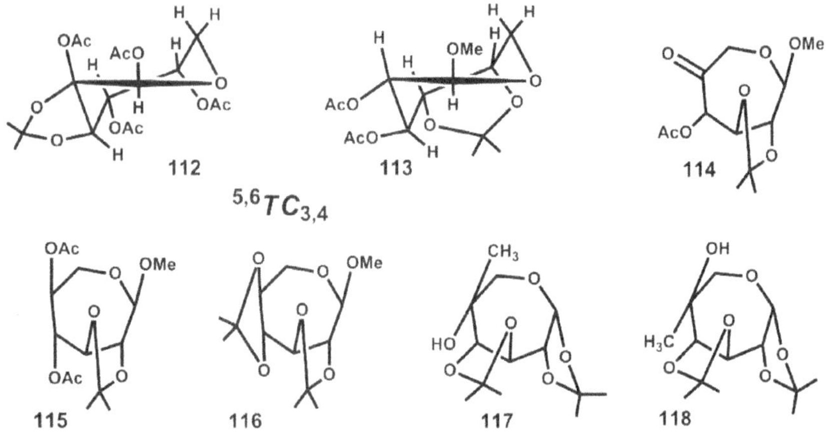

FIGURE 13.8 Solution-phase conformation $^{5,6}TC_{3,4}$ of **112–118**.

of >18.56 Hz was observed, whereas $^{6,0}C_3$ conformation of **111** was deduced from $J_{5,6a} = 2.95$ Hz and $J_{5,6b} = 10.50$ Hz [54–56].

Proton–proton coupling constants showed that septanosides adopted various twist-chair conformations depending on the nature of the protecting groups. Possible twist-chair conformations were estimated to be one of $^{5,6}TC_{3,4}$, $^{4,5}TC_{6,0}$, $^{1,2}TC_{6,0}$, $^{6,0}TC_{4,5}$, $^{3,4}TC_{1,2}$, and $^{2,3}TC_{3,4}$. Figure 13.8 shows a list of various septanoside derivatives adopting $^{5,6}TC_{3,4}$ conformation. The conformation $^{5,6}TC_{3,4}$ of **112–118** was deduced on the basis of the observed coupling constants $J_{2,3} > 9.0$ Hz and $J_{3,4} > 9.0$ Hz. The observed coupling constants $J_{2,3}$ of 7.32 Hz with **117** $J_{2,3}$ of 6.71 Hz with **118** were 2–3 Hz lower than the generally observed coupling constant. Other abundant conformations of septanoside in solution phase are $^{4,5}TC_{6,0}$ and $^{1,2}TC_{6,0}$. Septanoside derivatives **119–122** that adopt $^{4,5}TC_{6,0}$ conformation in solution are given in Figure 13.9. The conformation $^{4,5}TC_{6,0}$ was determined from the observed coupling constant $J_{5,6a} > 2.2$ Hz and $J_{6b,6a} > 14.0$ Hz for these derivatives [54–56].

The solution-state conformation of $^{1,2}TC_{6,0}$ of the bicyclic septanoside **123** and **124** was deduced from experimentally observed coupling constants (Figure 13.10). In both of the cases, the observed coupling constants $J_{5,6a} > 0.8$ Hz and $J_{5,6b} > 8.6$ Hz fulfil the condition for adopting this conformation. The *trans*-fused *O*-isopropylidene moiety restricted to $^{1,2}TC_{6,0}$ conformation for **123**, whereas long-range coupling

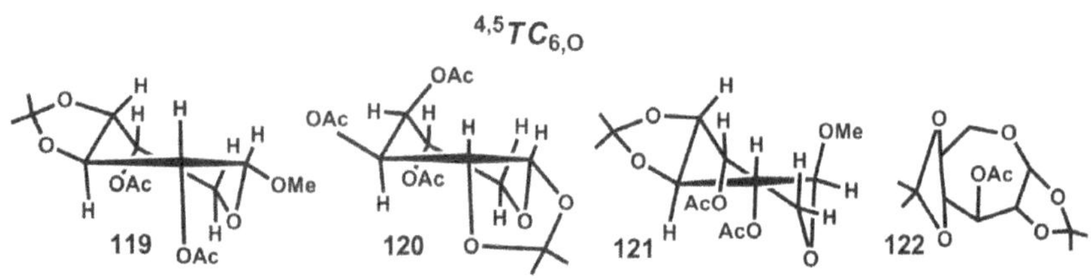

FIGURE 13.9 Solution-state conformation $^{4,5}TC_{6,0}$ of **119–122**.

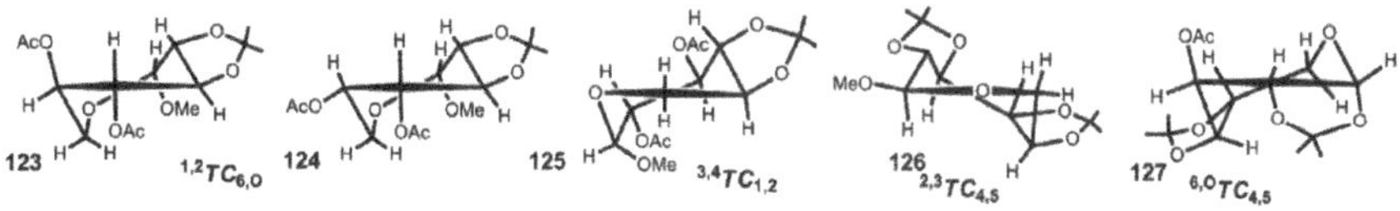

FIGURE 13.10 Solution-phase conformations of **123–127**.

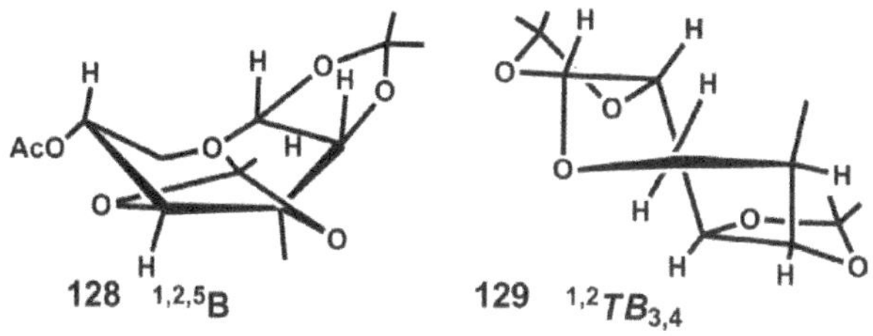

FIGURE 13.11 Solution-phase conformations of **128** and **129**.

constants $^4J_{4,6b}$ = 1.64 Hz and $^4J_{1,6a}$ = 0.56 Hz also supported the conformation for **124**. $^{3,4}TC_{1,2}$ conformation of septanoside **125** was determined by observing the following coupling constants: $J_{5,6b}$ = 9.48 Hz, $J_{5,6a}$ = 5.68 Hz, $J_{1,2}$ = 2.30 Hz, and $J_{2,3}$ = 8.01 Hz. Antiperiplanar orientation of H5 and H6a was attributed from the coupling constant $J_{5,6b}$ = 9.48 Hz. The observed coupling constants $J_{4,5}$ = 8.4 Hz, $J_{5,6a}$ = 10.0 Hz, and $J_{5,6b}$ = 2.5 Hz for **126** indicated $^{2,3}TC_{4,5}$ conformation with stable *O*-isopropylidene ring formation. Solution-state conformation **127** was determined from the observed coupling constants $J_{3,4}$ = 1.5 Hz and $J_{2,3}$ = 3.0 Hz. The analysis of the observed coupling constants and dihedral angles of ring protons led to conclude that it adopted a $^{6,O}TC_{4,5}$ conformation [54–56].

The solution-phase conformational analysis also reveals instances of boat and twist-boat conformations septanosides, examples of which are given in Figure 13.11. The coupling constants $J_{3,4}$ = 7.61 Hz, $J_{4,5}$ = 9.97 Hz, $J_{5,6a}$ = 6.28 Hz, $J_{5,6b}$ = 10.66 Hz, and $J_{6a,6b}$ = 11.35 Hz ascertained a boat conformation for **128**, whereas coupling constant $J_{3,4}$ = 7.61 Hz indicated $^{1,2,5}B$ conformation [56].

The solution-phase conformation of **129** showed that it adopted a twist-boat $^{1,2}TB_{3,4}$, (Figure 13.10), which was deduced from the observed coupling constants $J_{6a,6b}$ = 13.5 Hz, denoting an antiperiplanar arrangement of O5 with one of the *geminal* hydrogens of C6 [56].

13.4 CONCLUSIONS

Ring-opening reactions of bicyclic [4.1.0] carbohydrate derivatives provide an excellent route to synthesize seven-membered ring sugars, namely, septanoses and septanosides. The reactions promoted under acidic or basic catalysis are sensitive, depending on the cyclopropylated sugar derivative undergoing ring opening of the cyclopropyl moiety. Glycal-derived and oxyglycal-derived cyclopropane ring opening is a case in point, wherein the former affords exocyclic ring opening, leading to pyranoside formation eventually, with exocyclic substituent, and the latter undergoes endocyclic ring opening to a ring-expanded septanoside derivative, both the reactions mediated under base catalysis. On the contrary, differences among the reactivities between the above two cyclopropylated sugar derivatives are distinct under acid-catalyzed reaction conditions. Thus, cyclopropylated glycal derivative provides five-membered furanose derivatives upon ring opening to a septanose intermediate that subjects itself to further rearrangements. Under similar acidic conditions, cyclopropylated oxyglycal derivatives afford ring-expanded septanoside exclusively. These observations point to the possibility that synthesis of fully hydroxylated septanoside derivatives requires cyclopropanated oxyglycal as the most suitable precursor.

Ferrier-rearrangement-based ring-opening reactions of cyclopropylated sugar derivatives also find attractiveness, especially in the case of functionalized oxepane formation. With the emergence of the methods described in Section 13.2.8, in addition to cyclization, methodologies of linear precursors offer now a reasonable number of synthetic methods to prepare oxepanes and septanoses.

Interest in septanoses and septanosides is at a point where their biological roles await to be uncovered. Requirements that their structures, conformations, and intermolecular interactions are understood are important in this direction. General observations are that septanosides reside largely in twist-chair and chair conformations from solid-state, solution-phase, and computational studies. Such observations indicate that protein receptor sites for meaningful biological interactions may be probed within the constraints of these conformations. These and more studies involving septanoside unnatural sugars might be anticipated in future advancements in this specific area of carbohydrates.

ACKNOWLEDGMENTS

Authors thank the Department of Science and Technology, New Delhi, India, for the financial support. Council of Scientific and Industrial Research, New Delhi, is acknowledged for research fellowships to S.D. and N.V.G.

REFERENCES

[1] Kantorowski, E. J.; Kurth, M. J. *Tetrahedron* **2000**, *56*, 4317–4353.

[2] Leemans, E.; D'hooghe, M.; De Kimpe, N. *Chem. Rev.* **2011**, *111*, 3268–3333.

[3] Yu, M.; Pagenkopf, B. L. *Tetrahedron* **2005**, *61*, 321–347.

[4] Schleyer, P. v. R.; Williams, J. E.; Blanchard, K. R. *J. Am. Chem. Soc.* **1970**, *92*, 2377–2386.

[5] (a) Harvey, J. E.; Hewitt, R. J.; Moore, P. W.; Somarathne, K. K. A. *Pure Appl. Chem.* **2014**, *86*, 1377–1399. (b) Kleinke, A. S.; Webb, D.; Jamison, T. F. *Tetrahedron* **2012**,

68, 6999–7018. (c) Snyder, N. L.; Haines, H. M.; Peczuh, M. W. *Tetrahedron* **2006**, *62*, 9301–9320. (d) Pakulski, Z. *Pol. J. Chem.* **2006**, *80,* 1293–1326.

[6] Dey, S.; Jayaraman, N. *Pure Appl. Chem.* **2014**, *86*, 1401–1419.

[7] Micheel, F.; Suckfüll, F. *Ann. Chem.* **1933**, *502*, 85–98.

[8] Micheel, F.; Spruck, W. *Ber.* **1934**, *67B*, 1665–1667.

[9] (a) Stevens, J. D. *Chem. Commun.* **1969**, 1140–1141. (b) Stevens, J. D. *Carbohydr. Res.* **1972**, *21*, 490–492. (c) Ng, C. J.; Stevens, J. D. *Carbohydr. Res.* **1996**, *284*, 241–248.

[10] Anet, E. F. L. J. *Carbohydr. Res.* **1968**, *8*, 164–174.

[11] Contour, M. O.; Fayet, C.; Gelas, J. *Carbohydr. Res.* **1990**, *201*, 150–152.

[12] Tauss, A.; Steiner, A. J.; Stütz, A. E.; Tarling, C. A.; Withers, S. G.; Wrodnigg, T. M. *Tetrahedron Asymmetry* **2006**, *17*, 234–239.

[13] Sizun, G.; Dukhan, D.; Griffon, J.; Griffe, L.; Meillon, J.; Leroy, F.; Storer, R.; Gosselin, G. *Carbohydr. Res.* **2009**, *344*, 448–453.

[14] Castro, S.; Peczuh, M. W. *J. Org. Chem.* **2005**, *70*, 3312–3315.

[15] Enright, P. M.; Tosin, M.; Nieuwenhuyzen, M.; Cronin, L.; Murphy, P. V. *J. Org. Chem.* **2002**, *67*, 3733–3741.

[16] Baschang, G. *Ann. Chem.* **1963**, *663*, 167–173.

[17] Butcher, M. E.; Ireson, J. C.; Lee, J. B.; Tyler, M. J. *Tetrahedron* **1977**, *33*, 1501–1507.

[18] (a) Fukami, H.; Koh, H.-S.; Sakata, T. Nakajima, M. *Tetrahedron Lett.* **1967**, *14*, 1701–1704. (b) Wang, Z.; Miller, S. M.; Anderson, O. P.; Shi, Y. *J. Org. Chem.* **1999**, *64*, 6443–6458.

[19] Köver, A.; Matheu, M. I.; Díaz, Y.; Castillón, S. *ARKIVOC* **2007**, *4*, 364–379.

[20] Ovaa, H.; Leeuwenburgh, M. A.; Overkleeft, H. S.; van der Marel, G. A.; van Boom, J. H. *Tetrahedron Lett.* **1998**, *39*, 3025–3028.

[21] Peczuh, M. W.; Snyder, N. L. *Tetrahedron Lett.* **2003**, *44*, 4057–4061.

[22] (a) Schmidt, B.; Biernat, A. *Chem. Eur. J.* **2008**, *14*, 6135–6141. (b) Schmidt, B.; Biernat, A. *Org. Lett.* **2008**, *10*, 105–108.

[23] (a) Alcázar, E.; Pletcher, J. M.; McDonald, F. E. *Org. Lett.* **2004**, *6*, 3877–3880. (b) Boone, M. A.; McDonald, F. E.; Lichter, J.; Lutz, S.; Cao, R.; Hardcastle, K. I. *Org. Lett.* **2009**, *11*, 851–854.

[24] Castro, S.; Johnson, C. S.; Surana, B.; Peczuh, M. W. *Tetrahedron* **2009**, *65*, 7921–7926.

[25] (a) Gómez, A. M.; Lobo, F.; de las Vacas, D. P.; Valverde, S.; López, C. J. *Chem. Commun.* **2010**, *46*, 6159–6161. (b) Lobo, F.; Gómez, A. M.; Miranda, S.; López, C. J. *Chem. Eur. J.* **2014**, *20*, 10492–10502.

[26] (a) Hoberg, J. O. *J. Org. Chem.* **1997**, *62*, 6615–6618. (b) Batchelor, R.; Hoberg, J. O. *Tetrahedron Lett.* **2003**, *44*, 9043–9045. (c) Hoberg, J. O.; Bozell, J. J. *Tetrahedron Lett.* **1995**, *36*, 6831–6834.

[27] Venukumar, P.; Sudharani, C.; Sridhar, P. R. *Chem. Commun.* **2014**, *50*, 2218–2221.

[28] Yu, M.; Pagenkopf, B. L. *Tetrahedron* **2005**, *61*, 321–347.

[29] Ganesh, V.; Kundu, T.; Chandrasekaran, S. *Tetrahedron* **2014**, *70*, 7268–7282.

[30] (a) Hewitt, R. J.; Harvey, J. E. *J. Org. Chem.* **2010**, *75*, 955–958. (b) Hewitt, R. J.; Harvey, J. E. *Chem. Commun.* **2011**, 421–423.

[31] (a) Ganesh, N. V.; Jayaraman, N. *J. Org. Chem.* **2007**, *72*, 5500–5504. (b) Ganesh, N.

V.; Jayaraman, N. *J. Org. Chem.* **2009**, *74*, 739–746. (c) Ganesh, N. V.; Raghothama, S.; Sonti, R.; Jayaraman, N. *J. Org. Chem.* **2010**, *75*, 215–218.

[32] Dey, S. Jayaraman, N. *Carbohydr. Res.* **2014**, *389*, 66–71.

[33] Dey, S.; Jayaraman, N. *Beilstein J. Org. Chem.* **2012**, *8*, 522–527.

[34] Stoddart, J. F. *Stereochemistry of Carbohydrates*; John Wiley and Sons, Inc.: New York, 1971.

[35] Dione, P.; St-Jacques, M. *Can. J. Chem.* **1989**, *67*, 11–20.

[36] (a) Désilets, S.; St-Jacques, M. *J. Am. Chem. Soc.* **1987**, *109*, 1641–1648. (b) Entrena, A.; Campos, J. M.; Gallo, M. A.; Espinosa, A. *ARKIVOC* **2005**, *6*, 88–108.

[37] Jackobs, J.; Sundaralingam, M. *J. Chem. Soc., Chem. Commun.* **1970**, 157–159.

[38] Pallister, E. T.; Stephenson, N. C.; Stevens, J. D. *J. Chem. Soc., Chem. Commun.* **1972**, 98–99.

[39] Foster, S. J.; James, V. J.; Stevens, J. D. *Acta. Crystallogr.* **1983**, *C39*, 610–612.

[40] Hendrickson, J. B. *J. Am. Chem. Soc.* **1961**, *83*, 4537–4547.

[41] Hendrickson, J. B. *J. Am. Chem. Soc.* **1967**, *89*, 7036–7043.

[42] (a) Bocian, D. F.; Pickett, H. M.; Rounds, T. C.; Strauss, H. L. *J. Am. Chem. Soc.* **1975**, *97*, 687–695. (b) Bocian, D. F.; Strauss, H. L. *J. Am. Chem. Soc.* **1977**, *99*, 2876–2882.

[43] Wood, R. A.; James, V. J.; Rae, A. D.; Stevens, J. D.; Moore, F. H. *Aust. J. Chem.* **1983**, *36*, 2269–2277.

[44] McConnell, J. F.; Stevens, J. D. *J. Chem. Soc., Perkin Trans. 2*, **1974**, *3*, 77–81.

[45] Dey, S.; Basuroy, K.; Jayaraman, N. *Carbohydr. Res.* **2014**, *393*, 37–42.

[46] (a) Sutor, D. J. *Nature* **1962**, *195*, 68–69. (b) Sutor, D. J. *J. Chem. Soc.* **1963**, 1105–1110.

[47] (a) Desiraju, G. R. *J. Chem. Soc., Chem. Commun.* **1989**, 179–180. (b) Steiner, T. *J. Chem. Soc., Chem. Commun.* **1994**, 2341–2342.

[48] Bhadbhade, M.; Craig, D. C.; Ng, C. J.; Stevens, J. D. *Carbohydr. Res.* **2012**, *353*, 86–91.

[49] Beale, J. P.; Stephenson, N. C.; Stevens, J. D. *Acta Crystallogr.* **1972**, *B28*, 3115–3121.

[50] Bailey, C. J.; Craig, D. C.; Grainger, C. T.; James, V. J.; Stevens, J. D. *Carbohydr. Res.* **1996**, *284*, 265–270.

[51] Dey, S.; Basuroy, K.; Jayaraman, N. *Carbohydr. Res.* **2015**, *410*, 9–14.

[52] James, V. J.; Stevens, J. D. *Carbohydr. Res.* **1980**, *82*, 167–174.

[53] Craig, D. C.; James, V. J.; Stevens, J. D. *Aust. J. Chem.* **1990**, *43*, 2083–2086.

[54] Stevens, J. D. *Aust. J. Chem.* **1975**, *28*, 525–557.

[55] Ng, C. J.; Craig, D. C.; Stevens, J. D. *Carbohydr. Res.* **1996**, *284*, 249–263.

[56] Driver, G. E.; Stevens, J. D. *Aust. J. Chem.* **1990**, *43*, 2063–2081.

14

REARRANGEMENTS IN CARBOHYDRATE TEMPLATES TO THE WAY TO PEPTIDE-SCAFFOLD HYBRIDS AND FUNCTIONALIZED HETEROCYCLES

BERNARDO HERRADÓN, IRENE DE MIGUEL, AND ENRIQUE MANN
Instituto de Química Orgánica General, CSIC, c/ Juan de la Cierva 2, 28006 Madrid, Spain

14.1 INTRODUCTION

Carbohydrate derivatives possess a copious number of functionalized stereogenic centers [1]. Because of its rich and varied functionality as well as its stereochemical signature, they are suitable chiral starting materials for enantiomerically pure compound synthesis [2] as demonstrated in many literature examples [3–7]. In addition, when the carbohydrate is in a cyclic structure, especially in the pyranose ring (**A**; Figure 14.1), it has a conformational bias, due to combined steric and stereoelectronic effects [8], that allows to predict and control the stereochemical outcome of many reactions on the carbohydrate template (Figure 14.1).

Although the presence of multiple functionality is one of the main, potentially useful, features of carbohydrates as building blocks for organic synthesis, it can be also a drawback, in some cases, since most of the functional groups have very similar reactivity, which lowers the chemoselectivity of carbohydrate reactions, although several strategies have been developed to achieve this goal [9, 10].

Interesting alternatives to fully oxygenated pyranose (**A**) are the unsaturated counterparts, namely 3,4-dihydro-2*H*-pyran (glycal, **B**), 6-alkoxy-3,6-dihydro-2*H*-pyran

Domino and Intramolecular Rearrangement Reactions as Advanced Synthetic Methods in Glycoscience, First Edition. Edited by Zbigniew J. Witczak and Roman Bielski.

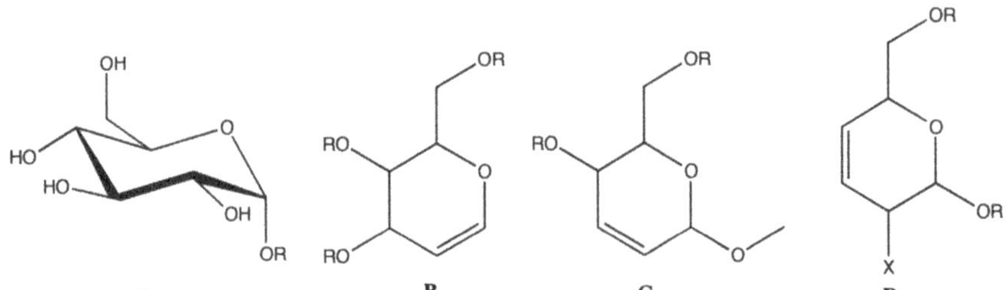

FIGURE 14.1 Generic structures of an α-D-glucoside in its pyranose form (**A**), 3,4-dihydro-2*H*-pyran (glycal, **B**), 6-alkoxy-3,6-dihydro-2*H*-pyran (2-hexenopyranoside, **C**) and 2-alkoxy-3,6-dihydro-2*H*-pyran (3-hexenopyranoside, **D**).

(2-hexenopyranoside, **C**), and 2-alkoxy-3,6-dihydro-2*H*-pyran (3-hexenopyranoside, **D**) (Figure 14.1), that can be obtained from the parent hexose in straightforward manners [11, 12].

In connection with our longstanding interest on the preparation of carbohydrate-like molecules [13–16] as well as the applications of carbohydrates to the synthesis of natural [17, 18] and unnatural products [19–21], we have had the opportunity to use unsaturated carbohydrates for the synthesis of peptide–carbohydrate hybrids [22–26], as well as functionalized heterocycles and alkaloids [27–29], and an outline of our recent results is reported herein.

14.2 SYNTHESIS OF THE CHIRAL BUILDING BLOCKS: APPLICATIONS OF THE CLAISEN–JOHNSON AND OVERMAN REARRANGEMENTS

Commercially available triacetyl-D-glucal (**1**) is a suitable starting material for the synthesis of functionalized 3,6-dihydro-2*H*-pyrans. Thus, the reaction of **1** with an alcohol under catalysis of Brønsted or Lewis acid gives a 6-alkoxy-3,6-dihydro-2*H*-pyran (2-hexenopyranoside) through a Ferrier rearrangement [30]. In our work [25, 26], we have used allyl alcohol, benzyl alcohol, and (±)-2-phenethyl alcohol as nucleophiles to give **2**, **3**, and **4**, respectively, as a mixture of diastereoisomers (Figure 14.2). The diacetates **2** and **3** were transformed into the corresponding diols **5** and **6** by mild hydrolysis. In terms of yield and stereoselectivity, the most suitable compound is the allyl glycoside **2**, and most of the remaining transformations were performed from this material.

[3,3]-Sigmatropic rearrangements are useful transformations because of its stereocontrolled nature due to its ordered mechanism [31, 32]. This kind of reactions is given by suitable derivatives of allylic alcohols. We have previously used two of these variants, namely Claisen [33–35] and Overman [36, 37] rearrangements, to nitrogenated heterocycles [24, 26, 38, 39] and carbohydrate derivatives [22–27]. The synthesis of the starting material and the reaction products of both rearrangements are indicated in Figure 14.3.

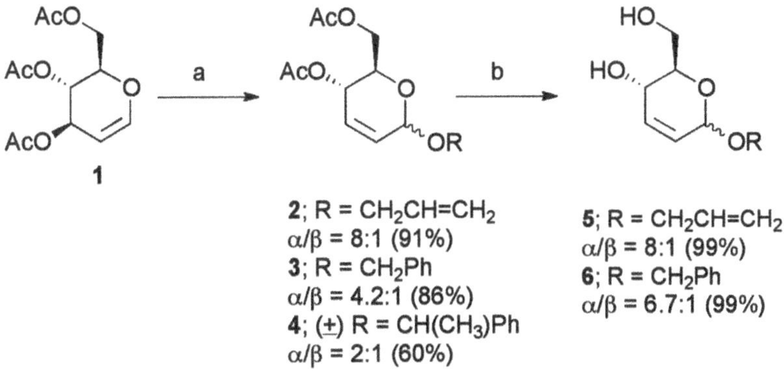

FIGURE 14.2 Synthesis of 3,6-dihydro-2*H*-pyrans 2-6. Reagents and conditions: (a) $FeCl_3$, R-OH, CH_3CN, rt (room temperature), and (b) KOH, MeOH, rt.

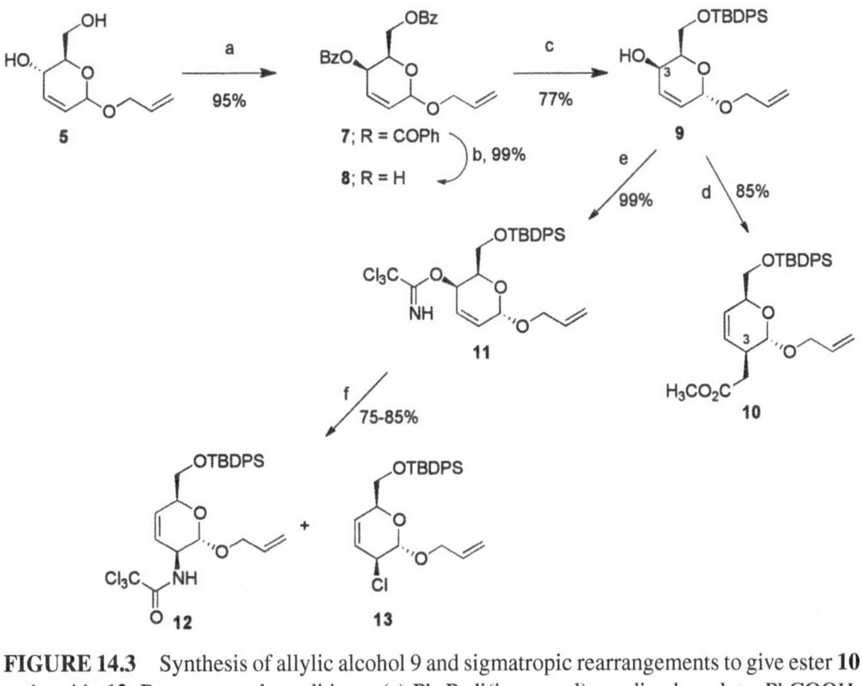

FIGURE 14.3 Synthesis of allylic alcohol 9 and sigmatropic rearrangements to give ester **10** and amide **12**. Reagents and conditions: (a) Ph_3P, di(isopropyl) azodicarboxylate, PhCOOH, THF, rt; (b) NaOMe, MeOH, rt; (c) TBDPSCl, Et_3N, DMAP, CH_2Cl_2, rt; (d) $CH_3C(OCH_3)_3$, $CH_3CH_2CO_2H$ (cat.), hydroquinone, 140°C; (e) Cl_3CN, 1,8-diazabicicloundec-7-ene (DBU), CH_2Cl_2, 0°C to rt; and (f) K_2CO_3, xylene, reflux.

The diol **5** was submitted to a double Mitsunobu reaction [40] with benzoic acid to give dibenzoate **7** (95% yield), with inversion of the configuration at the position 3 of the pyran. The dibenzoate **7** was treated with sodium methoxide to give diol **8** quantitatively, whose primary hydroxy group was selectively protected with *tert*-butyldiphenylsilyl chloride (TBDPSCl) to give **9**. At this stage, the two anomers were separated by chromatography, isolating the major in 77% yield. The allylic alcohol **9** was the starting material for the sigmatropic rearrangement. We found that the most convenient method to achieve Claisen rearrangement is the Johnson variant [41], which was performed by refluxing alcohol **9** in triethyl orthoacetate in the presence of propanoic acid and hydroquinone [42] to provide the cognate γ,δ-unsaturated ester **10** as the only product in a high yield (85%). The reaction took place in a completely stereoselective way, as it is expected for such suprafacial allyl rearrangements. The stereochemistry in compound **10** was confirmed by ^{1}H NMR spectroscopic analysis. The most significant data is the singlet pattern of the anomeric proton (H-2 in the pyran system) in agreement with a *trans*-relationship between the vicinal (H-2 and H-3) protons [43]. The *pseudo*-axial disposition of the free hydroxy group at C-3 (pyran numbering) in **9** plays an important role in the outcome of the rearrangement; since the epimeric alcohol gives a slow reaction and a poor yield in the rearrangement reaction [25, 44].

In a similar manner, from alcohol **9**, we performed the Overman rearrangement through the trichloroacetimidate **11** to give the trichloroacetamide **12** along with a small amount of the allylic chloride **13** [23, 27].

The sigmatropic rearrangement products **10** and **12** and some of the intermediates shown in the previous figures have been used in the synthesis of peptide–scaffold hybrids.

14.3 PEPTIDE–SCAFFOLD HYBRIDS

Peptide–scaffold hybrids are compounds where a peptide fragment is linked to a nonpeptidic portion (the *scaffold*), which is designed with the objective to modulate the properties of the peptide fragment, namely the conformation, the biological activity, and the pharmacological profile (pharmacokinetic, metabolism, etc.) [45]. It is well known that many peptides possess interesting biological activities [46–48] but they frequently suffer from limited stability due to hydrolysis by peptidase as well as poor transport through cell membranes [49] that limit their values as pharmaceuticals. To circumvent these inconveniences, some modifications of the peptides (generally termed as *peptidomimetics*) have been developed [50], although they have met with limited success in the pharmaceutical area. On the contrary, the biological activity of a peptide has been linked to the preferred conformation (secondary structure) of the peptide [51], which has been fueling an intense research activity in the field of conformationally restricted peptidomimetics [50].

With the objective to improve the transport and the peptidase resistance of peptides, we have been engaged in the synthesis of the peptide-scaffold hybrids. In the course of our longstanding interest on these molecules, we have prepared a panoply of

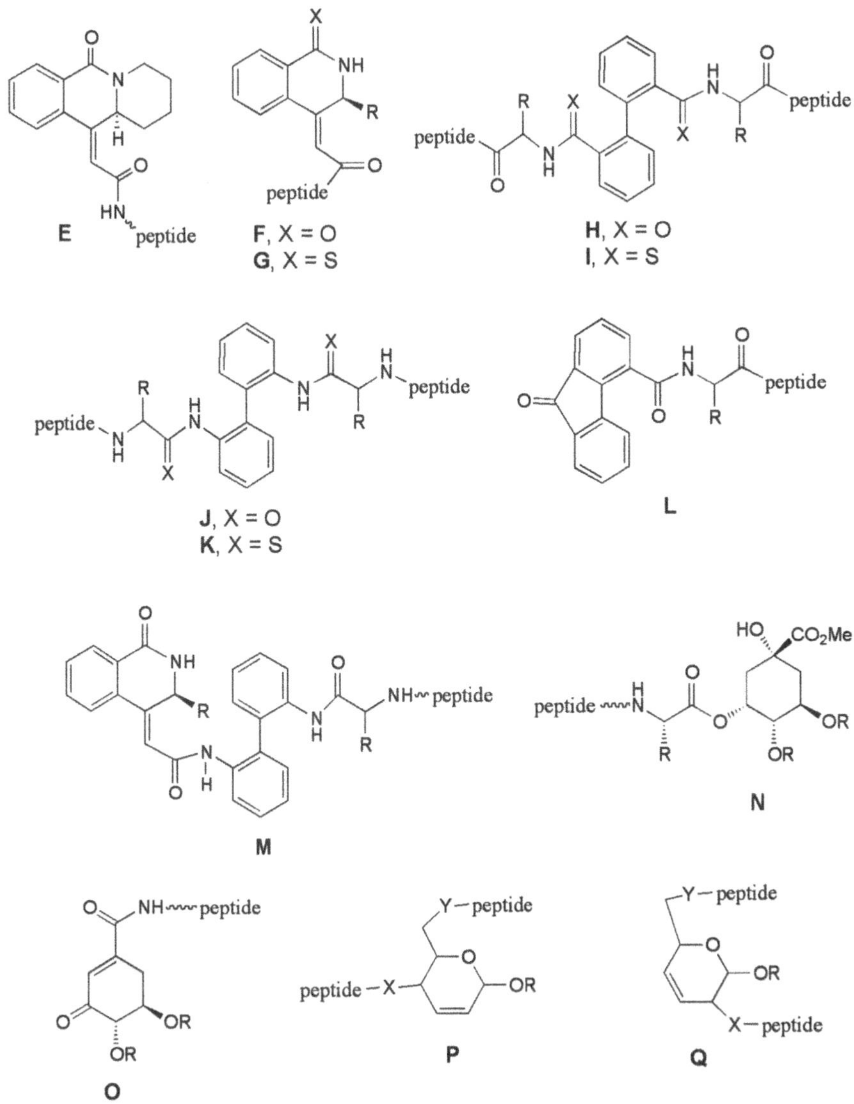

FIGURE 14.4 Generic structures of several peptide–scaffold hybrids prepared in our group.

compounds where the peptide fragment has been linked to a variety of scaffolds (Figure 14.4) such as partially reduced aromatic heterocycles (**E**-**G** and **M**) [52–55], biaryls (**H**-**K** and **M**) [56–61], fluorenones (**L**) [25], cyclohexanes (**N** and **O**) [62], and 3,6-dihydro-2*H*-pyrans (peptide–carbohydrate hybrids, **P** and **Q**) [22,23]. Some of heterocycles (**E**-**G**), biaryls (**H**-**K**), and the combined structure (**M**) have been found to be inhibitors of calpain, a protease involved in a variety of pathological conditions [63].

This last class of peptide–scaffold hybrids (**Q**) is readily synthesized from the Claisen and Overman rearrangements products. Some examples of peptide–carbohydrate hybrids are shown in Figure 14.5.

14.4 SEQUENTIAL REACTIONS FOR THE SYNTHESIS OF POLYANNULAR HETEROCYCLES

Nitrogenated heterocycles are frequently found in natural and synthetic molecules with interesting biological properties, which sometimes have been used as pharmaceuticals [64] and as tools to unveil important biochemical processes [65, 66].

Our group has been interested in the synthesis of alkaloids [67] and related compounds [68, 69]. In connection with our research on peptide–scaffold hybrids (see Section 14.3), we have synthesized a variety of partially reduced aromatic heterocyles [70–73]. As a step further, we have recently developed a methodology based in the sequential transformations of azide–alkene containing an electrophilic center, which is initiated by a 1,3-dipolar cycloaddition, followed by nitrogen extrusion, and latter transformation to an enamine that, in turn, reacts as nucleophile with the electrophile present in the molecule (Figure 14.6).

In this way, we have been able to synthesize a variety of heterocycles [29, 74, 75] (Figure 14.7).

14.5 THE FIRST TOTAL SYNTHESIS OF AMPHOROGYNINE C

A modification of this strategy has been employed for the synthesis of amphorogynine C, a pyrrolizidine alkaloid. This kind of alkaloids is present in more than 6000 plants of different families. As in most of secondary metabolites [76], its exact biological role in the plant is unknown, although they are likely deterrent for potential predators. The biological activity of pyrrolizidine alkaloids in mammals has been studied, finding that many of them are hepatotoxic [77], causing intoxications in both livestock and human being. On the contrary, some beneficial activities have been also found for pyrrolizidine alkaloids, among which, we can mention antiviral and antibacterial activities [78] as well as antagonists of the muscarinic acetylcholine receptor [79].

Païs and coworkers isolated in 1998 four related pyrrolizidine alkaloids, named amphorogynine A-D (Figure 14.8), from the plant *Amphorogyne spicata* (Santalaceae) from New Caledonia [80]. Structurally, these compounds are characterized by substitution only at positions C-1 and C-6, not previously identified in pyrrolizidine alkaloids.

Although two syntheses of amphorogynine A (of 17 and 21 steps with 3% and 5% overall yields) [81, 82] and one synthesis of amphorogynine D [83] have been reported, the preparation of amphorogynine C was not described up to our work [28].

In our approach to the synthesis of amphorogynine C, we used the Claisen rearrangement product **10** as a starting material and an intramolecular olefin–azide cyclization as a key step. The retrosynthetic analysis is shown in Figure 14.9. The

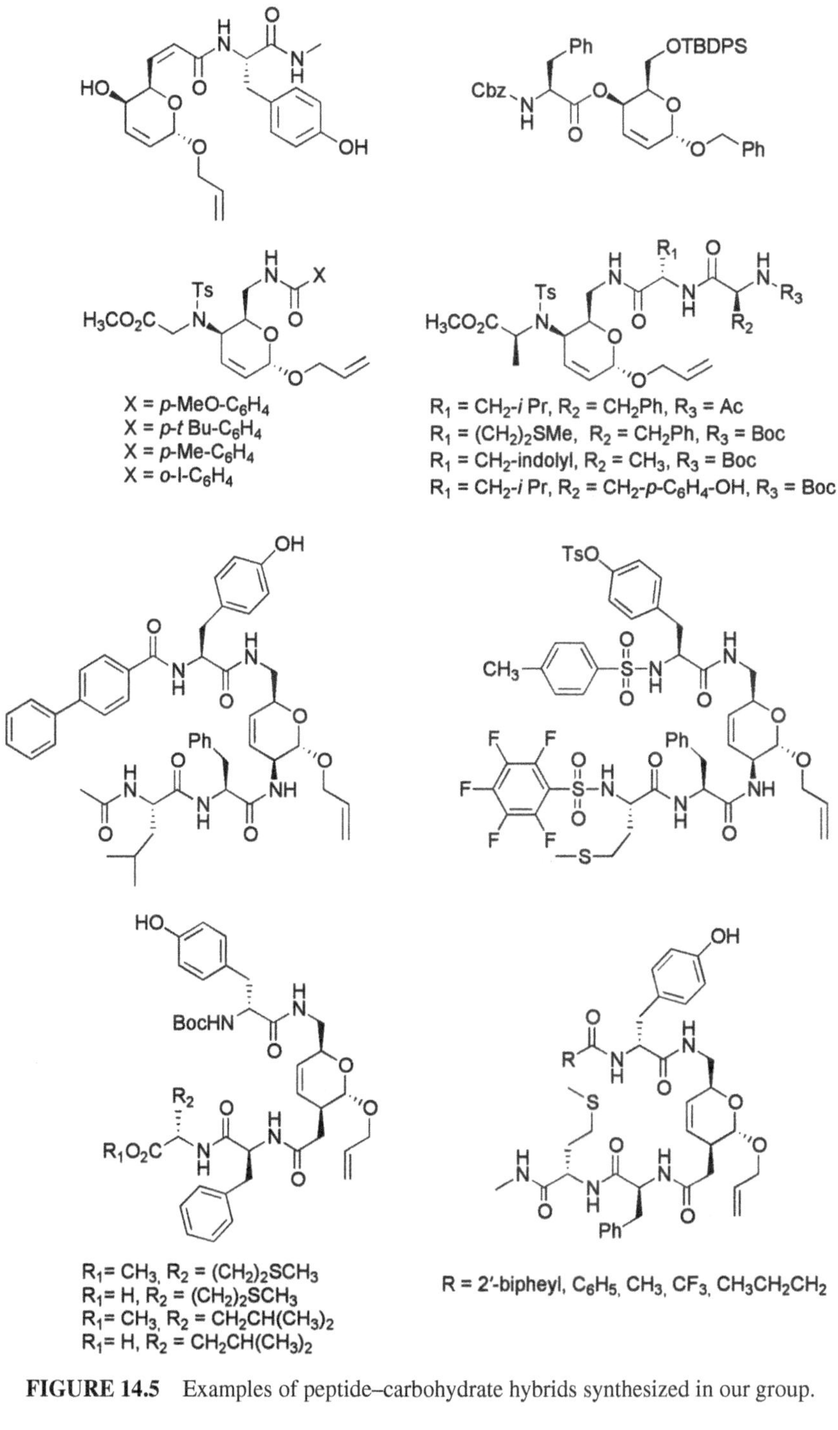

FIGURE 14.5 Examples of peptide–carbohydrate hybrids synthesized in our group.

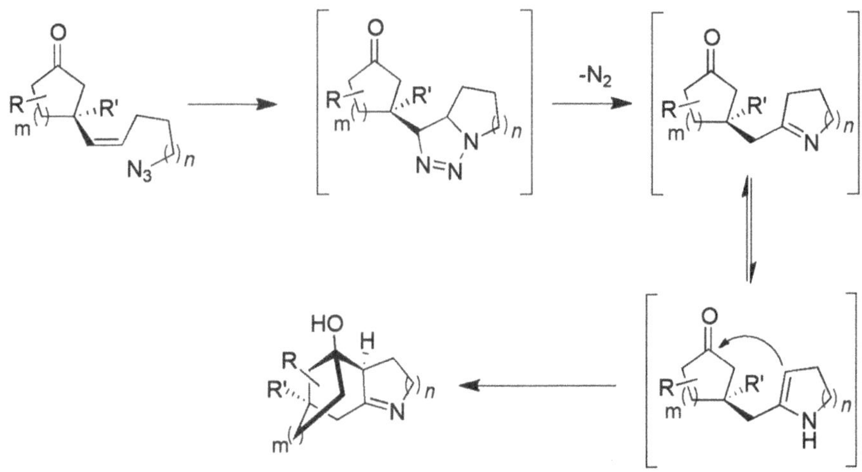

FIGURE 14.6 Strategy for the use of the sequential intramolecular azide–alkene cycloaddition/nitrogen extrusion/imine–enamine isomerization/nucleophilic addition.

natural product could be prepared by reaction of a derivative of dihydroferulic acid with the pyrrolizidine alcohol **14**, which in turn might be obtained from the bromolactone **15** by opening of the lactone to a hydroxy ester and intramolecular nucleophilic substitution. The lactone **15** could be generated by functional group manipulation from the acetal **16**, which might be synthesized by the azide–alkene cycloaddition and nitrogen extrusion from compound **17**, which, in turn, could be obtained from the γ,δ-unsaturated ester **10**. In the key conversion of **17** into **16**, we forecast the epimerization at the position C-3a of the pyrano[4,3-*b*]pyrrole system since compound **13** is more stable than the epimeric imine by around 2.3 kcal mol^{-1}, as computed by quantum chemical calculations carried out at the HF-DFT hybrid method with B3LYP functional and 6-31G(d) basis set.

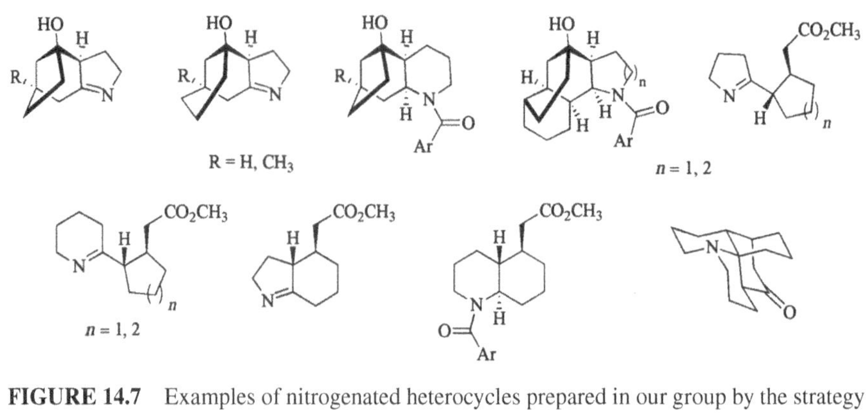

FIGURE 14.7 Examples of nitrogenated heterocycles prepared in our group by the strategy indicated in Figure 14.6.

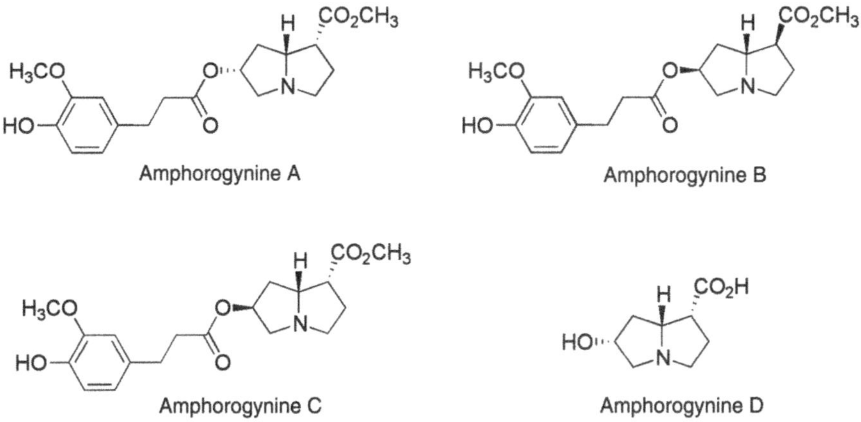

FIGURE 14.8 Structures of the pyrrolizidine alkaloids isolated from *Amphorogyne spicata*.

The synthesis of the target azide–alkene **17** was readily obtained in a three-step sequence with a nearly quantitative yield from the methoxycarbonylmethyl-substituted 3,6-dihydro-2*H*-pyran **10** (Figure 14.10).

With a convenient route to azide–alkene **17** in hand, we tested the key intramolecular 1,3-cycloaddition/nitrogen extrusion to obtain the fused pyrano[4,3-*b*]pyrrole **16**, which was attained along with the expected tricyclic aziridine **18**, as a minor product. The experiments were realized by heating a solution of **17** in either a polar solvent [dimethylformamide (DMF)] or an apolar solvent (toluene) in both sealed tubes (at 140°C for both solvents) or under microwave radiation (at 140°C in DMF and at 120°C in toluene). In the four experiments, the ratio between the pyrroline and the aziridine is approximately the same (4.3:1), which indicates that the observed ratio

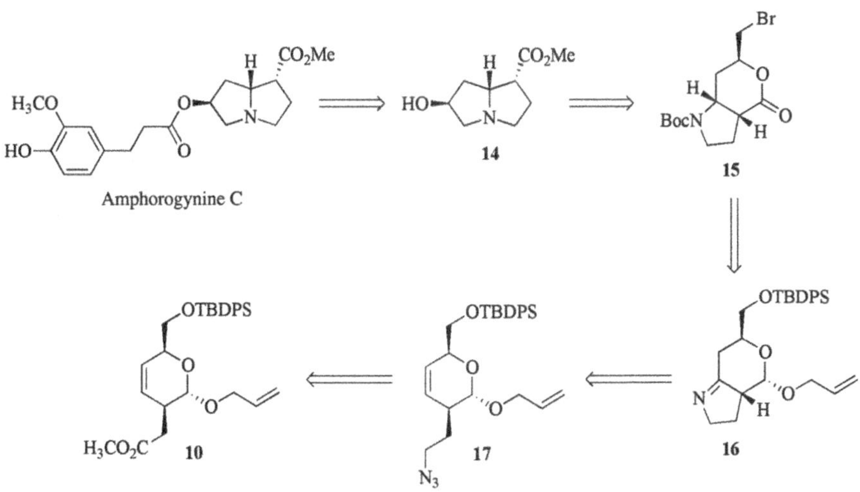

FIGURE 14.9 Retrosynthetic analysis of amphorogynine C from the ester **10**.

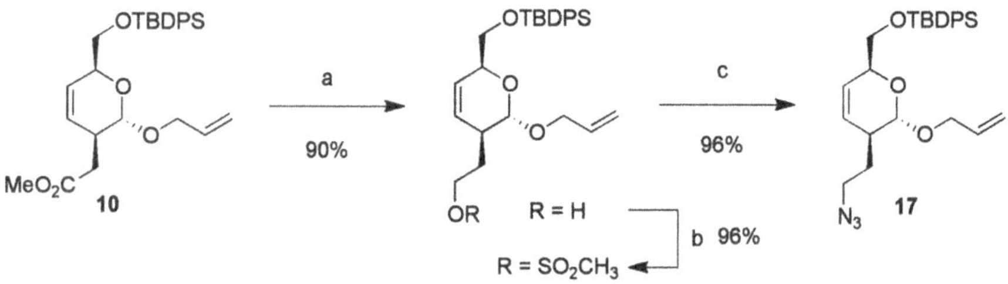

FIGURE 14.10 Conversion of ester **10** into the azide **17**. (a) $LiBH_4$, THF, MeOH, rt; (b) MsCl, Et_3N, CH_2Cl_2, 0°C to rt; and (c) NaN_3, DMF, 60°C.

is the thermodynamic one. The best conditions in terms of yield is the reaction of **17** in toluene at 140°C in a sealed tube that affords the imine **16** in 68% isolated yield for the two sequential reactions (cycloaddition/nitrogen extrusion) along with a 16% isolated yield of tricyclic aziridine **18** (Figure 14.11).

As expected, the epimerization at the position C-3a of the pyrano[4,3-*b*]pyrrole system was observed. This reaction is the result of the thermodynamic favorably isomerization at the vicinal position to the imino group through an imine/enamine tautomerization. The stereochemistry at C-3a of the pyrano[4,3-*b*]pyrrole ring system was readily assessed by measuring the coupling constant of the anomeric position of the carbohydrate and the vicinal proton (H-3a and H-4 of the bicyclic system) in the ^{1}H-NMR spectra [43]. It is well known that when both protons are in *trans*-disposition, the coupling constant J is nearly zero ($J \approx 0$ Hz), as happens with compound **17** (Figure 14.12) and the tricyclic aziridine **18** (Figure 14.13). However, the coupling constant between these two protons in the bicyclic imine **16** is $J = 4.7$ Hz (Figure 14.14), which indicates a *cis*-disposition of both protons. This stereochemical assignment was confirmed by the presence of NOE between the protons H-3a and H-4 in the NOESY spectra as well as by x-ray diffraction structure of the *bis*-(3,5-dinitrobenzoyl) derivative **19** (Figure 14.15) prepared from the imine **16** as indicated in Figure 14.16.

The imine **16** possesses all the carbon and nitrogen atoms of the pyrrolizidine fragment of amphorogynine C as well as the correct absolute configuration at C-3 and C-6 of the natural product. Although the transformation of **16** into amphorogynine C seems quite straightforward, the last steps of the synthesis required considerable experimentation.

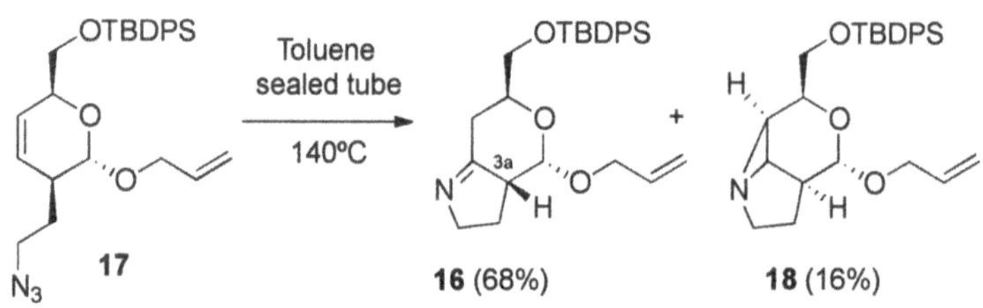

FIGURE 14.11 Intramolecular azide–alkene cycloaddition and nitrogen extrusion from **17**.

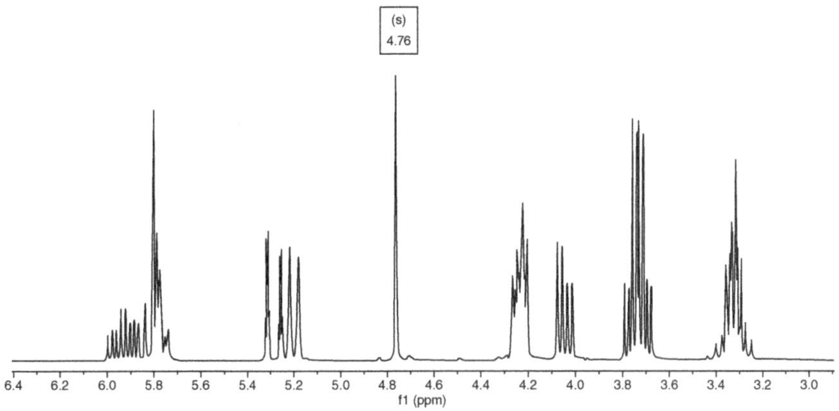

FIGURE 14.12 Portion of the ^{1}H-NMR spectrum of the azide **17**.

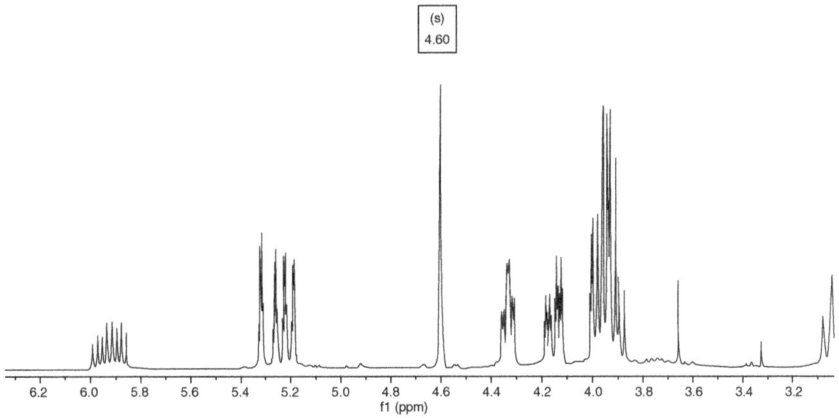

FIGURE 14.13 Portion of the ^{1}H-NMR spectrum of the tricyclic aziridine **18**.

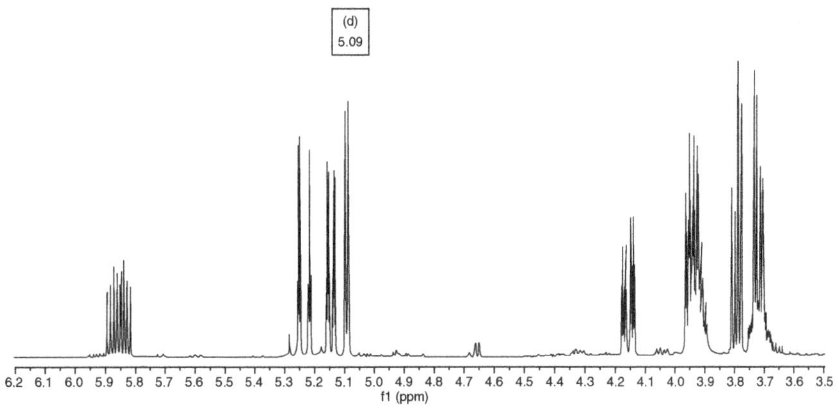

FIGURE 14.14 Portion of the ^{1}H-NMR spectrum of the bicyclic imine **16**.

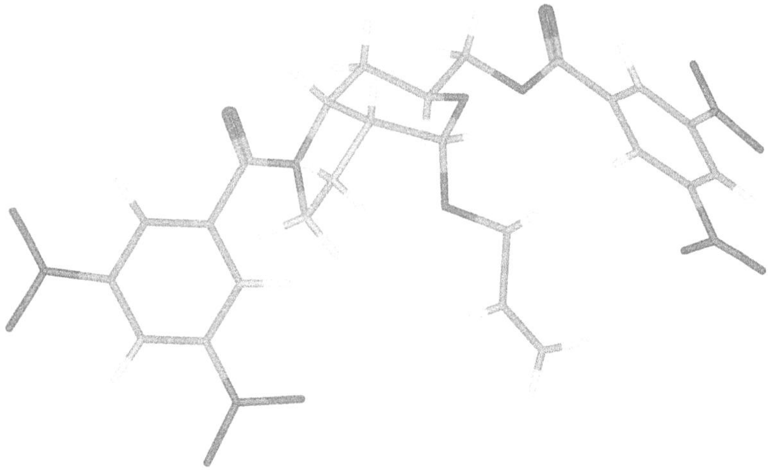

FIGURE 14.15 Solid-state structure of the *bis*-(3,5-dinitrobenzoyl) derivative **19**.

We first attempted the hydrolysis of the allyl acetal of **16**, using catalysis by either $PdCl_2$ or $Pd(PPh_3)_4$ under different experimental conditions; nevertheless, we did not obtain the expected lactol, but only starting material or decomposition products when forcing the conditions. We reasoned that the imine might coordinate the palladium atom hampering the deallylation reaction. In order to circumvent this drawback, we first reduced chemoselectively the imino group to give an amine that, without purification, was protected as tertbutylcarbamate to give **20** in a high yield (Figure 14.17). The all *cis*-configuration of **20** was corroborated by the presence of NOEs between the three protons H-3a, H-4, and H-7 of the pyrano[4,3-*b*]pyrrole system.

The next step was the deprotection of the lactol of **20** using $PdCl_2$ in methanol. Under this condition, the allyl group was unmasked; however, we did not obtain the lactol but the methyl acetal **21**. Changing the solvent to a 9:1 mixture of THF and water gave the expected lactol **22** as an 85:15 mixture of diastereoisomers that could be separated by chromatography and characterized (Figure 14.18), although the mixture of anomers was used in the next step.

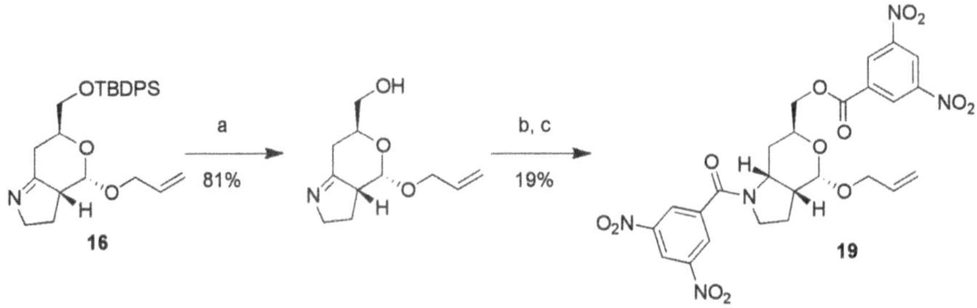

FIGURE 14.16 Synthesis of the *bis*-(3,5-dinitrobenzoyl) derivative **19** from imine **16**. Reagents and conditions: (a) tetrabutylammonium fluoride (TBAF), THF, rt; (b) $NaBH_4$, MeOH, 0°C; and (c) $3,5\text{-}(NO_2)_2C_6H_3COCl$, Et_3N, CH_2Cl_2, 0°C to rt.

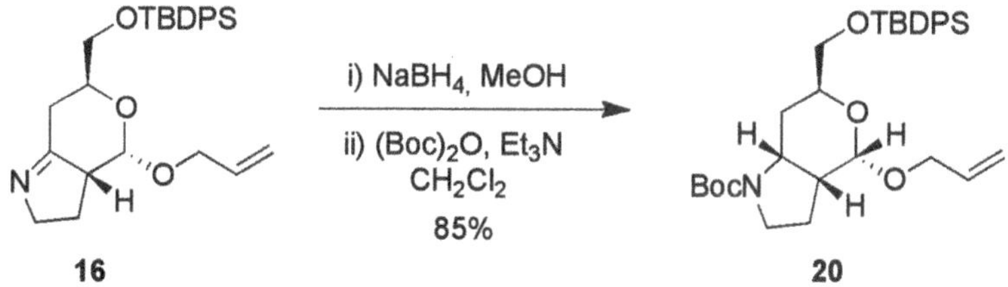

FIGURE 14.17 Synthesis of carbamate **20**.

The planned synthesis was continued with the oxidation of the lactol **22** to the corresponding lactone **23** (Figure 14.19). However, this reaction was quite difficult to realize since most of the tested oxidation reagents (Dess–Martin periodinane, pyridinium chlorochromate, MnO_2, and tetrapropylammonium perruthenate/*N*-methylmorpholine-*N*-oxide) gave starting material even after prolonged reaction times. Only the Jones oxidation (CrO_3/H_2SO_4 in acetone) gave the lactone **23**, albeit in a poor yield (20%). Since the observed low yield was due to the partial hydrolysis of the *tert*-butyldiphenylsilyl group (which goes with a velocity comparable to the oxidation of the lactol, even at the low temperature), we decided to perform the Jones oxidation followed by treatment with tetrabutylammonium fluoride to give the hydroxy-lactone **24**, which in turn was transformed to the bromolactone **15**, which by sequential treatment with trifluoroacetic acid (to hydrolyze the carbamate) and sodium methoxide in methanol (opening of the lactone to the alkoxy-substituted methyl ester and intramolecular nucleophilic substitution either directly or through an intermediate epoxide) to give the highly polar hydroxy-pyrrolizidine **14**, without purification, was condensed with *O*-benzylhydroferulic acid **25** in the presence of 1-ethyl-3-(3-dimethylaminopropyl)carbodiimide (EDC) and 4-dimethylaminopyridine (DMAP) to provide *O*-benzylamphorigynine C in a 50% yield from bromide (three

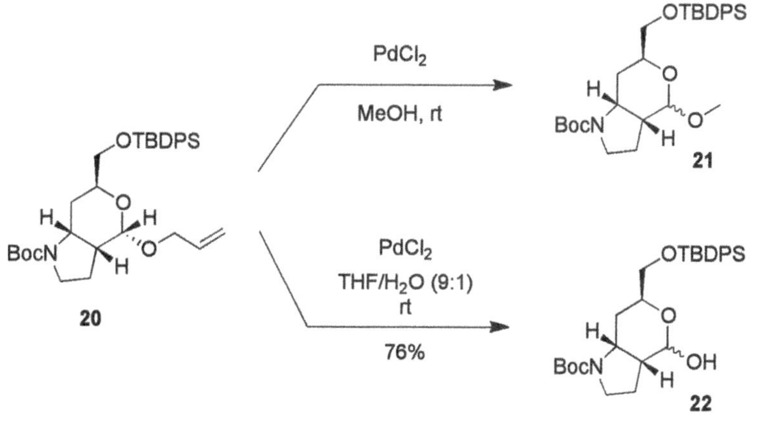

FIGURE 14.18 Synthesis of lactol **22**.

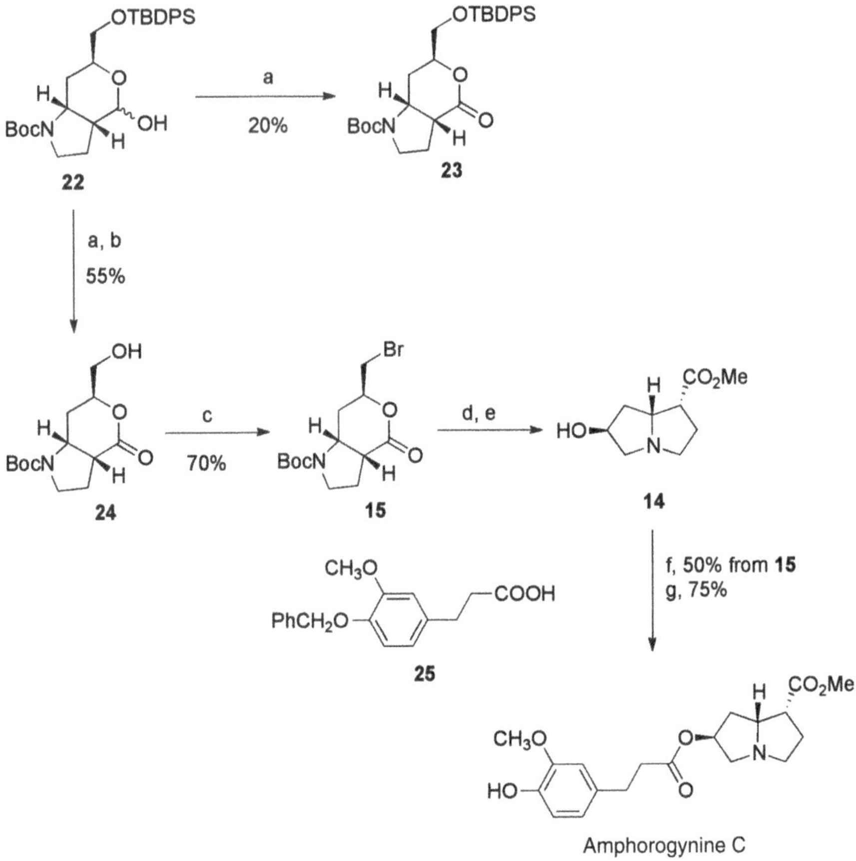

FIGURE 14.19 Final steps in the total synthesis of amphorogynine C. Reagents and conditions: (a) CrO_3, H_2SO_4, acetone, 0°C to rt; (b) TBAF, $3H_2O$, THF, rt; (c) CBr_4, PPh_3, CH_2Cl_2, 0°C to rt; (d) CF_3CO_2H, CH_2Cl_2, rt; (e) NaOMe, MeOH, rt; (f) **25**, EDC, DMAP, CH_2Cl_2, rt; and (g) H_2 (1 atm), Pd/C, AcOEt, rt.

steps, four reactions), and which was debenzylated to the natural product amphorigynine C in 75% (Figure 14.19).

The relative configuration of the natural product was confirmed by x-ray diffraction analysis (Figure 14.20) and the absolute configuration was assigned from the starting material (D-glucal).

Overall, the route goes in 20 steps from commercially available triacetyl-D-glucal in a 3% overall yield (84% as a mean yield for each step). This route allows confirming the relative and absolute configuration of the natural metabolite as well as to prove the synthetic potential of Claisen rearrangement and azide–alkene cycloaddition in carbohydrate chemistry. In addition, the reported methodology provides interesting synthetic intermediates for biological testing. Work along these lines is in progress.

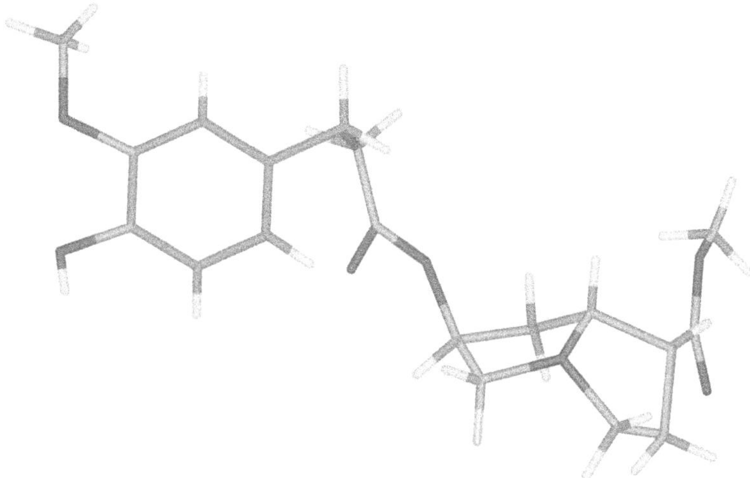

FIGURE 14.20 Solid-state structure of amphorogynine C.

ACKNOWLEDGMENTS

Our research has been generously supported by a project by the Spanish Government (CTQ2010-19295) and, in the initial steps of the peptide–carbohydrate research, by Janssen-Cilag. We thank previous members of our groups, namely Mercedes Alonso, Esperanza Benito, Antonio Chana, Roberto Chicharro, Carlos Miranda, Ana Montero, and Marina Velado for the collaboration.

REFERENCES

[1] Stick, R. V.; Williams, S. *Carbohydrates. The Sweet Molecules of Life*, 2nd edition; Elsevier, **2008**.

[2] Seebach, D.; Hüngerbihler, E. In *Modern Synthetic Methods*, Vol. 2; Scheffold, R. Ed.; Salle + Sauerländer, **1980**; pp 91–173.

[3] Hanessian, S. *Total Synthesis of Natural Products: The "Chiron" Approach*; Pergamon Press, **1983**.

[4] Bols, M. *Carbohydrate Building Blocks*; John Wiley & Sons, Ltd.: Chichester, **1996**.

[5] Boons, G.-J.; Hale, K. J. *Organic Synthesis with Carbohydrates*; John Wiley & Sons, **2000**.

[6] El Ashry, E. S. H., Ed., *Heterocycles from Carbohydrate Precursors*; Springer, **2007**.

[7] Hanessian, S.; Giroux, S.; Merner, B. L. *Design and Strategy in Organic Synthesis. From the Chiron Approach to Catalysis*; John Wiley & Sons, **2013**.

[8] Deslongchamps, P. *Stereoelectronic Effects in Organic Chemistry*; Pergamon Press, **1987**.

[9] Morcuende, A.; Valverde, S.; Herradón, B. *Synlett* **1994**, 89–91.

[10] Herradón, B.; Morcuende, A.; Valverde, S. *Synlett* **1995**, 455–458.

[11] Le, T. X.; Papin, C.; Doisneau, G.; Beau, J.-M. *Angew. Chem. Int. Ed.* **2014**, *53*, 6184–6187.

[12] Mal, K.; Sharma, A.; Das, I. *Chem. Eur. J.* **2014**, *20*, 11932–11945.

[13] Valverde, S.; Herradón, B.; Martín-Lomas, M. *Tetrahedron Lett.* **1985**, *26*, 3731–3734.

[14] Valverde, S.; Martín-Lomas, M.; Herradón, B.; García-Ochoa, S. *Tetrahedron* **1987**, *43*, 1895–1901.

[15] Valverde, S.; Martín-Lomas, M.; Herradón, B. *J. Carbohydr. Chem.* **1987**, *6*, 685–690.

[16] Sánchez-Sancho, F.; Valverde, S.; Herradón, B. *Tetrahedron Asymmetry* **1996**, *7*, 3209–3246.

[17] Valverde, S.; Herradón, B.; Rabanal, R. M.; Martín-Lomas, M. *Can. J. Chem.* **1987**, *65*, 339–342.

[18] Valverde, S.; Hernández, A.; Herradón, B.; Rabanal, R. M.; Martín-Lomas, M. *Tetrahedron* **1987**, *43*, 3499–3504.

[19] Valverde, S.; Herradón, B.; Rabanal, R. M.; Martín-Lomas, M. *Can. J. Chem.* **1987**, *65*, 332–338.

[20] Valverde, S.; Gómez, A. M.; Hernández, A.; Herradón, B.; López, J. C. *J. Chem. Soc., Chem. Comm.* **1995**, 2005–2006.

[21] Valverde, S.; Gómez, A. M.; López, J. C.; Herradón, B. *Tetrahedron Lett.* **1996**, *37*, 1105–1108.

[22] Montero, A.; Mann, E.; Herradón, B. *Eur. J. Org. Chem.* **2004**, 3063–3073.

[23] Montero, A.; Mann, E.; Herradón, B. *Tetrahedron Lett.* **2005**, *46*, 401–405.

[24] Mann, E. *Síntesis y estructura de nuevos híbridos péptido-heterociclo y péptido-areno*, PhD Dissertation; Universidad Autónoma, Madrid, 2002.

[25] Montero, A. *Síntesis, estructura y actividad biológica de híbridos péptido-carbohidrato y péptido-bifenilo*, PhD Dissertation; Universidad Autónoma, Madrid, 2004.

[26] Chicharro, R. *Síntesis, estructura y actividad biológica de heterociclos, híbridos péptido-heterociclo y péptido-carbohidrato*, PhD Dissertation; Universidad Autónoma, Madrid, 2007.

[27] Montero, A.; Benito, E.; Herradón, B. *Tetrahedron Lett.* **2010**, *51*, 277–280.

[28] de Miguel, I.; Velado, M.; Herradón, B.; Mann, E. *Eur. J. Org. Chem.* **2012**, *354*, 4347–4356.

[29] de Miguel, I., *Reacción de cicloadición 1,3-dipolar azida-alqueno intramolecular: Aplicaciones en síntesis de heterociclos y productos naturales*, PhD Dissertation; Universidad Autónoma, Madrid, 2013.

[30] Ferrier, R. J.; Zubkov, O. A. *Org. React.* **2003**, *62*, 569–736.

[31] Woodward, R. B.; Hoffmann, R. *Angew. Chem. Int. Ed. Engl.* **1969**, *8*, 781–853.

[32] Fleming, I. *Molecular Orbitals and Organic Chemical Reactions*; Wiley-VCH: Weinheim, **2009**.

[33] Claisen, L. *Ber. Deuts. Chem. Ges.* **1912**, *45*, 3157–3166.

[34] Hiersemann, M.; Nubbemeyer, U. *The Claisen Rearrangement*; Wiley-VCH: Weinheim, **2007**.

[35] Rehbein, J.; Hiersemann, M. *Synthesis* **2013**, *45*, 1121–1159.

[36] Overman, L. E. *J. Am. Chem. Soc.* **1974**, *96*, 597–599.

[37] Overman, L. E.; Carpenter, N. E. *Org. React.* **2005**, *66*, 1–107.

[38] Sánchez-Sancho, F. *Síntesis y transformaciones selectivas de heterociclos quirales*, PhD Dissertation; Universidad Autónoma, Madrid, 1999.

[39] Sánchez-Sancho, F.; Mann, E.; Herradón, B. *Synlett* **2000**, 509–513.

[40] Mitsunobu, O. *Synthesis* **1981**, 1–28.

[41] Johnson, W. S.; Wertheman, L.; Bartlett, W. R.; Brocksom, T. J.; Li, T.-T.; Faulkner, D. J.; Peterson, M. R. *J. Am. Chem. Soc.* **1970**, *92*, 741–743.

[42] Bao, R.; Valverde, S.; Herradón, B. *Synlett* **1992**, 217–219.

[43] Ferrier, R. J.; Vethaviyansar, N. *J. Chem. Soc. Perkin Trans 1* **1973**, 1791–1793.

[44] Krohn, K.; Flörke, U.; Gehle, D. *J. Carbohydrate Chem.* **2002**, *21*, 431–443.

[45] Hecht, S.; Huc, I. *Foldamers. Structure, Properties, and Applications*; Wiley-VCH: Weinheim, **2007**.

[46] Groner, B., Ed., *Peptides as Drugs. Discovery and Development*; John Wiley & Sons, **2009**.

[47] Sewald, N.; Jakubke, H.-D. *Peptides: Chemistry and Biology*; 2nd edition, Wiley-VCH: Weinheim, **2009**.

[48] Ling, L. L.; Schneider, T.; Peoples, A. J.; Spoering, A. L.; Engels, I.; Conlon, B. P.; Mueller, A.; Schäberle, T. F.; Hughes, D. E.; Epstein, S.; Jones, M.; Lazarides, L.; Steadman, V. A.; Cohen, D. R.; Felix, C. R.; Fetterman, K. A.; Millett, W. P.; Nitti, A. G.; Zullo, A. M.; Chen, C.; Lewis, K. *Nature* **2015**, *517*, 455–459.

[49] Hewitt, W. M.; Leung, S. S. F.; Pye, C. R.; Ponkey, A. R.; Bednarek, M.; Jacobson, M. P.; Lokey, R. S. *J. Am. Chem. Soc.* **2015**, *137*, 715–721.

[50] Hanessian, S.; McNaughton-Smith, G.; Lombart, H.-G.; Lubell, W. D. *Tetrahedron* **1997**, *53*, 12789–12854.

[51] Giannis, A.; Rübsam, F. *Adv. Drug Res.* **1997**, *29*, 1–77.

[52] Mann, E.; Chana, A.; Sánchez-Sancho, F.; Puerta, C.; García-Merino, A.; Herradón, B. *Adv. Synth. Catal.* **2002**, *344*, 855–867.

[53] Salgado, A.; Mann, E.; Sánchez-Sancho, F.; Herradón, B. *Heterocycles* **2003**, *60*, 57–71.

[54] Chicharro, R.; Alonso, M.; Mazo, M. T.; Arán, V. J.; Herradón, B. *ChemMedChem.* **2006**, *1*, 710–714.

[55] Alonso, M.; Chicharro, R.; Miranda, C.; Arán, V. J.; Maestro, M. A.; Herradón, B. *J. Org. Chem.* **2010**, *75*, 342–352.

[56] Mann, E.; Montero, A.; Maestro, M. A.; Herradón, B. *Helv. Chim. Acta* **2002**, *85*, 3624–3638.

[57] Montero, A.; Mann, E.; Chana, A.; Herradón, B. *Chem. Biodiver.* **2004**, *1*, 442–457.

[58] Montero, A.; Albericio, F.; Royo, M.; Herradón, B. *Org. Lett.* **2004**, *6*, 4089–4092.

[59] Herradón, B.; Montero, A.; Mann, E.; Maestro, M. A. *CrystEngComm* **2004**, *6*, 512–521.

[60] Montero, A.; Alonso, M.; Benito, E.; Chana, A.; Mann, E.; Navas, J. M.; Herradón, B. *Bioorg. Med. Chem. Lett.* **2004**, *14*, 2753–2757.

[61] Montero, A.; Albericio, F.; Royo, M.; Herradón, B. *Eur. J. Org. Chem.* **2007**, 1301–1308.

[62] Herradón, B. unpublished results.

[63] Huang, Y.; Wang, K. K. W. *Trends Mol. Med.* **2001**, *7*, 355–362.

[64] Wermuth, C. G., Ed., *The Practice of Medicinal Chemistry*, 3rd edition; Elsevier, **2008**.

[65] Villar, E. A.; Beglov, D.; Chennamadhavuni, S.; Porco, J. A., Jr; Kozakov, D.; Vajda, S.; Whitty, A. *Nature* **2014**, *10*, 723–732.

[66] Mann, E. *SEBBM* **2011**, *169*, 12–16; http://bit.ly/1ygAzuX (visited on January 13, 2015).

[67] Sánchez-Sancho, F.; Herradón, B. *Tetrahedron Asymmetry* **1998**, *9*, 1951–1965.

[68] Ors, M.; Morcuende, A.; Jiménez, M. I.; Valverde, S.; Herradón, B. *Synlett* **1996**, 449–451.

[69] Morcuende, A.; Ors, M.; Valverde, S.; Herradón, B. *J. Org. Chem.* **1996**, *61*, 5264–5270.

[70] Sánchez-Sancho, F.; Mann, E.; Herradón, B. *Adv. Synth. Catal.* **2001**, *343*, 360–368.

[71] Mann, E.; Mahía, J.; Maestro, M. A.; Herradón, B. *J. Mol. Struct.* **2002**, *641*, 101–107.

[72] Sánchez-Sancho, F.; Herradón, B. *Heterocycles*, **2003**, *60*, 1843–1854.

[73] Chicharro, R.; Alonso, M.; Arán, V. J.; Herradón, B. *Tetrahedron Lett.* **2008**, *49*, 2275–2279.

[74] de Miguel, I.; Herradón, B.; Mann, E. *Adv. Synth. Catal.* **2012**, *354*, 1731–1736.

[75] de Miguel, I.; Velado, M.; Herradón, B.; Mann, E. *Adv. Synth. Catal.* **2013**, *355*, 1237–1242.

[76] Krishnaswamy, N. R. *Chemistry of Natural Products. A Unified Approach*, 2nd edition; CRC Press, **2010**.

[77] Dreger, M.; Stanislawska, M.; Krajewska-Patan, A.; Mielcarek, S.; Mikolajczak, P. L.; Buchwald, W. *Herba Pol.* **2009**, *55*, 127–146.

[78] Asano, N.; Nash, R. J.; Molyneux, R. J.; Fleet, G. W. J. *Tetrahedron Asymmetry*, **2000**, *11*, 1645–1680.

[79] Hahn, K. N.; Fadeyi, O. O.; Cho, H. P.; Lindsley, C. W. *Tetrahedron Lett.* **2012**, *53*, 3577–3580.

[80] Huong, D. T. T.; Martin, M.-T.; Litaudon, M.; Sévenet, T.; Païs, M. *J. Nat. Prod.* **1998**, *61*, 1444–1446.

[81] Yoda, H.; Egawa, T.; Takabe, K. *Tetrahedron Lett.* **2003**, *44*, 1643–1646.

[82] Roche, C.; Delair, P.; Greene, A. *Org. Lett.* **2003**, *5*, 1741–1744.

[83] Roche, C.; Kadlecikova, K.; Veyron, A.; Delair, P.; Philouze, C.; Greene, A. E. *J. Org. Chem.* **2005**, *70*, 8352–8363.

15

PALLADIUM- AND NICKEL-CATALYZED STEREOSELECTIVE SYNTHESIS OF GLYCOSYL TRICHLOROACETAMIDES AND THEIR CONVERSION TO α- AND β-UREA GLYCOSIDES

Nathaniel H. Park, Eric T. Sletten, Matthew J. McKay, and Hien M. Nguyen

Department of Chemistry, University of Iowa, Iowa City, IA, USA

15.1 INTRODUCTION

The advancement of new methods for the stereoselective synthesis of naturally or unnaturally occurring glycosidic bonds is of immense importance for the development of potent, carbohydrate-based therapeutics. Development of such technology would not only enable easier access to compounds difficult to synthesize, but also facilitate selective modification of biologically active compounds to promote their enzymatic and chemical stability [1]. In this regard, the glycosyl urea has emerged as an important functional group for the replacement of labile *N*- and *O*-linked glycosidic linkages in the construction of *N*-linked glycopeptide mimetics [2, 3], neoglycoconjugates [4], and pseudooligosaccharides [5, 6]. The glycosyl urea is also observed in nature as a key structural feature of cinodine [7] and coumamidine [8] antibiotics (Figure 15.1). These antibiotics display a broad spectrum of activity against gram-negative bacteria. Instead of inhibiting protein synthesis, like aminoglycosides, these

Domino and Intramolecular Rearrangement Reactions as Advanced Synthetic Methods in Glycoscience, First Edition. Edited by Zbigniew J. Witczak and Roman Bielski.

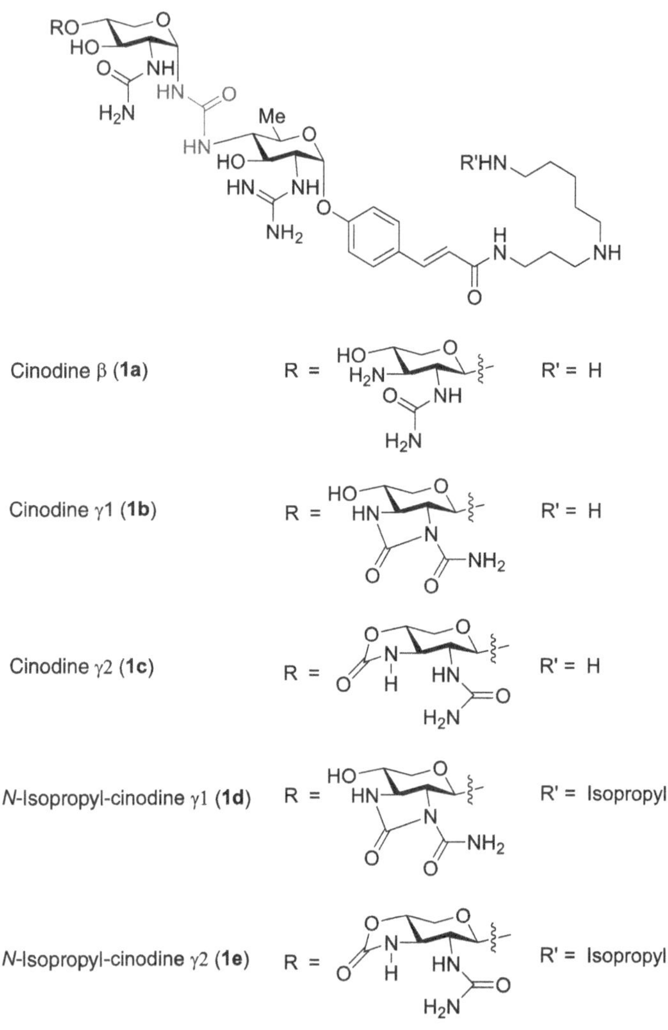

FIGURE 15.1 Cinodine antibiotics and semisynthetic derivatives.

compounds bind directly to bacterial DNA and the DNA organizational enzyme, DNA gyrase B [9–12].

Currently, there are limited methodologies available for the selective synthesis of glycosyl ureas and, in particular, α-glycosyl urea (Figure 15.2). These approaches often rely upon neighboring group participation to achieve the desired selectivity and they typically require numerous steps [13,14]. One approach, as reported by Ichikawa and coworkers, involves the generation of the glycosyl isocyanate *in situ*, which is trapped by the amine to form the corresponding glycosyl urea [2–4, 15] (Figure 15.3, eq. (a)). Although Ichikawa's approach retains the anomeric stereochemistry present in the starting material, the low selectivity (4:1 α:β) in the starting glycosyl azide, in

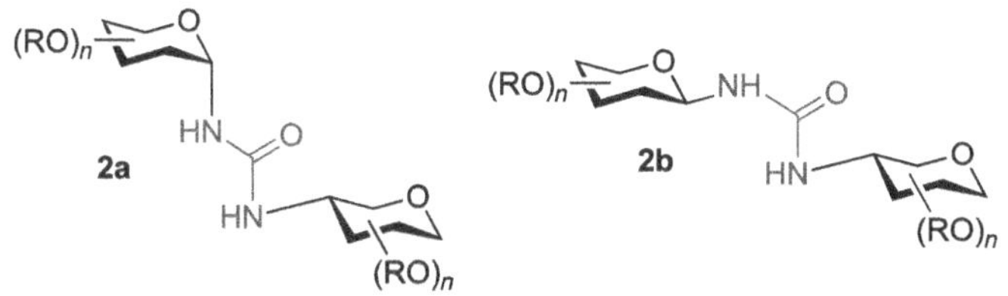

FIGURE 15.2 General structures of α- and β-glycosyl ureas.

combination with the numerous steps needed to access the isocyanate intermediate, limits the overall utility of this method [2–4]. In contrast, Bernandi and coworkers have developed a one-pot transformation in which glycosyl azide **3d** undergoes a modified Staudinger reduction with trimethylphosphine to generate the corresponding iminophosphorane **3e**, which is then trapped by an isocyanate to produce the corresponding urea **3f** [16] (Figure 15.3, eq. (b)). While this method is able to give

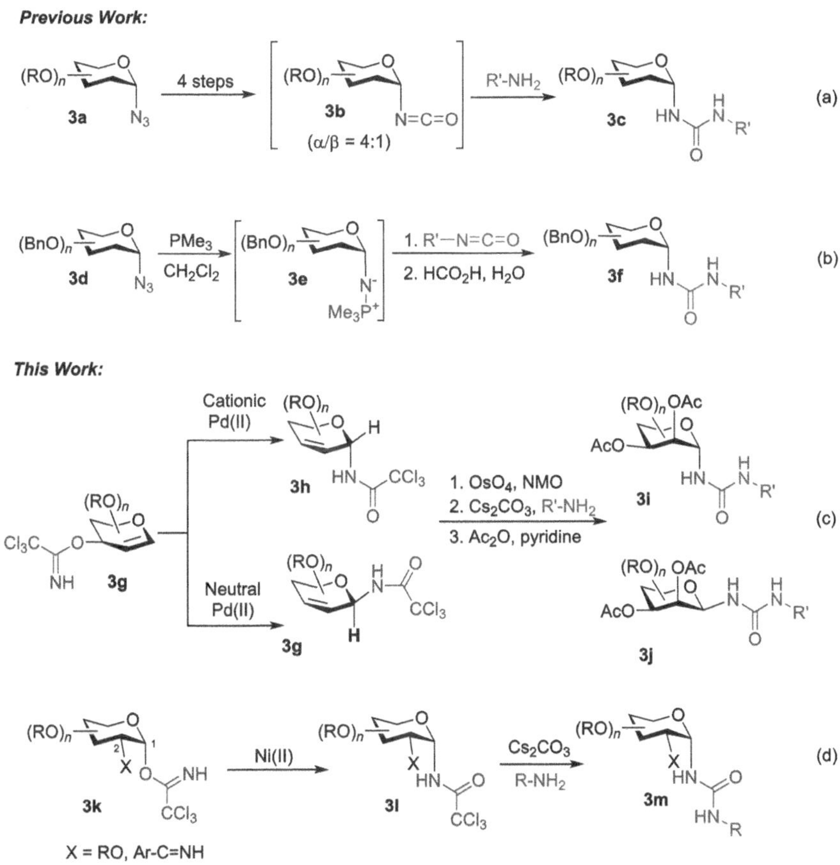

FIGURE 15.3 Methods for selective formation of glycosyl ureas.

the desired α-urea glycoside in a single step, the low reactivity of the iminophosphorane and the need to use the highly air-sensitive and pyrophoric trimethylphosphine hampers the overall applicability of this approach.

To overcome the challenges of synthesizing glycosyl ureas and provide high selectivity and the capability of being applied to a broad substrate scope, a new transition metal-catalyzed approach for the preparation of glycosyl ureas was conceived. Here, the use of a palladium or nickel catalyst in combination with a glycal or glycosyl trichloroacetimidate (**3g** or **3k**; Figure 15.3) promoted a stereoselective transformation to the corresponding trichloroacetamide (**3h, 3g,** or **3l**; Figure 15.3), which could then be readily modified to produce the desired α- or β-urea glycoside (**3i, 3j,** or **3m**; Figure 15.3).

15.2 DEVELOPMENT OF THE PALLADIUM(II)-CATALYZED GLYCAL TRICHLOROACETIMIDATE REARRANGEMENT

While Overman has extensively studied the [3,3]-sigmatropic rearrangement of allylic trichloroacetimidates [17–19], this method has never been utilized for the control of the anomeric selectivity in carbohydrate synthesis. In this case, there are two potential pathways for rearrangement that the glycal trichloroacetimidate may undergo when reacted with a palladium catalyst. In one path, the cationic palladium(II) catalyst would act as a Lewis acid and coordinate the trichloroacetimidate nitrogen functionality (**4b**; Figure 15.4), promoting the ionization of the glycal trichloroacetimidate to the corresponding ion-pair **4c** [20]. The ion pair can then undergo a stereoelectronically favored recombination to form α-trichloroacetamide **4d** (Figure 15.4).

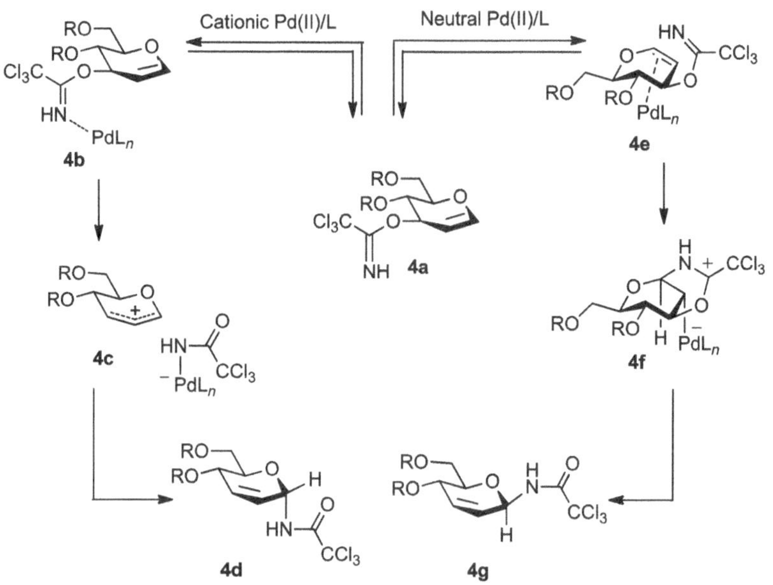

FIGURE 15.4 Proposed mechanism for palladium(II)-catalyzed rearrangement of glycal trichloroacetimidates.

In contrast, the use of a neutral palladium(II) catalyst would favor coordination of the glycal double bound to form the π-complex intermediate **4e** (Figure 15.4). This would activate the double bond toward nucleophilic attack by the trichloroacetimidate nitrogen function, facilitating generation of the intermediate σ-complex **4f** (Figure 15.4); its fragmentation should provide β-trichloroacetamide **4g** (Figure 15.4). The success of either pathway relies on the development of a suitable catalyst system that can facilitate and favor either the ionization recombination or cyclization-induced rearrangement pathways. Thus, the choice of a suitable supporting ligand for palladium is critical for determining the success of the overall transformation.

To investigate the feasibility of this proposed transformation, glucal trichloroacetimidate **5a** was treated with 2.5 mol % $Pd(PhCN)_2Cl_2$ in CH_2Cl_2 at 25°C for 2 h (Figure 15.5). The corresponding glucal trichloroacetamide **6a** was obtained in a 60% yield as a 1:1 mixture of α:β anomers (entry 1; Figure 15.5). Addition

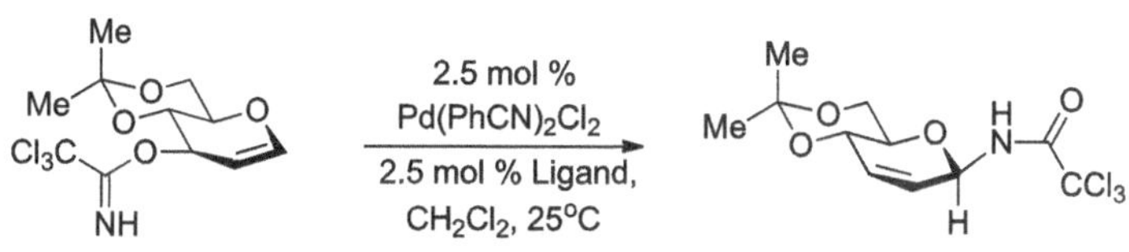

5a **6a-β**

Entry	Phosphine Ligand	Additive	Time	Yield[a]	α:β[b]
1	none	none	2 h	60%	1:1
2	none	4Å MS	2 h	75%	1:1
3	Ph_3P	none	16 h	83%	1:2
4	BINAP	none	20 h	85%	1:2
5	RuPhos	none	16 h	77%	1:3
6	*t*-BuX-Phos	none	25 h	73%	1:4
7	JohnPhos	none	23 h	75%	1:7
8	**TTMPP**	**none**	**16 h**	**89%**	**1:7**

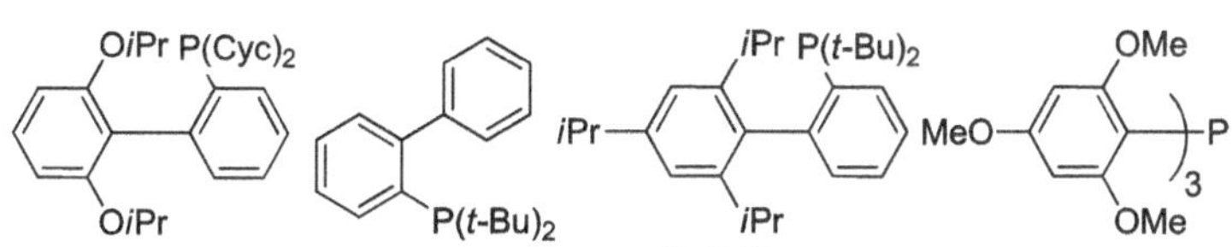

RuPhos JohnPhos *t*-BuX-Phos TTMPP

[a] Isolated yield. [b] ^1H NMR ratio.

FIGURE 15.5 Neutral Pd(II)-catalyzed stereoselective formation of β-*N*-glycosyl trichloroacetamide.

of 4Å molecular sieves increased the yield to 75%, but no improvement in selectivity was observed (entry 2; Figure 15.5). To modulate the steric and electronic properties of the palladium(II) catalyst and potentially increase the selectivity, phosphine ligands were applied. Triphenylphosphine and 2,2′-bis(diphenylphosphino)-1,1′-binaphthyl were able to improve the α:β selectivity to 1:2 with a slight increase in yield (entries 3 and 4; Figure 15.5). Switching the ligand to the more sterically encumbered biaryl phosphine ligands did improve the selectivity, with JohnPhos providing the best result (1:7 α:β, 75% yield, entry 7; Figure 15.5). Alternatively, using tris(trimethoxyphenyl)phosphine (TTMPP) as the ligand increased the yield to 89%, while maintaining the same 1:7 α:β selectivity at the anomeric center (entry 8; Figure 15.5).

Despite the moderate to good selectivities obtained with the phosphine ligands, the long reaction times suggested that the rate of nucleophilic attack of the trichloroacetimidate nitrogen onto the palladium–olefin complex might be slow. This likely allows increased competition from the ionization pathway (as shown in Figure 15.4), resulting in an erosion of β-selectivity. Thus, increasing the electrophilicity of the olefin via modulation of the electronic properties of the palladium catalyst could provide for both increased selectivity and faster reaction times. To validate this hypothesis, several σ-donor ligands were investigated. Addition of 10 mol % salicylaldehyde (L-1) [20] to reactions containing *t*BuXPhos and TTMPP leads to significantly decreased reaction times, while improving β-selectivity (entries 1 and 2; Figure 15.6). Increasing or decreasing the temperature leads to diminished selectivity, suggesting that an equilibration of both pathways occurs (entries 3 and 4; Figure 15.6). Performing the reaction in the absence of the TTMPP phosphine ligand resulted in a decreased selectivity, demonstrating that both the TTMPP and L-1 were needed to achieve high β-selectivity (entry 5; Figure 15.6). Next, the electronic properties of the σ-donor ligands were evaluated. The more electron-donating L-2 and L-3 resulted in a similar selectivity as L-1 but required longer reaction times (entries 6 and 7; Figure 15.6). In contrast, the use of the more electron-deficient L-4 and L-5 resulted in greatly improved selectivity and shorter reaction times (entries 8 and 9; Figure 15.6). A control experiment was conducted with 4-hydroxybenzaldehyde (L-6) as a ligand (entry 10; Figure 15.6). This led to a lower yield of the desired product and low α:β selectivity, suggesting that chelation of the σ-donor ligands to the palladium center is crucial for high selectivity.

As noted earlier, a more cationic palladium complex will likely favor the ionization-recombination pathway, resulting in a stereoelectronic preference for the α-anomer. To test this hypothesis, the more cationic $Pd(CH_3CN)_4(BF_4)_2$ catalyst [21] was utilized, which—in the absence of any added ligand—provided the desired product **7a** in a 73% yield and with 9:1 α:β selectivity (entry 1; Figure 15.7). The selectivity could be further improved via an addition of L-1 to the reaction mixture, which afforded the desired α-anomer **7a** in 13:1 to 14:1 α:β selectivity (entries 2 and 3; Figure 15.7). To determine whether small amounts of HBF_4 generated from the catalyst could be responsible for the observed course of the reaction, 2,6-di-*tert*-butylphosphine was added (entry 5; Figure 15.7) and the desired product was isolated

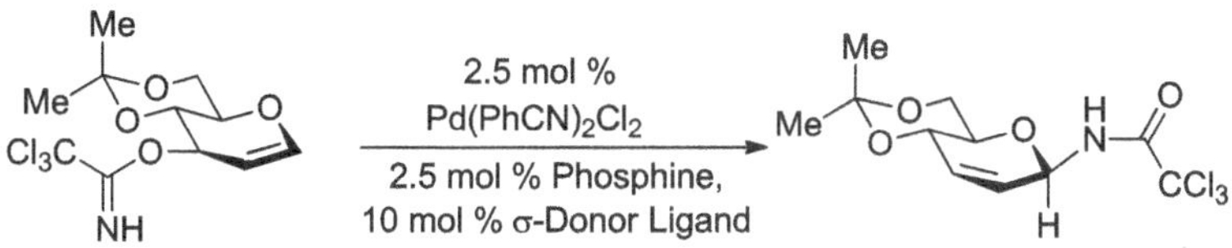

5a **6a-β**

Entry	Phosphine Ligand	σ-Donor Ligand	Time	Temp	Yield[a]	α:β[b]
1	*t*-BuX-Phos	L-1	6 h	25°C	70%	1:7
2	**TTMPP**	**L-1**	**4 h**	**25°C**	**86%**	**1:9**
3	TTMPP	L-1	8 h	0°C	30%	1:6
4	TTMPP	L-1	2 h	40°C	75%	1:4
5	None	L-1	1 h	25°C	71%	1:2
6	TTMPP	L-2	24 h	25°C	82%	1:8
7	TTMPP	L-3	7 h	25°C	70%	1:7
8	**TTMPP**	**L-4**	**2.5 h**	**25°C**	**95%**	**1:12**
9	**TTMPP**	**L-5**	**2.5 h**	**25°C**	**88%**	**1:16**
10	TTMPP	L-6	3 h	25°C	63%	1:3

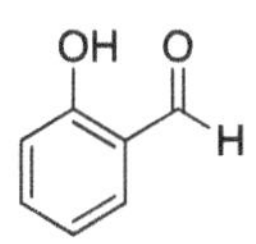

L-1
Salicylaldehyde

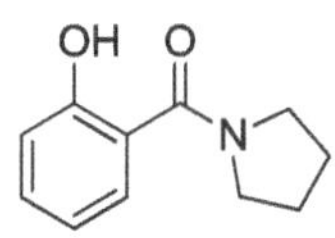

L-2
N-(2-Hydroxylbenzoyl)-pyrrolidine

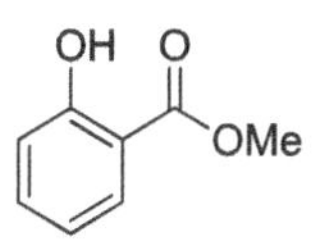

L-3
Methyl Salicylate

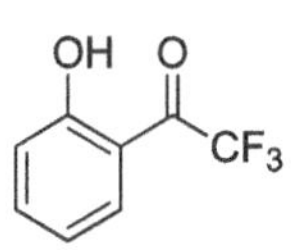

L-4
2-Trifluoroacetylphenol

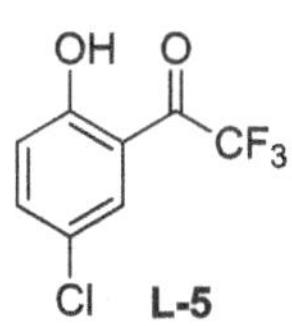

L-5
4-Chloro-2-trifluoroacetylphenol

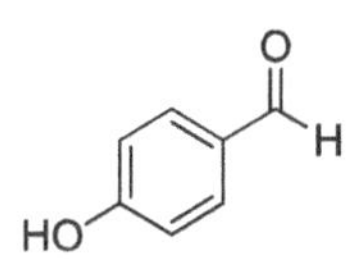

L-6
4-hydroxybenzaldehyde

[a] Isolated yield. [b] ^1H NMR ratio.

FIGURE 15.6 Optimization for stereoselective synthesis of β-*N*-glycosyl trichloroacetamide with σ-donor ligands.

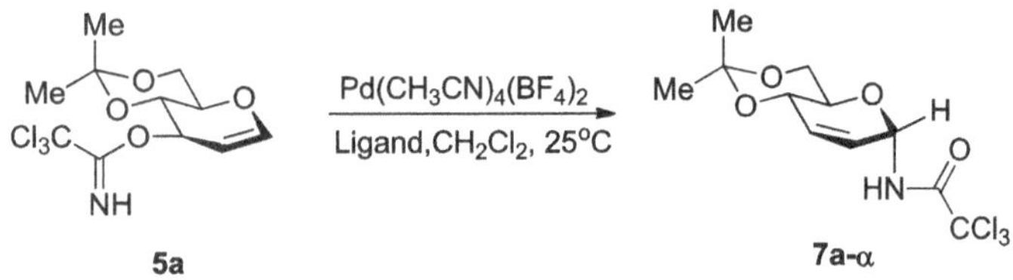

Entry	Palladium	Ligand	Additive	Time	Yield[a]	α:β[b]
1	2.5 mol %	None	None	45 min	73%	9:1
2	2.5 mol %	10 mol % L-1	None	1 h	80%	14:1
3	0.5 mol %	2 mol % L-1	None	1 h	82%	13:1
4	0.1 mol %	0.4 mol % L-1	None	2 h	78%	9:1
5	0.5 mol %	2 mol % L-1	DTBP	1 h	79%	9:1
6	**0.5 mol %**	**2 mol % L-4**	**None**	**45 min**	**93%**	**14:1**
7	0.5 mol %	2 mol % L-5	None	45 min	83%	15:1
8	0.5 mol %	2 mol % L-6[c]	None	45 min	58%	10:1

[a] Isolated yield. [b] ^{1}H NMR ratio. [c] L6= 4-hydroxybenzaldehyde

FIGURE 15.7 Cationic Pd(II)-catalyzed stereoselective formation of α-*N*-glycosyl trichloroacetamide.

in comparable yields and selectivity as in entries 1–4 (Figure 15.7). As with the neutral palladium-catalyzed transformation, switching to L-4 and L-5 resulted in both increased yield and selectivity (entries 6 and 7; Figure 15.7). A control experiment was also conducted with 4-hydroxybenzaldehyde (L-6), while the desired product **7a** was obtained in a 10:1 α:β selectivity; its yield was, however, much lower (entry 8; Figure 15.7). This is likely the result of acid-promoted decomposition of the starting material.

With the optimized conditions for neutral and cationic palladium-catalyzed rearrangement identified, their applicability was investigated in a series of glycal trichloroacetamides **8a–8g** (Figure 15.8). For the neutral palladium conditions, both L-1 and L-5 proved to be suitable in promoting the transformation, with L-5 typically providing the desired trichloroacetamides with higher β-selectivities (Figure 15.8). Overall, glycal substrates containing the cyclic 4,6-acetal protecting groups (**8a–8e**; Figure 15.8) were produced in an excellent β-selectivity. This is likely due to the restriction in conformation that favors the formation of the β-anomer while limiting ionization. In the cases where acyclic protecting groups were used (**8f** and **8g**; Figure 15.8), a mixture of anomers was obtained, with little to no β-selectivity. Under the cationic palladium conditions, high levels of α-selectivity were obtained for nearly all substrates **9a–9h** examined (Figure 15.9) in the rearrangement. Here, the use of L-4

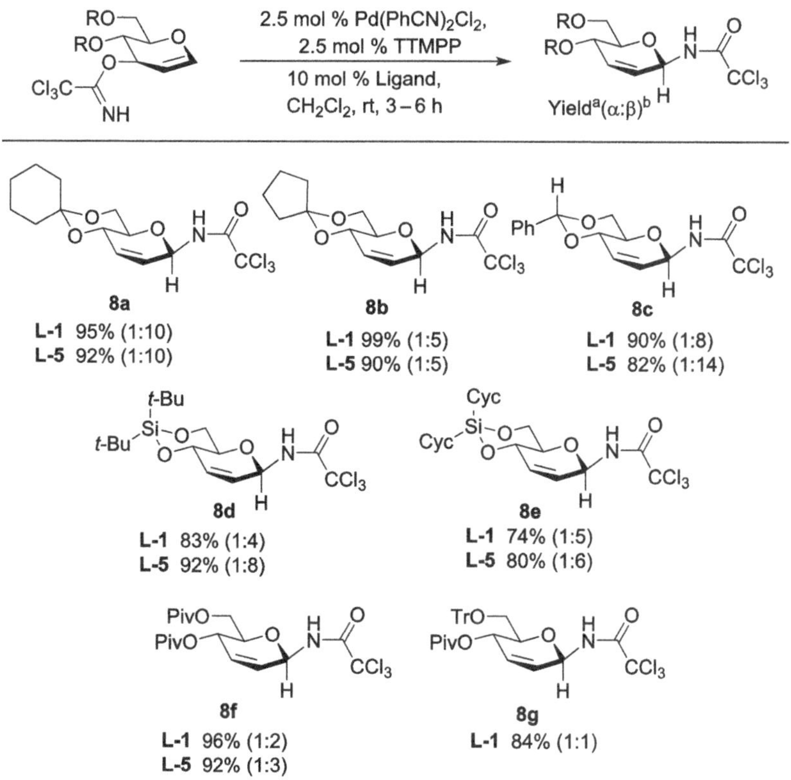

[a] Isolated yield. [b] ^{1}H NMR ratio.

FIGURE 15.8 Synthesis of β-*N*-glycosyl trichloroacetamides.

typically resulted in the improved selectivity and yield compared with L-1. The lack of the influence of different protecting groups on the observed α-selectivity implies that the stereoselectivity was under catalyst control.

In addition to evaluating the scope of the reaction with different glycal substrates, the reaction was investigated to determine whether it could be performed with a catalytic amount of Lewis acid or undergo thermally induced rearrangement. By using 0.5 mol % trimethylsilyl trifluoromethanesulfonate (TMSOTf) with glucal trichloroacetimidate **5a** at 0°C, none of the desired products was generated in the reaction and only decomposition of the starting material was observed (Figure 15.10). Under thermal conditions, the complete rearrangement was observed in 6 h at 120°C, providing **6a** as a single anomer with complete β-selectivity as determined by ^{1}H NMR analysis (Figure 15.10). This outcome is consistent with the concerted nature of the [3,3]-sigmatropic rearrangement, although the necessity of utilizing high temperatures makes it less desirable approach to installing the β-C1-trichloroacetamide.

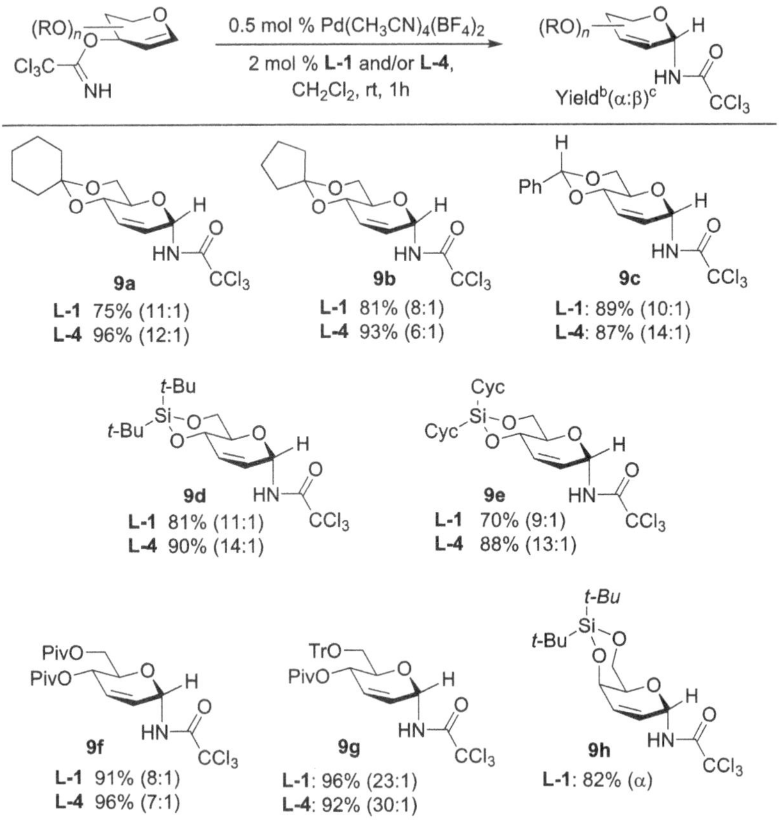

[a] Isolated yield. [b] ^{1}H NMR ratio.

FIGURE 15.9 Synthesis of α-*N*-glycosyl trichloroacetamides.

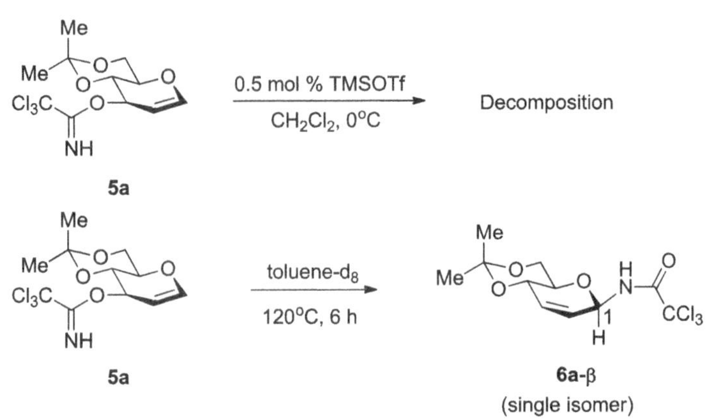

FIGURE 15.10 Comparative studies against cationic Pd(II)-catalyzed [3,3]-sigmatropic rearrangement.

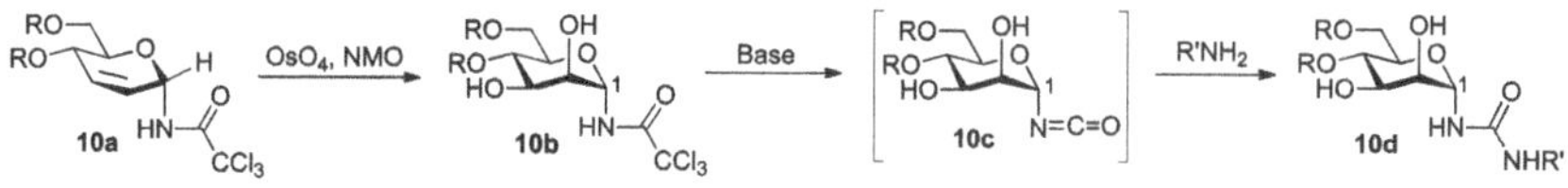

FIGURE 15.11 Strategy for the synthesis of glycosyl ureas.

15.3 STEREOSELECTIVE SYNTHESIS OF GLYCOSYL UREAS FROM GLYCAL TRICHLOROACETIMIDATES

Having successfully established conditions for the stereoselective formation of both α- and β-glycal trichloroacetamides via palladium catalysis, we turned our attention to their conversion into the corresponding glycosyl ureas. This strategy relies on performing a catalytic dihydroxylation of glycal **10a** to diol **10b** (Figure 15.11), which can be subsequently treated with a base to generate *in situ* the glycosyl isocyanate intermediate **10c** (Figure 15.11). Trapping **10c** with an amine nucleophile should produce the desired glycosyl urea **10d** (Figure 15.11). This approach should exclude the epimerization at the C1-anomeric center, thereby retaining the α- or β-selectivity obtained from the palladium-catalyzed rearrangement reaction. One major limitation of this strategy, however, is that the trichloroacetamide may undergo hydrolysis to generate glycosyl amine in the presence of a base.

To determine whether this approach is feasible, trichloroacetamide **11b** was treated with several different bases in combination with using *n*-butylamine (**12**) as the nucleophile (Figure 15.12). The use of NaH or K_2CO_3 in tetrahydrofuran (THF) did not induce any reaction (entry 1 and 2; Figure 15.12). Changing the solvent to DMF and using K_2CO_3 as the base promoted hydrolysis of **11b**, which then underwent epimerization to the β-anomer (entry 3; Figure 15.12). In contrast, by using Cs_2CO_3 as the base, trichloroacetamide **11b** was readily converted to the desired α-glycosyl urea **11c** with no erosion of stereochemical integrity at the anomeric center (entry 4; Figure 15.12). These conditions were applicable to a wide array of primary and secondary aliphatic amines, generating the corresponding α-glycosyl ureas in 50–70% yields (Figure 15.13). These conditions were also suitable in synthesizing unsymmetrical

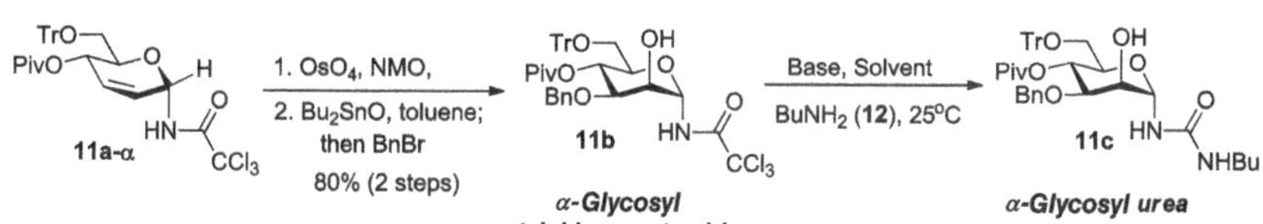

Entry	Base	Solvent	Results
1	NaH	THF	No reaction
2	K_2CO_3	THF	No reaction
3	K_2CO_3	DMF	Hydrolysis
4	**Cs_2CO_3**	**DMF**	**74%**

FIGURE 15.12 Transformation of *N*-glycosyl trichloroacetamide into glycosyl urea.

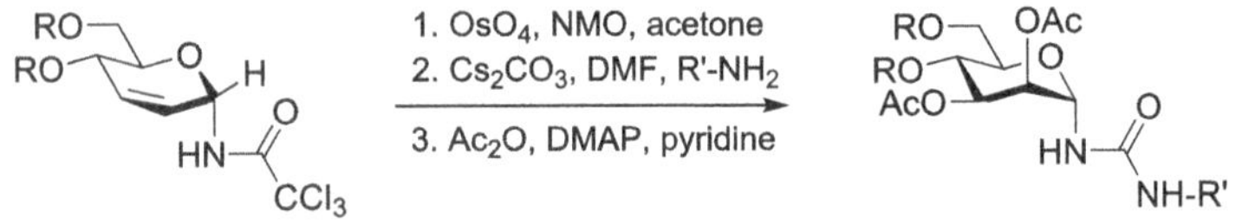

Entry	α-Trichloroacetamide	Amine	Urea	Yield (3 steps)
1	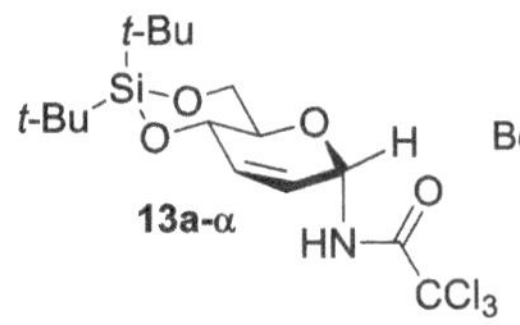 13a-α	Benzylamine	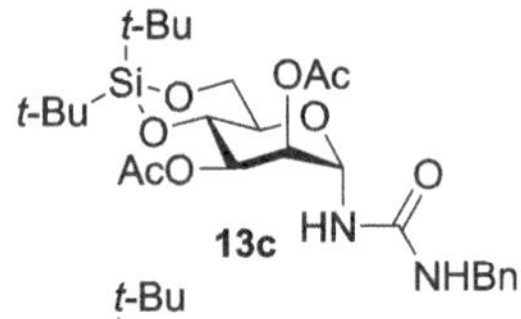 13c	50%
2	13a-α	*n*-Butylamine	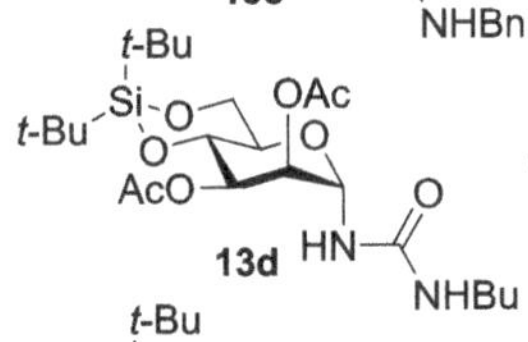 13d	58%
3	13a-α	Cyclohexylamine	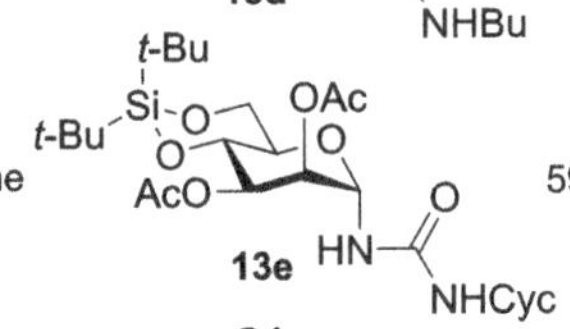 13e	59%
4	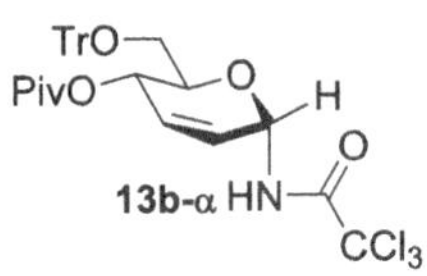 13b-α	Pyrrolidine	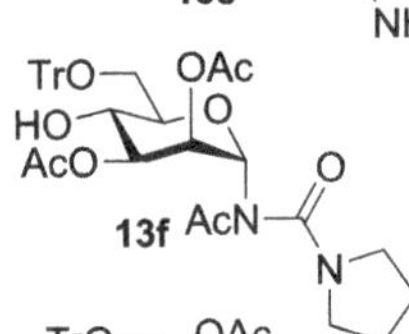 13f	58%
5	13b-α	Piperidine	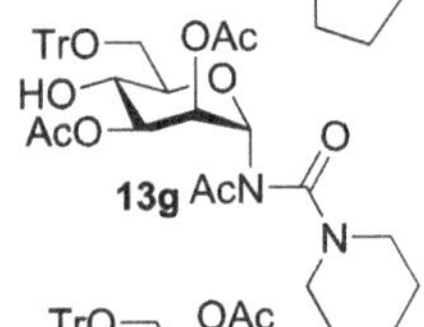 13g	61%
6	13b-α	Methylpiperazine	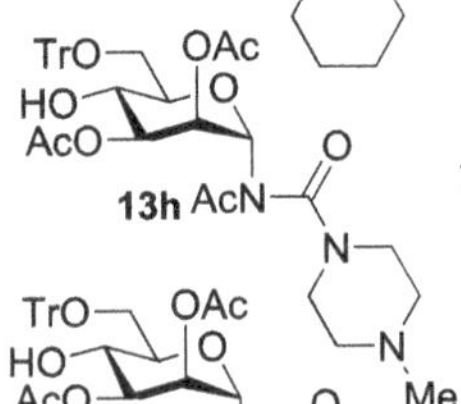 13h	70%
7	13b-α	Pyridylpiperazine	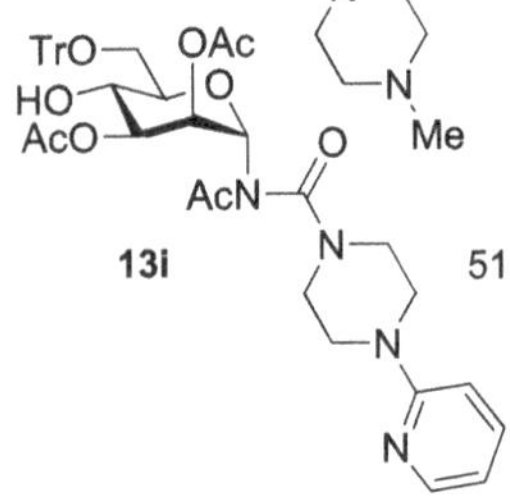 13i	51%

FIGURE 15.13 Stereoselective synthesis of α-glycosyl ureas.

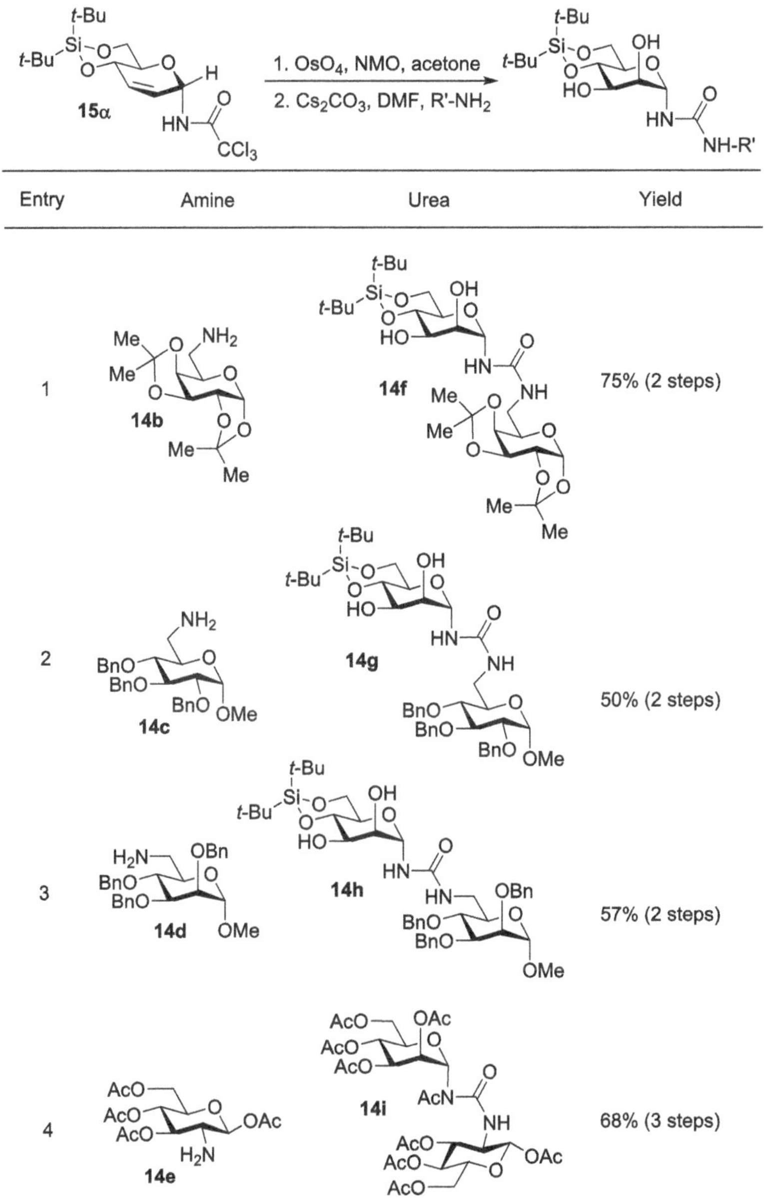

FIGURE 15.14 Synthesis of unsymmetrical urea-linked disaccharides.

urea-linked disaccharides **14f–14i**, tolerating many different glycosyl amine nucleophiles, including the more hindered C2 glucosamine nucleophile **14e** (Figure 15.14). Although an excess of the carbohydrate amine nucleophile (2–3 equivalents) was needed to achieve high conversions to the desired product, the unreacted amine could be recovered during purification. Finally, β-glycosyl ureas could be prepared in the

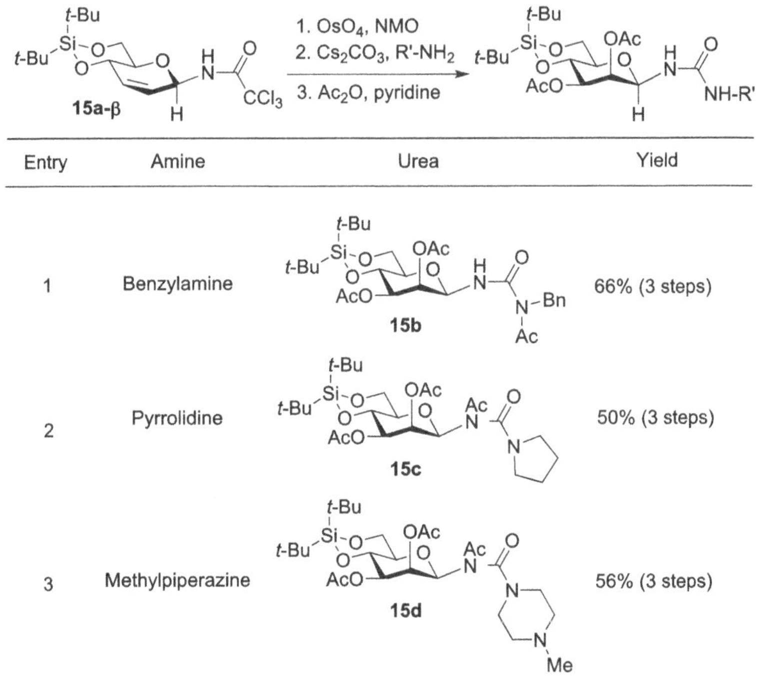

FIGURE 15.15 Stereoselective synthesis of β-glycosyl ureas.

same manner from the corresponding 1,2-unsaturated trichloroacetamides in 50–66% yield over three steps (Figure 15.15).

Having established the generality of these conditions for preparing glycosyl ureas, the potential of this methodology was evaluated by synthesizing the urea-linked pseudotrisaccharide **16g** (Figure 15.16). Starting from mannose trichloroacetimidate **16a**, a stereoselective glycosylation was performed to afford disaccharide **16c** in a 70% yield, exclusively as the α-anomer (Figure 15.16). Next the C3-acetyl group was removed and the resulting allylic alcohol was converted to the corresponding trichloroacetimidate **16d** in a 91% yield over two steps (Figure 15.16). The palladium-catalyzed rearrangement of **16d–16e** proceeded smoothly, affording the desired product in a 91% yield and ~5:1 α:β selectivity (Figure 15.16). Subsequent dihydroxylation and urea formation provided the desired urea-linked pseudotrisaccharide **16g** in a 60% yield over two steps and no epimerization at the anomeric C1-center (Figure 15.16).

15.4 DEVELOPMENT OF THE STEREOSELECTIVE NICKEL-CATALYZED TRANSFORMATION OF GLYCOSYL TRICHLOROACETIMIDATES TO TRICHLOROACETAMIDES

Although the palladium(II)-catalyzed rearrangement of 1,2-unsaturated trichloroacetamides to the corresponding trichloroacetamides was successful in accessing in

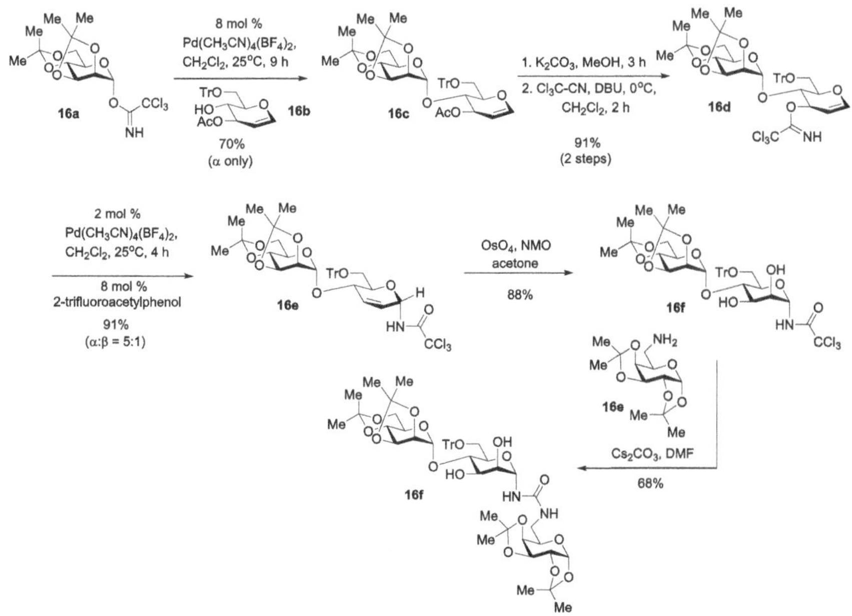

FIGURE 15.16 Synthesis of α-urea linked pseudotrisaccharide.

both α- and β-glycosyl ureas, the strategy is inherently limited for the generation of the urea-linked mannoside derivatives due to the necessity for the dihydroxylation (Figures 15.3 and 15.11). To overcome this limitation and expand the substrate scope to include a broader array of carbohydrates, a new route was developed to access C1-glycosyl trichloroacetamides. In this new concept, a transition-metal catalyst was envisioned to promote the conversion of a C1-glycosyl trichloroacetimidate **17a** to the corresponding trichloroacetamide **17b** in the absence of an external nucleophile (Figure 15.17). On the basis of previous results obtained with the palladium catalyst, it was hypothesized that a more cationic transition-metal catalyst would be able to promote the ionization and subsequent recombination to provide the stereoelectronically favored α-anomer.

With these factors in mind, the investigation commenced by treating the glucose trichloroacetimidate **18a** with 5 mol % $Pd(CH_3CN)BF_4$ at room temperature (Figure 15.18). After 5 h, only a trace amount of the desired product **18b** was observed in the reaction, suggesting that the catalyst may not be Lewis acidic enough to promote the

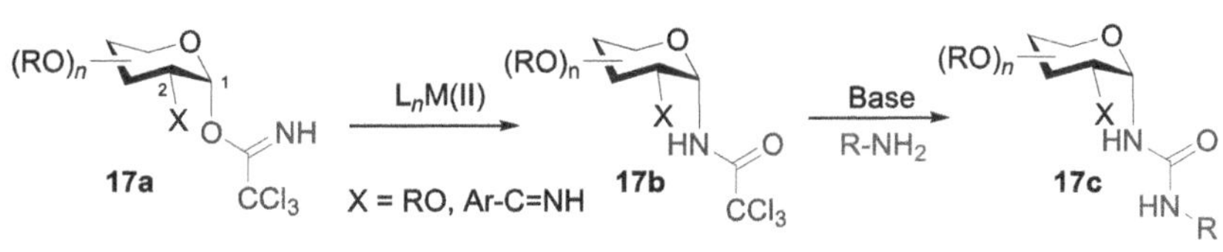

FIGURE 15.17 Formation of urea pseudosaccharides by cationic metal catalyzed rearrangement of trichloroacetimidates.

desired transformation (entry 1; Figure 15.18). To improve the ionization process, the catalyst was switched to the more cationic $Pd(PhCN)_2(OTf)_2$, which can be generated *in situ* from $Pd(PhCN)_2Cl_2$ and AgOTf. Using 5 mol % of this catalyst smoothly provided glycosyl trichloroacetamide **18b** in 1 h with 10:1 α:β selectivity and an 86% yield (entry 2; Figure 15.18). Reducing the catalyst loading to 2 mol % had no effect on the selectivity or yield of the transformation (entry 3; Figure 15.18). Switching from a cationic palladium catalyst to a cationic nickel catalyst, $Ni(PhCN)_4(OTf)_2$ afforded a similar rate and selectivity, albeit with a notable increase in conversion rate (entry 4; Figure 15.18). Adjusting the electronic properties of the benzonitrile ligands had no noticeable effect upon the rate, yield, or stereoselectivity of the transformation (entries 5 and 6; Figure 15.18). By changing the nitrile ligands to the bidentate phosphine ligand, 1,2-bis(diphenylphosphino)ethane (dppe), a significant increase in α-selectivity, while maintaining a similar yield as the other palladium and nickel catalysts was noted (entry 7; Figure 15.18). In comparison, the $Ni(dppe)Cl_2$ complex did not facilitate the reaction at all, indicating the necessity for using the more cationic triflate

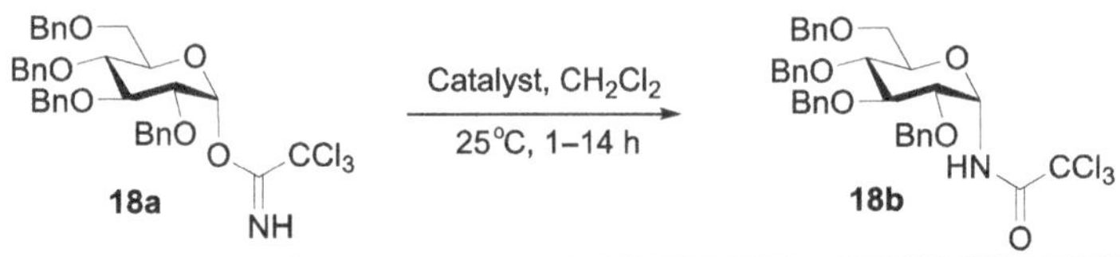

Entry	Catalyst	Catalyst Loading (mol %)	Time (h)	Yield[a] (%)	α:β[b]
1	$Pd(CH_3CN)_4(BF_4)_2$	5	5	NR	—
2	$Pd(PhCN)_2(OTf)_2$	5	1	86	10:1
3	$Pd(PhCN)_2(OTf)_2$	2	1	85	10:1
4	$Ni(PhCN)_4(OTf)_2$	2	0.25	84	11:1
5	Ni(*p*-F-PhCN)$_4$(OTf)$_2$	2	0.25	88	10:1
6	Ni(*p*-MeO-PhCN)$_4$(OTf)$_2$	2	0.25	90	10:1
7	$Ni(dppe)(OTf)_2$	2	1	85	30:1
8	$Ni(dppe)Cl_2$	2	6	NR	—
9	$Ni(dppp)(OTf)_2$	2	1.75	70	8:1
10	$Ni(dppb)(OTf)_2$	2	1.16	49	9:1
11	$Ni(PPh_3)(OTf)_2$	2	1.16	61	12:1
12	$Ni(OTf)_2$	2	0.25	57	19:1
13	AgOTf	6	14	72	5:1
14	TMSOTf	4	2	72	8:1
15	$BF_3.OEt_2$	4	6	65	4:1

[a] Isolated yield. [b] ^{1}H NMR ratio.

FIGURE 15.18 Screening of catalysts and conventional Lewis acids.

complex to achieve the desired transformation (entry 8; Figure 15.18). Other nickel catalysts using similar bidentate ligands, such as 1,3-bis(diphenylphosphino)propane or 1,4-bis(diphenylphosphino)butane, resulted in diminished yields and selectivity, implying that adjustments of the bite angle of the ligand [22] had a detrimental effect on catalyst performance (entries 9 and 10; Figure 15.18). The use of triphenylphosphine ligand provided a catalyst system that produced good α-selectivity, but a much lower yield of the desired product **18b** (entry 11; Figure 15.18). Next, $Ni(OTf)_2$ was examined to determine whether the phosphine ligand was a necessary component of the catalyst. While the α-selectivity of the reaction was high (19:1), the low yield and significant amounts of decomposition of the starting material suggested that the phosphine ligand plays an important role in the modulation of catalyst reactivity (entry 12; Figure 15.18). Traditional Lewis acids, including AgOTf, $BF_3•OEt_2$, and TMSOTf, were also evaluated to determine whether the observed catalyst reactivity and selectivity were a result of residual AgOTf or simple Lewis acid promoted transformation. In all cases, longer reaction times were necessary and the desired glycosyl trichloroacetamide **18b** was isolated with modest selectivity (entries 13–15; Figure 15.18).

Having identified Ni(dppe)OTf_2 as the optimal catalyst system for the transformation, the effect of solvent was investigated. More polar solvent such as THF, methyl *tert*-butyl ether, or dioxane, resulted in longer reaction times, lower yields, and lower stereoselectivity (entries 1–3; Figure 15.19). The same results came from less polar solvents such as toluene and trifluorotoluene; the yields and selectivity were decreased (entries 5 and 6; Figure 15.19). Thus, dichloromethane was selected as the solvent of choice as it provided the best yield, selectivity, and shortest reaction time.

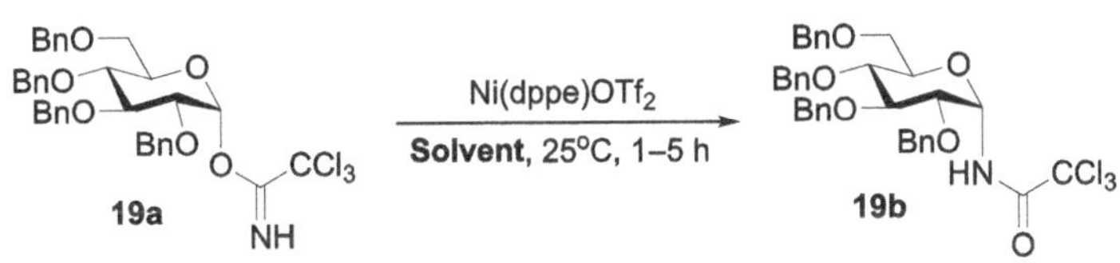

Entry	Catalyst	Solvent	Time (h)	Yield[a] (%)	α:β[b]
1	$Ni(dppe)(OTf)_2$	CH_2Cl_2	1	85	30:1
2	$Ni(dppe)(OTf)_2$	THF	3	49	19:1
3	$Ni(dppe)(OTf)_2$	MTBE	1.5	50	16:1
4	$Ni(dppe)(OTf)_2$	Dioxane	5	67	15:1
5	$Ni(dppe)(OTf)_2$	Toluene	2	75	20:1
6	$Ni(dppe)(OTf)_2$	Trifluorotoluene	1.25	67	13:1

[a] Isolated yield. [b] ^1H NMR ratio.

FIGURE 15.19 Screening of solvents.

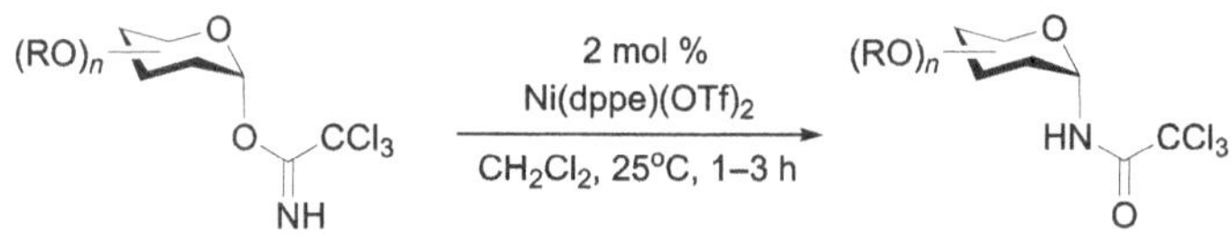

Entry	Imidates	Time (h)	Products	Yield (%)[a] (α:β)[b]
1	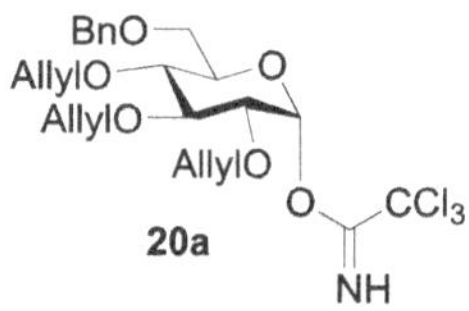 **20a**	1	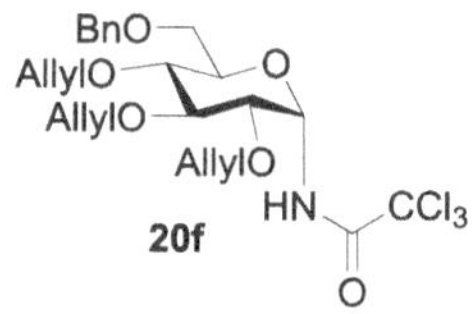 **20f**	91 (10:1)
2	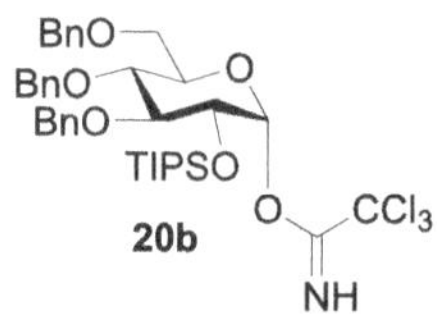 **20b**	10	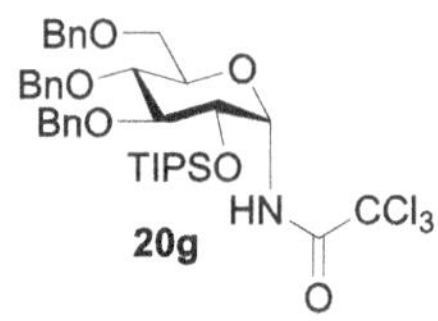 **20g**	90 (20:1)
3	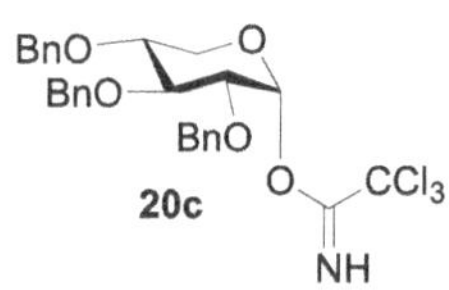 **20c**	0.5	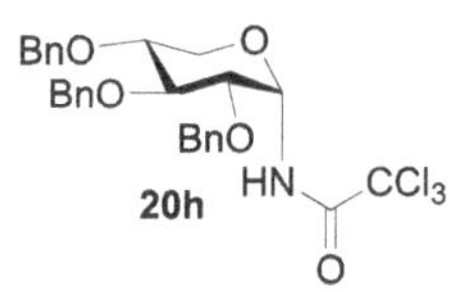 **20h**	87 (15:1)
4	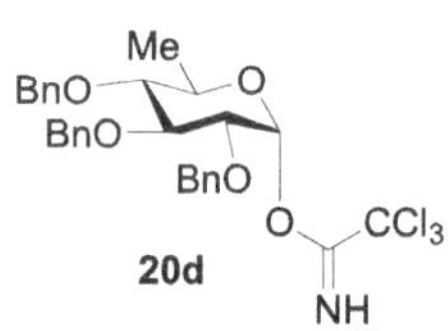 **20d**	0.5	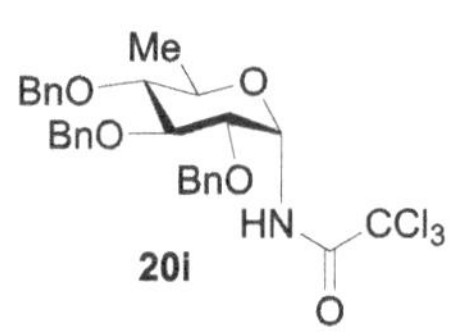 **20i**	91 (α only)
5	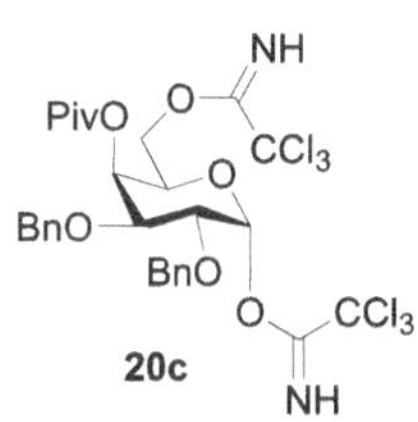 **20c**	12	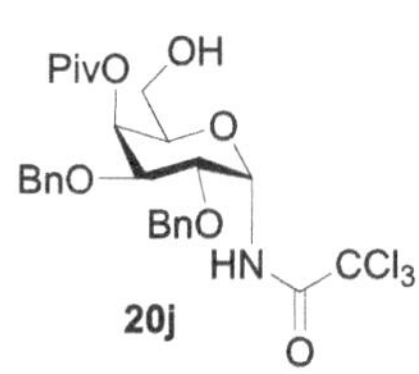 **20j**	72 (α only)

[a] Isolated yield. [b] ^{1}H NMR ratio.

FIGURE 15.20 Nickel-catalyzed transformation of monosaccharide trichloroacetimidates.

With the optimal conditions in hand, a variety of monosaccharide trichloroacetimidate substrates were then investigated (Figure 15.20). Incorporation of allyl or C2-triisopropylsilyl ether on the glucose trichloroacetimidates **20a** and **20b** was well tolerated, providing the desired products **20f** and **20g** (entries 1 and 2; Figure 15.20) in high yields (90–91%) and α-selectivity (α:β = 10:1–20:1). Xylose and quinovose

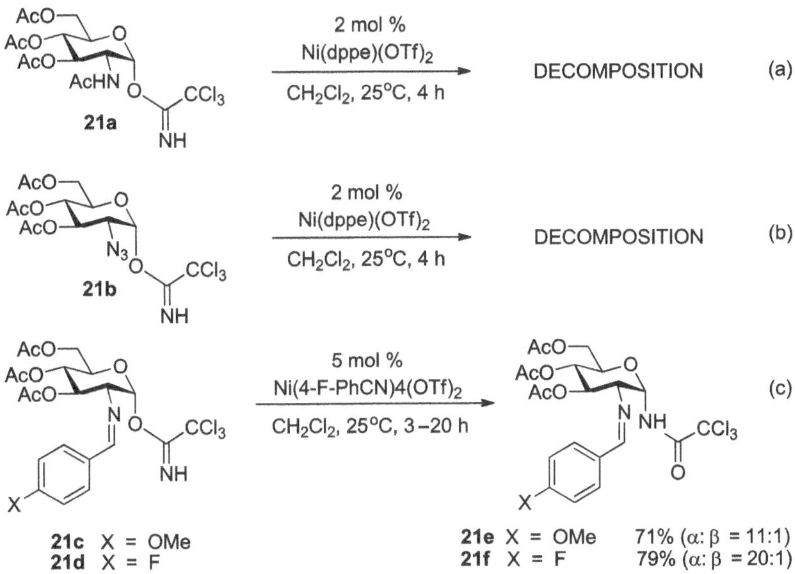

FIGURE 15.21 Transformation of C2-amino trichloroacetimidates.

trichloroacetimidates (**20c** and **20d**; Figure 15.20) that lack the C6-hydroxyl group were also examined. In both cases, the transformation proceeded smoothly, producing the desired trichloroacetamides **20h** and **20i** (entries 3 and 4; Figure 15.20) in 87–91% yields and excellent α-selectivity (α:β > 15:1). Finally, galactose trichloroacetimidate containing an additional trichloroactimidate group at the C6-hydroxyl group was studied (**20e**; Figure 15.20). Although the conversion proceeded slowly, the corresponding trichloroacetamide **20j** was isolated as a single α-anomer in a 72% yield (entry 5; Figure 15.20). The C6-trichloroacetimidate group was also cleaved under the reaction conditions affording the free hydroxyl group at that position, which could be potentially used in subsequent transformations as a glycosyl acceptor (entry 5; Figure 15.20).

In addition to C2-hydroxyl-containing monosaccharides, the robustness of this method was evaluated on C2-aminosugar substrates that are not compatible with current strategies [15, 16]. Glucose trichloroacetimidate substrates bearing the C2 *N*-acetyl and C2-azido groups were initially examined (**21a** and **21b**; Figure 15.21). In both cases, only decomposition of the starting material was observed under the optimized reaction conditions. Changing the C2 group to the *N*-benzylideneamino group allowed for a conversion of the starting material into the desired product (Figure 15.21). In this case, *para*-fluoro-*N*-benzylideneamino trichloroacetimidate **21d** proved to be superior to the *para*-methoxy-*N*-benzylidenamino **21c**, affording the corresponding glycosyl trichloroacetamide **21f** in a 79% yield and 20:1 α:β selectivity (Figure 15.21).

As the developed nickel catalyst system proved to be excellent for the stereoselective transformation of monosaccharide trichloroacetimidate substrates, its

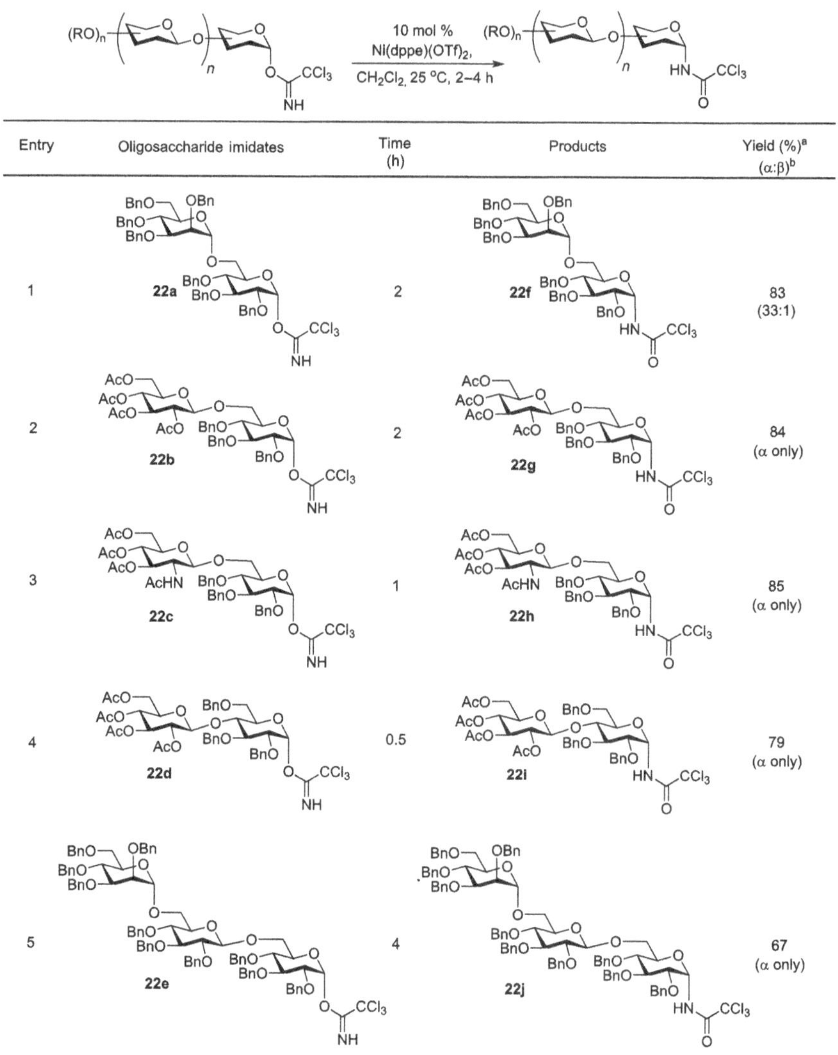

[a] Isolated yield. [b] ^{1}H NMR ratio.

FIGURE 15.22 Nickel-catalyzed transformation of oligosaccharide trichloroacetimidates.

compatibility with larger oligosaccharide trichloroacetimidates was subsequently probed. The investigation commenced with disaccharide substrate **22a**, which underwent the desired conversion to the corresponding trichloroacetamide **22f** (entry 1; Figure 15.22) in an 83% yield and with excellent α-selectivity (α:β = 33:1). While higher catalyst loadings were required to achieve the desired transformation, the high-yielding and excellent α-selectivity encouraged us to further evaluate with other disaccharide trichloroacetimidates **22b–22e**. In the cases of disaccharides **22b–22d**, high

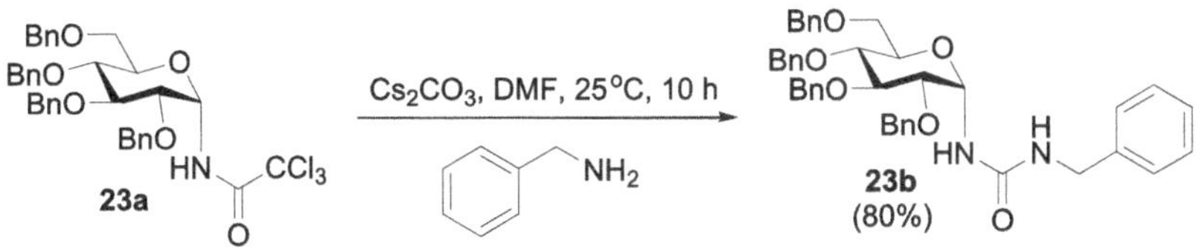

FIGURE 15.23 Conversion of trichloroacetamide to glycosyl urea.

yields of the corresponding trichloracetamides were obtained as single α-anomers (entries 2 and 4; Figure 15.22). Finally, rearrangement of trisaccharide **22e** produced the desired product **22j** as exclusively the α-anomer, however in a lower yield (entry 5; Figure 15.22).

15.5 TRANSFORMATION OF GLYCOSYL TRICHLOROACETIMIDATES INTO α- AND β-UREA GLYCOSIDES

The glycosyl trichloroacetamides could readily be transformed into the corresponding ureas using the previously developed conditions without any perturbation of the stereochemical integrity of anomeric C1-carbon (Figure 15.23). Secondary aliphatic amines were well tolerated as nucleophiles, and the monosaccharide trichloracetamides (**24a–24c**; Figure 15.24) were readily converted into the corresponding α-urea glycosides **24d–24f** in an excellent yield (88–93%). Amino acids such as valine and glycine or a proline derivative also served as competent amine nucleophiles, providing the corresponding α-glycoconjugates in 85–90% yields (entries 1–3; Figure 15.25). The unsymmetrical α-urea-linked disaccharides **25d** and **25e** could be accessed by employing a carbohydrate amine as a nucleophile, which afforded the desired products in 66–81% yield (entries 4 and 5; Figure 15.25). Finally, β-urea glycosides **26a–26c** could also be formed under the same conditions using both primary and secondary amine nucleophiles (Figure 15.26).

15.6 MECHANISTIC STUDIES ON THE NICKEL-CATALYZED TRANSFORMATION OF GLYCOSYL TRICHLORACETIMIDATES

To gain a better understanding of the origin of the nickel-catalyzed transformation of glycosyl trichloroacetimidates with high levels of α-selectivity, we decided to study in detail the mechanism of this process. Our initial mechanistic hypothesis began with a simultaneous coordination of the equatorial C2-ether and the imidate nitrogen by the nickel catalyst (**27b**; Figure 15.27). The trichloroacetimidate is then ionized, generating the corresponding oxocarbenium cation with the nickel catalyst remaining coordinated to the C2-ether (**27c**; Figure 15.27). Here, the redelivery of the coordinated nitrogen may occur directly (pathway A) or first dissociate to form

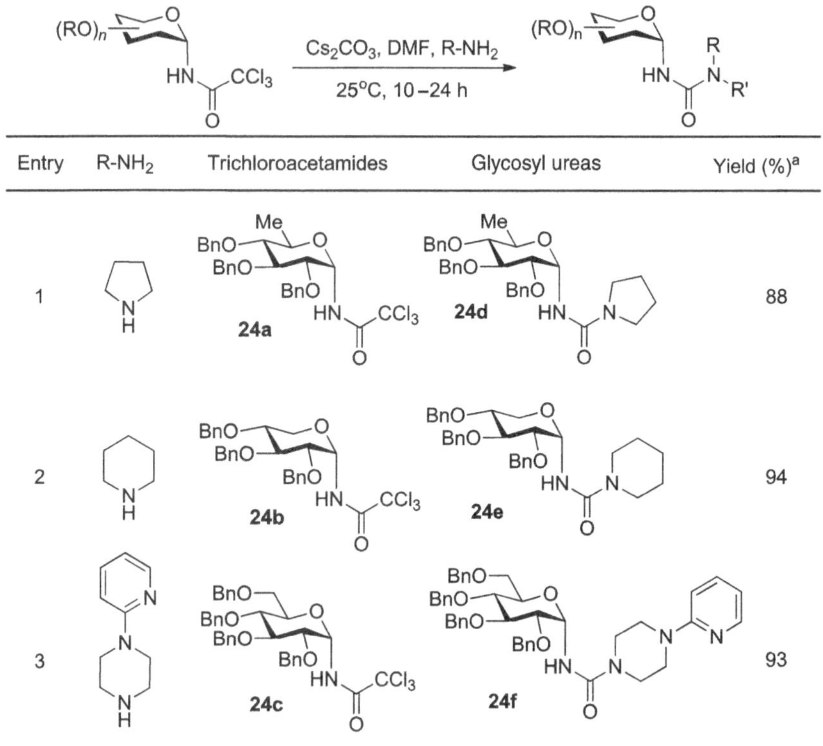

[a] Isolated yield.

FIGURE 15.24 Conversion of trichloroacetamides to urea with secondary amines.

a tight ion-pair **27d** and subsequently recombine in a stereoelectronically favored manner to provide the α-glycosyl urea.

As the first step in our hypothesis involved dual coordination by the C2-ether and imidate nitrogen (**27b**; Figure 15.27), it was determined that the α-orientation of the trichloroacetimidate was important for the overall reactivity and success of the transformation. To prove this mechanistic step, β-glucose trichloroacetamide **28a** was subjected to standard reaction conditions (Figure 15.28). The conversion was much longer (12 h) than that in case of the corresponding α-anomer and produced a 50% yield of desired trichloroacetamide **28b** (20:1 α:β; Figure 15.28) along with substantial decomposition products. While this result demonstrates that the β-trichloroacetamide is less reactive than the α counterpart, the high levels of α-selectivity suggests that the tight-ion pair pathway (Figure 15.27) is the major operative pathways in this reaction. To further test the importance of the C2-ether, 2-deoxyglucose substrate **29a** and β-2-deoxy-2-benzylideneamino imidate **29b** were investigated (Figure 15.29). Reaction of **29a** leads only to decomposition of the starting material, while no conversion of **29b** was noted, in contrast to the corresponding C2-benzyl ether **28a** (Figures 15.28 and 15.29).

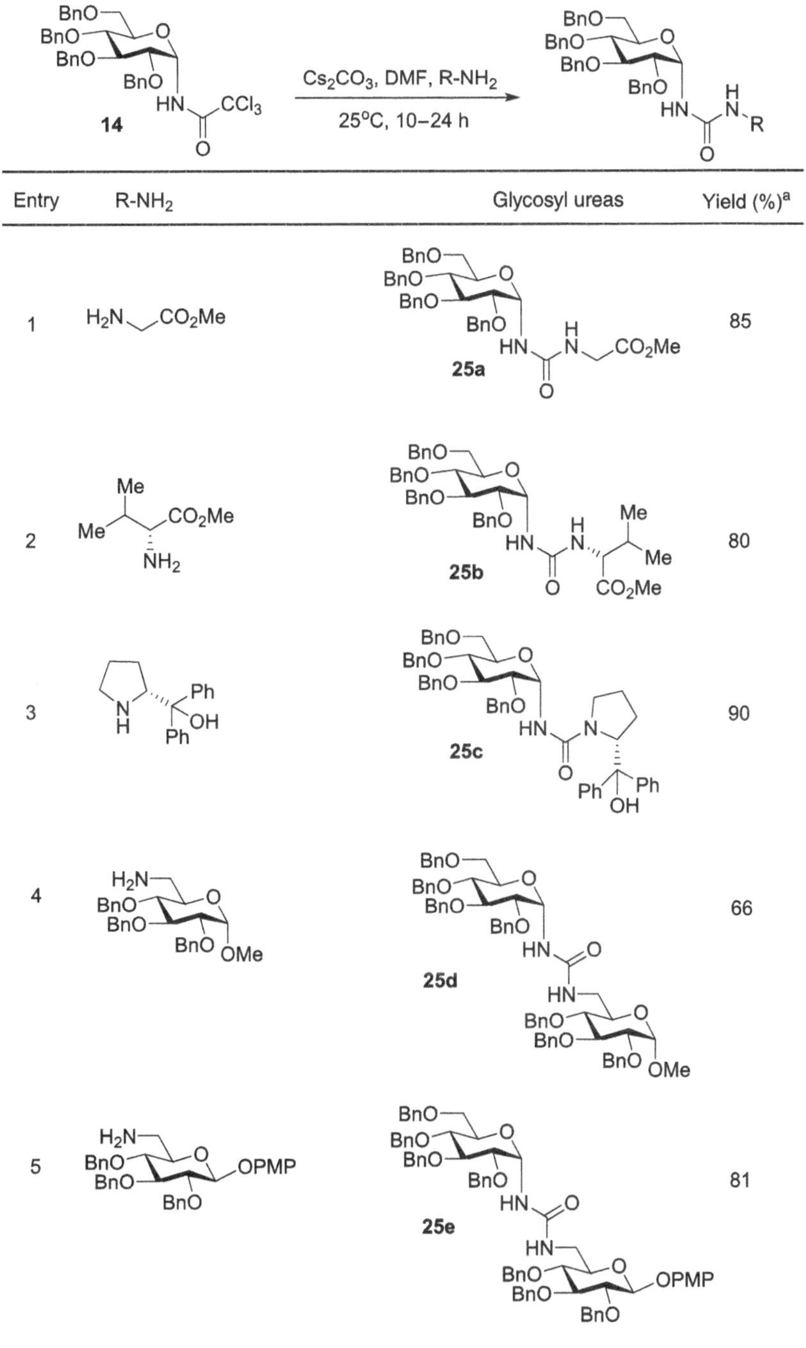

[a] Isolated yield.

FIGURE 15.25 Conversion of trichloroacetamide to urea with amino acid and carbohydrate amine nucleophiles.

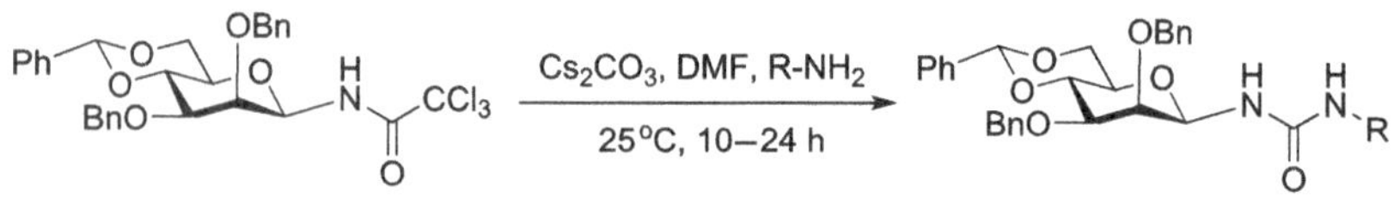

Entry	R-NH2	Glycosyl ureas	Yield (%)[a]
1	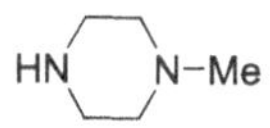	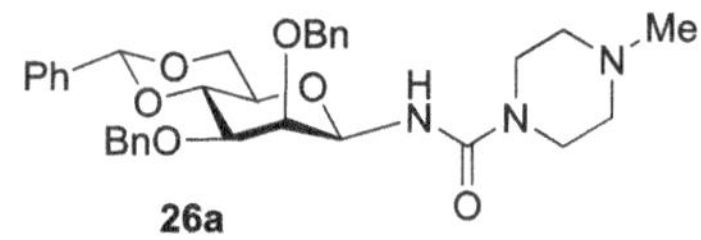 26a	75
2	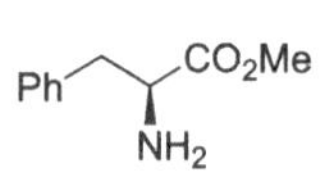	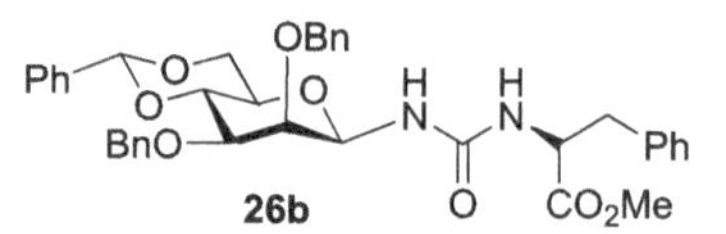 26b	88
3	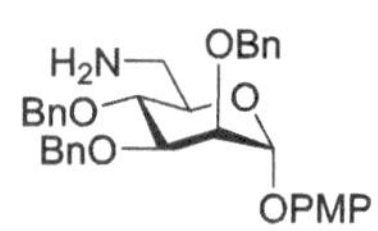	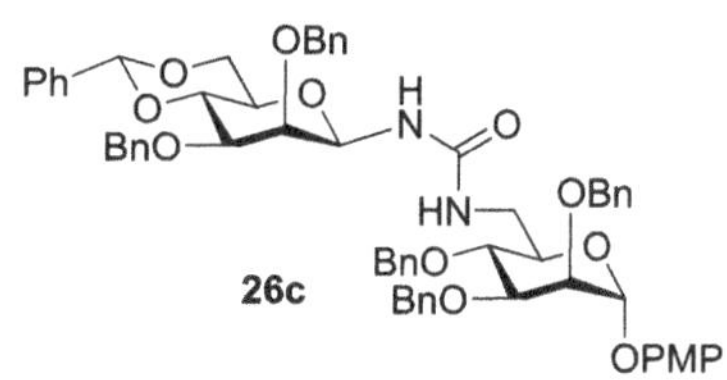 26c	76

[a] Isolated yield.

FIGURE 15.26 Stereoselective synthesis of β-urea-linked glycosides.

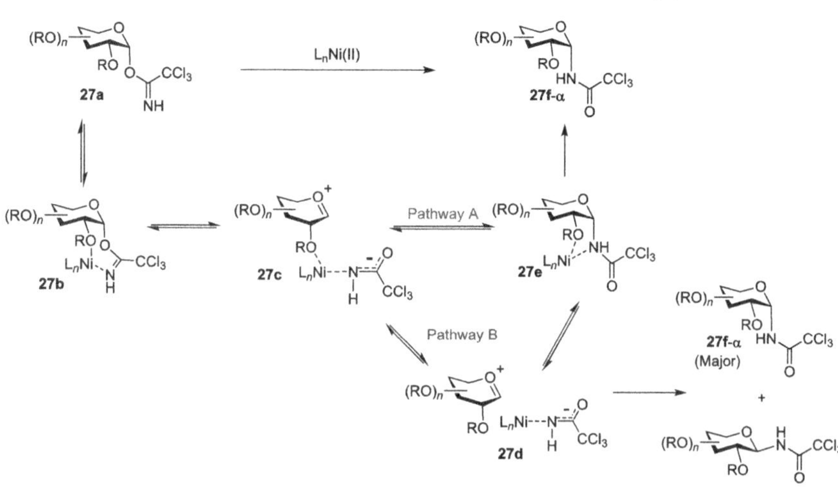

FIGURE 15.27 Proposed mechanism of nickel-catalyzed transformation of trichloroacetimidate.

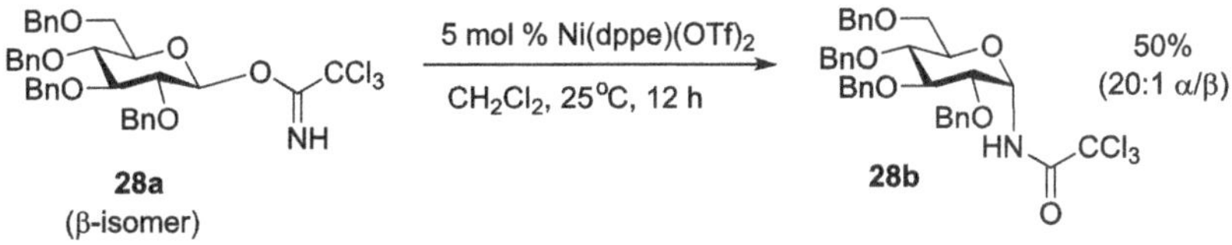

FIGURE 15.28 Rearrangement of β-trichloroacetimidate.

Tetrabenzyl mannose trichloroacetimidate **29c** was also tested to determine whether the axial or equatorial orientation of the C2-oxygen is critical for a stereoselective transformation (Figure 15.29). Under the standard reaction conditions, the corresponding trichloroacetamide **29d** was obtained in an 86% yield as a 1:1 mixture of α:β anomers (Figure 15.29). This implies that the axial C(2)-ether of the mannose substrate is too far from the α-trichloroacetimidate nitrogen to allow simultaneous coordination to the nickel catalyst and subsequent redelivery of the trichloroacetamide group [22–26]. In contrast, when the 4,6-benzylidene protecting group (**29e**; Figure 15.29) is used, high levels of β-selectivity are observed [26]. It may arise from the twist-boat conformation adopted by the intermediate oxocarbenium ion [27], allowing for joint coordination of the ether and the trichloroacetimidate nitrogen with the nickel catalyst on the β- face.

Having established the importance of the C2-ether in the transformation process, it was necessary to determine whether the reaction proceeded exclusively in an intramolecular fashion or if an intermolecular reaction could occur. Thus, a crossover

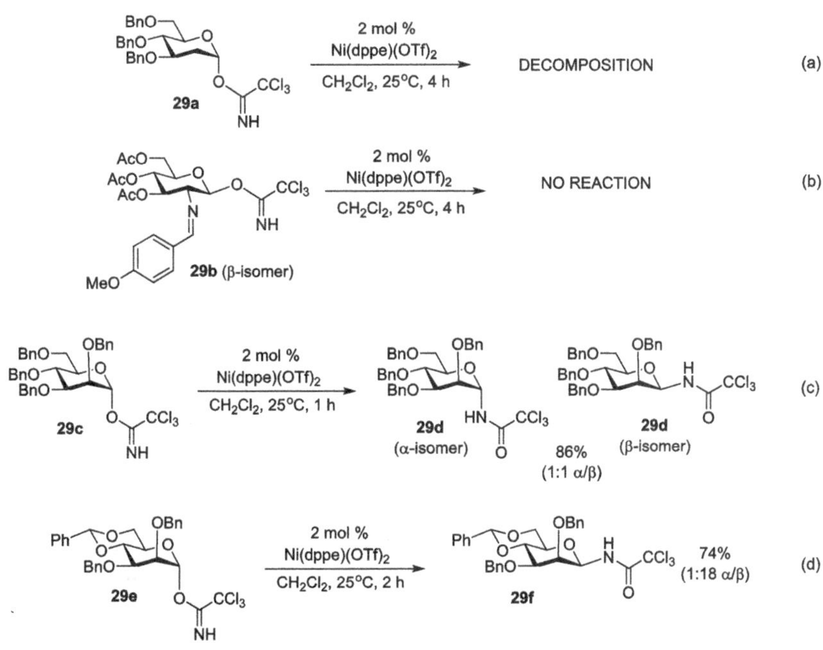

FIGURE 15.29 Substrates lacking equatorial C2-ether group.

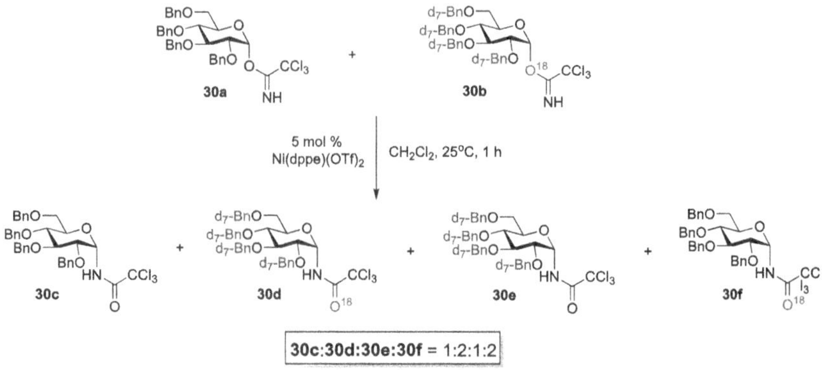

FIGURE 15.30 Crossover-labeling experiments.

labeling experiment was performed using equal amounts of labeled and unlabeled per-*O*-benzylated glucose (**30a** and **30b**, respectively (Figure 15.30). Upon treatment under the standard reaction conditions, two glycosyl trichloroacetamide products **30c** and **30d**, arising from their corresponding starting materials **30a** and **30b**, were observed in addition to two partially labeled products **30e** and **30f** (Figure 15.30). Therefore, the experiment demonstrates that the rearrangement may proceed via an intermolecular pathway.

To determine whether the nickel-catalyst facilitated delivery of the trichloroacetimidate nitrogen specious to the anomeric center, conversion of **31a** was conducted in the presence of an extra equivalent of trichloroacetamide (Figure 15.31). Instead of obtaining the normal 30:1 α:β selectivity, a 5.4:1 α:β mixture was observed in product **31b** (Figure 15.31). This suggests that the S_N2 displacement of the trichloroacetimidate leaving group by the external trichloroacetamide was likely occurring, generating more of the β-anomer during the course of the reaction.

Overall, the mechanistic experiments provide significant insight into the origin of the selectivity in the nickel-catalyzed transformation of glycosyl trichloracetimidates into the corresponding trichloroactamides. The equatorial C2-ether or benzylideneamino functionality in combination with the α-trichloroacetimidate at C1 is crucial for the high levels of selectivity and reactivity in the transformation. In addition, these results illustrate that both the intramolecular pathway and the intermolecular ion-pair pathway are operative during the reaction and both contribute directly to the formation of the trichloroacetamide product.

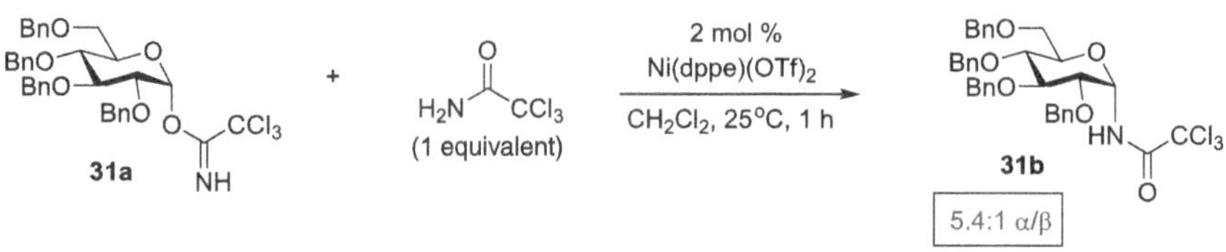

FIGURE 15.31 Rearrangement in the presence of external trichloroacetamide.

15.7 CONCLUSIONS

The stereoselective synthesis of α- and β-urea glycosides has traditionally been a challenging problem in the field of carbohydrate chemistry. Many of the current methods have suffered from epimerization of the anomeric center, required numerous steps, and have a narrow substrate scope [2–4, 13–16]. To overcome these challenges, the operationally simple and robust transition-metal-catalyzed process to access α- and β-trichloroacetamides selectively was developed [28–30]. The established conditions can readily be applied to a wide array of both mono- and oligosaccharides, including C2-aminosugars. These trichloroacetamide intermediates can then be directly converted to the corresponding urea without epimerization at the anomeric center. A large variety of amine nucleophiles can be utilized in the urea-forming reaction, including the methyl esters of amino acids and various amine-containing carbohydrates, allowing efficient access to both glycopeptides and unsymmetrical urea-linked disaccharides. The results of these studies have demonstrated the generality and versatility of the developed palladium and nickel catalyst systems. Future work will be aimed at further improving the method and utilizing it in more challenging settings such as the development of new urea-linked glycopeptide mimetics, pseudooligosaccharides, and antibacterial agents.

REFERENCES

[1] Pigman, W. In *Carbohydrates: Chemistry and Biochemistry*, Part II A; Horton, D., Ed.; Academic Press: New York, **1970**; pp 37–110.

[2] Ichikawa, Y.; Nishiyama, T.; Isobe, M. *Synlett* **2000**, 1253–1256.

[3] Ichikawa, Y.; Ohara, F.; Kotsuki, H.; Nakano, K. *Org. Lett.* **2006**, *8*, 5009–5012

[4] Ichikawa, Y.; Nishiyama, T.; Isobe, M. *J. Org. Chem.* **2001**, *66*, 4200–4205.

[5] Prosperi, D.; Ronchi, S.; Panza, L.; Rencurosi, A.; Russo, G. *Synlett* **2004**, 1529–1532.

[6] Prosperi, D.; Ronchi, S.; Lay, L.; Rencurosi, A.; Russo, G. *Eur. J. Org. Chem.* **2004**, 395–405.

[7] Ellestad, G. A.; Cosulich, D. B.; Broschard, R. W.; Martin, J. H.; Kunstmann, M. P.; Morton, G. O.; Lancaster, J. E.; Fulmor, W.; Lovell, F. M. *J. Am. Chem. Soc.* **1978**, *100*, 2515–2524.

[8] Fernandes, P.; Swanson, R.; Hardy, D.; Hanson, C.; McDaniel, D.; Beyer, J.; Chen, R. *J. Antibiot.* **1989**, *42*, 538–541.

[9] Osburne, M.; Maiese, W.; Greenstein, M. *Antimicrob. Agents Ch.* **1990**, *34*, 1450–1452.

[10] Dobashi, K.; Nagaoka, K.; Watanabe, Y.; Nishida, M.; Hamada, M.; Naganawa, J.; Takita, T.; Takeuchi, T.; Umezawa, H. *J. Antibiot.* **1985**, *38*, 1166–1170.

[11] Nishiyama, T.; Isobe, M.; Ichikawa, Y. *Angew. Chem. Int. Ed.* **2005**, *44*, 4372–4375.

[12] Nishiyama, T.; Kusumoto, Y.; Okumura, K.; Hara, K.; Kusaba, S.; Hirato, K.; Kamiya, Y.; Isobe, M.; Nakano, K.; Kotsuki, H.; Ichikawa, Y. *Chem. Eur. J.* **2010**, *16*, 600–610.

[13] McKay, M. J.; Nguyen, H. M. *Carbohydr. Res.* **2014**, *385*, 18–44.

[14] Spanu, P.; Ulgheri, F. *Curr. Org. Chem.* **2008**, *12*, 1071–1092.

[15] Ichikawa, Y.; Nishiyama, T.; Isobe, M. *Tetrahedron* **2004**, *60*, 2621–2627.

[16] Bianchi, A.; Ferrario, D.; Bernandi, A. *Carbohyd. Res.* **2006**, *341*, 1438–1446.

[17] Overman, L. E. *J. Am. Chem. Soc.* **1974**, *96*, 597–599.

[18] Anderson, C. E.; Overman, L. E. *J. Am. Chem. Soc.* **2003**, *125*, 12412–12413.

[19] Watson, M.; Overman, L. E.; Bergman, R. G. *J. Am. Chem. Soc.* **2007**, *129*, 5031–5044.

[20] Miller, K. J.; Bagg. J. H.; Abu-Omar, M. M. *Inorg. Chem.* **1999**, *38*, 4510–4514.

[21] Wayland, B. B.; Schramm, R. F. *Inorg. Chem.* **1969**, *8*, 971–976.

[22] Nukakda, T.; Berces, A.; Wang, L.; Zgierski, M. Z.; Whitfield, D. M. *Carbohydr. Res.* **2005**, *340*, 841–852.

[23] Amat, L.; Carbo-Dorca, R. *J. Chem. Inf. Comput. Sci.* **2000**, *40*, 1188–1198.

[24] Winkler, D. A.; Holan, G. *J. Med. Chem.* **1989**, *32*, 2084–2089.

[25] Winkler, D. A. *J. Med. Chem.* **1996**, *39*, 4332–4334.

[26] Lucero, C. G.; Woerpel, K. A. *J. Org. Chem.* **2006**, *71*, 2641–2647.

[27] Crich, D.; Chandrasekera, N. S. *Angew. Chem. Int. Ed.* **2004**, *43*, 5386–5389.

[28] Mercer, G. J.; Yang, J.; McKay, M. J.; Nguyen, H. M. *J. Am. Chem. Soc.* **2008**, *130*, 11210–11218.

[29] Park, N. H.; Nguyen, H. M. *Org. Lett.* **2009**, *11*, 2433–2436.

[30] McKay, M. J.; Park, N. H.; Nguyen, H. M. *Chem. Eur. J.* **2014**, *20*, 8691–8701.

INDEX

Domino and Intramolecular Rearrangement Reactions as Advanced Synthetic Methods in Glycoscience, First Edition. Edited by Zbigniew J. Witczak and Roman Bielski.